ADVANCES IN PSYCHONEUROIMMUNOLOGY

HANS SELYE SYMPOSIA ON NEUROENDOCRINOLOGY AND STRESS

Series Editors:

Sandor Szabó, *Brigham and Women's Hospital, Boston, Massachusetts*
Yvette Taché, *UCLA, Los Angeles, California*
Beatriz Tuchweber-Farbstein, *University of Montreal, Montreal, Quebec, Canada*

Volumes published by Plenum Press:

Volume 2—NEUROENDOCRINOLOGY OF GASTROINTESTINAL
 ULCERATION
  Edited by Sandor Szabó and Yvette Taché

Volume 3—ADVANCES IN PSYCHONEUROIMMUNOLOGY
  Edited by Istvan Berczi and Judith Szélenyi

ADVANCES IN PSYCHONEUROIMMUNOLOGY

Edited by

Istvan Berczi

Department of Immunology
The University of Manitoba
Winnipeg, Manitoba, Canada

and

Judith Szélenyi

National Institute of Haematology,
  Blood Transfusion and Immunology
Budapest, Hungary

PLENUM PRESS • NEW YORK AND LONDON

Library of Congress Cataloging-in-Publication Data

On file

QP
356
.47
.A385
1994

ISBN 0-306-44883-1

© 1994 Plenum Press, New York
A Division of Plenum Publishing Corporation
233 Spring Street, New York, N. Y. 10013

Printed in the United States of America

FOREWORD

I greatly appreciate being asked to write a foreword to this book. Unfortunately, illness precluded me from participating in and attending this evidently successful meeting. Apparently, the fact that I was born in Budapest encouraged the editors to ask me to write something for the book, and I am pleased to do so.

The first time that I was introduced to this area was through the seminal work of Ader and Cohen. By that time, I was certainly familiar with the extraordinary body of work associated with Hans Selye, that remarkable Hungarian scientist who chose to work in Canada.

It is of more than passing interest to observe that while most of our scientific, medical and other colleagues are convinced that the expression of emotions or grief or even anger is accompanied by physical manifestations of lacrimation, nasal edema and so on, there appear to be difficulties in appreciating the fact that the brain may regulate other physiological functions. Similarly, there is no problem understanding that anger, frustration, depression and so on can influence the normal physiological homeostasis of the stomach, especially since this has been described so beautifully in the work of Wolf and Wolff. These same authors subsequently went on to describe similar changes in the nose which included not only observations on relative vascularity, edema, and secretion, but also evidence for the regulation of leukocyte emigration into nasal secretions as a consequence of emotional change.

We know now that stress can give rise to the formation of ulcers in the stomach. The more recent observations that ulcers in the stomach are pathophysiologically associated with colonization by an organism known as Helicobacter pylori, and that this ulceration may be attenuated or cured by the elimination of these organisms as a result of antibiotic treatment, raise the intriguing relationship between stress (the brain), bacteria and the immune system which would, under other circumstances, be expected to eradicate this infection. Apparently, the lack of eradication is the eventual cause of ulceration. This is an example of psychoneuromicroimmunology.

Clearly, the immune system can have input into the nervous system. Communication by any of a large number of chemicals or mediators made by immune cells, could be expected to have afferent input to the brain. Effector responses may be the end result of such interactions, stimulated as a result of antigen-immune cell-nervous interactions via axon or spinal reflexes. Bidirectional activity between the immune system and the nervous system may be expected, and in many instances has been shown to actually occur.

The term integrative physiology has been coined and introduced into the literature to cover the extraordinarily complex interactions at a tissue level, the net result of which can be read out as the physiological state at that time. The interactions between the nervous system, the immune system, the structural tissue cells as well as the endocrine systems are so complex as to bewilder most investigators. Nevertheless, it is foolish to consider that it is possible to study single events or single populations of cells in test tubes and arrive at accurately predictive conclusions as to how these events will be expresed in situ in tissues, without taking these complex interactions into account. Psychoneuroimmunology, a term coined by Ader, is not a discipline, but a descriptive term of the conceptual approach used to begin to explore the interactions between the nervous and the immune systems. As can be seen quite readily from the contents of this book, this ranges from psychological experiments such as those involving Pavlovian conditioning to the description of molecular interactions between neuropeptides and immune cells and cells of the nervous system. It is encouraging to see the extent and diversity of the scientific interests and disciplines being brought to bear in this exciting and rapidly developing field.

John Bienenstock

PREFACE

Observations indicating the interaction of the neuroendocrine and immune systems date back to the beginning of this century. There were numerous attempts in the 1930s and 1940s to study the effect of various hormones on the hemopoietic and immune systems, which did not yield conclusive results and have long been forgotten. The interaction of the nervous, endocrine and immune systems is only now being considered seriously. This field represents a novel, multidisciplinary approach in Biological Sciences. Even the name of the field has not been settled as yet and there are debates going on with regards to the proper term. We have adopted "Psychoneuroimmunology," which was originally coined by Ader, as a simple term, which implies the involvement of higher nervous system activity in the regulatory interactions of the neuroendocrine and the immune systems. The alternative terms neuroimmunomodulation, neuroendocrinimmunology and immunoendocrinology, etc., refer to the same field or to certain areas of the field.

During the past decade or so, there have been approximately a dozen conferences held in the West covering various aspects of psychoneuroimmunology. These meetings were hardly accessible to scientists in the East. This was the first international conference held with the participation of scientists from Eastern Europe. The conference and this volume pay tribute to scientists of Hungarian descent who made seminal contributions to the field, i.e. Hans Selye, who was the first to point out the interaction between the neuroendocrine and immune systems; Andor Szentivanyi, who demonstrated for the first time that allergic reactions are regulated by the central nervous system; and Miklos Jancso and coworkers, who discovered neurogenic inflammation. It is only now that we begin to fully appreciate the significance of these important discoveries which were made all too early so that they could not be analyzed and understood by the contemporary scientific community. Modern science is equipped with powerful research tools which make it feasible to advance quickly in this complex multidisciplinary field, with the aim of understanding the whole organism, rather than trying to analyze restricted areas. The developments are spectacular, indeed, and the new insights gained through studies on psychoneuroimmunology have already advanced our understanding of certain human diseases, such as autoimmune disease, inflammatory diseases, nervous and endocrine abnormalities and the influence of behavioral factors and of aging on the immune response and disease. We sincerely hope that this volume will contribute to the understanding and acceptance of this brave new area of modern scientific enquiry.

Istvan Berczi
Judith Szelenyi

ACKNOWLEDGEMENTS

The editors are indebted to members of the Scientific Advisory Committee: R. Ader, H. Besedovsky, J. Bienenstock, M. Dardenne, C.J. Grossman, K.W. Kelley, K. Kovacs, C.A. Ottaway, J.P. Revillard, T. Roszman, S. Szabo, G. Wick, and C.R. Wira, for their valuable contribution regarding the selection of the contents and the authors for this volume. The devoted secretarial and excellent editorial assistance of Jean Sylwester is gratefully acknowledged. Financial support for the conference and for the production of this volume was provided in part by: Ciba Geigy (Switzerland), The Council for Tobacco Research (USA), The Hans Selye Foundation (Canada), Immunobiology Research Institute (USA), Marion Merrell Dow, Inc. (USA), The Medical Research Council of Canada, Research Life (Italy), Sandoz (Canada), Psychoneuroimmunology Foundation (Hungary).

CONTENTS

Immunoregulation by the ACTH-Adrenal Axis

Neuroendocrine and Metabolic Effects of Cytokines

Steroid Hormones

ABBREVIATIONS

A	Adrenaline
A431	Human epidermal cells
A549	Human lung epithelial cells
Ab	Antibody
ACh	Acetylcholine
AChE	Acetylcholine esterase
ACTH	Adrenocorticoytropic hormone
ADCC	Antibody dependent cellular cytotoxicity
Adrx	Adrenalectomy
ALL	Acute lymphocytic leukemia
AMDGF	Alveolar macrophage-derived growth factor
AML	Acute myelogenous leukemia
AMP	Adenosine monophosphate
ANS	Autonomic nervous system
anti-SRBC	Antibody against sheep red blood cells
AS	Adrenal steroids
ATP	Adenosine triphosphate
AVP	Arginine vasopressin
BAL	Bronchoalveolar lavage
BCG	Bacille Calmette-Guerin
BE	Beta-endorphin
BRC	Bromocriptine
BW	Body weight
Ca^{++}	Calcium
CAChl	Radioactive acetylcholine
CAM	Cell adhesion molecule
cAMP	Cyclic 3',5'-guanosine monophosphate
Capr	Caproate
CD	Cluster designation (of cell surface molecules)
CDF	Cholinergic differentiation factor
CDFR	Cholinergic differentiaton factor receptor
CG	Chorionic gonadotropin
CGRP	Calcitonen gene related peptide
CH	Constant portion of heavy chain
ChAT	Choline acetyltransferase
CIF	Colony inhibitory factor
CL	Constant portion of light chain
CM	Calmodulin
CML	Chronic myelogenous leukemia
CNS	Central nervous system
CNTF	Ciliary neurotropic factor
CNTFR	Ciliary neurotropic factor receptor
Con-A	Concanavalin-A
CP	Chlorpromazine
CP 96,345	Substance P antagonist (Pfizer)
CR	Conditioned response
CRF	Corticotropin releasing factor
CRH	Corticotropin-releasing hormone

CS	Conditioned stimulus
CSA	Cyclosporine A
CSF	Colony stimulating factor
CST-SMG	Cervical sympathetic trunk-submandibular gland
CTL	Cytotoxic T lymphocytes
DDT_1MF_2	Hamster smooth muscle cells
DEAE	Diethylaminoethanol
DES	Diethylstilbestrol
DHA	$[^3H]$dihydroalprenolol
DHT	Dihydrotestosterone
DMEM	Dulbecco modified minimal essential medium
dsRNA	Double stranded RNA
DTH	Delayed type hypersensitivity
E1	Estrone
E2	Estrogen or estradiol
E3	Estriol
EAE	Experimental allergic encephalomyelitis
Ec	Endothelial cells
ECF-A	Eosinophil chemotactic factor of anaphylaxis
EE	Ethinyl estradiol
EEG	Electroencephalogram
EGF	Epidermal growth factor
EGTA	Ethylene glycol-bis (beta-aminoethyl ether)-N,N,NI,NI-tetraacetic acid
EMG	Electromyogram
β-END	β-Endorphin
EPR	Erythropoietin receptor
ER	Estrogen receptor
ESF	Erythropoietin-stimulating factor
ESR	Electron spin resonance
E/T	Effector to target ratio
FACS	Fluorescent cell sorter
FcRII	IgE receptor
FCS	Fetal calf serum
FDC	Follicular dendritic cells
FEV	Forced expiratory volume
FGF	Fibroblast growth factor
FSH	Follicle stimulating hormone
GABA	Gamma-aminobutyric acid
GALT	Gut-associated lymphoid tissue
G-CSF	Granulocyte colony stimulating factor
G-CSF-R	Granulocyte colony stimulating factor receptor
GDP	Guanosine diphosphate
GH	Growth hormone
GHRF	Growth hormone releasing factor
GHRH	Growth hormone releasing hormone
GlyCAM-1	Cell adhesion molecule, ligand for L-selectin, on B cells
GM-CSF	Granulocyte-monocyte colony stimulating factor
GM-CSF-R	Granulocyte-monocyte colony stimulating factor receptor
GS	Gonadal steroids
GTP	Guanosine triphosphate
H7	Protein kinase C inhibitor
HEPES	N-2-hydroxyethylpiperazine-N'-2-ethanesulfonic acid
HETE	5-Hydroxyeicosatetraenonic acid
hGH	Human growth hormone
HIV	Human immunodeficiency virus
^{3}H-NMS	^{3}H N-methyl scopolamine
HPA	Hypothalamic-pituitary adrenal axis
HPLC	High performance liquid chromatography
^{3}H-QNB	^{3}H-Quinuclidinyl benzoate
HRF	Histamine-releasing factor
H/S	Helper/suppressor ratio

HYP	Hypothalamus
Hypox	Hypophysectomy
IBD	Inflammatory bowel disease
^{125}I-Btx	^{125}I-α-bungarotoxin
ICAM-1	Intercellular adhesion molecule-1
i.c.v.	Intra-cerebro-ventricular
IEL	Intraepithelial lymphocytes
IFN	Interferon
IFN-R	Interferon receptor
Ig	Immunoglobulin
IGF-I,II	Insulin-like growth factor I,II
IGFBP	IGF binding protein
IgSF	Immunoglobulin superfamily
IL	Interleukin
IL1RA	Interleukin-1 receptor antagonist
IL-R	Interleukin receptor
i.p.	Intra-peritoneal
I/R	Ischemia/reperfusion
ISC	Immunosuppressive cytokine
Isc	Intestinal short-circuit current
KCS	Keratoconjunctivitis sicca
LA	Lipid A
LAK	Lymphokine activated killer
LDL	Low density lipoprotein
Leu-M3 (CD14)	Monocyte-, macrophage surface molecule
LGL	Large granular lymphocytes
LH	Luteinizing hormone
LH-RH	Luteinizing hormone-releasing hormone
LIF	Leukemia inhibitory factor
LIFR	Leukemia inhibitory factor receptor
LPS	Lipopolysaccharide
MABP	Mean arterial blood pressure
mAChR	Muscarinic acetylcholine receptor
MadCAM-1	Adhesion molecule of B lymphocytes
MB-35	Thymic peptide
MBH	Median basal hypothalamus
MDP	N-acetylmuramyl-L-alanyl-D-isoglutamine
MECA-79	Cell adhesion molecule or addressin of T cells
MHC	Major histocompatibility complex
MLA	Levator ani muscle
MLN	Mesenteric lymph node
MMCP II	Mouse mast cell protease II
MP	Muramyl peptide
MPO	Myeloperoxidase
mRNA	Messenger RNA
MSH	Melanocyte stimulating hormone
MW	Molecular weight
NA	Noradrenaline
nAChR	Nicotinic acetylcholine receptor
N-CAM	Neural cell adhesion molecule
NCF	Neutrophil chemotactic factor
NDF	Neuromal differentiation factor
NDV	Newcastle disease virus
NE	Norepinephrine
NGF	Nerve growth factor
NK	Natural killer
NMDA	N-methyl-D-aspartate
NMK	Neuromedin K
NPY	Neuropeptide Y
NREMS	Non-rapid eye movement sleep
2-OCH3 E1	2-Methoxyestrone

6-OHDA	6-Hydroxydopamine
2-OH E1	2-Hydroxyestrone
OS	Obese strain
OVLT	Organum vasculosum laminae terminalis
OVX	Ovariectomy
P	Progesterone
PAF	Platelet activating factor
PBL	Peripheral blood lymphocytes
PBMC	Peripheral blood mononuclear cells
PBS	Phosphate buffered saline
Pc	Pericytes
PC	Phosphatidyl choline
PCL	Picryl chloride
PCR	Polymerase chain reaction
PCV	Postcapillary venules
PDGF	Platelet derived growth factor
PD(QS)	Palladium di(sodium alizarine monosulfonate)
PE	Phosphatidyl ethanol-amine
PG	Prostaglandin
PGE$_2$	Prostaglandin E$_2$
PHA	Phytohemagglutinin
PHA-PT	Phytohemagglutinin and pertussis toxin
Phe	Phentolamine
PIT	Pituitary
PITA	Pituitary adenoma
Pit-hGH	Pituitary-derived human growth hormone
PKC	Protein kinase C
PL	Plasma cell
PLA	Phospholipase A
PMA	Phorbol 12-myristate 13-acetate
PMN	Polymorphonuclear leukocytes
POA	Preoptic area
POMC	Proopiomelanocortin
PP	Peyer's Patches
PR	Progesterone receptor
PRL	Prolactin
PRL-R	Prolactin receptor
Pro	Propranolol
PVN	Paraventricular nucleus
PWM	Pokeweed mitogen
REMS	Rapid-eye movement sleep
RES	Reticuloendothelial system
RMCP II	Rat mast cell protease II
RNA	Ribonucleic acid
RRBC	Rat red blood cells
RT-PCR	Reverse transcription polymerase chain reaction
SC	Secretory component
SCG	Superior cervical ganglia
SCGx	Superior cervical ganglionectomy
SDAT	Senile dementia of Alzheimer type
SDS-PAGE	Sodium dodecyl sulfate-polyacrylamide gel electrophoresis
SK	Substance K
SLE	Systemic lupus erythematosus
SMA	Superior mesenteric artery
SMG	Submandibular gland
SMGx	Sialadenectomy
SMS 202,995	Somatostatin analogue (Sandoz)
SOM	Somatostatin
SP	Substance P
SPG	Syngeneic pituitary graft
SRBC	Sheep red blood cells

SRIF	Somatostatin
SRS-A	Slow reactive substance of anaphylaxis
STAI	State-Trait-Anxiety-Inventory
STH	Somatotrophic hormone
SV	Seminal vesicles
SWS	Slow-wave sleep
T	Testosterone
T3	Triiodothyronine
T4	Thyroxine
TBL	Tuberal lesions
TCA	Trichloro-acetic acid
TEC	Thymic epithelial cells
TF	Tissue factor
TFR	Tissue factor receptor
TGF	Transforming growth factor
Th	T helper cell
Thy-1	T cell surface antigen
TLC	Thin layer chromatography
TNF	Tumor necrosis factor
TNP	Trinitrophenol
TPA	Phorbol ester
TRH	Thyrotropin-releasing hormone
TSH	Thyroid stimulating hormone
TX	Tamoxifen
UC	Ulcerative colitis
UCR	Unconditioned response
UCS	Unconditioned stimulus
VCAM	Vascular adhesion molecule
VIP	Vasoactive intestinal peptide
VLA	Very late activation antigen
VP	Ventral prostate
W	Wakefulness
XLA	X-linked agammaglobulinemia

STRESS AND DISEASE: THE CONTRIBUTION OF HANS SELYE TO PSYCHONEUROIMMUNOLOGY. A PERSONAL REMINISCENCE

Istvan Berczi

Department of Immunology, Faculty of Medicine
University of Manitoba
Winnipeg, Manitoba, R3E 0W3, Canada

When the news came from Montreal in the summer of 1967 that I had been accepted by Dr. Selye to work in his Institute, I was madly in love with a girl who later became my wife. So I told her about the legendary Dr. Selye and outlined carefully that my highest desire was to work with him. After having listened carefully to the story, she asked me what is going to happen to our marriage plans now? "Well," I said, "would you like to get married before I go, or after I come back?" The answer was, without much hesitation, that we should get married before, so we did on September 3, 1967, and off I went a few weeks later to Montreal. The prospect was that we would spend an entire year, the duration of my planned stay in Montreal, separated as the common practice of the Hungarian authorities at the time was not to allow spouses or family members to follow individuals on study trips or other kinds of missions to the West.

My journey to Montreal by plane impressed me very much in that I flew across the ocean and set foot on another continent on the same day. I remembered my teenage readings about all the troubles, disease and death that Columbus and other sailors went through during the same trip. At the airport in Montreal one of my fellow countrymen, Dr. Pal Vegh, was waiting for me and escorted me to my room nearby the University which was rented by the Institute so that I would have a place to come to. The next morning after receiving directions from my French Canadian landlady, whose English was much better than mine, I set out to find the University on foot, and did so without much trouble. There it was, majestically standing on the top of a hill, like a cathedral of knowledge and science. As I learned later, not very far from the University there was indeed, on the hill a Roman Catholic cathedral as well.

Although I had learned a lot about Dr. Selye and his Institute from my colleague and good friend, Dr. Lorand Bertok, who was the first to go from Hungary, being there in person was not short of surprises. On the main corridor there were dozens of pictures of scientists, most of which I had only heard about by name, but had no idea what they looked like. My greatest surprise was that there was even a

Hans Selye

Stress

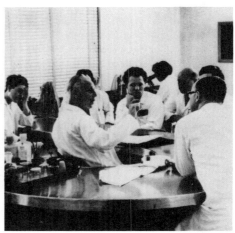

Discussion in the autopsy room

portrait of Kossa, who was a professor of my alma mater, and I knew that he pioneered histochemistry by the development of a staining method for the deposition of calcium in animal tissues, but I had never seen his portrait before. At the main entrance, near the elevator, there was the inscription:

> *"Neither the prestige of your subject, and the power of your instruments, nor the extent of your learnedness and the precision of your planning, can substitute for the originality of your approach and the keeness of your observation."*

I had never thought about these problems consciously before, but during my stay in the Institute, the message became engraved in my mind through numerous discussions about great discoveries, the history and psychology of science and of personal characteristics of scientists. In the corridor there was also a world map marked with little flags of all the countries from which students or visiting scientists came to the Institute. At my arrival there were students/scientists from Russia (even though this was at the height of the cold war after the Cuban crisis), from Korea, Japan, Egypt, Turkey, Italy, Germany, Czechoslovakia, Poland, Argentina, and of course, Canada. I couldn't help noticing that whenever Dr. Selye met someone in the corridor, frequently he spoke to him/her in his/her native language as he was fluent in Hungarian, German, English, French, and Italian and knew a fair bit of Spanish, Czech, Russian, and several other languages.

Soon after my arrival, Dr. Selye saw me in his office as he always found the time to greet in a very personal manner newly arrived students/scientists and visitors this way. After telling him about my trip we settled my project, which was to do experiments on calcifilaxis. Then he pointed to a bouquet of snow lillies (adelweiss) tied with a red, white and green ribbon, which was framed and hung on the wall, and said: "You see, this symbolizes my heritage and native land, which is Austria/Hungary. My father was Hungarian and a surgeon in the army and my mother was Austrian. So, by ancestry, I am neither Hungarian nor Austrian; I am Austro-Hungarian." I often wondered, why he called my attention to these things right at the beginning. Perhaps it was because I heard from a number of my colleagues while still at home that he was Hungarian, but then later on I also came across people who were convinced that he was Austrian or German. In any case, later on I was fortunate and privileged to have numerous private conversations with him through which I have learned a great deal about his personal views, character and attitude towards life in general. The other thing he pointed out to me in his office was a picture of desperate people that was hanging above the door of his office, and said, "You see, this is stress."

I was given an office in the Institute shared by Dr. Arpad Somogyi down the corridor and began the work. First I had to learn about the phenomenon of calcifilaxis quickly, which was by reading and by attending the daily rounds with Dr. Selye, when we examined the experimental animals collectively and each investigator had to present his/her animals, especially if symptoms of interest and/or importance were noticeable. The round finished in the autopsy room where Dr. Selye, with our participation, would look at every experimental animal that was terminated on that day. Again, the investigator had to call attention to pathological changes of possible importance.

After finishing with the autopsy, plans for new experiments were presented by the various investigators to Dr. Selye for his approval. Each experiment had to be justified on theoretical grounds and this frequently led to vivid discussions of scientific problems, theories and views. Each autopsy session also had a slide puzzle which was placed under the microscope and was to be viewed by everybody and diagnosed by a secret vote as to what it was. To my surprise, I noticed that some of the slides were coded in Hungarian which gave the correct answer. Apparently this strategy worked very well until some of us who understood the code arrived. One day a chicken experiment was evaluated and Dr. Selye remarked while looking at the carcasses that the lymph nodes along the neck were normal. Well, I had to tell him with all due respect that those nodules, though looking exactly like lymph nodes, were really the thymus, and that the chicken did not have any real lymph nodes. This came so quickly in compliance with my character that I hadn't even considered the possible consequences of contradicting this world famous great scientist in public, or even worse, in front of his students and collaborators. Intuitively, I felt that he would appreciate the information, which came from my training in comparative anatomy, and would not be insulted or take this as a criticism. This is exactly what happened. During a similar session weeks later he remarked that we all profit from these discussions and that he learned from me the other day that the chicken has no lymph nodes.

Being newly wed and separated and determined to be faithful I went to the Institute also on weekends. I had a lot to learn about my project so that I could propose decent plans for experiments and interpret the results, write up the papers, all of which was required if I was to last for some time in the Institute. To my surprise and great satisfaction, the Institute had a fantastic library and collection of reprints and I could hardly ask anything from the librarian, Mr. Krzyzanowsky, which was not on my desk by the end of the day. This was in sharp contrast with the situation in Budapest, where I had to travel frequently to the Central Medical Library, and sometimes wait in line for a particular book or journal for hours, or occasionally I had to return on another day in order to see the article I wanted. Once I got to see the article I was not allowed to take it out of the building, but had to sit down and very quickly take notes of the salient findings and conclusions. All these problems were solved by the quick availability of the material and by the use of the Xerox machine which was a wonder for me in the western world. Among the other wonders, I was most fascinated by the efficiency of the telephone which allowed one to phone instantly all over the world. The 15 months I spent in the Institute, not only was one of the most memorable periods of my life, but it was also very productive.

It didn't take long for Dr. Selye to find out that I was there every day as he also was there every day. For his weekdays he worked 12 hours a day (from 6:00 in the morning until 6:00 in the evening), and during the weekend he took it easy and stayed from 6:00 in the morning until only 4:00 in the evening or sometimes even until 2:00 in the afternoon. Soon he became a constant visitor to my office during the weekend for personal chats which was most exciting and memorable and I felt very honoured and privileged by this relationship. We discussed everything from politics to religion, history, personal life, even telling jokes. He called my attention to the fact that contrary to the prevailing opinion in Hungary, Ferenc Deak, who was responsible for entering into confederation with Austria, was not a traitor, but in fact, a very wise politician. He said that all I had to do was look around in Budapest and determine when the House of the Opera and a whole list of other beautiful public buildings were erected. "You will find," he said, "that much was built around the turn of the century between 1867 and 1914." I was astonished by this logical argument and asked myself, why I didn't think of this before? He also told me how angry he was about the two world wars, both of which he lived through; the terrible suffering and damage to Europe and, indeed, to mankind, these wars inflicted, and that one of the reasons why he escaped to North America was that science and creativity were virtually paralyzed in pre-Second World War Europe. "The only aristocracy I respect," he said, "is the aristocracy of intellect. Politicians," he pointed out, "are trying to guide their countries into the future while thinking in terms of the past, which is based on nationalism."

During our discussions he never missed the opportunity to point out that science is international and that the only thing which should count is the importance of the contribution. He had the greatest admiration for Claude Bernard, who was the first to recognize that the internal environment, which he called "*millieu interieur*", is under sophisticated physiological regulation.[1] Walter Cannon was the other scientist whom he regarded as his immediate predecessor in terms of thinking.[2] He also had high esteem for Pasteur and Koch and for many of his contemporaries that were invited to the Institute as "Claude Bernard Professors."

The visit of the Claude Bernard Professors to the Institute was usually for two days to deliver one or two lectures, participate in rounds, and to withstand the "roasting party" from members of the Institute, which was a very exciting event. The party was held at Dr. Selye's house in the evening to which, in keeping with good Hungarian customs, all the staff and students were invited. After having something to eat and one or two drinks to raise our spirits, we sat down around our guest and asked questions which one would not ask under any other circumstances. Almost anything was allowed, but the topic usually circled around the human factors that drive scientists, about the philosophy and psychology of science, the question of originality, discoveries, and so on. I met Dr. Aurelia Jancso this way and learned about the discovery of neurogenic inflammation, listened to George Palade, who later on received the Nobel Prize for his work, and got a chance to have some very interesting conversations with Dr. Elwin Kabat, who, as I realized later, was one of the founders of modern immunology. I can only appreciate now the caliber of Claude Bernard Professors that visited the Institute over the years and the diversity of topics discussed. These included Sir Macfarlane Burnett, Baruj Benacerraf, and Pierre Grabar, just to name a few.

These invitations did not only reflect Selye's ability of judging scientific accomplishment, but also testified to the flexibility of his mind. Lecturers were invited, not only from the well established disciplines in medicine, but also from the area of psychology, as he was always very conscious about looking at the whole organism in its entire complexity. I will never forget his reaction to the publication by Christian Barnard of the first heart transplantation in man.[3] He said that transplantation is not the solution to the problem, but rather, prevention of the disease is the right approach. He also remarked that transplantation will never be suitable for the routine treatment of people with heart disease. It will be a treatment only for the rich, he said, even if all the technical problems are solved, and stated that this approach may even generate criminal activity. Over the years I have remembered this many times, admiring how well he saw the problem and how right he was.

No matter what the subject was, he always challenged your imagination and intellect, therefore touching your emotions. I don't remember meeting anybody who was neutral towards him. Some admired him constantly, others were in violent opposition with almost everything he stood for, and there were a fair number of people who had a love/hate relationship with him. I have to confess that I have had my differences and difficulties, and I didn't always agree with his opinion. As I said already, I was working on calcifilaxis, and made a lot of effort to catch up with the literature and to produce meaningful results. But the news of the first heart transplantation in man shifted his interest back to his old favourite subject which was experimental cardiopathies. Before long, within a few weeks, the most important problem became cardiopathies in the Institute and not calcifilaxis and related phenomena. Somehow he managed to quickly convince most everybody to switch to this problem, except me. I felt that I should stick to my original problem as I only had a year to stay, and I planned to do my doctoral work there. My doctoral thesis would have had to be on a coherent subject and not calcifilaxis and cardiopathy. In the end he understood my reasons and agreed that I continue with my project for a few more months, and even asked me to co-author a review article[4] with him on the subject which was a great honour.

As I became more and more familiar with Selye's research interests and achievements, it became clear to me that his conviction was, in fact, that neuroendocrine factors play major roles in most, if not all, diseases. I learned from him, and also from his popular books,[5-9] that the discovery of the stress syndrome was accidental during his attempts to isolate some hormones from the placenta. For a while it was thought that the adrenal enlargement and involution of lymphoid organs was specific for a particular hormone, but attempts to purify it always failed as the activity was lost. At some point it occurred to Selye that this, in fact, could be a nonspecific response to nocuous agents, and, indeed, when he performed the control experiments, that was the case. He published a short note about his findings in Nature in 1936.[10] During the same year a longer article was published by him in the British Journal of Experimental Pathology,[11] where he demonstrated that the involution of the thymus was in fact mediated by the adrenal gland as it was absent in adrenalectomized animals if stressed. His experiments in chickens revealed that the Bursa of Fabricius is also extremely sensitive to steroid hormones.[12]

A review was published by Selye in 1946,[13] where he already gives a comprehensive theory of the *general adaptation syndrome* which is supported by experimental facts. He also talks about the possibility that diseases of adapation do exist. He states that, after exposure to stress, initially there is shock, which is followed by a counter shock phase, and this gradually goes into a stage of resistance. If, however, the stressor persists, resistance may go into exhaustion and death may ensue. He points out that specific and nonspecific resistance follow the same course but this latter "cross resistance" will fall much sooner and stays below normal during the period of resistance. He also presents data of blood sugar and chlorine changes and points out that white blood cell counts rise invariably during stress, regardless of the stressor used. The changes in the adrenal cortex and of thymus involution are also illustrated histologically. The adrenal cortex becomes wider with loss of lipid granules and the border between the zona fasciculata and reticularis is no longer distinct. The thymus shows a depletion of cortical thymocytes. Nuclear debris is evident and pyknotic thymocyte nuclei are abundant. He notes that this "accidental involution" becomes most pronounced during the countershock phase when the adrenal cortex reaches its maximum development. Large macrophages engulf the dead thymus cells and carry them away through the lymphatics. At the same time he noted that thymic reticulum reverts to its origianl epithelial type and the cells become roundish or polygonal and rich in cytoplasm. When involution is most acute the entire organ is distended with jelly-like edema. He points out that lymph nodes, the spleen and other lymphatic organs are almost as markedly affected as the thymus, although they do not involute quite as rapidly and their involution cannot be completely prevented by adrenalectomy. His summary figure, which is a fairly accurate outline of the acute phase response, as we recognize it toay,[14] is reprinted here (Fig. 1).

Today we know that a variety of insults, including trauma and infection stimulate the release of chemotactic-, proinflammatory cytokines, and a whole host of other mediators from a variety of cells in the damaged area that include mast cells, endothelial cells, platelets. The released mediators attract blood borne leucocytes, such as neutrophilic granulocytes, monocytes/macrophages, lymphocytes, eosinophils and basophils that release additional mediators, and thus contribute to the inflammatory response. In some cases certain cytokines, such as interleukin-1 (IL-1), tumor necrosis factor-α (TNF-α) and interleukin-6 (IL-6), become detectable in the

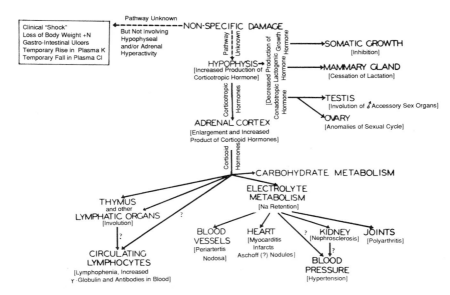

Figure 1. Functional interrelations during the general adaptation syndrome. Schematized drawing indicating that non-specific damage causes clinical shock, loss of body weight and nitrogen, gastro-intestinal ulcers, temporary rise in plasma potassium with fall in plasma Cl, through unknown pathways (nervous stimulus?, deficiency?, toxic metabolites?) but manifestly not through the stimulation of the hypophyseoadrenal mechanism. This is proven by the fact that the above manifestations are not prevented either by hypophysectomy or by adrenalectomy; they even tend to be more severe in the absence of either or both of these glands.

Non-specific damage, again through unknown pathways, also acts upon the hypophysis and causes it to increase corticotropic hormone production at the expense of a decreased elaboration of gonadotropic hormone production at the expense of a decreased elaboration of gonadotropic, lactogenic and growth hormones. The resulting corticotropic hormone excess causes enlargement of the adrenal cortex with signs of increased corticoid hormone production. These corticoids in turn cause changes in the carbohydrate (sugar active corticoids) and electrolyte metabolism (salt-active corticoids) as well as atrophy of the thymus and the other lymphatic organs. It is probable that the cardiovascular, renal, blood pressure and arthritic changes are secondary to the disturbances in electrolyte metabolism since their production and prevention are largely dependent upon the salt intake. The changes in γ-globulin, on the other hand, appear to be secondary to the effect of corticoids upon the thymicrolymphatic apparatus.

We do not know as yet, whether the hypertension is secondary to the nephrosclerosis or whether it is a direct result of the disturbance in electrolyte metabolism caused by the corticoids. Similarly, it is not quite clear, as yet, whether corticoids destroy the circulating lymphocytes directly, or whether they influence the lymphocyte count merely by diminishing lymphocyte formation in the lymphatic organs. Probably both these mechanisms are operative. (Taken from Ref.[13].)

blood and function as acute phase hormones. They act on the brain causing fever and other functional modifications (IL-1, TNF-α), release certain pituitary hormones and inhibit others (much of which is indicated in the graph above), promote general catabolism, (mediated primarily by TNF-α, also known as cachectin), stimulate the production of new serum proteins known as acute phase reactants in the liver (the joint action of IL-6, glucocorticoids and catecholamines), and also elevate the production of leucocytes in the bone marrow, the mechanism of which is not fully elucidated.[14,19] (For further reference, please see also the papers by Asa and Kovacs, Besedovsky and del Rey, Gaillard, and Nagy and Berczi in this volume.) Thus, with the recent discovery of cytokines and our increasing recognition of their functions, we have begun to fill in the gaps in Dr. Selye's adaptation syndrome outlined nearly half a century ago.

In 1949 Selye discovered that an inflammatory reaction, which can be induced in the rat by the parenteral administration of egg white, is inhibited by cortisone or by purified ACTH. On the other hand, desoxycorticosterone acetate, a mineral-corticoid compound, tends to aggravate the reaction.[15] These experiments initiated his interest in inflammation which became the most lasting topic in his research and led to the proposition later that diseases, like rheumatoid arthritis, anaphylaxis, etc. are in fact diseases of adaptation as stated in numerous publications. In his review article in Science,[16] entitled "*Stress and disease*", he shows a diagram of the stress response with inflammation clearly in mind (Fig. 2).

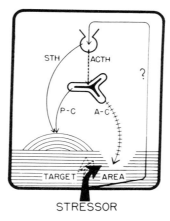

Figure 2. Diagram illustrating the principal pathways of the stress response.
A-C: glucocorticoids, also known as antiphlogistic corticoids; ACTH: adrenocorticotrop hormone; P-C: the mineralocorticoids, also known as prophlogistic corticoids; STH: somatotropic (growth) hormone. (Taken from Ref.[16])

The paragraphs in this article that contain important conclusions and theoretical considerations are as follows:

> "*Certain recent experiments suggest that, depending on the conditions, ACTH may cause a predominant secretion of one or the other type of corticoid. However, be this as it may, the 'growth hormone,' or somatotrophic hormone (STH), of the pituitary increases the inflammatory potential of connective tissue very much as the prophlogistic corticoids do; hence, it can sensitize the target area to the actions of the prophlogistic corticoids. . .*
> "*Among the derailments of the general adaptation syndrome that may cause disease, the following are particularly important: (i) an absolute excess or deficiency in the amount of adaptive hormones (for example, corticoids, ACTH, and STH) produced during stress; (ii) an absolute excess or deficiency in the amount of adaptive hormones retained (or 'fixed') by their peripheral target organs during stress; (iii) a disproportion in the relative secretion (or fixation) during stress of various antagonistic adaptive hormones (for example, ACTH and antiphlogistic corticoids, on the one hand, and STH and prophlogistic corticoids, on the other hand); (iv) the production by stress of metabolic derangements, which abnormally alter the target organ's response to adaptive hormones (through the phenomenon of 'conditioning'); and (v) finally, we must not forget that, although the hypophysis-adrenal mechanism plays a prominent role in the general-adaptation syndrome,*

other organs that participate in the latter (for exmple, nervous system, liver, and kidney) may also respond abnormally and become the cause of disease during adaptation to stress. . .

"Corticoid requirements during stress. During stress, the corticoid requirements of all mammals are far above normal. After destruction of the adrenals by disease (as after their surgical removal), the daily dose of corticoids, necessary for the maintenance of well-being at rest, is comparatively small, but it rises sharply during stress (for example, cold, intercurrent infections, and hemorrhage), both in experimental animals and in man. . .

"Anti-inflammatory effects of corticoids. The same antiphlogistic corticoids (cortisone and cortisol) that were shown to inhibit various types of experimental inflammations in laboratory animals exert similar effects in a human being afflicted by inflammatory diseaes (for example, rheumatoid arthritis, rheumatic fever, and allergic inflammations). . .

"Sensitivity to infection after treatment with antiphlogistic corticoids. In experimental animals, the suppression of inflammation by antiphlogistic hormones is frequently accompanied by an increased sensitivity to infection, presumably because the encapsulation of microbial foci is less effective and perhaps partly also because serologic defense is diminished. . .

"Psychological and psychiatric effects of corticoid overdosage. It has long been noted that various steroids - including desoxycorticosterone, cortisone, progesterone, and many others - can produce in a variety of animal species (even in primates such as the rhesus monkey) a state of great excitation followed by deep anesthesia. It has more recently been shown that such steroid anesthesia can also be produced in man, and, of course, the marked emotional changes (sometimes bordering on psychosis) that may occur in predisposed individuals during treatment with ACTH, cortisone, and cortisol are well known. Several laboratories reported furthermore that the electroshock threshold of experimentl animals and their sensitivity to anesthetics can be affected by corticoids."

He concludes as follows:

"If I may venture a prediction, I would like to reiterate my opinion that research on stress will be most fruitful if it is guided by the principle that we must learn to imitate - and if necessary to correct and complement - the body's own autopharmacologic efforts to combat the stress factor in disease."

The prediction by Selye that the pituitary gland has the capacity to both stimulate and inhibit inflammatory reactions is the subject of recent investigations and is proven correct.[17,18] The notion of prophlogistic steroids has not been studied to a great extent to date, but the antiinflammatory effect of glucocorticoids is firmly established and it is clear today that the adrenal gland plays an important physiological role in the regulation of immune and inflammatory responses.[19] The disproportion of hormones and other mediators, altered responsiveness in tissues and the significance of metabolic derangements during acute phase reactions related to sepsis, severe trauma and shock are the subject of current investigations and deemed to be highly relevant to prognosis. The involvement of the central nervous system, the liver and of other organs, such as the kidney, is also substantiated.[14] That "conditioning" may also play a role in host defence is also gaining ground.[20] Some hard evidence is forthcoming regarding the corticoid requirements during infection and other forms of stress.[14] The antiinflammatory effect of cortisone and cortisol are well recognized and are widely applied in medicine today. That corticosteroids increase the sensitivity to infection is of common knowledge. The phenomenon of stress related anesthesia is well recognized, but opioid peptides rather than steroid hormones are considered to be the mediators.

My colleague, Dr. Lorand Bertok, upon his return to Budapest from Selye's Institute, brought the news that the study of mast cells was in the focus of interest and he also brought a copy of the book written by Dr. Selye: *The mast cells.*[21] As it turned out, the inflammation Dr. Selye induced years earlier in rats by the injection of egg white was due to the discharge of mast cells. Because of my intentions to go to Dr. Selye's laboratory, I studied this book and was struck by the importance of mast cells in various pathological phenomena. At that time the role of mast cells in immune mechanisms had not been firmly established. However, a related cell type, the basophilic leucocyte's role in immune mechanisms was known already. Selye was interested in mast cells mainly because they play a major role in inflammation and he was puzzled by the powerful effects of mediators released by mast cells which could play a role in various pathological processes, such as inflammation, necrosis, calcification or thrombohemorrhagic phenomena.[4,21-24] After my arrival in Montreal, the name of Professor Jancso and coworkers came up frequently during discussions because of their discovery that the stimulation of sensory nerves induces inflammation.[25] By 1967 Professor Jancso had passed away so his wife and long time collaborator, Dr. Aurelia Jancso, was invited to the Institute as a Claude Bernard Professor. It was most exciting to hear the story of *neurogenic inflammation* from someone who participated in its discovery.

Looking at Dr. Selye's principal areas of research during his scientific career, which spanned half a century, one may observe that even if they seem to be very different, all of them were inspired by the profound conviction of Dr. Selye that neuroendocrine mechanisms play a role in what he called "*diseases of adapation.*" Once he decided to work on an area, he thoroughly surveyed the literature which was usually published then in the form of a book. In these books[21,26-34] a possible role of neuroendocrine mechanisms in relation to the subject was never neglected, even in the event of not much information being available. Then he presented related findings from his own laboratory, which also appeared in numerous publications (his lifetime output was 1,325 papers). His contributions to modern immunology are major indeed. He discovered that steroid hormones regulate lymphoid organs such as the thymus, spleen, lymph nodes and Bursa of Fabricius. He demonstrated that thymic atrophy is mediated by the ACTH-adrenal axis during stress with glucocorticoids being the final effector molecules. He also described the antiinflammatory action of adrenal steroid hormones. He and his coworkers made significant contributions to our understanding of the role of mast cells in various pathological phenomena. Selye made all these contributions without knowing the function of the thymus, lymph nodes or the Bursa of Fabricius. The function of these organs was understood only a few years prior to my arrival in Montreal. I followed these developments closely because my major interest has always been immunology.

One of Selye's major dilemmas was that he was never able to define stress. In his comprehensive review published in 1946[13] he talks about the alarm reaction that is comprised of shock and counter shock, the recovery phase which leads to resistance and eventually, if the stress lasts, to breakdown. However, 20 years later when I got to his laboratory he recognized already that not only damaging, unpleasant or even dangerous events/agents, but also pleasurable experiences evoke a neuroendocrine response that involves the ACTH-adrenal axis. As a matter of fact, I think he realized that the pituitary gland constantly reacts to our internal and external environment, which enables us to function normally. Therefore, he began to talk about the stress of life and divided stress into two subgroups: *eustress* and *distress*, the former referring to pleasurable and the latter referring to unpleasant or dangerous impulses.[35,36] Another difficulty was the realization that there is great variance in individual reactivity towards stressors and that the same stimulus which could be pleasurable to one individual may be very stressful to another. I remember him talking about the frustrated businessman who was forced to take a holiday by his family while he had so much to do. Perhaps this is why he showed me the picture of desperate people above the door in his office because that was the best definition he could provide. Looking at a picture we all formulate an idea of what it could mean though we are not absolutely certain, and it is likely that each person who looks at it will have somewhat different impressions.

The prediction by Dr. Selye of the pluricausal nature of most diseases is really the recognition that living organisms have evolved multiple mechanisms to defend themselves against harmful agents. For this reason, in most cases, it is necessary to interfere with these defense mechanisms at more than one point to cause disease. The redundancy of immune effector mechanisms[14,37] or the recent recognition that it is necessary to deregulate more than one gene to cause cancer,[38] certainly supports this view. In his last years he turned his attention to the protective power of certain hormones against various toxins and other noxious stimuli and created the term "*catatoxic steroids*"[29,35] for those hormones that have protective effect. A book of two volumes entitled "*Hormones and Resistance*"[27] has also been published by him on the same topic. That hormones are important in immunological and other forms of resistance is the subject of current scientific inquiry.[14,17,18]

Selye faced many criticisms and by the time I came to his laboratory he had been highly regarded only by psychologists, whereas most scientists in other disciplines had ignored his ideas. It was widely held that his experiments were highly artificial and hence not relevant to pathophysiological processes that occur in real life. It was also a problem that stress was not and could not be defined. In most disciplines scientists were busy working to discover and characterize basic interactions like the interaction of antigen presenting cells, T cells and B lymphocytes within the immune system, and were not concerned about stress. Another difficulty was that neither the basic knowledge nor the tools necessary to study and understand further the stress problem were available. Clearly, one could not possibly study the effect of interleukin-1 and of other cytokines on the release of pituitary hormones before these mediators were discovered.

He knew about this criticism already in 1955 and refers to it in his article in Science[16] as follows:

> "*Pasteur, Koch, and their contemporaries introduced the concept of specificity into medicine, a concept that has proved to be of the greatest heuristic value up to the present time. Each individual, well-defined disease, they held, has its own specific cause. It has been claimed by many that Pasteur failed to recognize the importance of the 'terrain,' because he was too preoccupied with the pathogen (microorganism) itself. His work on induced immunity shows that this is incorrect. Indeed, at the end of his life he allegedly said, 'Le microbe n'est rien, le terrain est tout.'*"

Only a scientist with very strong convictions of being right would maintain his position in the face of such mounting criticisms and he did so. Moreover, he dedicated his entire scientific career to furnishing more and more proof for the importance of neuroendocrine mechanisms in the development of disease. After his forced retirement at the age of 70 he organized the International Institute of Stress and remained active in promoting his cause.

Selye did not only present his findings in scientific journals, but also wrote popular books about the process of scientific research, about stress and related subjects.[5-9] These books have been translated into many languages. Of his popular books I was most impressed by the one having the title: *"In vivo."*[5] This is a collection of lectures dealing with his philosophy of the scientific process, with the human factors involved, and giving advice to scientists with regards to what is important and what is not so important. On the first page there is a quotation from Albert Szent-Gyorgyi:

> *"In studying life, you keep diving from higher levels to lower ones until somewhere along the way life fades out, leaving you empty handed. Molecules and electrons have no life."*

The main message of the book is that one always should try to view the entire organism in its complexity and not to get lost with tiny little details without even considering how it relates to the whole organism. He considers that although intellect is important in science, instinct is indispensable as well. He classifies scientists into problem finders and problem solvers as follows:

> *"In problem finding, the principal requirement is inspiration, perhaps with a certain amount of opportunism, a tendency to follow the line of least resistance rather than the steadfast pursuit of what we set out to find. On the other hand, problem solving is based on careful planning and persistence on a steady course until the aim is reached. It requires patience and the courage to resist all temptations to start on something new in the hope of quick returns. Here perspiration is often more effective than inspiration. Yet, even in problem solving, it is rarely possible to be guided merely by the laws of logic."*

The roles of logic, intellect and of hypotheses in science are summarized by him as cited below:

> *"No matter how distasteful this may be to many scientists, we must accept the fact that intellect is not always the safest approach to exploration and the acquisition of knowledge. The homing pigeon's 'knowledge' of geography, the bat's 'understanding' of radar are very effective though not intellectual; it is not through the logical rules of grammar that a child learns his native tongue. . .*

> *"Even the construction of hypotheses is much less dependent upon logical reasoning than most people think. No hypothesis can be arrived at by logical reasoning alone since it must be based on insufficient evidence, or else it is not a hypothesis at all but a factual conclusion. Indeed, the more a lack of facts forces a hypothesis to depend upon imagination, the more ingenious it is."*

The importance of originality is illustrated through the example of Mendel who had no laboratory and, as a matter of fact, was not even a scientist, yet he discovered the basic laws of genetics. Then he describes the discovery of penicillin, which is attributed to Fleming. He points out that the phenomenon described by him had, in fact, been described repeatedly before, namely that bacterial colonies will not grow around colonies of mold on solid agar containing medium. He speculates that there could have been a lot more people who saw the phenomenon but did not publish it as it indicates sloppiness in work. Normally, he remarks, mold is not supposed to grow in bacterial cultures. The importance of Fleming's discovery was fully realized after Florey and Chain demonstrated its practical value. Indeed, he says, it may be asked whether even Fleming himself fully appreciated all the potentialities of what he had found since he abandoned this field for many years in favour of less important investigations.

Discovery, he states, is the realization that something new exists. This necessarily means that the finding is of something unpredictable, but discovery does not necessarily imply importance. To be important the discovery must not only be unexpected, but also generalizable, i.e. applicable to many situations. Only this gives it real scope. In contrast, development is the further exploration of an already discovered fact. Because discoveries cannot be planned, intuition and "peripheral vision" play the most important role in this process.

In defence of *in vivo* research he remarks:

"It would hardly have been possible to discover anaphylaxis, yellow fever, and the phenomenon of homograft rejection by the use of electron microscope or of cell chemistry. . .

"You could never learn what a mouse is like by carefully examining each of its cells separately under the electron microscope any more than you could appreciate the beauty of a cathedral through the chemical analysis of each stone that went into its construction."

While he acknowledges that specialization is unavoidable in research, he emphasizes a need for "general practitioners" of science:

"Nowadays, you rarely meet a mature scientist who has retained at least the recent medical graduate's general knowledge of, say histology, physiology, biochemistry, pharmacology, clinical medicine and surgery. Most of them try to specialize and become experts in one field. This is undoubtedly the sound attitude for the vast majority whose primary interest is the solution of well-defined problems. But we shall always need at least a few general practitioners of medical research, men whose minds are open to the many things that come their way. We shall depend upon them in research just as we shall always require general practitioners of clinical medicine who can look at the patient as a whole and at least determine to which kind of specialist he ought to be sent."

While I was at the Institute of Experimental Medicine and Surgery, one of my colleagues, Dr. Cirpilli, who came from Turkey, remarked that Dr. Selye could have become equally as well a writer, a linguist, a philosopher, or a politician. Indeed, he was not only much ahead of his time in science, but also with regards to his views on politics and society. His efforts to popularize science and teach important general principles to his young colleagues was just as important as his scientific activity.

If you ask me what did I learn from him, the answer is very simple: a lot. I became convinced by his work that the immune system is under neuroendocrine regulation and never gave up my intentions to do research in this field, despite the funding difficulties which sometimes seemed insurmountable. I never forgot Selye's teaching about originality and that one must keep an open mind and adjust the theories to the facts found and not the other way around. Ever since we started to work on the role of the pituitary gland in immune function with my dear colleague, Dr. Eva Nagy, we trusted *in vivo* research and it was not disturbing for us that for years we were unable to demonstrate a repoducible effect of prolactin or growth hormone on *in vitro* immune reactions. We always reasoned that mother nature is more reliable than our primitive *in vitro* experiments and that if it does not work in the bottle, there must be some important detail that we don't know and this is why we cannot do it. Erythropoietin and some of the colony stimulating factors that regulate bone marrow function have already been produced by recombinant DNA technology, which could be considered as a triumph of *in vitro* research.[37] However, the role of pituitary hormones in hemopoiesis has never been uncovered by these experiments. In contrast, one can find good indication for such a role in the old literature[38] and we saw it very clearly in our *in vivo* experiments.[39] So again the *in vitro* experiments, although highly sophisticated and successful, did not give a complete picture and it was necessary to investigate hemopoiesis *in vivo* in the complexity of the organism in order to uncover further aspects of the regulatory network involved in bone marrow function.

Through my endeavours I have learnt over the years that there is much truth in the old literature. Many consider old papers to be useless and irrelevant, yet with proper insight one can gain a lot of information and even confidence from the results of early investigators. Consequently I cannot help but be fascinated and puzzled by the foresight and wisdom of those scientists who, with primitive methodology and with enormous gaps in knowledge, predicted biological laws which got forgotten only to be rediscovered later again. Hans Selye was one of these geniuses. He saw very clearly what his contemporaries were unable to see. It is only now that we are beginning to understand what he was really talking about. Indeed, I realized myself that I have evolved over the years to appreciate more and more of what he was saying, which deepened my admiration towards him. I feel very fortunate and privileged to have been associated with him.

ACKNOWLEDGEMENTS

I am much obliged to the Hans Selye Foundation, to Mme. Louise Drevet Selye and to Dr. Beatriz Tuchweber-Farbstein, who is president of the Foundation, for providing material and valuable information for this manuscript. I also thank Mrs. Jean Sylwester for her devoted work of processing this text.

REFERENCES

1. C. Bernard, Lecons sur les phenomenes de la vie communs aux animaux et aux vegetaux, *Balliere, Paris* 1878.
2. W.B. Cannon, The emergency function of the adrenal medulla in pain and the major emotions, *Am. J. Physiol.* 33:356 (1914).
3. C.N. Barnard, A human cardiac transplant, an interim report of a successful operation performed at Groote Schuur Hospital, Cape Town, *S. Afr. Med. J.* 41:1271 (1967).
4. H. Selye and I. Berczi, The present status of calciphylaxis and calcergy, *Clin. Orthop. Related Res.* 69:23 (1970).
5. H. Selye, "In Vivo: The Case for Supramolecular Biology," Livesight Pub. Co., New York (1967).
6. H. Selye, "The Story of the Adaptation Syndrome," Acta Inc. Med. Pub., Montreal (1952).
7. H. Selye, "The Stress of Life," McGraw Hill, New York (1956).
8. H. Selye, "From Dream to Discovery," McGraw Hill, New York (1964).
9. H. Selye, "The Stress of My Life: A Scientist's Memory," Van Nostrand Reinhold, New York (1979).
10. H. Selye, A syndrome produced by diverse nocuous agents, *Nature (Lond)* 138:32 (1936).
11. H. Selye, Thymus and adrenals in the response of the organism to injuries and intoxication, *Brit. J. Exp. Path.* 17:234 (1936).

12. H. Selye, Morphological changes in the fowl following chronic overdosage with various steroids, *J. Morph.* 73:401 (1943).
13. H. Selye, The general adaptation syndrome and the diseases of adaptation, *J. Clin. Endocrinol.* 6:117 (1946).
14. I. Berczi and E. Nagy, Neurohormonal control of cytokines during injury, *in:* "Brain Control of Responses to Trauma," N.J. Rothwell and F. Berkenbosch, eds., Cambridge University Press (1994) (in press).
15. H. Selye, Effect of ACTH and cortisone upon an "anaphylactoid reaction," *Can. Med. Assoc. J.* 61:553 (1949).
16. H. Selye, Stress and disease, *Science* 122:625 (1955).
17. I. Berczi, Immunoregulation by pituitary hormones, *in:* "Pituitary Function and Immunity," I. Berczi, ed., CRC Press, Boca Raton, FL (1986).
18. I. Berczi, The role of the growth and lactogenic hormone family in immune function, *Neuroimmunomodulation* (1994) (in press).
19. H.O. Besedovsky and A. del Rey, Immune-neuroendocrine circuits: integrative role of cytokines, *Front. Neuroendocrinol.* 13:61 (1992).
20. R. Ader and N. Cohen, The influence of conditioning on immune responses, *in:* "Psychoneuroimmunology II," R. Ader, D.L. Felten and N. Cohen, eds., Academic Press (1991).
21. H. Selye, "The Mast Cells," Butterworth, Washington (1965).
22. H. Selye, G. Gabbiani and B. Tuchweber, The role of mastocytes in the regional fixation of blood borne particles, *Brit. J. Exp. Path.* 44:38 (1963).
23. H. Selye, Mast cells and necrosis, *Science* 152:1371 (1966).
24. H. Selye, "Thrombohemorrhagic Phenomena," C.C. Thomas, Springfield (1966).
25. N. Jancso, A. Jancso-Gabor and J. Szolcsanyi, Direct evidence for neurogenic inflammation and its prevention by denervation and by pretreatment with capsaicin, *Brit. J. Pharmacol.* 31:138 (1967).
26. H. Selye, "Anaphylactoid Edema," Warren H. Green, St. Louis (1968).
27. H. Selye, "Hormones and Resistance," 2 vols., Springer-Verlag, New York (1970).
28. H. Selye, "The Pluricausal Cardiopathies," Charles C. Thomas Pub., Springfield (1961).
29. H. Selye, Catatoxic steroids, *Can. Med. Assoc. J.* 101:51 (1969).
30. H. Selye, "Stress," Acta Inc., Montreal (1950).
31. H. Selye, "The Chemicla Prevention of Cardiac Necroses," Ronald Press, New York (1958).
32. H. Selye, "Calciphylaxis," Univ. Chicago Press, Chicago (1962).
33. H. Selye, "Experimental Cardiovascular Disease," Springer-Verlag, New York (1970).
34. H. Selye, "Stress in Health and Disease," Butterworth, Boston (1976).
35. H. Selye, Hormones and resistance, *J. Pharm. Sci.* 60:1 (1971).
36. H. Selye, Forty years of stress research: principal remaining problems and misconceptions, *Can. Med. Assoc. J.* 115:53 (1976).
37. S.C. Clark and R. Kamen, The human hematopoietic colony-stimulating factors, *Science* 236:1229 (1987).
38. I. Berczi and E. Nagy, Effects of hypophysectomy on immune function, *in:* "Psychoneuroimmunology II," R. Ader, D.L. Felten and N. Cohen, eds., Academic Press (1991).
39. E. Nagy and I. Berczi, Pituitary dependence of bone marrow function, *Brit. J. Haematol.* 71:457 (1989).

HISTAMINE, CAPSAICIN AND NEUROGENIC INFLAMMATION. A HISTORICAL NOTE ON THE CONTRIBUTION OF MIKLÓS (NICHOLAS) JANCSÓ (1903-1963) TO SENSORY PHARMACOLOGY

Gábor Jancsó

Department of Physiology, Albert Szent-Györgyi Medical University
Dóm tér 10., H-6720 Szeged, Hungary

Miklós (Nicholas) "Jancsó was undoubtedly the greatest genius of post World War II Hungarian biomedical sciences," noted Professor Szentágothai in his introductory remarks to a symposium on "Capsaicin and the sensory system" held in Budapest in 1985.[1] It is not possible to give here even a superficial account of N. Jancsó's scientific achievements. Instead, the intention of the present author is to draw attention to some interesting aspects of his investigations and to show that, despite the apparent diversity of the fields of medical research in which he was involved, his work is characterized by a remarkable unbroken thematic continuity. This is also emphasized in the excellent book by Professor B. Issekutz senior,[2] in which he wrote about the life and work of Jancsó and his father, N. Jancsó senior, who was a renowned malaria researcher and Director of the Department of Internal Medicine at the Kolozsvár (now Cluj, Rumania) and later on at Szeged University. Professor Issekutz was himself an outstanding pharmacologist, who headed the Pharmacology Department at Szeged and later at Budapest University. He greatly appreciated the brilliance of his young colleague and supported him in his career in many ways. He established a chemotherapy division for Jancsó in the Pharmacology Department at Szeged University after he had completed a successful Fellowship at the Robert Koch Institute in Berlin in 1931. In the pre-war period of his scientific career, Jancsó made a significant contribution to the understanding of the mechanisms involved in the action of chemotherapeutic agents, as will be mentioned briefly in this article. However, there is little doubt that it is the discovery of the selective effect of capsaicin on sensory neurones for which he is mostly renowned today. In this short historical note, therefore, particular emphasis is given to his studies which led to the recognition of the unique pharmacological actions of capsaicin-type compounds and the mechanisms of neurogenic inflammatory responses.

In his early research, Jancsó was engaged in investigations into the mechanism of action of chemotherapeutic agents. He was the first to succeed in the development of a histochemical technique suitable for the demonstration of the tissue distribution

Advances in Psychoneuroimmunology, Edited by
I. Berczi and J. Szélenyi, Plenum Press, New York, 1994

Miklós Jancsó

of arsenobenzene compounds, including Salvarsan.[3] During his stay at the Robert Koch Institute in Berlin in 1930/31, he studied the action of trypanocide agents and made several important observations. He provided direct evidence in support of Ehrlich's suggestion that chemotherapeutic agents exert their effects via direct action on the microorganisms. Indeed, by means of fluorescence microscopy, he demonstrated the presence of acridine compounds within the trypanosomes isolated from the blood of infected mice. Furthermore, he showed that the trypanosomes resistant to acridine compounds failed to fluoresce. Hence, he concluded that the failure of the dye to accumulate in resistant trypanosomes is the cause of the resistance against these trypanocide agents.[4]

These studies led to the discovery of a group of new trypanocide drugs, the guanidines.[5] As pointed out by Gaddum,[6] this was one of the most interesting discoveries in chemotherapy. It was based on reasoning relating to the presumed mechanism of action of trypanocide drugs. Many studies have shown that the survival of trypanosomes outside the body is dependent on an adequate supply of glucose. It was therefore clear that any drug which deprived the trypanosomes of glucose, or which prevented them from using glucose, would be effective against them. In 1934, Jancsó came to the conclusion that germanin actually acted this way[7]. He therefore decided to try synthalin, because this causes a fall in blood sugar through its action on the liver and might be expected to deprive the trypanosomes of their main nutrient. Hence, the effect of synthalin on mice infected with trypanosomes was examined. The experiments were successful, and Jancsó claimed that this was the first discovery of a compound with chemotherapeutic activity through theoretical reasoning rather than by pure accident. He also pointed out that this was impressive evidence of the importance of investigation of the mode of action of drugs.[5] It was later shown that synthalin exerts direct action on the trypanosomes (hypoglycaemia on its own had little effect on them), and a number of similar compounds were synthesized, some of which were very effective against trypanosomes and other microbes.[8]

In the course of these experiments, Jancsó became interested in the role of the reticulo-endothelial system in the defence mechanisms of the body. He devised a new method involving systemic injection of electrocolloidal copper, which allowed a selective destruction of the cells of the reticulo-endothelial system, and in particular the Kupffer cells of the liver. This approach offered a possibility to estimate the contribution of the reticulo-endothelial system to the defence mechanisms of the organism and its role in chemotherapeutic action.[9] In the next twenty years or so, he continued his studies on the function of the reticulo-endothelial system. The results of these studies were summarized in a monograph [10] which contained mostly previously unpublished work. In the space available here, it is not possible to give even a superficial account of the work contained in this book. It includes the results of the early experiments on the effects of capsaicin, since these originated from studies on the effects of histamine on the function of the reticulo-endothelial system and on the vascular epithelium.

Studies on the reticulo-endothelial system led Jancsó to observations of the effects of histamine on the functions of the reticulo-endothelial cells and of vascular endothelial cells. In a paper published in Nature in 1947, he suggested that histamine is a physiological activator of the reticulo-endothelial system, and furnished convincing evidence for this. He showed that Kupffer cells of the isolated perfused rat liver phagocytose colloidal particles. If the perfusion fluid contained an antihistamine, phagocytosis by Kuppfer cells was inhibited. Similarly, the Kupffer cell function was strongly inhibited by previous in vivo histamine desensitization. The phagocytosing capacity was restored if histamine was added to the perfusion fluid.[11]

In the same paper he reported that painting the skin of rats, mice and guinea pigs with an alcoholic solution of histamine after a previous intravenous injection of a solution of Indian ink resulted in a black coloration of the affected skin area. In suitable histological preparations, it could be shown that the small blood vessels exhibit a characteristic labelling with carbon. This phenomenon is known as angiopexis. This could be prevented by antihistamines and also by histamine desensitization. [11]

Jancsó became particularly interested in the mechanism of the phenomenon of angiopexis, also known as vascular labelling. He showed that the histamine-induced accumulation of colloidal particles within the vascular wall can be inhibited by repeated local administrations of histamine of increasing concentrations, i.e., by local histamine desensitization. These findings therefore demonstrated that it is possible to achieve a significant localized change in the sensitivity towards the effects of histamine by repeated applications of the compound in increasing concentrations.

It seems that these studies on the effects of histamine played a crucial role in the discovery of the unique effects of capsaicin on sensory neurones. In further experiments he found that histamine is not the only stimulus which is able to produce the characteristic histological picture of vascular labelling. A number of irritants proved to be effective, including capsaicin, which is the pungent principle of red peppers. Thus, it was shown that the application of certain irritants, such as capsaicin, xylene or mustard oil, onto the skin produces the characteristic histological picture of angiopexis or vascular labelling. [12,13] The small blood vessels, and predominantly the venules, showed up clearly as a consequence of the presence of colloidal silver in their walls. This histological picture corresponds to that seen after topical histamine application. Therefore, Jancsó at first thought that the effect of capsaicin may be accounted for by a release of histamine. This assumption was reasonable, since desensitization to histamine inhibited the effect of several irritants in causing vascular labelling. Through a continuation of the reasoning in this line, the next step was logical; at least it seems especially obvious in retrospect. It was argued that, if it is possible to produce desensitization towards the effects of histamine by repeated administrations of histamine, it might be possible to achieve a similar desensitization towards the effects of capsaicin, provided it is administered to the animal repeatedly. The experiments initiated on this basis led to the discovery of the unique phenomenon of capsaicin desensitization and offered a new approach to the study of nociceptive and inflammatory responses.

The experiments with capsaicin began in the late thirties or early forties. The first results on the effects of capsaicin on the functions of the sensory nervous system were communicated by Jancsó and Aurelia Jancsó-Gábor to the Hungarian Physiological Society at the annual meeting in 1949. In this first report, the actions of capsaicin on sensory nerve endings are already clearly described. It was shown that, after repeated local or systemic administrations of capsaicin, rats, mice and guinea pigs become insensitive to the sensory irritant actions of capsaicin. It was pointed out in this early report that capsaicin interferes selectively with the functions of specific "pain receptors". The functions of receptors sensitive to innocuous and noxious mechanical stimuli were unaffected. The condition brought about by repeated injections of capsaicin and characterized by a selective insensitivity to chemogenic pain was termed capsaicin desensitization. The next reports on capsaicin appared only ten years later. These described the effects of capsaicin on chemosensitive pain receptors and on hypothalamic warmth receptors. [14-16]

To illustrate the selective effects of capsaicin on chemosensitive pain receptors, it seems appropriate to recall a few sentences from Jancsó's paper published in the Bulletin of the Millard Filmore Hospital in 1960.

"To cite a few examples to describe the behavior of desensitized animals: The desensitized guinea pig shows no signs of irritation even in a strongly lacrimating mist produced by aerosolization of ammonia or chloraceptophenone solution and behaves calmly. This is in sharp contrast to the control animals who reacted to this situation by blepharospasm, intensive lacrimation and violent scratching of their noses. Formalin, nicotine, veratrine, allyl alcohol solution or hypertonic saline or glucose solution can be instilled in the eye of the desensitized animal without causing lacrimation or chemosis. The nose of the desensitized animals can then be smeared with mustard oil or formic acid without resulting in scratching, lacrimation, or sneezing. Even dousing the nose with concentrated capsaicin solution failed to elicit sneezing, whereas the untreated guinea pigs sneezed from 10 to 20 times consecutively. Similarly, no pain and no defensive reflex is elicited following rubbing the skin with xylene, mustard oil or chloracetophenone solution. In sharp contrast, the corneal and sneezing reflexes can easily be elicited by tactile stimulus; the touch receptors function perfectly, and the sensory threshold for pain elicited by pinching, pricking, heat or electric current remains unaltered."

These findings showed that capsaicin desensitization resulted in a loss of protective reflexes and behavioral reactions evoked by chemical, but not mechanical stimuli. The significance of these findings was clearly recognized by Jancsó. He concluded that separate populations of sensory fibres, and in particular pain fibres, may be distinguished pharmacologically. In addition, he suggested that the transduction processes at these sensory receptors may also differ. Thus, using capsaicin as a highly selective pharmacological tool, it was possible to dissect separate populations of nociceptive afferents. Furthermore, the method of capsaicin desensitization allowed a selective, relatively long-lasting inhibtion of the function of this particular population of sensory neurones, and thereby the study of their functional significance.

Capsaicin applied onto the skin or mucous membranes or into the eye evokes not only nociceptive responses, but also inflammatory reactions. These responses can be particularly well studied on the eye or ear of guinea pigs or rats. The inflammatory responses of the conjunctiva or of the ears were quantitated by measuring the amount of silver contained in the inflamed tissue after a previous intravenous injection of a colloidal silver solution. Application of a solution of capsaicin into the eye resulted in a massive accumulation of silver in the inflamed tissue as compared with the control. It also caused clearly visible chemosis, i.e. oedema of the conjunctiva. Systemic capsaicin treatment practically completely abolished both silver accumulation and oedema. Similarly to systemic pretreatment with capsaicin, repeated local applications of capsaicin abolished the inflammatory response to capsaicin. These findings demonstrated that both systemic and repeated local administrations of capsaicin resulted in the development of a characteristic insensitivity towards its own action. These phenomena are known as local and systemic capsaicin desensitization, respectively. [17,18]

Further experiments revealed that a number of other irritants (e.g. xylene, mustard oil and chloroacetophenone) also cause an accumulation of silver in the affected tissues and that, most importantly, this can be largely inhibited by capsaicin pretreatment. The involvement of the nervous system in these inflammatory reactions was supported by the findings of experiments performed on denervated tissues. The extent of silver accumulation was not different from the control up to 24 hours after denervation. After a few days, however, silver accumulation could hardly be demonstrated in the denervated tissue. Obviously, this was related to the degeneration of the nervous structures, i.e. sensory nerve endings, which is complete only after a few days, and not merely after 24 hours.

These findings clearly indicated the existence of a specific form of inflammation which depends on a structurally and functionally intact sensory innervation. Direct evidence of the involvement of sensory nerves in neurogenic inflammatory responses was provided by experiments which showed that antidromic electrical stimulation of sensory nerves resulted in a characteristic inflammatory response. It emerged that stimulation at C-fibre strength is necessary to obtain an inflammatory response. This response could be conveniently demonstrated and measured by using a quantitative Evans blue technique introduced by Jancsó.[18]

Neurogenic inflammatory responses could also be demonstrated in man. This was shown first in a self-experiment performed by Jancsó on his facial skin.[17] The freshly shaved facial skin was painted repeatedly with a 0.5% solution of capsaicin and the skin temperature was recorded from both the treated and the untreated cheeks. Initially there was a significant difference between the temperatures of the two cheeks. The temperature of the treated skin considerably exceeded that of the control skin area. Applications were accompanied by marked hyperaemia and a burning pain. After repeated applications of capsaicin, however, the temperature difference between the two cheeks gradually decreased, as did the pain and hyperaemia; the skin became desensitized towards the actions of capsaicin. These findings demonstrated the existence of neurogenic inflammation in the human skin, and the involvement of capsaicin-sensitive nerves in that reaction. Subsequent studies established that the flare component of the triple response of the skin is mediated by capsaicin-sensitive nerves, since it can not be elicited in a capsaicin-treated skin area.[18]

Studies aimed at an understanding of the mechanisms of neurogenic inflammatory responses revealed that in the animal skin a neurogenic inflammatory response can invariably be elicited in a skin area treated with a local anesthetic. In the human skin, the flare component of the triple response cannot be evoked under local anesthesia, although a local response can be clearly observed at the site of contact with the irritant. On the basis of investigations on human and animal skin, the hypothesis was put forward that sensory nerve endings contain a vasoactive "neurohumor" which, upon release from these nerve terminals, causes vasodilatation and plasma extravasation. Release of this neurohumor can be elicited in two different ways: by antidromic stimulation of sensory nerves, and by orthodromic direct stimulation of pain receptors with irritants. The discharge of this substance cannot be inhibited with atropine, physostigmine, hexamethonium, phentolamine, antihistamines (H_1 antagonists) or serotonin antagonists. Desensitization with capsaicin, however, prevents the release of this mediator. Therefore, Jancsó introduced the term sensory neurone blocking agent[19] to denote this unique pharmacological character of capsaicin. Desensitization to the effects of capsaicin was explained by a (temporary) depletion of this neurohumor from these sensory nerve endings. He emphasized that these particular afferent fibres possess a dual function: they are involved in the transmission of nociceptive impulses towards the central nervous system; and through the release of a vasoactive neurohumor from their peripheral endings, they are involved in the mechanisms of antidromic vasodilatation and neurogenic plasma extravasation. This hypothesis offered by Jancsó[18] to explain the mechanisms involved in the processes served by capsaicin-sensitive neurones has needed little if any revision in the light of the experimental data subsequently obtained. The neurohumors released from these sensory nerve endings have now been identified and the morphological, neurochemical and functional traits of capsaicin-sensitive neurones are largely characterized.[20-24]

It was not my aim here to give an update concerning the field to which Miklós Jancsó made a pioneering contribution. Rather, I intended to give a short historical account of his achievements relating to the study of the mechanisms of pain sensation and neurogenic inflammatory responses, and to the discovery of the unique pharmacological actions of capsaicin-type compounds. In the past decade, the potential of capsaicin as a highly selective neurotoxin has been greatly exploited and we are now facing a complicated system of capsaicin-sensitive neurones of diverse morphological and functional character. Jancsó's original findings had a significant influence on this work and he may be regarded as a founder of a new field, sensory pharmacology.

REFERENCES

1. J. Szolcsányi, Miklós Jancsó, 1903-1966,*Acta Physiol. Hung.* 69:263 (1987).
2. B. Issekutz, sen., Id. Jancsó Miklós és ifj. Jancsó Miklós, a két orvostudós. (Nicholas Jancsó, sen., and Nicholas Jancsó, jr., the two medical scientists.) Akadémiai Kiadó, Budapest.
3. N. Jancsó, Eine neue histochemische Methode zur biologischen Untersuchung des Salvarsan und verwandter Arsenobenzolderivate, *Z. ges. exp. Med.* 61:63 (1928).
4. N. Jancsó, Wirkungsmechanismus der Chemotherapeutica bei Trypanosen, *Klin. Wschr.* 11:1305 (1932).
5. N. Jancsó and H. Jancsó, Chemotherapeutische Wirkung und Kohlenhydratstoffwechsel. Die Heilwirkung von Guanindinderivaten auf die Trypanosomeninfektion, *Z. Immunitatsforsch.* 86:1 (1935).
6. J.H. Gaddum, Discoveries in therapeutics, *J. Pharm. Pharmacol.* 6:497 (1954).
7. N. Jancsó and H. Jancsó, Mikrobiologische Grundlagen der chemotherapeutischen Wirkung. 1. Mitteilung: Wirkungsmechanismus des Germanins (Bayer 205) bei Trypanosomen, *Ztbl. Bakter.* 132:257 (1934).
8. H. King, E.M. Lourie and W. Yorke, New trypanocidal substances, *Lancet* II:1360 (1937).
9. H. Kroó and N. Jancsó, Die Bedeutung des Reticuloendothels für die Immunität and Chemotherapie, *Z. Hyg.* 112:544 (1931).
10. N. Jancsó, Speicherung; Stoffanreicherung im Retikuloendothel und in der Niere. Akadémiai Kiadó, Budapest (1955).
11. N. Jancsó, Histamine as a physiological activator of the reticulo-endothelial system, *Nature* 160:227 (1947).
12. N. Jancsó, Sichtbarmachung von Histaminwirkungen in den Geweben, *Ber. Physiol.* 126:475 (1941).
13. Z. Dirner, Wirkungsmechanismus von Hautreizstoffen, *Ber. Physiol.* 126:475 (1941).
14. N. Jancsó and A. Jancsó-Gábor, Dauerausschaltung der chemischen Schmerzempfindlichkeit durch Capsaicin, *Naunyn-Schmiedeberg's Arch. exp. Path. Pharmak.* 236:142 (1959).
15. J. Pórszász and N. Jancsó, Studies on the action potentials of sensory nerves in animals desensitized with capsaicine, *Acta Physiol. Sci. Hung.* 16:299 (1959).
16. N. Jancsó and A. Jancsó-Gábor, Die Wirkungen des Capsaicins auf die hypothalamischen Thermoreceptoren, *Naunyn-Schmiedeberg's Arch. exp. Path. Pharmak.* 251:136 (1965).
17. N. Jancsó, Role of the nerve terminals in the mechanism of inflammatory reactions, *Bull. Millard Fillmore Hosp.* (Buffalo, N.Y.) 7:53 (1960).
18. N. Jancsó, Desensitization with capsaicin and related acylamides as a tool for studying the function of pain receptors, *in*: "Pharmacology of Pain," R.K.S. Lim, ed., Pergamon Press, Oxford (1968).
19. N. Jancsó, Neurogenic inflammatory responses, *Acta Physiol. Acad. Sci. Hung.* Suppl. 23:3 (1964).
20. G. Jancsó, E. Király and A. Jancsó-Gábor, Pharmacologically induced selective degeneration of chemosensitive primary sensory neurones, *Nature* 270:741 (1977).
21. C.A. Maggi and A. Meli, The sensory-efferent function of capsaicin-sensitive nerves, *Gen. Pharmacol.* 19:1 (1988).
22. J. Szolcsányi, Capsaicin, irritation, and desensitization. Neurophysiological basis and future perspectives, *in:* "Irritation, Chemical Senses," Vol. 2, B.G. Green, J.R. Mason and M.R. Kare, eds., Marcel Dekker, New York and Basel (1990).
23. P. Holzer, Capsaicin: Cellular targets, mechanisms of action, and selectivity for thin sensory neurons, *Pharmacol. Rev.* 43:143 (1991).
24. G. Jancsó, Pathobiological reactions of C-fibre primary sensory neurones to peripheral nerve injury, *Exp. Physiol.* 77:405 (1992).

NEURAL REGULATION OF ALLERGY AND ASTHMA. THE DISCOVERIES AND SEMINAL CONTRIBUTIONS OF ANDOR SZENTIVANYI

Istvan Berczi

Department of Immunology
University of Manitoba
Winnipeg, Manitoba R3E 0W3, Canada

The aspirations of natural sciences are to discover and understand the laws of Mother Nature with the final goal of diminishing human suffering and maintaining human activity in harmony with our environment. The presentation of facts pertinent to establishing natural laws is the overwhelming consideration in the literature of natural sciences and the human and social factors seldom get presented even though they play a major role in the evolution of science. I often wonder what it was like to state for the first time that the earth is not the center of the universe, or that God did not create man, but human beings are the result of a process of evolution from simple organisms to more complicated ones, or to uncover that diseases may be caused by invisible microorganisms, just to mention a few of the major advances in natural sciences. Perhaps one can rephrase this question. How many people recognized that the earth is moving but were afraid to say so either because they did not trust their own judgement or, perhaps, because they were afraid of punishment for heretic thinking. In the Middle Ages, heretic thinking was punishable by death so it took an extreme amount of courage on the part of Galilei to publicize his discovery. Why did he do it? One can only resort to guesses to outline his reasons. First of all, he must have been absolutely convinced that the evidence that he found was sufficient to prove his point. Second, he must have trusted his judgement and was not discouraged from publicizing his conclusion by the possibility that no one would agree with him. He must have felt very strongly that it was so important to convince his fellow citizens of his discovery that he disregarded all the possible consequences, including severe personal punishment. Therefore, one may conclude that the true hallmark of a genius is to be able to discover new facts by keen observation and powerful reasoning and to have the confidence and courage to publicize the discovery. Although in modern times the presentations of true discoveries do not carry the risk of capital punishment, they may be dismissed by the scientific community for lengthy periods of time and thus deny recognition from the scientist, perhaps for a lifetime, which may have dreadful consequences to the individual. However, it is inevitable that ultimately the discovery

Advances in Psychoneuroimmunology, Edited by
I. Berczi and J. Szélenyi, Plenum Press, New York, 1994

Andor Szentivanyi

will be supported by irrefutable evidence and generally appreciated by the scientific community.

We are fortunate enough to present a personal account by Andor Szentivanyi of his discovery that immune reactions are regulated by the nervous system and to learn some of the historical facts and human factors that shaped his career throughout the years. The story begins in 1951 when he was a resident and was called to a patient having a severe asthmatic attack. He observed to his great disappointment that treatment with adrenalin was ineffective to alleviate the attack and he was not able to save the life of this individual. Did other people experience this phenomenon before? The answer is a definite "yes" as he found out from his professor the next day.

Did anybody else wonder why should it be that some patients react very well to treatment with adrenalin which could be lifesaving, whereas some do not? Perhaps, but these individuals either did not see beyond the problem or, even if they recognized that there might be a difference in the neural regulation of asthma in these individuals, they lacked the talent, ambition, and/or courage to try to solve the problem. In contrast, Andor Szentivanyi, after seeing at the bedside that this is a matter of life or death, made this problem the dominant project of his scientific enquiry, which still continues today.

Ingeniously Szentivanyi decided to use anaphylactic shock in the guinea pig as an animal model, which was known at that time to be due to systemic antigen-antibody reactions. This knowledge and the fact that serum antibodies are able to agglutinate antigenic particles (red blood cells, bacteria, etc.) or to precipitate soluble antigen sum up the basic knowledge of immunology of the time. Antibodies were presumed to be produced by the "reticuloendothelial system" and not by lymphocytes; the function of the spleen and of lymph nodes was presumed to be filtering of lymph and blood, and the thymus was regarded as a peculiar organ with lymphoid morphology and no known function. It was not recognized as yet that in addition to antibody mediated (humoral) immunity, other immune reactions are executed by thymus derived (T) lymphocytes (cell mediated immunity). Immune reactions were regarded as normal if they provided the host with protection (prophylaxis) and abnormal if the result was harmful (anaphylaxis) rather than protective. It was also known that some animals/patients were unable to respond to immunization, and thus they were categorized as "anergic", whereas those responding normally were "euergic", and the ones exhibiting deleterious reactions were considered "allergic". That some forms of hypersensitivies are mediated by a special subclass of antibodies (IgE), which are fixed to the surface of tissue mast cells and of basophils in the blood, was beyond consideration as was that other forms of hypersenstivity reactions are mediated by T lymphocytes or by immune complexes. To put this in a little more perspective, imagine that the morphological interaction of nerve fibers with tissue mast cells was found in 1951 (and probably was). Today this is considered to be a major evidence for neuroendocrine-immune interaction, whereas at that time this information was uninterpretable.

One can only understand the ingenuity of Szentivanyi after considering how little was known about immune reactions, asthma at the time and that the possible interaction of the nervous system with the immune system was beyond consideration as the latter hadn't even been defined. He had to design his experiments on the basis of scattered and confused knowledge and of suspicions and gut feelings. Clearly, it took special talent, unique insights, powerful reasoning and unusual courage to go ahead and try to solve the problem that he had identified at the bedside. What he found is truly remarkable: major discoveries that laid down the foundation to understanding how the immune system with all its powers of destruction functions in homeostasis and harmony of the organism. These discoveries pave the road to the clarification of pathophysiological pathways that are the basis of diseases that are due to the malfunction of the immune system. Most of these diseases are ill understood and poorly controllable today.

Once the problem had been identified it did not take long for the young Szentivanyi to get into action. He quickly learned the technique of inducing brain lesions in animals with great precision and within 6 months of observing the patient with adrenalin resistance he knew already that brain lesions inhibit the anaphylactic reaction in guinea pigs. The first report on these findings appeared under the title "Anaphylaxis and nervous system" in Acta Medica Hungarica in 1952.[1] This was followed by publications on the subject in German and also in English journals. The salient findings of these investigations may be summarized as follows: (i) Tuberal lesions (TBL) of the hypothalamus in preimmunized guinea pigs and rabbits inhibited the anaphylactic reaction elicited by intravenous application of the specific antigen. (ii) Antibody production was also inhibited if the lesions were induced prior to immunization. (iii) The reaction of antibodies with the specific antigen was not affected by TBL. (iv) TBL had no effect on the liberation of tissue material mediating anaphylaxis. (v) TBL increased the resistance of the animals to biologically active substances such as histamine, which was a temporary effect. (vi) TBL inhibited the anaphylactic reaction even when the animals were provided with passively transferred antibody which elicited lethal shock in normal animals after systemic challenge with the antigen. (vii) The Schultz-Dale test, which was performed with small pieces of intestine of the animals *in vitro*, was also inhibited if the donors were subjected to TBL. (viii) The Arthus reaction, turpentine induced inflammation and the Sanarelli-Schwartzmann phenomenon were uneffected by TBL. (ix) Lesions inflicted in other areas of the hypothalamus or the central nervous system had no influence on any of the above listed parameters. (x) Instead of inflicting lesions, electrical stimulation of the mamillary region of the hypothalamus had an inhibitory effect on the anaphylactic response and increased the resistance of animals to histamine.[2-5]

These findings attest that the following discoveries have been made: (a) The central nervous system (CNS) has a significant influence on immune reactions as measured by antibody formation. (b) Immediate hypersensitivity reactions (which are now known to be elicited by IgE antibodies fixed to tissue mast cells or basophils, and were measured by Szentivanyi using anaphylaxis and the Schultz-Dale test) are regulated by the CNS to a large extent, whereas other inflammatory responses due to immune complexes (Arthus reaction) to sensitized T lymphocytes (turpentine sensitivity) or to a thrombohemorrhagic reaction (Sanarelli-Schwartzmann phenomenon) are not readily influenced by brain lesions. (c) The reaction to histamine (a major mediator of immediate type hypersensitivity reactions) is also altered by TBL. These are major discoveries, indeed, done with remarkable experimental planning, and keen observations while contemporary confusion and misconceptions prevailed about the immune system and of immunologically mediated conditions.

More discoveries were to follow. Szentivanyi and coworkers showed that in mice the hyperglycemia which could be induced by histamine or serotonin administration was mediated by adrenalin-stimulated hydrocortisone release from the adrenal gland. The observation that histamine or serotonin treatment failed to increase blood sugar in pertussis-sensitized mice was interpreted as a failure of adrenalin action. This was confirmed by the treatment of these animals with adrenalin which was ineffective. A resistant state to the action of histamine and serotonin similar to those of pertussis-sensitized animals, could be induced by the treatment of mice with drugs capable of blocking β -adrenergic receptors. Pertussis-sensitized mice showed an increased incorporation of radioactive glucose into tissues and a hypersensitivity to histamine and serotonin.[6,7] It was also shown that the treatment of various strains of mice and guinea pigs by β -adrenergic blocking agents sensitized the animals to anaphylaxis or to its pharmacological mediators (histamine). A significant strain difference has also been observed in this respect and in some experiments an additive blockage of β - or simultaneous stimulation of α -adrenergic receptors could break the strain resistance to anaphylactic sensitization. Hyperglycemia and treatment with aminophylline also counteracted the histamine-sensitizing effect of β -adrenergic blockage in mice.[8]

The discoveries coming from these experiments are as follows: (a) Immunization has a significant metabolic effect as measured by serum glucose levels and incorporation into tissues. (b) Both the nervous and endocrine systems are involved (i.e. adrenalin and glucocorticoids) in the induction of these metabolic changes. (c) The function of β -adrenergic receptors is altered as the consequence of immunization which leads to increased sensitivity to inflammatory mediators such as histamine and serotonin. That immunization could alter metabolism or significantly affect neural transmission was unheard of at the time.

By 1968 Szentivanyi had synthesized the implications of all his findings with those of others which was presented in the *Journal of Allergy* under the title: "The β - adrenergic theory of the atopic abnormality in bronchial asthma."[9] In this article he argues that bronchial asthma, whether it is due to "extrinsic" or "intrinsic" causes, is ultimately elicited by the same mediators such as histmaine, serotonin, acetylcholine, slow reacting substances and plasma kinins. Catecholamines are also released during asthmatic reactions and thus could be considered as an additional group of mediators. He states that in many tissues and in most species the catecholamines are natural antagonists of the inflammatory mediators. He postulates that the atopic abnormality in asthma is due to the abnormal functioning of the β -adrenergic system irrespective of what the triggering event may be (Figure 1). He explains as follows:

> *"The β -adrenergic theory regards asthma not as an 'immunological disease' but as a unique pattern of bronchial hypersensitivity to a broad spectrum of immunological, psychic, infectious, chemical, and physical stimuli. This view gives to the antigen-antibody interaction the same role as that of a broad category of nonspecific stimuli which function only to trigger the same defective homeostatic mechanism in the various specialized cells of bronchial tissue."*

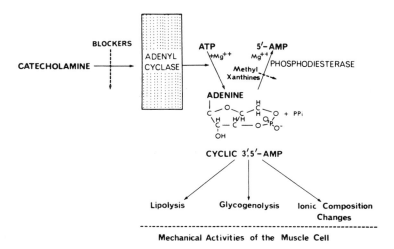

Figure 1. Schematic representation of the biochemical pharmacology of adrenergic action. (Reprinted from Ref. 9.)

Szentivanyi uses powerful arguments derived from his own experiments and those of others in a logical and compelling manner to support his thesis, as follows: (i) asthmatic persons show a remarkable bronchial hypersensitivity to the mediators of antigen-antibody responses even when the patients are completely asymptomatic; (ii) the pharmacological blockage of β -adrenergic receptors is capable of inducing a hypersenstivity to histamine, serotonin, acetylcholine, bradykinin, and slow reacting substance regardless of whether or not these substances are of endogenous or exogenous origin; (iii) β -adrenergic blockage increases bronchial reactivity to inhaled allergens in patients with seasonal allergic rhinitis without a previous history of asthma and increased airway resistance is also present in normal subjects up to 30 minutes after drug administration; (iv) α -adrenergic blockage ameliorates asthmatic attacks and reduces bronchial hypersensitivity; (v) bacterial products are capable of eliciting changes similar to the pharmacological blockage of β receptors; (vi) abnormal immune responses leading to the excessive production of mast cell sensitizing antibody may have a pathogenic role; (vii) epinephrine has an antiproliferative effect mediated by the β receptor and receptor dysfunction could conceivably produce enhanced lymphoproliferation; and (viii) β blockers can lead to a selective hyperplasia of lymphoid tissues and may function as immunological adjuvants.

The role of infectious agents in the induction of asthma is discussed at length with endotoxin given special consideration. Endotoxins are potent releasers of amine mediators, sensitizers of adrenergic target cells, and condition animals for pathological lesions such as hemorrhagic necrosis (Sanarelli-Schwartzmann reaction) at the site of epinephrine injection. In relation to viral infections he notes that killed influenza virus which does not elicit local inflammatory reactivity also produces pharmacological hypersensitivity. Evidence is cited that eosinophilia, which is frequently associated with asthma, may also be the consequence of β -adrenergic malfunction.

The beneficial effects of aminophylline and methylxanthines on asthma is attributed by Szentivanyi to the inhibition of phosphodiesterase, an enzyme responsible for the inactivation of 3',5'-AMP (Figure 1). The action of these agents is bypassing the biochemical site of the postulated β -adrenergic lesion and lower receptive threshold to catecholamine action. The moderating effect of glucocorticoids on asthma is explained by their general antiinflammatory effect. He notes that corticosteroids support the normality of target cell responses to catecholamines and that in the absence of glucocorticoids, receptor threshold for catecholamines may be raised occasionally up to the point of complete unresponsiveness. Finally, the reason for susceptibility to a variety of unrelated factors that can precipitate asthma is considered. He summarized the problem as follows:

> "In addition to the antigen-antibody interaction, asthmatic episodes are known to be triggered by a large variety of stimuli such as infection, various synthetic and natural chemicals, conditioned reflexes, psychic stimuli, changes in atmospheric pressure, inhalation of cold air, nonantigenic dust, fumes, and other irritants, etc. Any molecular interpretation of a susceptibility to such a large variety of unrelated stimuli would appear to necessitate the postulate that the primary lesion be connected with a final common pathway operating through a biologically unusually broad messenger system. The adenylcyclase-3',5'-AMP system is the first known messenger system capable of responding to a wide variety of neural, humoral, and hormonal agents subserving homeostasis."

The ingenuity of this article is that it considers a wide spectrum of facts which are seemingly unrelated and puts them into logical order and with powerful reasoning. The conclusion derived is that a neuroendocrine regulatory defect must be involved in the pathogenesis of asthma and of atopic diseases in general. This article has been much appreciated by clinical scientists, but is seldom considered in authoritative textbooks of immunology in relation to allergic diseases. In these texts the role of IgE antibodies to "allergens" and the inflammatory mediators released by allergen-IgE reaction from mast cells and basophils is presented as the major mechanism. The fact that the level of IgE antibodies frequently does not correlate with clinical symptoms is grossly ignored and asthma or atopic reactions elicited by nonimmunological factors are regarded as a clinical curiosity of minor importance, if any.

Ever since the presentation of the β-adrenergic theory of atopy and asthma, Szentivanyi remained faithful to the idea that the major pathomechanism underlying these diseases is related to the malfunction of the β-adrenergic signalling pathway. This common thread runs through virtually all of his numerous papers, reviews, book chapters and it is the subject of some of the 18 books he published. The major topics he engaged himself in were as follows: (a) studies on the nature and function of adrenergic (especially β-adrenergic) and cholinergic receptors;[1-34] (b) studies on adenyl cyclase, cyclic AMP and on signal transduction;[35-59] (c) the isolation, characterization and pharmacological modulation of phosphodiesterase;[43,60-85] (d) systemic effects of immunization and endotoxin, on the adrenergic and cholinergic systems, on metabolism, and on sensitivity to inflammatory mediators, etc.;[37,86-109] (e) clinical studies on asthma and related conditions.[12,15,22,25,28,49,51-53,72,110-121] He discovered that β-adrenergic subsensitivity does exist in patients with atopic dermatitis who never received adrenergic medication and, therefore, the therapeutic desensitization with adrenalin cannot account for the dysfunctional β-system.[16] It was also established that the β-adrenergic reactivity of lung tissue of lymphocytes and adipocytes from patients with atopic asthma was abnormal and various patterns of drug-vs-disease-induced subsensitivities could be recognized.[11,12,14,122-124] It was also concluded that the bronchial hyperreactivity to cholinergic agents in asthma is not mediated through cholinergic mechanisms but is caused by the adrenergic abnormality which is analogous to the so-called "denervation supersensitivity".[125-128] Lymphocytes of asthmatic patients showed a significant decrease in ligand binding to β-adrenergic receptors independent of therapy, whereas β receptors of polymorphonuclear cells were unaffected.[12,14,124] Lymphocytes from asthmatic patients showed a shift from α_1 to α_2 adrenoreceptors, which was also present in adipocytes from asthmatic patients. These abnormalities were absent from patients with asymptomatic triad asthma.[129]

In 1983 Krzanowski and Szentivanyi[115] summarized the pathogenesis of β-adrenergic subsensitivity in asthma as follows:

"An acute asthmatic episode may be seen as the dramatic culmination of an internal struggle for homeostasis. Initiated by a remarkable range of environmental and internal factors, the struggle probably begins at the site of an inherited or acquired intrinsic defect or imbalance at the molecular level. . . The abnormality may be: (1) acquired by a functional regulatory shift caused by hormonal changes, infection, allergic tissue injury, etc., (2) genetically determined, and (3) caused by autoimmune disease. In case of a given atopic disorder, one, or a combination of two, or all three of these major mechanisms may be operative."

This statement is supported by findings from the author's laboratory and of others, indicating that (i) there is a broad genetic variation in man with regards to autonomic reactivities; (ii) endotoxin is capable of altering adrenoreceptor reactivity in human bronchial smooth muscle; (iii) killed influenza virus vaccine increases the sensitivity of asthmatic patients to aerosolized methacholine; (iv) upper respiratory infections reduced β-adrenergic activity, cyclic AMP synthesis and lysosomal enzyme release in human granulocytes; (v) patients with chronic bronchitis with no previous history of asthma may develop frank asthmatic attacks during therapy with β-adrenergic blocking drugs given for some unrelated reason; (vi) all hormonal agents that have been studied so far, including corticosteroid, thyroid hormones, insulin and estrogen, are capable of producing an α- or β-oriented shift in adrenoreceptor numbers; and (vii) finally, cases of autoimmune reactions against adrenergic and cholinergic receptors are cited.

Other areas of scientific enquiry by Szentivanyi are related to inflammation and the effect of glucocorticoids.[130-133] More recently he focused his attention on the influence of cytokines on adrenergic receptors as it may be related to asthma and some other diseases such as cystic fibrosis. In the following chapter[134] he observes that: (i) inflammation, bronchial hyperresponsiveness and the influx of lymphocytes to the site are inseparable features of the pathogenesis of asthma; (ii) bronchial hyperresponsiveness correlates with T lymphocyte derived products and not directly with IgE antibodies; (iii) lymphocytes are capable of secreting mediators that alter β adrenergic function. In the hope that the puzzle that got him to venture to solve the pathogenesis of asthma over 40 years ago can now be brought to final conclusion he is working on these problems more feverishly than ever before.

Szentivanyi is truly the founder of the brave new area of scientific enquiry that may be designated as psychoneurobiology of which a more restricted area is psychoneuroimmunology. From the moment he identified the problem at the bedside he was committed to this area of investigation regardless of any other circumstances, such as the immense difficulties of approaching the problem at a time when immunology and experimental medicine as a whole were in their infancy. Most of the salient observations he made with his colleagues was done in wartorn Hungary during the early 50s which must have contributed significantly to the obstacles he had to overcome. While in the United States he continued his work which culminated in the presentation of the β-adrenergic theory for atopy and asthma followed by the systematic collection of experimental and clinical evidence for its validation. Although the work has not yet been concluded, and for this reason, he is not quite satisfied; it must be rewarding for him to see that the neural regulation of immune and inflammatory reactions is now taken very seriously. In fact there is compelling evidence to support almost all of the conclusions he laid out way before this field was considered a valid area of scientific enquiry. Now there is good evidence that adrenergic and cholinergic nerves and a whole host of neuropeptides have a significant influence on immune reactions, on inflammation, and play a role in the pathogenesis of hypersensitivity reactions, including atopy and asthma. Moreover, there is good evidence to indicate that immune inflammatory processes affect neural regulatory processes and elicit a systemic metabolic effect as he has shown in his pioneering experiments. We congratulate him and pay homage for his remarkable achievements. We wish him well with his future endeavours.

REFERENCES

1. G. Filipp, A. Szentivanyi and B. Mess, Anaphylaxis and nervous system, *Acta Med. Hung.* Tomus III, Fasciculus 2:163 (1952).
2. A. Szentivanyi and G. Filipp, Anaphylaxis and the nervous system. Part II, *Ann. Allergy* 16:143 (1958).
3. G. Filipp and A. Szentivanyi, Anaphylaxis and the nervous system. Part III, *Ann. Allergy* 16:306 (1958).
4. A. Szentivanyi and J. Szekely, Effect of injury to, and electrical stimulation of, hypothalamic areas on anaphylactic and histamine shock of the guinea pig. A preliminary report, *Ann. Allergy*

14:259 (1956).

5. A. Szentivanyi and J. Szekely, Anaphylaxis and the nervous system. Part IV, *Ann. Allergy* 16:389 (1958).

6. A. Szentivanyi, C.W. Fishel and D.W. Talmage, Adrenaline mediation of histamine and serotonin hyperglycemia in normal mice and the absence of adrenaline-induced hyperglycemia in pertussis-sensitized mice, *J. Infect. Dis.* 113:86 (1963).

7. C.W. Fishel and A. Szentivanyi, The absence of adrenaline-induced hyperglycemia in pertussis-sensitized mice and its relation to histamine and serotonin hypersensitivity, *J. Allergy* 34:439 (1963).

8. R.G. Townley, I.L. Trapani and A. Szentivanyi, Sensitization to anaphylaxis and to some of its pharmacological mediators by blockade of the beta adrenergic receptors, *J. Allergy* 39:177 (1967).

9. A. Szentivanyi, The beta adrenergic theory of the atopic abnormality in bronchial asthma, *J. Allergy* 42:203 (1968).

10. S. Katsh, D.G. Halkias and A. Szentivanyi, Analysis of the structural requirements for the histamine-sensitizing activity of the beta adrenergic blocking agents, *J. Allergy* 43:171 (1969).

11. A. Szentivanyi, The conformational flexibility of adrenoceptors and the constitutional basis of atopy, *Triangle* 18:109 (1979).

12. A. Szentivanyi, O. Heim and P. Schultze, Changes in adrenoceptor densities in membranes of lung tissue and lymphocytes from patients with atopic disease, *Ann. N.Y. Acad. Sci.* 332:295 (1979).

13. A. Szentivanyi, P. Schultze, O. Heim, J. Szentivanyi and J.B. Polson, Two different patterns of adrenoceptor mechanisms in drug versus disease-induced beta adrenergic subsensitivity in membranes of T and B lymphocytes, *J. Reticuloendoth. Soc.* 26:291 (1979).

14. A. Szentivanyi, La flexibilite de conformation des adrenocepteurs et la base constitutionelle du terrain allergique, *Rev. Franc. Allergol.* 19:205 (1979).

15. A. Szentivanyi, J. Szentivanyi and H. Wagner, Measurement of numbers of adrenoceptors in lymphocytes and lung tissue of patients with reversible obstructive airways disease, *Clin. Pharmacol. Ther.* 27:193 (1980).

16. A. Szentivanyi, O. Heim, P. Schultze and J. Szentivanyi, Adrenoceptor binding studies with [^{3}H] dihydroalprenolol with [^{3}H] dihydroergocryptine on membranes of lymphocytes from patients with atopic disease, *Acta Dermato-Venerol. Suppl.* 92:19 (1980).

17. S. Asai, J.J. Krzanowski, D.F. Martin, R.F. Lockey, S.C. Bukantz and A. Szentivanyi, Ptychodiscus brevis toxin stimulates a new receptor site in activating parasympathetic nerve axonal voltage-sensitive sodium channels, *J. Allergy Clin. Immunol.* 69:10 (1982).

18. A. Szentivanyi, Adrenergic receptors, *Int. J. Immunoparmacol.* 4:292 (1982).

19. A. Szentivanyi and J. Szentivanyi, Altered adrenoceptor function in disease states, *Pathobiol. Annu.* 12:340 (1982).

20. A. Szentivanyi and J. Szentivanyi, Some selected aspects of the immunopharmacology of adrenoceptors, *Adv. Immunopharmacol.* 2:269 (1983).

21. J.B. Polson, R.F. Lockey, S.C. Bukantz, S. Lowitt, J.J. Krzanowski and A. Szentivanyi, Effects of ketotifen on the responsiveness of human peripheral blood lymphocyte beta adrenergic receptors, *The Pharmacologist* 29:132 (1987).

22. J.B. Polson, R.F. Lockey, S.C. Bukantz, S. Lowitt, J.J. Krzanowski and A. Szentivanyi, Responsiveness of lymphocyte beta adrenergic receptors in patients treated with ketotifen, *Clin. Pharmacol. Therap.* 43:137 (1988).

23. J.B. Polson, R.F. Lockey, S.C. Bukantz, S. Lowitt, J.J. Krzanowski and A. Szentivanyi, Effects of ketotifen on the responsiveness of peripheral blood lymphocyte beta adrenergic receptors, *Int. J. Immunopharmacol.* 10:657 (1988).

24. J.J. Krzanowski, M.E. Schwartz, S. Reiner, O. Heim, C. Abarca and A. Szentivanyi, Adrenergic influences on the induction of histidine decarboxylase synthesis in hematopoietic progenitor cells, *J. Leukocyte Biol.* 46:294 (1989).

25. A. Szentivanyi, M.E. Schwartz, O. Heim, G. Filipp and C. Abarca, Ketotifen treatment restores the beta adrenergic modulation of the N-formylmethionyl-leucyl-phenylalanine (FMLP) induced respiratory burst in polymorphonuclear leukocytes (PMN) in asthma, *J. Clin. Pharmacol.* 29:859 (1989).

26. A. Szentivanyi, M.E. Schwartz, S. Reiner, O. Heim, C. Abarca, S. Robicsek and J.J. Krzanowski, Requirements for central and peripheral adrenergic input in the induction of histidine decarboxylase synthesis by interleukin 3 (IL-3) and granulocyte-macrophage colony

stimulating factor (GM-CSF) in hematopoietic progenitor cells, *Cytokine* 1:133 (1989).

27. A. Szentivanyi, S. Reiner, M.E. Schwartz, O. Heim, J. Szentivanyi and S. Robicsek, Restoration of normal beta adrenoceptor concentrations in A549 lung adenocarcinoma cells by leukocyte protein factors and recombinant interleukin-1a (IL-1a), *Cytokine* 1:118 (1989).

28. G. Kunkel, R. Paus, J.B. Polson and A. Szentivanyi, Beta-adrenergic receptors in nasal tissue with reference to their role in asthma and atopy, *Allergol. Immunopathol.* 17:67 (1989).

29. A. Szentivanyi, S. Reiner, O. Heim, S. Robicsek and J.F. Hackney, Studies on the nature of the lymphocytic protein factors upregulating beta-adrenergic receptors, *Clin. Pharmacol. Therap.* 149:139 (1991).

30. A. Szentivanyi, P. Schultze, O. Heim, S. Reiner, S. Robicsek, J. Zority, K.J. Hurt, J.F. Hackney, E.G. Calderon, J.J. Dwornik and R.F. Lockey, The elution profile of the A549 beta adrenergic (bAR) regulating activity of lymphocyte conditioned medium (LCM) of IM9 cells developed by DEAE ion exchange HPLC, *Int. J. Immunopharmacol.* 13:68 (1991).

31. A. Szentivanyi, E.G. Calderon, O. Heim, P. Schultze, H. Wagner, J. Zority, R.F. Lockey, J.J. Dwornik and S. Robicsek, Adrenergic and histamine receptor concentrations in bone marrow cells, *Int. J. Immunopharmacol.* 13:740 (1991).

32. A. Szentivanyi, O. Heim, P. Schultze, H. Wagner, J. Zority, E. Calderon, J.J. Dwornik and R.F. Lockey, Sensitization of bone marrow hemopoietic progenitor cells to the interleukin-3 (IL-3) and granulocyte macrophage colony stimulating factor (GM-CSF) induced de novo synthesis of histidine decarboxylase through the activation of beta adrenoceptors, *Int. J. Immunopharmacol.* 13:73 (1991).

33. J. Szentivanyi, S. Reiner, O. Heim, J.F. Hackney, S. Robicsek, R.F. Lockey, J.J. Dwornik and A. Szentivanyi, Impaired capacity of lymphocyte proteins to upregulate beta-adrenoceptors in A549 human lung adenocarcinoma cells in respiratory and cutaneous atopic disease, *Int. J. Immunopharmacol.* 13:743 (1991).

34. E.G. Calderon, O. Heim, A. Szentivanyi, P. Schultze, H. Wagner, J. Zority, R.F. Lockey, J.J. Dwornik and S. Robicsek, Adrenergic and histamine receptor concentrations in bone marow cells, *J. Allergy Clin. Immunol.* 89:277 (1992).

35. R.A. Ortez, T.W. Klein and A. Szentivanyi, Impairment of the adenyl cyclase system in spleens of mice sensitized bacterially or by beta pharmacological blockade, *J. Allergy* 45:111 (1970).

36. R.A. Ortez, T.W. Klein and A. Szentivanyi, Cyclic AMP levels in spleens of pertussis-sensitized mice, *Fed. Proc.* 29:640 (1970).

37. J.J. Krzanowski, J.B. Polson and A. Szentivanyi, Histamine, isoproterenol, and epinephrine induced in vivo activation of adenyl cyclase in mouse lung following Bordetella pertussis, *The Pharmacologist* 16:212 (1974).

38. J.B. Polson, J.J. Krzanowski and A. Szentivanyi, Effects of methacholine, histamine, and atropine on pulmonary guanosine-3',5'-cyclic monophosphate in hypersensitive mice, *The Pharmacologist* 16:212 (1974).

39. J.B. Polson, J.J. Krzanowski and A. Szentivanyi, Histamine-induced changes in pulmonary guanosine-3',5'-cyclic monophosphate and adenosine-3',5'-cyclic monophosphate levels in mice following sensitization by Bordetella pertussis and/or propranolol, *Res. Commun. Chem. Pathol. Pharmacol.* 9:243 (1974).

40. J.B. Polson, J.J. Krzanowski and A. Szentivanyi, Elevation of pulmonary cyclic GMP levels caused by histamine and methacholine in mice simulating atopic disease, *Adv. Cyclic Nucleotide Res.* 5:815 (1974).

41. J.J. Krzanowski, J.B. Polson and A. Szentivanyi, In vivo alterations in adenosine-3',5'-cyclic monophosphate levels following adrenergic activation in animal models of atopic disease, *Adv. Cyclic Nucleotide Res.* 5:819 (1975).

42. J.J. Krzanowski, J.B. Polson and A. Szentivanyi, Pulmonary patterns of adenosine-3',5'-cyclic monophosphate accumulations in response to adrenergic or histamine stimulation in Bordetella pertussis-sensitized mice, *Biochem. Pharmacol.* 25:1631 (1976).

43. J.B. Polson, J.J. Krzanowski and A. Szentivanyi, Pulmonary guanosine-3',5'-cyclic monophosphate in sensitized mice: Accumulations produced by H-1 receptor activation and evidence of altered phosphodiesterase activity, *J. Allergy Clin. Immunol.* 57:265 (1976).

44. J.J. Krzanowski, J.B. Polson and A. Szentivanyi, Reduction in levels of pulmonary cyclic guanosine-3',5'-monophosphate induced by theophylline and isoproterenol, *J. Allergy Clin. Immunol.* 57:265 (1976).

45. D.F. Fitzpatrick and A. Szentivanyi, The effects of protein kinase and cyclic nucleotides on calcium (Ca^{2+}) uptake by rabbit aortic microsomes, *The Pharmacologist* 18:182 (1976).

46. A.L. Goldman, T.A. Ebel, J.J. Krzanowski, J.B. Polson and A. Szentivanyi, Lung cyclic nucleotides

in bronchospastic and nonreversible obstructive lung disease, *Am. Rev. Resp. Dis.* 115:330 (1977).

47. J.J. Krzanowski, J.B. Polson and A. Szentivanyi, Pulmonary cyclic nucleotide lowering after ganglionic blockade and prostaglandin synthetase inhibition, *The Pharmacologist* 19:204 (1977).

48. D.F. Fitzpatrick and A. Szentivanyi, Stimulation of calcium uptake into aortic microsomes by cyclic AMP and cyclic AMP-dependent protein kinase, *Naun. Schmied. Arch. Pharmacol.* 298:255 (1977).

49. J.J. Krzanowski, J.B. Polson, A.L. Goldman, T.A. Ebel and A. Szentivanyi, Cyclic nucleotide levels in human lung from patients with bronchospastic airway disease, *The Pharmacologist* 19:203 (1977).

50. J.B. Polson, J.J. Krzanowski, D.F. Fitzpatrick and A. Szentivanyi, Studies on the inhibition of phosphodiesterase-catalyzed cyclic AMP and cyclic GMP breakdown and relaxation of canine tracheal smooth muscle, *Biochem. Pharmacol.* 27:254 (1978).

51. J.B. Polson, J.J. Krzanowski, A.L. Goldman, T.A. Ebel and A. Szentivanyi, Cyclic nucleotide phosphodiesterase activity in patients with reversible and nonreversible airway obstruction, *Adv. Cyclic Nucleotide Res.* 9:757 (1978).

52. J.J. Krzanowski, J.B. Polson, A.L. Goldman, T.A. Ebel and A. Szentivanyi, Cyclic nucleotide levels in human lung from patients with bronchospastic airway disease, *Adv. Cyclic Nucleotide Res.* 9:767 (1978).

53. J.J. Krzanowski, J.B. Polson, A.L. Goldman, T.A. Ebel and A. Szentivanyi, Reduced adenosine 3',5'-cyclic monophosphate levels in patients with reversible obstructive airways disease, *Clin. Exp. Pharmacol. Physiol.* 6:111 (1979).

54. J.J. Krzanowski, J.B. Polson and A. Szentivanyi, Divergent effects of prostaglandin synthesis inhibitors on pulmonary cyclic 3',5'-adenosine monophosphate and cyclic 3',5'-guanosine monophosphate levels in untreated and histamine-sensitized mice, *Arch. Int. Pharmacodyn.* 241:324 (1979).

55. J.J. Krzanowski, A. Urdaneta-Bohorquez, J.B. Polson, Y. Sakamoto and A. Szentivanyi, Studies of two smooth muscle relaxants (adenosine and theophylline) both capable of influencing tissue cyclic AMP levels on canine tracheal smooth muscle, *Adv. Cyclic Nucleotide Res.* 17A:49 (1984).

56. R.A. Duncan, J.J. Krzanowski, J.S. Davis, J.B. Polson, R.G. Coffey, T. Shimoda and A. Szentivanyi, Polyphosphoinositide metabolism in canine tracheal smooth muscle in response to a cholinergic stimulus, *Biochem. Pharmacol.* 36:307 (1987).

57. R. Duncan, J.J. Krzanowski, J. Polson, J.B. Coffey, T. Shimoda and A. Szentivanyi, The kinetics of the acetylcholine induced inositol phosphate formation in isolated contracting canine tracheal smooth muscle, *J. Allergy Clin. Immunol.* 79:196 (1987).

58. J.J. Krzanowski, A. Urdaneta-Bohorquez, J.B. Polson, Y. Sakamoto and A. Szentivanyi, Effects of adenosine and theophylline on canine tracheal smooth muscle tone, *Arch. Int. Pharmacodyn. Ther.* 287:224 (1987).

59. R.A. Duncan, J.J. Krzanowski, J. Davis, R.G. Coffey, J.B. Polson, T. Watanabe and A. Szentivanyi, The dose-response relationship between inositol phosphate formation and contractile resposnes in canine tracheal smooth muscle, *The Pharmacologist* 29:117 (1987).

60. J.J. Krzanowski, J.B. Polson and A. Szentivanyi, Theophylline attenuation of histamine-induced increases in pulmonary adenosine-3',5'-cyclic monohosphate, *J. Allergy Clin. Immunol.* 55:133 (1975).

61. J.B. Polson, J.J. Krzanowski and A. Szentivanyi, Theophylline, theobromine and caffeine inhibition of cAMP and cGMP hydrolysis by respiratory smooth muscle phosphodiesterase, *J. Allergy Clin. Immunol.* 55:133 (1975).

62. J.B. Polson and A. Szentivanyi, Theophylline, theobromine, caffeine, and ICI 58,301 inhibition of adenosine-3',5'-cyclic phosphate and guanosine-3',5'-cyclic phosphate hydrolysis by canine tracheal smooth muscle phosphodiesterase, *Fed. Proc.* 34:261 (1975).

63. J.J. Krzanowski, A. Szentivanyi and D.B. Tyler, Reduction of the histamine-induced increase in pulmonary adenosine-3',5'-cyclic monophosphate levels by theophylline, *Fed. Proc.* 34:332 (1975).

64. D.F. Fitzpatrick, J.B. Polson, J.J. Krzanowski and A. Szentivanyi, Correlation of tracheal smooth muscle relaxation with phosphodiesterase inhibition, *The Pharmacologist* 17:271 (1975).

65. J.F. Williams, J.B. Polson and A. Szentivanyi, Investigation of the pharmacological activity of the metabolites of the methylxanthines, *The Pharmacologist* 18:181 (1976).

66. J.B. Polson, J.J. Krzanowski, A.L. Goldman, T.A. Ebel and A. Szentivanyi, inhibition of human

pulmonary phosphodiesterase by therapeutic levels of theophylline, *The Pharmacologist* 19:203 (1977).

67. J.F. Williams, S. Lowitt, J.B. Polson and A. Szentivanyi, Pharmacological and biochemical activities of some monomethylxanthine and methyluric acid derivatives of theophylline and caffeine, *Biochem. Pharmacol.* 27:1545 (1978).

68. J.B. Polson, J.J. Krzanowski, A.L. Goldman and A. Szentivanyi, Inhibition of human pulmonary phosphodiesterase activity by therapeutic levels of theophylline, *Clin. Exp. Pharmacol. Physiol.* 5:553 (1978).

69. J.P. Polson, J.J. Krzanowski, W.H. Anderson, D.F. Fitzpatrick, D.P.C. Hwang and A. Szentivanyi, Analysis of the relationship between pharmacological inhibition of cyclic nucleotide phosphodiesterase and relaxation of canine tracheal smooth muscle, *Biochem. Pharmacol.* 28:1391 (1979).

70. J.B. Polson, J.J. Krzanowski and A. Szentivanyi, Stimulation of cyclic AMP phosphodiesterase activity by the heated supernatant fraction of canine tracheal smooth muscle, *Allergol. Immunopathol.* 7:297 (1979).

71. J.B. Polson, J.J. Krzanowski and A. Szentivanyi, Separation of the cyclic nucleotide phosphodiesterase of canine tracheal smooth muscle, *Fed. Proc.* 40:665 (1981).

72. J.B. Polson, J.J. Krzanowski, A.L. Goldman and A. Szentivanyi, Cyclic nucleotide phosphodiesterase activity in patients with obstructive airways disease, *Allergol. Immunopathol.* 10:101 (1982).

73. J.B. Polson, J.J. Krzanowski and A. Szentivanyi, Inhibition of a high affinity cyclic AMP phosphodiesterase and relaxation of canine tracheal smooth muscle, *Biochem. Pharmacol.* 31:3403 (1982).

74. J.J. Krzanowski, J.B. Polson, M. McPherson, A. Urdaneta-Bohorquez and A. Szentivanyi, Adenosine-theophylline interactions in canine tracheal smooth muscle strips, *J. Allergy Clin. Immunol.* 71:131 (1983).

75. J.B. Polson, J.J. Krzanowski and A. Szentivanyi, Cyclic nucleotide phosphodiesterases partially purified from canine tracheal smooth muscle, *Adv. Cyclic Nucleotide Res.* 17A:41 (1984).

76. J.B. Polson, J.J. Krzanowski and A. Szentivanyi, Correlation between inhibition of a cyclic GMP phosphodiesterase and relaxation of canine tracheal smooth muscle, *Biochem. Pharmacol.* 34:1875 (1985).

77. S. Robicsek, J.B. Polson, J.J. Krzanowski, R.F. Lockey and A. Szentivanyi, Study of human lymphocyte phosphodiesterase using high pressure liquid chromatography, *J. Allergy Clin. Immunol.* 79:168 (1987).

78. S. Robicsek, J.B. Polson, J.J. Krzanowski, R.F. Lockey and A. Szentivanyi, High pressure liquid chromatography of human T-lymphocyte phosphodiesterases, *Clin. Pharmacol. Ther.* 43:167 (1988).

79. S. Robicsek, J.B. Polson, J.J. Krzanowski, R.F. Lockey and A. Szentivanyi, Separation of lymphocyte cyclic nucleotide phosphodiesterases by ion exchange high pressure liquid chromatography, *Adv. Sec. Mess. Phosphopro. Res.* 21A:24 (1988).

80. S. Robecsek, J.J. Krzanowski, A. Szentivanyi and J.B. Polson, Analysis of human T-lymphocyte cyclic nucleotide phosphodiesterase activity by high pressure liquid chromatography and type III inhibitors, *FASEB J.* 3:1295 (1989).

81. J.B. Polson, S.A. Robicsek, J.J. Krzanowski and A. Szentivanyi, Paritculate cyclic AMP phosphodiesterase in human T-lymphocytes, *Clin. Pharmacol. Ther.* 47:169 (1990).

82. S. Robicsek, D.K. Blanchard, J. Djeu, J.J. Krzanowski, A. Szentivanyi and J.B. Polson, Attenuation of human T-lymphocyte blastogenesis by the selective inhibition of cAMP-phosphodiesterases, *FASEB J.* 4:A337 (1990).

83. S. Robicsek, D.K. Blanchard, J. Djeu, J.J. Krzanowski, A. Szentivanyi and J.B. Polson, Synergistic effect on T-lymphocyte blastogenesis by selective cAMP-phosphodiesterase inhibitors, *FASEB J.* 4:2036 (1990).

84. S.A. Robicsek, D.K. Blanchard, J.Y. Djeu, J.J. Krzanowski, A. Szentivanyi and J.B. Polson, Multiple high affinity cAMP-phosphodiesterases regulate proliferation in human thymic lymphocytes, *Biochem. Pharmacol.* 42:869 (1991).

85. S. Robicsek, D.K. Blanchard, J. Djeu, J.J. Krzanowski, A. Szentivanyi and J.B. Polson, Attenuation of IL-2 and Tac receptor expression by selective PDE inhibitors in human T-lymphocytes, *Int. J. Immunopharmacol.* 13:734 (1991).

86. R.A. Ortez, C.W. Fishel and A. Szentivanyi, The reduced sensitivity of Bordetella pertussis vaccinated murine spleen and platelet adenyl cyclase to epinephrine, *Fed. Proc.* 31:748 (1972).

87. C.W. Fishel, A. Szentivanyi and T.W. Klein, Effect of epinephrine on plasma cyclic adenosine monophosphate levels of Bordetella pertussis-vaccinated and of adrenergically blocked mice, *Fed. Proc.* 32:1009 (1973).

88. J.B. Polson, J.J. Krzanowski and A. Szentivanyi, Effect of histamine on the pulmonary levels of cyclic nucleotides in normal mice and under conditions of pharmacologic or bacterial sensitization, *J. Allergy Clin. Immunol.* 53:100 (1974).

89. J.J. Krzanowski, J.B. Polson and A. Szentivanyi, Energy mtabolism and patterns of behavior of adenosine-3',5'-cyclic monophosphate in lung from control and bacterially sensitized mice under conditions of rest and adrenergic activation, *J. Allergy Clin. Immunol.* 53:100 (1974).

90. T.W. Klein, A. Szentivanyi and C.W. Fishel, Effects of serotonin on platelets of normal and Bordetella pertussis-injected mice, *Proc. Soc. Exp. Biol Med.* 147:681 (1974).

91. R.A. Ortez, D. Seshachalam and A. Szentivanyi, Alterations in adenyl cyclase activity and glucose utilization of Bordetella pertussis sensitized mouse spleen, *Biochem. Pharmacol.* 24:1297 (1975).

92. J.F. Williams and A. Szentivanyi, Effect of Bordetella pertussis vaccine on the drug-metabolizing enzyme system of mouse liver, *Fed. Proc.* 34:261 (1975).

93. C.W. Fishel, D.G. Halkias, T.W. Klein and A. Szentivanyi, Characteristics of cells present in peritoneal fluids of mice injected intraperitoneally with Bordetella pertussis, *Infect. Immun.* 13:263 (1976).

94. J.F. Williams and A. Szentivanyi, Depression of hepatic drug-metabolizing enzyme activity by Bordetella pertussis vaccination, *Eur. J. Pharmacol.* 43:281 (1977).

95. J.F. Williams, S. Lowitt and A. Szentivanyi, Effect of endotoxin and phenobarbital on heme enzymes of rat liver, *The Pharmacologist* 21:232 (1979).

96. J.F. Williams, S. Lowitt and A. Szentivanyi, Involvement of a heat-stable and heat-labile component of Bordetella pertussis in the depression of murine hepatic mixed-function oxidase system, *Biochem. Pharmacol.* 29:1483 (1980).

97. J.F. Williams, S. Lowitt and Z. Szentivanyi, Endotoxin depression of hepatic mixed function oxidase system in C3H/Hej and C3H/H3N mice, *Immunopharmocology* 2:285 (1980).

98. J.F. Williams, A.L. Winters, S. Lowitt and A. Szentivanyi, Depression of hepatic mixed-function oxidase activity by Bordetella pertussis in splenectomized and athymic nude mice, *Immunopharmacology* 3:101 (1981).

99. A. Szentivanyi and J. Szentivanyi, Neuester Stand der Rezeptoren-Theorie bei Atopie, *Allergologie* 6:155 (1983).

100. J.F. Williams and A. Szentivanyi, Continued studies on the effect of interferon inducers on the hepatic microsomal mixed-function oxidase system of rats and mice, *J. Interferon Res.* 3:211 (1983).

101. H. Friedman and A. Szentivanyi, Microbial vaccines - specific and non-specific modifiers of immunity, *Immunol. Allergy Prac.* 7:40/17 (1985).

102. J.F. Williams and A. Szentivanyi, Effect of carbon tetrachloride on hepatic cytochrome P-450 activity in endotoxin tolerant and polymixin B treated rats, *The Pharmacologist* 27:251 (1985).

103. J.F. Williams and A. Szentivanyi, Induction of tolerance in mice and rats to the effect of endotoxin to decrease the hepatic microsomal mixed-function oxidase system. Evidence for a possible macropahge-derived factor in endotoxin effect, *Int. J. Immunopharmacol.* 7:501 (1985).

104. H. Friedman and A. Szentivanyi, Antibacterial immunity, vaccines, and allergy, *Allergologie* 8:357 (1985).

105. H. Friedman, T. Klein, A. Szentivanyi and A. Nowotny, Immunomodulation by bacterial products: Role of intermediary soluble factors, *Int. J. Immunother.* 2:6164 (1986).

106. A. Nowotny, K. Blanchard, C. Newton, T.W. Klein, W. Stewart II, A. Szentivanyi and H. Friedman, Interferon induction by endotoxin-derived nontoxic polysaccharides, *J. Interferon Res.* 7:371 (1987).

107. A. Szentivanyi, S. Reiner, O. Heim, G. Filipp and C. Abarca, The effect of sympathetic ablation [6-hydroxydopamine hydrobromide (6-OHDA); axotomy] on endotoxin induced adrenergic mechanisms, *The Pharmacologist* 31:118 (1989).

108. A. Szentivanyi, S. Reiner, O. Heim, G. Filipp and C. Abarca, In vivo and in vitro studies on adrenergic mechanisms induced by Escherichia coli endotoxin, *J. Leukocyte Biol.* 46:328 (1989).

109. A. Szentivanyi, J.J. Krzanowski, J.B. Polson and C.M. Abarca, The pharmacology of microbial modulation in the induction and expression of immune reactivities. I. The

pharmacologically active effector molecules of immunologic inflammation, immunity, and hypersensitivity, *Immunopharmacol. Rev.* 1:159 (1990).

110. J.F. Williams and A. Szentivanyi, Implications of hepatic drug metabolizing activity in the therapy of bronchial asthma, *J. Allergy Clin. Immunol.* 55:125 (1975).

111. S. Lowitt, J.F. Williams and A. Szentivanyi, Dexamethasone induction to tryptophan oxygenase activity in vivo and in isolated rat hepatic parenchymal cells: Effect of bacterial endotoxin, *The Pharmacologist* 20:157 (1978).

112. A. Szentivanyi and E. Middleton Jr., Asthma pharmacotherapy and why, *J. Respirat. Therap.* 1:10 (1980).

113. J.F. Hackney and A. Szentivanyi, Response of isolated guinea pig tracheal muscle to glucocorticoid and non-glucocorticoid succinates, *Arch. Int. Pharmacodyn. Therap.* 244:4 (1980).

114. L. Kirkman, A. Goldman, J.J. Krzanowski, J.B. Polson, W. Anderson, A. Richman and A. Szentivanyi, Enhanced in vitro sensitivity to propranolol-induced beta adrenergic blockade in COPD patients, *Am. Rev. Resp. Dis.* 123:79 (1981).

115. J.J. Krzanowski and A. Szentivanyi, Invited editorial: Reflections on some aspects of current research in asthma, *J. Allergy Clin. Immunol.* 72:433 (1983).

116. Y. Sakamoto, J.J. Krzanowski, R.F. Lockey, J.B. Polson and A. Szentivanyi, The potential of ergonovine to precipitate bronchial asthmatic attacks via serotonin receptors, *J. Allergy Clin. Immunol.* 73:129 (1984).

117. S. Robicsek, J.J. Krzanowski, J.B. Polson and A. Szentivanyi, Reevaluation of theophylline for asthma? *J. Clin. Pharmacol.* 29:859 (1989).

118. A. Szentivanyi, M.E. Schwartz, O. Heim and G. Filipp, Ketotifen and the N-formylmethionyl-leucyl-phenylalanine (FMLP) induced respiratory burst in asthma, *Clin. Pharmacol. Therap.* 47:137 (1990).

119. A. Szentivanyi, M.E. Schwartz, O. Heim, G. Filipp and C. Abarca, Ketotifen and the n-formylmethionyl-leucyl-phenylalanine (FMLP) induced respiratory burst in asthma, *J. Clin. Pharmacol.* 30:92 (1990).

120. A. Szentivanyi, S. Reiner, O.Heim, C. Abarca, J. Szentivanyi, S. Robiecsek and J. Hackney, Impaired lymphokine regulation of b-adrenoceptors in asthma and atopic dermatitis, *J. Leukocyte Biol.* Suppl.1:183 (1990).

121. J. Szentivanyi, O. Heim, P. Schultze, H. Wagner, E.G. Calderon, R.F. Lockey, S. Robicsek and A. Szentivanyi, Similarities and differences in patterns of beta-adrenergic regulation by lymphocytic proteins in respiratory and cutaneous atopy versus cystic fibrosis, *J. Allergy Clin. Immunol.* 89:162 (1992).

122. A. Szentivanyi and D.F. Fitzpatrick, The altered reactivity of the effector cells to antigenic and pharmacological influences and its relation to cyclic nucleotides. II. Effector reactivities in the efferent loop of the immune response, *in*: "Pathomechanismmus und Pathogenese Allergischer Reaktionen," G. Filipp, ed., Werk-Verlag Dr. Edmund Banachewski, Grafelfing bei Munchen (1980).

123. A. Szentivanyi, J.B. Polson and J.J. Krzanowski, The altered reactivity of the effector cells to antigenic and pharmacological influences and its relation to cyclic nucleotides. I. Effector reactivities in the efferent loop of the immune response, *in*: "Pathomechanismus und Pathogenese Allergischer Reaktionen," G. Filipp, ed., Werk-Verlag, Dr. Edmund Banachewski, Grafelfing bei Munchen (1980).

124. A. Szentivanyi, J.J. Krzanowski, J.B. Polson and W.H. Anderson, Evolution of research strategy in the experimental analysis of the beta adrenergic approach to the constitutional basis of atopy, *in*: "Advances in Allergology and Clinical Immunology," A. Oehling, E. Mathov, I. Glazer and C. Arbesman, eds., Pergamon Press, Oxford (1980).

125. A. Szentivanyi, Effect of bacterial products and adrenergic blocking agents on allergic reactions, *in*: "Textbook of Immunological Diseases," M. Samter, D.W. Talmage, B. Rose, W.B. Sherman and J.H. Vaughan, eds., Little, Brown and Co., Boston, MA (1971).

126. A. Szentivanyi, J.J. Krzanowski and J.B. Polson, The autonomic nervous system: Structure, function, and altered effector responses, *in*: "Allergy: Principles and Practice," E. Middleton, C.E. Reed and E.F. Ellis, eds., C.V. Mosby Co., St. Louis, MO (1978).

127. A. Szentivanyi and J.F. Williams, The constitutional basis of atopic disease, *in*: "Allergic Diseases of Infancy, Childhood, and Adolescence," C.W. Bierman and D.S. Pearlman, eds., W.B. Saunders Co., Philadelphia, PA (1980).

128. A. Szentivanyi, Adrenergic and cholinergic receptor studies in human lung and lymphocytic membranes and their relation to bronchial hyperreactivity in asthma, *in*: "Patient Care

Publications," Darien, CT (1982).

129. A. Szentivanyi and J. Szentivanyi, Anti-allergic drugs and beta adrenoceptors, *in*: "Proceedings of the Congress of the European Academy of Allergology and Clinical Immunology," Clermont-Ferrand, France (1982).

130. J.F. Hackney and A. Szentivanyi, The specificity of glucocorticoids in the relaxation of respiratory smooth muscle in vitro, *J. Allergy Clin. Immunol.* 55:123 (1975).

131. J.F. Hackney and A. Szentivanyi, The unique action of glucocorticoid succinates on respiratory smooth muscle in vitro, *The Pharmacologist* 17:271 (1975).

132. S. Lowitt, A. Szentivanyi and J.F. Williams, Endotoxin inhibition of dexamethasone induction of tryptophan oxygenase in suspension culture of isolated rat parenchymal cells. II. Effect of in vivo pretreatment of rats with endotoxin, *Biochem. Pharmacol.* 31:693 (1982).

133. A. Szentivanyi and J. Szentivanyi, Mechanisms of action of corticosteroids, *in*: "Proceedings of the International Symposium on Allergy and Immunoloty," Lima, Peru (1985).

134. A. Szentivanyi, The immune-neuroendocrine circuitry--The next, and possibly, the last frontier of vertebrate immunity, *in*: I. Berczi and J. Szelenyi, eds., "Advances in Psychoneuroimmunology," Plenum Press (1994).

THE IMMUNE-NEUROENDOCRINE CIRCUITRY--THE NEXT, AND POSSIBLY, THE LAST FRONTIER OF VERTEBRATE IMMUNITY*

Andor Szentivanyi and Christine M. Abarca

Departments of Internal Medicine, Neurology, Pharmacology, and Environmental and Occupational Health
University of South Florida Colleges of Medicine and Public Health
Tampa, Florida 33612 U.S.A.

David Wilson Talmage, one of the most important pioneers of the molecular biology of immunoregulation, in discussing recently (1988) what is beyond molecular immunology, poses the following questions:

> *"What will we study after we understand all the genes and their products that impact on the immune system? Surely, in a few years we will understand how lymphocytes differentiate and are activated, how they synthesize, secrete, kill, migrate. What will we study then? My prediction is that immunology will then have its greatest challenge. How can we put it all together for the benefit of mankind? How will we prevent allergies, autoimmunity, transplant rejection, and how will we cure cancer? The coming era of immunology will be a stage of synthesis-a return to Stage 1 and the immunology of the whole animal."*[1]

This return to stage one, the immunology of the whole animal, has already begun as reflected by this chapter.

Indeed, of the various evolving views on immunologic inflammation, immunity and hypersensitivity, this paper discusses the irrevocable shift and turnabout in our concepts of immunoregulation as connected with our growing understanding of the

---

*Dedicated to the memory of Leo Szilard.

Advances in Psychoneuroimmunology, Edited by
I. Berczi and J. Szélenyi, Plenum Press, New York, 1994

immune-neuroendocrine circuitry. This network, powerful enough both conceptually and in *de facto* functioning to bring about a radical change in our perceptions of the human immune, endocrine, and nervous systems, has already enlisted the minds, hearts and resources of a large number of our leading laboratories in many areas of life sciences on an international scale.

The immune-neuroendocrine circuitry represents an immensely complex, powerful and wide-ranging charter of human physiological and pathologic possibilities that, among others, is working its way to the creation of a new immunology based on a vastly enlarged vision of immunologic potential in health and disease.

THE DISCOVERY OF THE IMMUNE-NEUROENDOCRINE CIRCUITRY AND THE CONCEPTS OF PREVAILING IMMUNOLOGIC THOUGHT THAT IMPEDED THE TIMELY RECOGNITION OF ITS ROLE IN IMMUNE HOMEOSTASIS

The integrative center of this awesome edifice is the hypothalamus. The latter is a small, anatomically complex region of the diencephalon which in a variety of ways contributes to a number of regulatory systems. The functional and anatomic complexity of the hypothalamus results in part from its role as a nodal region for (1) convergence of input from the limbic system which contribute to an association between visceral and behavioral functions, (2) bidirectional bundles of nerve fibers with their cell bodies (perikarya) in the telencephalon or brainstem, and (3) local neurons coordinating distant organ system activities through the effector functions of the endocrine and autonomic nervous systems.

The role of hypothalamic influences in the induction and expression of immuno-logic inflammation, immunity, and hypersensitivity was first discovered in my laboratory in the fall of 1951 at the University of Debrecen School of Medicine in Hungary. The circumstances that led to this discovery could be stated as follows. As a first year resident on call in the First Department of Internal Medicine in Debrecen, I was called one night in late June 1951 to a hospitalized 52-year-old male patient with a twelve- year history of chronic asthma. The patient was in severe status asthmaticus requiring the repeated administration of epinephrine which in the post-war years in Hungary was just about the only antiasthmatic agent in the therapeutic armamentarium of the hospital. In the beginning, epinephrine provided symptomatic relief but had to be administered repeatedly and the extent and length of these reliefs between two administrations became gradually shorter and less adequate ultimately resulting in a complete loss of effectiveness. I thought that this was due to massive mechanical obstruction in the airways and as such should not be surprising. The patient died and I requested an autopsy to be held in the morning. Macroscopic and later histologic examination of the airways showed a limited degree of obstruction that by no means could account for the refractoriness to epinephrine. I concluded that there was something unusual about this patient's asthma, or there was something unusual about the disease itself, at least as far as responsiveness to epinephrine was concerned. I decided to go to see the chairman of our department, Professor Bela Fornet, who at that time was the foremost expert on bronchial asthma in Hungary. He said that I should "calm down" since what I observed was nothing unusual but a frequent experience with the disease; in fact, status asmaticus was defined by some as a condition that is refractory to epinephrine. Embarrassed by my ignorance about asthma, I thanked Professor Fornet for the clarification and left.

In the next couple of days, however, I could not "calm down." On the contrary, I became more and more agitated since I felt that if epinephrine refractoriness is indeed a characteristic and not an uncommon feature of asthma, then this characteristic may be of pathognomonic significance. Such a possibility brought into focus the question that how is it possible that highly experienced and knowledgeable scholars of asthma did not seem to associate this strange tolerance to a "drug" with the pathogenesis of the condition. This was all the less comprehensible since epinephrine was not a "drug" but a natural substance with enormous importance in mammalian physiology--one of the three catecholamines representing the adrenergic nervous system together with norepinephrine and dopamine.

I went to see Professor Fornet again and I presented my argument as follows. It is established that the main catecholamine store of the mammalian body is the chromaffin tissue including the adrenal medullae, paraganglia, and various other extraadrenal chromaffin cells. The precursors of these chromaffin cells are derived from the neural crest of the embryo as primordial sympathetic ganglion cells; only after migrating outside the nervous system do some of the ganglion cells become further differentiated as adrenal medullary cells. This ganglionic ancestry is further exemplified by the unusual anatomic relationship of these cells to their innervating autonomic fibers. The latter, reaching the adrenal gland via the lesser splanchnic nerve and through the direct branches from the lumbar sympathetic trunk, terminate among the medullary chromaffin cells without the interposition of postganglionic neurons. Both cholinesterase activity and the synaptic vesicles, the characteristic storage organelles of the cholinergic neurotransmitter (acetylcholine), are confined to the presynaptic terminals of the splanchnic fibers, whereas the catechol-containing storage granules are localized postsynaptically in the medullary chromaffin cells. This arrangement is consistent with the concept that the nerve fibers represent the preganglionic cholinergic innervation of the medullary cells. The limiting membrane of the latter stands for the postganglionic membrane, whereas the cells themselves substitute for the terminals of the postganglionic adrenergic neurons. The commonly shared characteristics in electrophysiologic properties and sensitivity to various ganglion blocking agents further testifies to the specific relationship between their limiting membrane and the postganglionic adrenergic membrane. This is well illustrated by the excellent agreement between the sensitivity of both membranes to methonium compounds of various chain lengths; in both cases decamethonium (^{c}10) is ineffective, hexamethonium (^{c}6) is the most potent. The adrenal medullary cells, therefore, may be regarded as the embryologic, anatomic, and functional analogues of the adrenergic neurons and their products, the catecholamines, represent the effector molecules of the adrenergic division of the autonomic nervous system. I further argued that the asthmatic response to epinephrine may reflect a dysfunctional adrenergic system which may contribute to the development of asthma through a resulting autonomic imbalance. I suggested that such an animal model could be established through various hypothalamic manipulations in the anaphylactic guinea pig and I asked Professor Fornet to contact Professor Szentagothai to inquire whether I could go for a study leave in his Institute of Anatomy at the University of Pecs which, at that point, was the most authoritative center in hypothalamic research in Europe. I also mentioned that I had had discussions about such a plan with Dr. Janos Hankiss, a resident colleague in our department, and it would be helpful if both of us could go. Professor Fornet made the arrangements with Professor Szentagothai for a three-month stay in his institute for the purpose of learning hypothalamic methodology. For reasons I no longer remember, Dr. Hankiss lost interest in this plan. In his place Dr. Geza Filipp (later Professor of Internal Medicine at the University of Saarland, Germany) joined us and it was Dr. Bela Mess (currently Professor and Chairman at the Second Department of Anatomy at the University of Pecs in Hungary) who was designated by Szentagothai to teach us the Horsley-Clarke stereotaxic methodology. This is how our exploration of the immune-neuroendocrine circuitry started and with some interruptions continued up to the writing of this chapter. In addition to G. Filipp and B. Mess, my associates in various phases of our work included I. Legeza, B. Fornet, L. Vegh, M. Keszthelyi, P. Demeny, A. Haraszthy, J. Szekely, O. Heim, S. Reiner, P. Schultze, D.W. Talmage, J. Radovich, R.J. Weiler, D. Hofstra, R. Blaisdell, Z. Houban, R.A. Ortiz, T.W. Klein, W.E. Linaweaver, D. Hurst, R.G. Townley, I.L. Trapani, S. Katsh, D.G. Halkias, C.W. Fishel, J. Szentivanyi, K. Haberman, M.E. Schwartz, H. Wagner, E. Wertheimer, and more recently J.J. Krzanowski, J.B. Polson, D.F. Fitzpatrick, J.F. Hackney, J.F. Williams, S. Lowitt, W. Anderson, S. Robicsek, K. Hurt, J. Zority, E. Calderon, C. Abarca, R.F. Lockey, S. Brooks and K. Ali.

The rationale behind the decision for a systematic exploration of the hypothalamus was as follows. Historically, the interpretation of the symptomatology and the underlying reaction sequence of human asthma was patterned after those of the anaphylactic guinea pig. However, the range of atopic responsiveness in asthma includes a variety of stimuli that are non-immunologic in nature. Foremost among these is a broad range of pharmacologically active mediators that today could be considered as the chemical organizers of central and peripheral autonomic regulation. Therefore, I believed that anaphylaxis could not be used as a model for the investigation of the constitutional basis of atopy in asthma. It was postulated that such a model, if it were to be meaningful, must be able to imitate both the immunologic and autonomic abnormalities of the disease. Since at that early time (1951) none of the currently existing neuroactive agents (agonists, antagonists, etc.) which one could conceivably use as an experimental tool to induce an autonomic imbalance were in existence, it was concluded that the best chance to develop such a condition could be through various manipulations at the neuroendocrine regulatory level of the hypothalamus. Consequently, hypothalamically imbalanced anaphylactic animals were used. These were produced by the electrolytic lesion, and conversely, the electrical stimulation of various nuclear groupings in the hypothalamus through permanently implanted depth electrodes placed stereotaxically into the hypothalamus. The resulting cumulative findings obtained on such a model indicated that the hypothalamus has a modulatory influence on all cellular and humoral immune reactivities and that both neural as well as endocrine pathways are required for hypothalamic modulation of immune responses.[2-41]

Concurrently with these developments, however, an unparalleled expansion of information on the basic aspects of immunology, in general, and on the nature of antibody diversity, in particular, started to occupy the center stage of immunologic interest. Most importantly, the new perceptions surrounding the nature of antibody diversity began to surface in the late 1950s with the first conclusive genetic studies having been completed and a new set of concepts defined. These circumstances led to a total transformation of prevailing immunologic thought ultimately leading to the replacement of instructionist theories by the selective theories as advanced by Talmage (spring 1957) and Burnet (fall 1957).

Three major postulates were implicit in these theories of heritable cellular commitment: (1) the antigen receptor site and the antibody combining site whose synthesis that cell controls are identical and are derived at least partially from the same structural gene; (2) the condition guaranteeing the correspondence of the immunoglobulin synthesized with the antigen is that they are limited to the same cell that is the cell specialized for the synthesis of a single antibody; (3) the cell specialization stipulated in item two is inherited and therefore, clonal (the clonal selection theory of acquired immunity). Subsequently, it became established that virtually all antibody diversity and specificity encoded in the immune system can be accounted for in genetic terms and thus the controls for the antibody response must reside largely within the major histocompatibility complex where different genes appear to code for immune response, suppression, and cell interaction. Another impediment in the timely recognition of the significance of the immune-neuroendocrine circuitry in immune homeostasis was the ability to have immune reactivities proceed *in vitro*. This further supported the concept that the immune system is a totally autonomous and self-regulating unit (this view has overlooked the rich neurohormonal milieu in which most *in vitro* immune responses occur). Sporadic refutations of these postulates have occurred and continue to surface in the literature, but the great bulk of the evidence is supportive of the clonal selection theory. The theory's sheer eloquence, however, has probably been most responsible for its dominant role in immunologic thought and its acceptance as dogma since the early '60s. In any case, these concepts and the large body of supportive evidence have so permeated the field that it became difficult, if not impossible, to think of immunology outside of this framework.[42]

In the late '50s and throughout the '60s the two conceptual centers of these ideas were under the leadership of D.W. Talmage at the University of Chicago and later at Colorado, and the group at Walter and Eliza Hall Institute in Australia under M. Burnet. Because of the close association with Talmage, extending over a period of ten years, I was very much under the influence of these views and developed reservations against the significance of the above mentioned hypothalamic findings in immunoregulation. Nevertheless, by 1966 in a chapter of a German text[16] through an extensive analysis of our findings and the dominant immunologic concepts, I came to articulate the following conclusions: (1) the significance of the immunopharmacologic mediators of immune manifestations in normal mammalian physiology is that they are the chemical organizers of central and peripheral autonomic action; (2) the preceding suggests the inseparability of the immune response system from the neuroendocrine system; (3) such inseparability indicates the *de facto* existence of immune-neuroendocrine circuits and the necessity for a bidirectional flow of information between the two systems; (4) one must distinguish between the concepts of autoregulation as one that primarily revolves around one effector molecule of immunity, the antibody, and satisfies the requirements of antibody diversity and specificity in contrast to the more complex requirements of immune homeostasis; (5) in contrast to autoregulation that is always self-contained, homeostatic control is always beyond the constraints of one single cell or tissue system; and (6) thus, immune homeostasis must represent a far more sophisticated level of control than autoregulation, and is based on immune-neuroendocrine circuits. Indeed, as Gail Schechter[43] points out, no bodily system is as simple, sacred, or singular as once thought. Instead, as in any good relationship, the separate components strive for sensitivity, synchrony, and synergy. Recognition and communication among the immune, endocrine, and nervous systems exemplify the formula for harmony and homeostasis.

While the manifold similarities between the immune and nervous systems are fully realized (see below), the immune system has a major additional level of complexity over that of the nervous system. Although the nervous system with its spectacular masses of much-revealing and well-defined projection patterns is well-moored in the body in a static web of axons, dendrites, and synapses, the elements of the immune system are in a continuously mobile phase incessantly scouring over and percolating through the body tissues, returning via an intricate system of lymphatic channels and then blending again in the blood. This dynamism is relieved only by scattered concentrations called lymphoid organs. These circumstances would appear to indicate that the functional plasticity of the immune system is far greater than that of the nervous system, and consequently, its regulation must require a more complex and sophisticated level of control. For these reasons, I have raised the frivolous question in the above-mentioned text in 1966 whether the immune system is more "intelligent" than the brain.

When these conclusions were reached (in the '60s) our understandings of cellular immunology were in an early phase. The '70s saw the discovery of the lymphokines, monokines and a broad range of other effector molecules of immunologic inflammation, immunity, and hypersensitivity. But it was only in the '80s that we recognized that the cells of the immune system, primarily the lymphocytes, synthesize, store, and release neurotransmitters, hypothalamo-hypophyseal hormones, etc. and by all criteria serve as neuroendocrine cells "par excellence."[16-19,40,44] We shall return to the lymphocyte below.

DEVELOPMENTAL INTERRELATIONSHIPS AMONG THE CELLULAR AND HUMORAL COMPONENTS OF THE IMMUNE-NEUROENDOCRINE CIRCUITRY

These interrelationships may be briefly stated through a discussion of (1) the cells involved in the synthesis, storage, secretion and/or release of the effector molecules of immunologic reactivities; (2) neural crest interactions in the development of the immune system; (3) cerebral dominance or lateralization and immune disorders; (4) the ancient superfamily of immune recognition molecules and the neural cell adhesion molecule; (5) coevolution of cytokine receptor families in the immune and nervous systems; (6) shared molecular mechanisms in the development of immune and neuronal memories; and (7) the unique recognition and communication powers of the immune and nervous systems as shared characteristics.

The Cells Involved in the Synthesis, Storage, Secretion, and/or Release of the Effector Molecules of Immunologic Reactivities

The functions of the immune system are the properties of cells distributed throughout the body. They include (1) free or circulating cells of the blood, lymph and intravascular spaces; (2) similar cells collected into units that allow for close interaction with lymph or circulating blood--lymph nodes, spleen, liver, and bone marrow; and (3) two major control organs for the system, the thymus gland and the hypothalamic-pituitary-adrenal complex. The cells involved in the synthesis, storage, secretion, and/or release of the effector molecules of immunologic inflammation, immunity,and immunologically-based hypersensitivity (allergy) represent a continuous spectrum of related cell types specialized in the production and storage of various physiopharmacologically active effector substances in variable proportions, i.e., of cells that have a common developmental origin with differentiation being determined by the specific requirements of the local neurohumoral regulation. Accounting only for those effector molecules for which the cell type has been identified, this incomplete spectrum of cells and effector substances includes macrophages and lymphocytes (interleukins 1-12, interferons, tumor necrosis factors, lysosome and complement components, prostaglandins, leukotrienes, acid hydrolases, neutral proteinases, arginase, nucleotide metabolites, various neuroactive immunoregulatory peptides including ACTH, CRF-like activity, bombesin, endorphins, enkephalins, TSH, growth hormone, prolactin, neurotensin, chorionic gonadotropin, VIP, tachykinin neuropeptides including substance P, substance K, neuromedin K, somatostatins, mast cell growth factor, etc.), neutrophil leukocytes (SRS-A, ECF-A, enzymes, PAF and other vascular permeability factors, kinin-generating substances, a complement-activating factor, histamine-releasers, a neutrophil inhibitory factor, VIP, 5-HETE, etc.), basophilic leukocytes (histamine, SRS-A, ECF-A, NCF, PAF, SP, SOMs, etc.), murine basophilic leukocytes (the same as in humans plus serotonin), eosinophilic leukocytes (PAF, 8,15-diHETE, SRS-A, eosinophil peroxidase, major basic protein, etc.), serosal, connective tissue or TC mast cells (histamine, SRS-A, ECF-A, NCF, PAF, VIP, SP, SOMs, etc.), mucosal or T mast cells (histamine, SRS-A, ECF-A, NCF, PAF, VIP, SP, SOMs, etc.), "chromaffin positive" mast cells (dopamine in ruminants; in other species possibly norepinephrine and neuropeptide Y), the so-called P-cells (histamine, serotonin), enterochromaffin cells (serotonin), chromaffin cells (epinephrine, norepinephrine, dopamine, neuropeptide-Y, IL-1α, etc.), platelets (depending on species, histamine, serotonin, catecholamines, prostaglandins, 12-HETE), neurosecretory cells (histamine, serotonin, catecholamines, acetylcholine, prostaglandins, and other eicosanoids, kinins, the various hypothalamic substances that release or inhibit the release of the anterior pituitary hormones, and the group of neuroactive immunoregulatory peptides including ACTH, bombesin, neurotensin, endorphins, enkephalins, TSH, growth hormone, prolactin, chorionic gonadotropin, VIP, the tachykinin neuropeptides, somatostatins, neuropeptide-Y, IL-1α, IL-1β, IL-6, etc.), medullary thymic epithelial cells and the Hassals corpuscles (thymosins and other thymic factors), SIF cells (dopamine), and other nerve cells including essentially all the effector molecules listed under the neurosecretory cells. For more detailed information on all the foregoing cell types and effector molecules see Szentivanyi et al., 1990.[41]

Many of these cell types possess different morphologic, physicochemical and general biologic characteristics. Nevertheless, in passing from one member of this cell spectrum to another, obvious transitions are seen in all these characteristics. Furthermore, when one surveys their properties and their probable physiologic function in the higher organism, significant cohesive features become apparent that set them apart from other body constituents as a distinct single class of cells that must be included in current concepts of neurosecretion.

There are some workers who postulated that the cellular components of the immune-neuroendocrine circuitry and their effector molecules could be viewed as two different divisions of this network. The two major divisions according to these workers may be defined as involved in neurovascular immunology and neuroendocrine immunology. Neurovascular immunology is concerned with immune response-related actions of vasoactive neurotransmitter substances that function as potent, short-lived local "hormones" first identified and studied as mediators of immunologic inflammation and hypersensitivity. They also play important roles in blood flow, vascular permeability, and pain transmission. These soluble effector molecules are simple compounds (i.e., amine mediators), short-chain peptides (i.e., kinins, substance P, etc.) and short-chain lipids (i.e., prostaglandins, leukotrienes) and have a long evolutionary history in biological defense. They use parakine and synaptic signalling on their effector cells. The second major division, defined above as involved in neuroendocrine immunology, represents all the immune response-related hypothalamic, pituitary, and other hormones which use endocrine signalling and are primarily immunomodulatory in character. This perhaps convenient but arbitrary functional separation of these two divisions is intrinsically incorrect as discussed in Szentivanyi et al.[45]

Neural Crest Interactions in the Development of the Immune System

In discussing neural crest interactions in the development of the immune system at first we must briefly characterize the developmental biology of this structure. The neural crest is produced from ectodermal cells which are released from the apical portions of the neural folds at about the time fusion occurs to form the neural tube and a separate overlying ectodermal layer. The basement membrane underlying the neural crest cells breaks down, the cellular characteristics change, and they become separated from the other components of the neural fold. There is a change in relative spatial relationship, or migration, of the neural crest cells to varied associations and destinies.[46]

We have known for some time that neural development is conducted in several distinct, consecutive steps but it is only recently that we have begun to understand the nature of the molecular mechanisms underlying these steps. The first step is primarily controlled by inductive mechanisms, i.e., by established diffusible factors of vertebrate embryonic induction. This is followed by intracellular regulation of a set of transcriptional factors containing for instance, Hox and Pou domains.[47,48] When the neural tube is formed, neural crest cells, i.e., multipotent stem cells, start to differentiate into the various cell types that comprise the nervous system. This is the stage at which neuronal immunoglobulin gene superfamilies (IgSFs) and cytokine receptors play important roles. Three processes take place at this stage to form the nervous system: cell migration, pathway finding, and determination of cell lineage. Recent studies suggest that the molecules involved in these three processes may have originated from the same protein module that has also evolved to generate major molecules in the immune system (see discussion of coevolution of cytokine receptors in the immune and nervous systems below).

The portion of the neural crest pertinent to this discussion is that which is closely associated with the developing brain, specifically, the hind brain. Crest cells in this cranial portion (anterior to the fifth somite) differentiate into mesenchyme, in addition to other connective tissue, muscular and nervous components. Neural crest cells migrate ventrolaterally through the bronchial arches and contribute mesenchymal cells to a number of structures. It is this mesenchyme that forms the layers around the epithelial primordia of the thymus.[49]

The full significance of the foregoing will be even more appreciated when viewed in the context of three additional considerations: (1) the thymus is formed by contributions from different sources which must interact in a precisely timed sequence for proper development; (2) ablation of small portions of neural crest prevents or alters the development of the thymus; and (3) formation of the thymus precedes that of the more secondary, peripheral lymphoid tissues reflecting a critical thymic role already in the early development of the immune system. Taken together, development of the immune system is inherently linked to the neural crest and any aberration in this link results in defective immune development such as that seen for instance in the DiGeorge syndrome.

The Ancient Superfamily of Immune Recognition Molecules and the Neural Cell Adhesion Molecule

The ancient superfamily of immune recognition molecules and the N-CAM represent another aspect of the interrelationships among the cellular and molecular components of the immune-neuroendocrine circuitry. Most of the glycoproteins that mediate cell-cell recognition or antigen recognition in the immune system contain related structural elements, suggesting that the genes that encode them have a common evolutionary history. Included in this Ig superfamily are antibodies, T-cell receptors, MHC glycoproteins, the CD2, CD4, and CD8 cell-cell adhesion proteins, some of the polypeptide chains of the CD3 complex associated with T cell receptors and the various Fc receptors on lymphocytes and other white blood cells--all of which contain one or more Ig-like domains (Ig homology units). Each Ig homology unit is usually encoded by a separate exon, and it seems likely that the entire supergene family evolved from a gene coding for a single Ig homology unit similar to that encoding Thy-1 or β_2-microglobulin which may have been involved in mediating cell-cell interactions. Since a Thy-1-like molecule has been isolated from the brain of squids, it is probable that such a primordial gene arose before vertebrates diverged from their invertebrate ancestors some 400 million years ago. New family members presumably arose by exon and gene duplications, and similar duplication events probably gave rise to the multiple gene segments that encode antibodies and T cell receptors.[50]

An increasing number of cell-surface glycoproteins that mediate Ca^{2+}-independent cell-cell adhesion in vertebrates are being discovered to belong to the Ig superfamily. One of these is the so-called N-CAM, which is a large, single-pass transmembrane glycoprotein (about 1,000 amino acid residues long). N-CAM is expressed on the surface of nerve cells and glial cells and causes them to stick together by a Ca^{2+}-independent mechanisms. When these membrane proteins are purified and inserted into synthetic phospholipid vesicles, the vesicles bind to one another, as well as to cells that have N-CAM on their surface; the binding is blocked if the cells are pretreated with monovalent anti-N-CAM antibodies. Thus, N-CAM binds cells together by a homophilic interaction that directly joins two N-CAM molecules.[51]

Anti-N-CAM antibodies disrupt the orderly pattern of retinal development in tissue culture and when injected into the developing chick eye, disturb the normal growth pattern of retinal nerve cell axons. These observations suggest that N-CAM plays an important part in the development of the central nervous system by promoting cell-cell adhesion. In addition, the neural crest cells that form the peripheral nervous system have large amounts of N-CAM on their surface when they are associated with the neural tube, lose it while they are migrating, and then re-express it when they aggregate to form a ganglion suggesting that N-CAM plays a part in the assembly of the ganglion.

There are several forms of N-CAM, each encoded by a distinct messenger RNA (mRNA). The different mRNAs are generated by alternative splicing of an RNA transcript produced from a single large gene. The large extracellular post of the polypeptide chain (~ 680 amino acid residues) is identical in most forms of N-CAM and is folded into five domains characteristic of antibody molecules. Thus, N-CAM belongs to the same ancient superfamily of recognition proteins to which antibodies belong.[52]

I have mentioned earlier the guidance provided by the neural crest in the development of the thymus, that is, a central regulatory organ for the immune system. The converse also appears to be true: the immune system has a special role in the development of the nervous system. In this context, the most critical feature of the immune system is its tremendous polymorphism so that lymphocytes that are produced can recognize enormous numbers of different antigens. For this reason, the immune system is ideally suited to provide markers or "anchoring sites" that enable developing structures to be built up in precisely the correct form. In no organ system is this type of detailed anchorage mechanism as important as in the developing nervous system in which many millions of nerve fibers traverse great distances and establish connections with particular groups of target cells. Marking by means of histocompatibility antigens provides exactly such a system, as indeed the original function of the MHC antigens has been defined already in the 1970s as the general plasma membrane anchorage site of organogenesis-directing proteins.[53]

Coevolution of Cytokine Receptor Families in the Immune and Nervous Systems

Established members of neurotrophic factors (NTFs) and neuronal differentiation factors (NDFs) belong to two major groups: cytokines and growth factors. The family of cytokines may be divided into three subclasses based on structural homology of their receptors (Bazan[54,55]): (1) Class I receptors: IL-1R, IL-3R, IL-4R, IL-5R, IL-6R, IL-7R, granulocyte-macrophage colony stimulating factor receptor (GM-CSFR), erythropoietin receptor (EPR), cholinergic differentiation factor receptor (CDFR), leukaemia inhibitory factor receptor (LIFR), ciliary neutrophic factor receptor (CNTFR); (2) Class II receptors: interferon-alpha receptor (IFN-α R), interferon-gamma receptor (IFN-γ R), and tissue factor receptor (TFR); and (3) class III receptors, i.e., those receptors whose primary structures belong to immunoglobulin gene superfamilies (IgSFs) such as IL-1. Growth factors can be classified into gene families based on the homology of their structures: nerve growth factor (NGF) family, fibroblast growth factor (FGF) family, epidermal growth factor (EGF) family, insulin growth factor (IGF) family, etc.

Class I cytokine receptors can be further divided into at least two subtypes, Class Ia and Class Ib, based on structural differences[56]. Class Ia includes IL-2R, IL-3R, IL-5R, and GM-CSFR[57] and consists of the common cytokine domain with four conserved cysteines and a Trp-Ser-X (variable residues)-Trp-Ser motif near the transmembrane domains. Class Ib contains IL-6R, CNTFR, LIFR, and G (granulo-cyte)-CSFR, and has additional elements, i.e., three fibronectin-like domains near the transmembrane domain and an Ig-like domain at the N-terminal end in the extracellular sites. It is the group of cytokines in Class Ib that is of special interest to us in the context of this discussion. As mentioned briefly before, a number of NTFs and NDFs have been identified recently together with their receptors at molecular levels. It was shown that NTFs enhance survival of certain types of neurons whereas NDFs affect neuronal phenotype without having any influence on neuronal survival. Cholinergic differentiation factor (CDF) that has a potent NDF activity on sympathetic neurons is identical to leukaemia inhibitory factor (LIF) that plays an important role in the immune system[58]. Ciliary neutrophic factor (CNTF) with an NTF activity on ciliary ganglia neurons also possesses NDF activity as that of CDF/LIF by inducing cholinergic differentiation of sympathetic neurons in culture[59]. Another member of this group (Class Ib), IL-6, an established immunoregulatory molecule, improves the survival of mesencephalic catecholaminergic and septal cholinergic neurons from postnatal rats (two weeks old) in cultures[60], and triggers the association of its receptor (IL-6R) with the signal transducer gp130[61]. Generation of sensory neurons is stimulated by LIF[62] and LIFR is structurally related to the IL-6 signal transducer gp130[63]. Furthermore, there is evidence that not only LIF but also CNTF act on neuronal cells via a shared signal pathway that involves the IL-6 signal transducing receptor component[64].

In view of the very similar NTF and NDF activities of all cytokine members of Class Ib, Yamamori[65] proposed that their receptors are likely to have a closely similar structure. This suggestion was considerably strengthened by findings in a number of laboratories showing that their receptors belong to a subclass of receptors for cytokines with both neuronal and immunoregulatory activities.[66-70] This close structural relationship between the nervous and immune systems at the molecular level may be explained by evolutionary mechanisms in agreement with the proposal of Edelman[71] that the structural resemblance of immunoglobulin and neural cell surface molecules may have resulted from a common evolutionary origin, and with the analysis of Bazan[54] of the structural design and molecular evolution of the cytokine receptor superfamily. Finally, Yamamori and Sarai[56] in an attempt to trace the evolutionary origin of the cytokine receptors, constructed a higher subdomain structure of the receptor for CDF/LIF based on its known primary structure. In their study, the receptor appears to contain immunoglobulin and fibronectin-like domains in addition to common domains of the cytokine receptor, similar to those cell surface molecules of the neural IgSFs. Taken together, these and other above-mentioned findings indicate that the Class Ib cytokine receptor evolved as a consequence of fusion of the genes for a more primitive cytokine receptor in Class Ia and for the IgSF, and similarly that a large number of molecules regulating neural development and vertebrate immunoregulation have a common evolutionary background. Thus, the origin of cytokine receptors represents a classical example of molecular coevolution in the immune and nervous systems.[56]

CEREBRAL DOMINANCE OR LATERALIZATION AND IMMUNE DISORDERS

Leo Szilard

This section requires the introduction of a living legend of the world of international science, Professor Leo Szilard. Here we must go back to the saddest chapter of twentieth century European history. When Hitler arrived in 1933, the remarkable German tradition of scholarship was destroyed almost overnight. Germany, and later the entire European continent, was no longer hospitable to the imagination--and not just the scientific imagination. A whole conception of culture was in retreat: the conception that human knowledge is personal and responsible, an unending adventure at the edge of uncertainty. And since evil does not take place in isolation but is a product of amorality by consensus, silence fell on the entire European continent, as after the trial of Galileo. The great men went out into a threatened world. Max Born. Erwin Schrödinger. Albert Einstein. Sigmund Freud. Thomas Mann. Arthur Koestler. Stephan Zweig. Bertolt Brecht. Arturo Toscanini. Bruno Walter. Marc Chagal. Enrico Fermi. And the incomparable Hungarian quintet: John Von Neumann, Eugene Wigner, Dennis Gabor, Edward Teller, and Leo Szilard who arrived finally after many years at the University of Chicago.

This group of Hungarians has had an inordinate impact on our culture: Von Neumann (invention of the computer), Wigner (Nobel prize for contributions in theoretical physics), Gabor (Nobel prize for holography), and Teller (the father of the hydrogen bomb). But what of Szilard? He was the "genius in the shadows," as his recent biography by Lanoutte[72] is most appropriately entitled, who was behind these Hungarian contributions and was associated with many of the most noteworthy scientific and historical events of our century.

Leo Szilard was a Hungarian whose university life was spent in Germany as a nuclear physicist. In 1929, he had published a pioneer paper on Information Theory, the relation between knowledge, nature, and man. He was also the first to realize that if you hit an atom with one neutron and it happens to break up and release two, you then would have a chain reaction. He wrote a specification for a patent that contained the word "chain reaction" which was filed in 1934. In an attempt to keep the patent secret and thus prevent science from being misused, Szilard assigned the patent to the British Admiralty so that it was not published until after the war. But meanwhile the war in Europe was becoming more and more threatening and the march of progress in nuclear physics, and the march of Hitler went step by step, pace by pace, to its inexorable and incomprehensibly evil end.

On 2 December 1942, on the squash court, West Stands, Stagg Field, University of Chicago, the exponential graphite-uranium pile designed by the group under Enrico

Fermi and Leo Szilard went into operation. Thus, man achieved the first self-sustaining nuclear chain reaction, and thereby initiated the controlled release of nuclear energy. The feasibility of such an achievement had already been predicted in 1939 independently by Szilard, Joliot Curie in France and Niels Bohr in Denmark. In the same year, Szilard wrote to Joliot Curie asking him whether one could make a prohibition on publishing in the field of atomic physics. Finally on 2 August 1939, Szilard prepared a letter for the signature of Albert Einstein to be sent to President Roosevelt informing him of current progress in atomic bombs. This was the letter that started the development of the Manhattan project. When in 1945 the European war had been won and Szilard realized that the bomb was now about to be made and used on the Japanese, Szilard marshalled protests everywhere he could and wrote memorandum after memorandum. He always wanted the bomb tested openly before the Japanese and an international audience, so that the Japanese should know its power and surrender before people died. Szilard failed and with him the community of scientists failed. He did what a man of integrity could do. He resigned as professor of nuclear physics and turned to biology. It was at this point that he sought out David W. Talmage at the University of Chicago whom he knew was in the process of paving the way to developing a new view of immunoregulation--the view that we have discussed earlier (for more extensive coverage of the subject the reader is referred to Szentivanyi and Friedman, 1994[73]). It was Talmage who introduced me to Szilard but while at the University of Chicago I had no personal relationship with him. It was only after Talmage moved to the University of Colorado Medical School in Denver and asked me to go with him that a more personal and close relationship developed between us. This relationship became an outgrowth of our friendship with Szilard's only sister, Rose, and her husband Roland Detre, by that time a nationally recognized painter, who resided in Denver. Because of these circumstances, Szilard frequently visited Denver, leading gradually to more meetings between us and ultimately resulting in a friendship that lasted until the day of his death on May 20, 1964 in LaJolla, California. On that day, I was in LaJolla at an NIH research conference; it was a very sad day.

On one of his visits to Denver, I told Szilard that one of my old research interests while still in Hungary had to do with an observation that epilepsy occurs more frequently in patients with disorders of atopic allergy (asthma, atopic dermatitis, hay fever, etc.), especially if the patients are left-handed. I explained that this was about the time when electroencephalography emerged as a new tool in neurophysiology. Further, the first laboratory of this kind was established at the University of Marseille under the direction of Professor Henry Gastaut where Raphael Panzani, an allergist, was in the process of studying precisely the same strange relationship between left-handedness, epilepsy, and atopic disease. I requested authorization for a study-leave at the University of Marseille but the Hungarian authorities did not approve my exit visa and I could not go. Nevertheless, these findings, i.e., a relationship between atopic disease and left-handedness, remained a source of considerable fascination for me for the main reason that they did not make any sense at all.

Szilard looked at me and said, "On the contrary, they do make a great deal of sense even in nuclear physics." I still see him, sitting before me, a man ready on short notice with memoranda outlining strategies for international peace or economic recovery, charming, impulsive, restless, and a very able agent of seduction from the straight path of immunology and immunological disease to such bizarre relationships as left-handedness and atopic allergy. To Szilard's statement I remarked that I always thought that it was a central and proven tenet of the laws of physics that elementary particles are symmetric. Szilard said that these concepts had been proven to be incorrect a long time ago. He pointed to the example of beta decay in which an electron is emitted along with a newly formed particle, the neutrino, that spins at right angles to the direction of movement. In the old concepts of symmetry, there would be equal numbers of neutrinos with left-sided and right-sided spins. In reality, however, the neutrinos so produced are all "left-handed," i.e., as they move toward the observer they all spin in the clockwise direction while the emitted electrons spin in the reverse direction. The asymmetric spins of particles emitted in beta decay are not restricted to the physics of elementary particles. In fact, they are present in all physical, chemical, and biological systems. This nonconservation of parity, continued Szilard, may very well apply to the problems of cerebral dominance or lateralization and the development of immune disorders and it may explain our observations on the relationship between left-handedness and atopic disease (for a more current and comprehensive account of these issues see Szentivanyi et al., 1994[45]).

Lateralization and Immune Disorders

The recognition of cerebral lateralization grew out of the discovery in the last century of cerebral dominance, that is, the superior capacity of each side of the brain to acquire particular skills. Over the past 120 years, it was believed that hemispheric dominance was based on functional asymmetry--on the differences in function of the two sides of the brain and of specific regions within them. In the face of the prevalent belief that cerebral dominance lacked an anatomic correlate, work over the past three decades has conclusively established that cerebral dominance is based on asymmetries of structure. An example of this is the early detectable asymmetry in the human brain that involves the upper surface of the posterior portion of the left temporal lobe, the *planum temporale*. The larger size of the left *planum temporale* reflects the greater extent of a particular temporoparietal cytoarchitectonic area on the left. There are other asymmetries in the human brain and the same applies to the findings throughout the animal kingdom. In addition to genetics, several other factors in the course of development, both prenatal and postnatal, influence the direction and extent of these structural differences.[74]

Associations of anomalous cerebral dominance include not only developmental disorders such as dyslexia, autism, stuttering, mental retardation, and learning disorders, as well as some extraordinary musical, mathematic, athletic and other talents**, but also alterations in many bodily systems including the immune system. In these situations, the same influences that modify structural asymmetry in the brain also modify other systems such as the immune system. The suspected molecular mechanisms involved in the influence of structural asymmetry on the development of immune reactivities are discussed in more detail in two larger reviews.[75,76] Here I shall only mention that animal experiments in the past ten years provided some insight into the nature of the association between anomalous hemispheric dominance and immune reactivities.

Thus, beginning in 1980 and continuing throughout the decade, Renoux and coworkers showed that immunomodulation of the T-cell lineage in rodents can be a phenomenon of hemispheric lateralization. In 1980, the initial observation presented data indicating that lesioning the left cerebral neocortex depresses T-cell mediated responses in mice without affecting B-cell responses. These observations were extended in experiments where animals with a right cortical lesion served as controls

---

**It is difficult to speak in some of these cases of extraordinary talents as the "pathology of superiority" but that is what the evidence dictates.

for animals with a left cortical ablation. The findings demonstrated a balanced brain asymmetry in which the right hemisphere controls the inductive influence on T-cells of signals emitted by the left hemisphere. In addition, most recent studies found that ditiocarb sodium (Imuthiol), an immunostimulant specifically active on the T-cell lineage, can replace the signals emitted by the left neocortex, since mice without a left neocortex were stimulated to increased T-cell-dependent responses by treatment with ditiocarb sodium, whereas the agent did not modify the responses already increased in right decorticates. B-cell-dependent and some macrophage-dependent responses are not affected by either neocortical ablation or ditiocarb sodium. This lateralization of cortical influences on immune function in rodents is likely to be predictive of an even greater influence in humans with more profound and complex cortical functions.[77]

IMMUNE AND NEURONAL MEMORY

The immune system and the nervous system possess short- or long-term memory, or both. The latter may be defined as the recording of experiences that can modify behavior. This general definition encompasses a broad spectrum of phenomena from the bacterial capacity of sensing chemical gradients to cognitive learning in humans.

The clonal selection theory of acquired immunity provides a useful conceptual framework for understanding the cellular basis of immunologic memory. According to this scheme, immunologic memory is generated during the primary immune response because (1) the proliferation of antigen-triggered virgin cells creates a large number of memory cells--a process known as clonal expansion; (2) the memory cells have a much longer life span than do virgin cells and recirculate between the blood and secondary lymphoid organs; and (3) each memory cell is able to respond more readily to antigen than does a virgin cell.

One reason, if not the most important reason, for the increased responsiveness of memory B-cells is the higher affinity (avidity) of their antibody receptors for the homologous antigen. Thus, with the passage of time after immunization there is a progressive increase in the affinity of antibodies produced against the immunizing antigen. This phenomenon is known as affinity maturation and it is due to the accumulation of somatic mutations in variable (V)-region coding sequences after antigen stimulation of B-lymphocytes. The rate of somatic mutation in these sequences is estimated to be 10^{-3} per nucleotide pair per cell generation which is about a million times greater than the spontaneous mutation rate in other genes. This process is called somatic hypermutation. Since B-cells are stimulated to proliferate by the binding of antigen, any mutation occurring during the course of an immune response that increases the affinity of a cell surface antibody molecule will cause the preferential proliferation of the B-cell making the antibody, especially when antigen concentration decreases with increasing time after immunization. Thus, affinity maturation is the consequence of repeated cycles of somatic hypermutation followed by antigen-driven selection in the course of an antibody response.

Research in the field of neuronal memory is still in an early phase primarily because of methodologic difficulties and the validity of approaches currently used. The human brain is extraordinarily complex (10^{12} neurons) and intricate (an average neuron may have 10,000 dendrites interacting with other neurons), dictating the use of reductionist approaches which always require a correlation with the whole organism to verify the conclusions reached at the molecular level. Such relationship emphasizes the importance relating the biochemical events in single cells to the more complex organisms such as Aplysia, Drosophila, rodents, cats, and humans. The fact, however, that adjacent neurons are almost never identical means that the quantity of material needed to perform biochemical analyses necessitates a cell line approach. Both bacteria and neural cell lines provide a homogeneous population of cells that can be studied biochemically. Bacteria detect chemical gradients using a memory obtained by the combination of a fast excitation process and a slow adaptation process. This model system, which has the advantages of extensive genetic and biochemical information, shows no features of long-term memory.

To study long-term memory, other biologic systems that exhibit two phenomena associated with learning and memory, habituation and potentiation, must be used. Habituation is defined as the decreased responsiveness to a stimulus when it is presented repetitively over time. Potentiation, on the other hand, is defined as the increased responsiveness to a stimulus when that stimulus is presented repetitively over time. In the mammalian brain, the hippocampus plays a special role in learning: when it is destroyed on both sides of the brain, the ability to form new memories is largely lost although previous long established memories remain. The evidence obtained on hippocampal slices indicates that the biochemical changes in the synapse represent the molecular bases for long-term memory.[78,79] Despite the wealth of information provided by investigations in mammalian brain slices, it became increasingly clear that the study of cultured neural cells is more desirable because only in such a system could one be certain that the complete biochemical pathway, that is, the complete signal transduction pathway from stimulatory input to a behavioral output, could be analyzed. To study memory, however, some modifiable behavior needs to be observed. Because neurons communicate with each other chemically through the release of a neurotransmitter, the secretion of neurotransmitters (the output) evoked by various chemical stimuli (input) could be used to monitor the responsiveness of the cell. This experimental system was used to study the input-output properties of a particular neuron. Both habituation and potentiation could be demonstrated in neuronal cell lines indicating that they can serve as good model systems for the memory process, except that they do not possess synaptic connections. In the early phase of these studies the absence of synaptic connections posed a substantial problem for two major reasons: (1) as stated before, current evidence favors (in more organized neural tissue such as brain slices) the idea that the biochemical changes in the synapse are the molecular basis of long-term memory, and (2) the synapse is a unique anatomic association of two cells that occurs only in the nervous system and therefore represents a special *sui generis* neuronal feature.

Whether memories, however, are generally recorded in presynaptic changes or in postsynaptic changes, in synaptic chemistry or in synaptic structure, or indeed in synapses at all, are still open questions. Regardless of the validity of any of these questions, it appears that the biochemical features of all memory-forming processes (i.e., habituation, potentiation, and associative learning in invertebrates, mammalian brain slices, and cultured clonal neural cell lines) are highly similar (the PC 12 cells[80,81] and HT4 cells[82,83]). They can be characterized as follows: 1) a mono-amine (primarily serotonin) and a glutamate receptor (known as NMDA receptor because it is selectively activated by the artificial glutamate analog N-methyl-D-aspartate) are involved; (2) binding of the neurotransmitter (serotonin, glutamate) by these receptors initiates a cascade of enzymatic reactions; (3) the first step in this cascade is the activation of a G-protein which may either interact directly with ion channels or control the production of cyclic adenosine monophosphate (AMP) or Ca^{2+}; (4) the two second messengers in turn regulate ion channels directly or activate kinases that phosphorylate various proteins including ion channels; (5) at many synapses both channel-linked and non-channel-linked receptors are present, responding either to the same or to different neurotransmitters; (6) responses mediated by non-channel-linked receptors (serotonin) have a slow onset and long duration and modulate the efficacy of subsequent synaptic transmission providing the basis for memory formation; (7) channel-linked receptors that allow Ca^{2+} to enter the cell (NMDA receptor) also mediate long-term memory effects; and (8) either too much or too little cyclic AMP can interfere with memory formation.[84]

In all these processes, the interaction between serotonin and glutamate and their respective receptor systems can be illustrated by the studies carried out on the HT4 neural cell line.[82] The HT4 cells do not habituate to repetitive membrane depolarization but after exposure of these cells to various neurotransmitters, serotonin has the capacity to potentiate cellular responsiveness. Depending on the strength of the serotonin stimulus, both short- and long-term potentiation can be induced. For instance, a two-minute exposure to serotonin results in the transient increase in cellular responsiveness, where a five-minute presentation gives rise to a more permanent potentiation, the difference between the two involving the activation of NMDA receptors. Thus, the stronger (five minute) serotonin stimulus results in the endogenous release of excitatory amino acids with activation of NMDA receptors. Consistent with this mechanism, long-term secretory potentiation can also be produced with a two-minute stimulus of serotonin only if glutamate or NMDA is given simultaneously.

As I now begin a comparison of immunologic versus neuronal memory, I have to return to an earlier statement that immunologic memory is due to clonal selection and lymphocyte maturation. Although this is correct in cellular terms, in molecular terms the problem of clonal selection and expansion reduces to the issue of affinity maturation of the antibody on the surface of the lymphocyte. In other words, the entire antigen-driven selection of antibody-producing lymphocytes is based on the strength of the antibody-antigen interaction, which depends on both the affinity and the number of binding sites. The affinity of an antibody reflects the strength of binding of an antigenic determinant to a single antigen-binding site and it is independent of the number of sites. However, the total avidity of an antibody for a multivalent antigen, such as a polymer with repeating subunits, is defined as the total binding strength of all of its binding sites together. When a multivalent antigen combines with more than one antigen binding site on an antibody, the binding strength is greatly increased because all the antigen-antibody bonds must be broken simultaneously before the antigen and antibody can dissociate. Thus, a typical IgG molecule will bind at least 1,000 times more strongly to a multivalent antigen if both antigen-binding sites are engaged than if only one site is involved. For the same reason, if the affinity of the antigen-binding sites in an IgG and an IgM molecule is the same, the IgM molecule (with ten binding sites) will have a much greater avidity for a multivalent antigen than an IgG molecule (with two sites). This difference in avidity, often 10^4-fold or more, is important because antibodies produced early in an immune response usually have much lower affinities than those produced later. Because of its high total avidity, IgM--the major Ig class produced early in immune responses--can function effectively even when each of its binding sites has only a low affinity.[85]

In the late 1950s, an excellent correlation was shown between the body temperature of rabbits and the affinity and avidity of the antibody produced against the radiolabeled antigen as tested in equilibrium dialysis experiments.[86] The higher the body temperature was, the greater the affinity and avidity were. In the beginning of these studies commercially available *E. coli* endotoxin was used, but later the rise in temperature was reproduced by electrical stimulation of the posterior hypothalamus or hippocampus through stereotaxically implanted permanent depth electrodes in studies without endotoxin administration.[87] Discovery of the peptidoglycan and its derivatives as powerful immunologic adjuvants opened up a new window for the consideration of the relationship between immunologic and neuronal memory. The peptidoglycan, which is the basal layer of the bacterial cell wall, is a rigid macromolecule surrounding the cytoplasmic membrane. It is formed by the polymerization of a disaccharide tetrapeptide subunit; in the intact peptidoglycan, disaccharides form linear chains whereas peptides are linked by interpeptide linkages.[44] The recognition of the immunomodulating properties of peptidoglycans and peptidoglycan fragments is the result of the work aimed at identifying the structure responsible for the adjuvant activity of the mycobacterial cells in Freund's adjuvant.[88] Simple active molecules were soon produced by organic synthesis followed by a vast array of analogs and derivatives that can be classified into several categories. The one that is most pertinent to this discussion is the group of "simple muramyl peptides." Of these the smallest immunoactive synthetic muramyl peptide is N-acetylmuramyl-L-alanyl-D-isoglutamine (MDP).[89,90] This substance has a pyrogenic effect that originally was attributed to its ability to induce the release of endogenous pyrogen from mononuclear phagocytes. However, a direct central nervous system action could not be excluded since MDP was found to be active by the intracerebroventricular route (MDP was shown to cross the blood-brain barrier[91]), and in rabbits made leukopenic by nitrogen mustard treatment.[44] In addition, it was subsequently shown that MDP can also induce sleep, and the somnogenic effect can be separated from its pyrogenic activity. MDP's pyrogenic activity does not affect brain temperature changes that are tightly coupled to sleep states.[92] More importantly with respect to the direct central nervous system neuronal effects of MDP, this substance is capable of specific binding to serotonin receptors of synaptosomal membranes of brain tissue and competes with serotonin for these binding sites, and the kinetics of serotonin binding to brain homogenates is altered after sleep deprivation.[93] Additional findings on the capacity of MDP to act directly and specifically on central neurons include the following: MDP alters neuronal firing rates in different regions of the brain,[94] humoral antibody responses are enhanced by lowering serotonin levels in the brain,[95] administration of para-chlorophenylalamine which markedly decreases the level of brain serotonin completely abolishes the MDP-induced rise in body temperature and the somnogenic effect.[96] Finally, it has been established that immunization decreases the concentration of serotonin in the hypothalamus and the hippocampus.[97]

Although the foregoing evidence is fragmentary, it does establish a set of future reference points to begin to undertake a more informed comparison of the molecular mechanisms involved in immune and neuronal memory.

It appears that one of these future reference points may now be defined as a consequence of some important recent findings in our laboratory. They have to do with our demonstration that lymphocyte conditioned medium contains a group of adrenergically active lymphocytic proteins that participate in the physiologic regulation of beta$_2$-adrenergic receptors in human pulmonary epithelial cells which will be discussed later. What is pertinent for this discussion involves one of these adrenergically active proteins, an interleukin-1alpha (IL-1α) that is present in B-cells but among the T-cell subpopulations its presence is restricted to the CD4$^+$ CD45RO$^+$ subset of the T-memory cells. The only other cells where this IL-1α is found to be selectively accumulated are the hippocampal neurons which are the site of memory. It is added that we have shown that the hippocampal neurons have the capacity to synthesize, store, and release this agent as well as express specific receptors for the same. It is possible, therefore, that IL-1α is an effector molecule in a memory mechanism that is shared by both the immune and neuronal memory.[98-103]

THE UNIQUE RECOGNITION AND COMMUNICATION POWERS OF THE IMMUNE AND NEUROENDOCRINE SYSTEMS AS SHARED CHARACTERISTICS

Earlier I cited Schechter, pointing out that a good relationship between two biologic systems must be sensitive, synchronized, and synergistic.[43] There are no two biologic systems where such characterization of an ideal relationship would be more valid than in the case of the immune and neuroendocrine systems as reflected by their unique recognition and communication powers as shared characteristics. The latter are based on four critical features shared by both: (1) they are composed of extraordinarily large numbers of phenotypically distinct cells organized into intricate networks. Moreover, the size of this extensive cellular arsenal continuously increases as new sequence information becomes available and new members of the Ig superfamily of cell surface molecules appear each year. Within these cell networks, the individual cells can interact either positively or negatively and the response of one cell reverberates through the system by affecting many other cells; (2) cells of both systems synthesize, secrete and/or release the same effector molecules; (3) recognition of these effector molecules is realized by the same cellular receptors and second messenger mechanisms of both cell systems; and (4) these cellular and molecular determinants make a continuous, bilateral flow of information, the *sine qua non* of the unique interactions within the immune-neuroendocrine circuitry, possible.

A more amplified view on the basic biochemistry and molecular biology of receptor-effector coupling by G-proteins (i.e., the fundamental mechanism used by hormones, neurotransmitters, and the immunomodulatory cytokines for signal transmission by G- proteins) is presented by Szentivanyi[104] and by Lochrie and Simon,[105] and Birnbaumer and Brown.[106] Here I shall only mention that about 80% of all known neurohormones, neurotransmitters, immunomodulatory lymphokines and other autocrine and paracrine factors that regulate cellular interactions in the immune-neuroendocrine circuitry, called "primary" messengers, elicit cellular responses by combining with specific receptors that are coupled to effector functions by G-proteins. Although the primary messengers are many, the number of physicochemically and biologically distinct receptors that mediate their action is even larger. So far about 80 distinct receptors that recognize 40 hormones, neurotransmitters, and so on, can be identified. It is reasonable to assume that the total number of distinct receptors coupled by G-proteins will be 100 to 150. In contrast to receptors, the number of final effector functions regulated by these receptors and the number of G-proteins that provide for receptor-effector coupling are much lower, probably not much more than 15 each.

In mammals, a total of eight G-proteins have been purified essentially free from each other (Gt, Gs, Gi1, Gi2, Gi3, G01, G02, and Gz/x) and the cDNAs derived from a total of nine genes encoding G-alpha subunits have been cloned and designated $alpha_s$, $alpha_{i1}$, $alpha_{i2}$, $alpha_{i3}$, $alpha_0$, $alpha_{tr}$, $alpha_{tc}$, $alpha_{01\beta}$, and $alpha_{z/x}$, giving rise to 12 messenger RNAs because of the fourfold variation in the splicing of the $alpha_s$ precursor mRNA. In addition, there is evidence for the existence of at least seven additional G-alpha genes. Homology cloning has also revealed that there are at least four G-beta genes and three G-gamma genes.

The G-proteins are heterotrimeric membrane proteins (alpha, beta, gamma; 1:1:1), distinguished by unique alpha subunits, but sharing common beta subunits. Stated in a different way, G-proteins may be viewed to be composed of a unique, but homologous alpha subunit in reversible association with a complex comprised of a beta and a gamma subunit commonly shared by several different G-protein alpha subunits. Thus, the alpha subunit of Gs (the G-protein that stimulates adenylate cyclase) may share beta-gamma complex in common with the alpha subunits of the family of G-proteins that mediate inhibition of adenylate cyclase (G1), or other G-proteins like G0, Gz, and Gt. The alpha subunits bind and hydrolyze GTP, and are often the substrates for NAD^+-dependent ADP ribosylation by bacterial toxins (i.e., pertussis, cholera). Activated beta adrenoceptors catalyze the exchange of GTP for bound guanosine diphosphate (GDP) by the alpha subunit of the holoprotein, promoting the dissociation of the GTP-bound alpha subunit from the beta-gamma complex. It is the "free" GTP ligand alpha subunit of a G-protein that regulates the activity of the membrane-bound effector units such as adenylate cyclase.

The primary sequences of several G-protein-linked effectors, including adenylate cyclase and phospholipase C, have been determined and molecular cloning of phospholipase A_2 and Ca^{++} and K^+ channels are in an advanced stage in several laboratories. As mentioned earlier G-alpha $_s$ mediates the stimulation of adenylate cyclase and has been shown to regulate Ca^{++} channel activity. The G-proteins that mediate the inhibition of adenylate cyclase, termed Gi, constitute a family with at least three members, G-alpha $_{i1}$, G-alpha $_{i2}$, and G-alpha $_{i3}$, each the produce of a separate gene. Of these, it is G-alpha $_{i2}$ that mediates the inhibition of adenylate cyclase.[104]

At the time of writing, receptors for simple substances, such as the amine mediators and short-chain peptides as well as lipids, and for more than 20 different hypothalamopituitary peptides have been identified in the cells of the immune system, essentially in lymphocytes. In addition to the hypothalamopituitary hormones, lymphocytes also express receptors for peptides secreted from neurons together with other neurotransmitters. These neuropeptides take on added significance as immunomodulators, since it is now known that lymphoid organs are directly innervated with nerves secreting these agents. From the standpoint of the integration of information in the immune-neuroendocrine circuitry, future studies will have to examine these parallel signaling pathways in isolation. In other words, it will be necessary to determine how an individual cell completely processes and integrates information from these individual pathways. This is all the more remarkable because the cell is faced with the task of balancing the need to communicate with other cells with the need for growth and maintenance of the differentiated state while preserving adequate flexibility to support regulation, sensitivity, and gain. One early result of such inquiries is the demonstration of cross-regulation (cross-talk) between the various G-protein mediated signaling pathways. Thus, it was shown that in the cross-regulation between α_1 and ß $_2$-adrenergic receptor-mediated pathways, activation of ß $_2$-adrenergic receptors increased α_1-adrenergic receptor mRNA levels.[107] Conversely, activation of the $G_{1\alpha}$-mediated inhibitory pathway of adenylate cyclase cross-regulates the stimulatory ($G_{s\alpha}$-mediated) beta-adrenergic-sensitive adenylate cyclase system by (1) upregulating ß $_2$-adrenergic receptors and enhancing the activation of the stimulatory ($G_{s\alpha}$-mediated) adenylate cyclase pathway, and (2) downregulating elements of the inhibitory adenylate cyclase pathway, $G_{1\alpha2}$ and A_1-adenosine receptor binding, respectively.[108] It may be added that cross-regulation is also observed between signaling pathways that do not share the same effectors. Although much more work remains to be done to unravel the complexities of the coordinated regulation of information processing and integration by the cell, it is already possible to state that there is cross-regulation between neurally derived substances and lymphokines.

In the foregoing sections, discussions of the reciprocal, regulatory interplay between the immune and neuroendocrine systems have mainly covered the peripheral pathways by which the neuroendocrine influences are able to affect immune functions. However, as stated earlier, the flow of information in the immune-neuroendocrine circuitry is bidirectional, and there is conclusive evidence that products of the immune system are capable of modulating neuroendocrine processes. There are two lines of evidence indicating that the products of the immune system can influence the brain or the pituitary gland, or both. The first is provided by correlational studies which show that changes occur in the brain during the course of an immune response. Along this line, Korneva and Klimenko[109] recorded single unit activity in the hypothalamus showing significant changes in the neuronal firing patterns in the posterior, ventromedial, and supramaxillary nuclei during the course of an immune response. These observations were independently confirmed by Besedovsky and coworkers[110] who found a considerable increase in the firing rate of neurons in the ventromedial hypothalamus 1 to 5 days following sensitization to trinitrophenol (TNP)-hemocyanin. Srebro and associates[111] found a significant increase in the nuclear volume of neurosecretory cells in the supraoptic nucleus during skin allograft rejection. Changes in the serotonin levels occur in the hypothalamus and hippocampus following immunization with typhoid antigen,[97] whereas increases in dopamine-stimulated adenylate cyclase activity in caudate homogenates are found following bacille Calmette-Guérin (BCG) antigen administration.[112] In recent years, these observations have been expanded by the findings on the effects of Newcastle disease virus on the metabolism of cerebral biogenic amines,[113] and similar changes have also been observed with influenza virus.

The second line of evidence implicating the immune system in regulating physiologic processes at the level of the brain or the pituitary gland, or both, is derived from studies in which products of the cells of the immune system were administered to experimental animals or added to cultured neuronal or pituitary cells. IL-1 stimulation of ACTH secretion was first shown in a mouse pituitary cell line, AtT-20 cells[114] and subsequently confirmed on primary pituitary cells.[115,116] In addition, IL-1 alters the release of TSH, growth hormones, and prolactin;[117] stimulates astroglial proliferation following brain injury;[118] stimulates somatostatin synthesis in the fetal brain;[119] inhibits progesterone secretion in cultures of granulosa cells;[120] and elicits the production of CRF by the hypothalamus.[121,122] In these neuronal interactions IL-1 acts on specific receptors in the brain.[123] IL-1 shows complex, multitargeted effects on insulin secretion: (1) it has direct glucose-dependent inhibitory and stimulatory effects on pancreatic β-cell function;[124] (2) IL-1 induces hyperinsulinemia by a central action;[125] and (3) it acts as a hypoglycemic agent independently from effects on insulin release.[126] Other lymphokines also have effects on the neuroendocrine system. IL-2 stimulates oligodendroglial proliferation and maturation[127] and also induces ACTH secretion in pituitary cells.[128] TNF-α and IL-6 augment ACTH secretion *in vivo* together with numerous other effects on the neuroendocrine system,[123,129-132] and the thymic hormone, thymosin fraction 5, stimulates prolactin and growth hormone release from anterior pituitary cells.[133]

THE LYMPHOCYTE AS THE UNIFYING REGULATORY CELL COMPONENT OF THE IMMUNE-NEUROENDOCRINE CIRCUITRY AND ITS RELATION TO HUMAN ASTHMA

In the past ten years the lymphocyte in general, and in the past three years the T-lymphocyte in particular, emerged as both the basic regulatory unit of the immune-neuroendocrine network and as a fundamental cell in the pathogenesis of bronchial asthma. This view is connected with our growing realization of the T-lymphocyte's role in the development of the bronchial obstructive process not only as an immunoregulator but also as the inflammatory cell of possibly central significance, the cell with a key involvement in bronchial hyperresponsiveness, the effector cell of the delayed-type hypersensitivity (DTH) reaction in the genesis of airway hyperreactivity, and a genetically defective neuroendocrine cell that may be specifically responsible for the immunologic, inflammatory, and beta-adrenergic dysregulation in contrast to the large number of other factors that may and do contribute secondarily to the beta-adrenergic deficit in this disorder.[104]

The T-Lymphocyte as a Critical Cell of Bronchial Inflammation

The role of IgE in immunologically-based asthma has been known for a long time as was the demonstration that its production clearly depended on multiple subsets of lymphocytes. With the discovery of soluble immunoregulatory molecules (i.e., monokines and lymphokines) it has also been established that the T-cell-derived lymphokines, IL-4, IL-5, and IFNɣ, are intimately involved in the regulation of IgE production.[134] Until recently, however, the critical role of T-cell-derived lymphokines in the regulation and expression of the inflammation associated with allergy and asthma of both the extrinsic and intrinsic varieties had not been established. Our current understanding now clearly shows a coordinated regulation of immune and inflammatory responses by T-cell-derived lymphokines.[135] Indeed, T-cells appear to orchestrate the bronchial inflammatory response to inhaled allergens and other stimuli in asthma by the production of several lymphokines with widely varying attributes such as IL-1, IL-3, IL-5 and granulocyte-macrophage colony-stimulating factor (GM-CSF). Of these, IL-5 has selective and the most pronounced effects on eosinophils including chemokinetic, chemotactic, phagocytic, cytotoxic, superoxide, and complement receptor-inducing activities together with prolongation of survival and activation.[136-138] The chemotactic factors released by T-cells recruit eosinophils to the sites of airway inflammation, whereas other lymphokines activate eosinophils and indirectly regulate the release of their products.[139,140] Also, T-lymphocytes have chemotactic activity for neutrophils, basophil granulocytes as well as monocytes, and can activate or degranulate these effector cells.[141,142] Furthermore, production of eosinophils by bone marrow is dependent on T-cells and their lymphokines.[143] Of the latter, IL-4 may require special note since this substance promotes the synthesis of IgE by B-lymphocytes[109] and induces the production of an IgE-binding factor (a cleavage fragment of the IgE receptor, F_cRII[144]). Lymphocytes from asthmatic patients are capable of spontaneously producing the IgE-binding factor.[145]

T-lymphocytes also play a role in the proliferation and activation of mast cells, and in fact an intact immune system is a requirement for the development of a functional mucosal mast cell system. The latter cells require a T-cell-derived factor to divide after stimulation. T-cells also produce substances that enhance the release of histamine from mast cells. Thus, IFNɣ, IL-1, IL-3, and GM-CSF can prime mast cells to increase their release of histamine to appropriate stimulation. In addition, T-cells produce histamine-releasing factors (HRFs) as well as a histamine-release inhibitory factor, which directly or indirectly participate in the release of histamine from mast cells or in its inhibition, respectively.[146-148] The issue of these lymphocytic HRFs is further discussed below.

The Involvement of T-Lymphocytes in Bronchial Hyperresponsiveness in Asthma

Non-specific bronchial hyperresponsiveness is usually defined as an increased reactivity of the airways to physical, thermal, chemical, immunologic, microbial (both bacterial and viral), pharmacologic, and otherwise physiologic stimuli. This manifestation has been interpreted as a characteristic of asthma and associated with airway inflammation. In the past decade, the lymphocytes, more specifically the T-lymphocytes, have become conspicuous among the inflammatory cells infiltrating the bronchi in asthmatic patients.[149,150]

Two recently published extensive studies demonstrated the relationship of T-lymphocytes to other inflammatory cells and to bronchial hyperresponsiveness. The first of these studied the regulatory role of activated T-lymphocytes in eosinophilic inflammation by investigating T-cell activation and eosinophilia in blood and bronchoalveolar lavage (BAL) from patients with asthma not receiving steroid treatment.[151] Compared to that from normal individuals, BAL from asthmatics contained markedly increased numbers of both lymphocytes and eosinophils. The lymphocytosis consisted of increased numbers of both CD4[+] and CD8[+] T-cells, and the T-cell populations expressed elevated levels of T-cell activation markers such as IL-2 receptor (CD25), HLA-DR, and very late activation antigen-1 (VLA-1). Close correlation was found between CD4[+] IL-2R[+] T-cells, eosinophil concentrations, and the degree of bronchial hyperresponsiveness. In the second publication,[152] T-lymphocytes, eosinophils, mast cells, neutrophils and macrophages were studied in bronchial biopsy specimens from subjects with or without asthma and their relationship to bronchial hyperresponsiveness determined using immunohistochemical techniques and monoclonal antibodies at two airway levels. There were no significant differences in the numbers of mucosal-type or connective tissue-type mast cells, elastase positive neutrophils, or Leu-M3[+] cells in the airways in asthmatics compared to controls. Conversely, at both proximal and subsegmental biopsy sites, significantly more IL-2R-positive (CD25[+]) cells and "activated" (EG2[+]) eosinophils were present in the airways of patients with asthma. There were positive correlations between numbers of T-lymphocytes, activated (CD25[+]) cells, eosinophils, CD3 and EG2, CD3 and CD25, and CD25 and EG2[+] in the airways of asthmatic patients. Furthermore, the ratio of EG2[+] to CD45[+] cells correlated with the bronchial hyperresponsiveness as measured by the provocative concentration of methacholine that caused a 20% decrease of one-second forced expiratory volume (FEV_1). This supports the view that activated (CD25) T-lymphocytes release products that regulate recruitment of eosinophils into the airway wall and that the eosinophils may be responsible for airway mucosal damage, particularly shedding of epithelial cells, which are believed to be a factor in precipitating bronchial hyperreactivity.[153,154] These considerations are in harmony with findings indicating that asthmatic bronchial mucosal biopsy specimens contain T-cells expressing mRNA for IL-5 and their numbers correlate with those of CD25[+] cells and the number of eosinophils in the biopsy specimens.[155] Thus, these and other studies were interpreted to mean that the prominent eosinophilic infiltration in atopic asthma and the eosinophil-mediated damage result in bronchial hyperresponsiveness. It was further proposed that eosinophil accumulation is under the control of T-lymphocytes or mast cell products, or both, and that neutrophils and macrophages further amplify mucosal inflammation through release of their own mediators.

The T-Lymphocytes as the Effector Cells of the Delayed-Type Hypersensitivity (DTH) Reaction in the Genesis of Airway Hyperreactivity

In the early 1980s, a murine model of pulmonary DTH was established through immunization with picryl chloride (PCL)[156] and revealed a peribronchial and perivascular mononuclear cell infiltrate as well as an increase in mucus-producing cells. This pulmonary DTH was shown to be antigen-specific, T-cell- and serotonin-dependent, and associated with increased vascular permeability. Subsequent analysis of this model showed that mice with PCL-induced pulmonary DTH developed T-cell-dependent, antigen-specific airway hyperreactivity, as determined by the measurement of pulmonary resistance *in vivo* and of tracheal reactivity to carbachol *in vitro*.[157] It is not known whether inflammation found in late phase bronchial reactions represents a T-cell-mediated DTH response or whether some immediate hypersensitivity reactions might be triggered by non-IgE, antigen-specific factors analogous to the PCL-F identified in cutaneous DTH in mice. Nor is it clear whether the phenotypic subtypes of T-cells in murine pulmonary or cutaneous DTH are the same as those involved in manifestations discussed in the preceding sections. What is clear, however, is that all the available data taken together support a central role of T-lymphocytes in the pathogenesis of asthma and in the development of bronchial hyperresponsiveness.

Bronchial Hyperresponsiveness and Lymphokines

The foregoing considerations resulted in a perception that asthma (both extrinsic and intrinsic), bronchial inflammation, bronchial hyperresponsiveness, and the bronchial influx of lymphocytes are interrelated, in fact, inseparable, features in the genesis of asthma. These views, however, could not account for the following observations: (1) profound inflammation of the airways is present without airway hyperresponsiveness;[158,159] (2) prevention of airway inflammation does not block the development of airway hyperresponsiveness;[160-162] (3) bronchial hyperresponsiveness is not a common feature of other inflammatory conditions (i.e., sarcoidosis) that are characterized by lymphocyte infiltration of the lungs;[163] (4) the quantitatively increased and qualitatively altered end-organ sensitivity to a broad range of specific as well as non-specific stimuli is not restricted to asthma but is also present in allergic rhinitis and atopic dermatitis;[164] (5) airways hyperresponsive to methacholine challenge represent a frequent finding in patients with atopic dermatitis without airway inflammation or symptomatic asthma;[165] (6) there is evidence that bronchial hyperresponsiveness usually precedes the development of asthma;[166] (7) bronchial hyperresponsiveness correlates with the production of T-lymphocyte products and not directly with IgE antibody synthesis;[167] and (8) there is evidence in twins that bronchial hyperresponsiveness, total serum IgE, and skin test scores are separately controlled by genetic factors.[168,169]

Although additional investigations are needed to clarify further the profile and kinetics of appearance of lymphocytes in bronchial biopsy specimens and/or in the bronchial lavage fluid of patients with asthma, some important generalizations can already be made at this time. Thus, we know that these lymphocytes are "activated," thereby supporting a functional role for lymphocytes, or more specifically for their lymphokines, in airway walls and in lung lining fluid, in the pathogenesis of airway hyperreactivity associated with asthma. The "activated state" of these lymphocytes and their lymphokines must reflect a specifically altered functional capacity that is directly related to the developmental mechanism of bronchial hyperresponsiveness. This altered functional capacity of these lymphocytes and the role of their lymphokines in non-specific bronchial hyperresponsiveness are further supported by a series of observations, some of which have been already mentioned but require some further consideration. One of these is the lymphokine HRF that lymphocytes from patients with intrinsic and extrinsic asthma spontaneously produce and release even under *in vitro* conditions. HRF releases histamine from basophils and mast cells together with leukotriene C_4 and its production can be further enhanced by preincubating lymphocytes from patients with extrinsic asthma or intrinsic asthma, with skin test positive allergens, or antigenic bacterial antigens, respectively. Furthermore, the magnitude of its spontaneous production by lymphocytes *in vitro* correlates with bronchial hyperresponsiveness in asthma of both varieties.[120-122,170] The roles of two other lymphokines, IL-4 and IL-5, have already been adequately reviewed in preceding sections. In many viral respiratory infections precipitating asthma there is an increased IFNγ production by lymphocytes[171] which in turn releases PAF that is known to be involved in bronchial hyperresponsiveness. It has also been shown that IFNγ activation of alveolar macrophages leads to the release of more thromboxane B_2, prostaglandin $F_{2\alpha}$ and leukotriene B_4 which have the capacity to induce bronchial hyperresponsiveness. [172,173] As mentioned earlier, lymphocytes from asthmatic patients are capable of spontaneously producing an IgE-binding factor that increases IgE synthesis by B-cells and enhances antigen-induced histamine release. [119] Finally, when challenged with antigen, the lymphocytes from patients with asthma produce more IL-2 than those from the controls, and viral respiratory infections in asthmatics can further enhance IL-2 production after exposure to the antigen. [123] The role of the lymphokines in the developmental mechanism of bronchial hyperresponsiveness and in relation to the constitutional basis of asthma is further discussed below from the perspective of the lymphocyte as a neuroendocrine cell.

Emergence of the Lymphocyte as a Neuroendocrine Cell and Its Significance for Asthma

As pointed out in preceding sections, the entire concept of the nature and role of the lymphocyte in immunology, immunologic inflammation, and allergy has so radically changed in the past ten years that it again requires the listing of the effector molecules that must be expected to have a regulatory influence on both health and disease (primarily asthma) of the human airways. These include interleukins, interferons, tumor necrosis factors, nucleotide metabolites, various neuroactive immunoregulatory peptides including hypothalamic CRF-like activity, ACTH, TSH, growth hormone, prolactin, chorionic gonadotropin, bombesin, endorphins, enkephalins, neurotensin, and VIP; tachykinin neuropeptides including substance P, substance K, neuromedin K, somatostatins, mast cell growth factor, suppressin, and so on. In the past four years an additional group of lymphocytic proteins with highly potent adrenergic regulatory activities was also discovered. This vast array of effector molecules makes the lymphocyte a veritable mobile neuroendocrine and/or neurosecretory cell with a range of activities that are bound to have a dominant regulatory participation in both the pathogenesis as well as the constitutional basis of asthma. It is necessary to further discuss some important aspects and activities of the recently discovered adrenergically active lymphocytic proteins (see Szentivanyi [75]).

Three macromolecular fractions with adrenergic activity can be identified in lymphocyte conditioned medium by diethylaminoethanol (DEAE) ion exchange high performance liquid chromatography (HPLC), immunoneutralization, molecular mass, sequence analysis, and biologic characterizations. One of these fractions contains a secretory variant of beta-arrestin and an IL-1α antagonist, both of which downregulate β_2-adrenergic receptors in A549 human lung epithelial cells. The two other fractions represent protein components that upregulate β_2-adrenergic receptors. One of these contains a mixture of IL-1α and IL-1β, whereas the adrenergically active component(s) of the remaining fraction is currently being characterized. The first question that may be asked in the context of this chapter is whether the adrenergically highly active IL-1α obtained from the corresponding macromolecular fraction of lymphocyte conditioned medium has a specific receptor on airway cells. Recent studies designed to explore this question used human bronchial epithelial cells isolated and cultured from the normal bronchi of patients undergoing surgery (for standard clinical reasons) essentially as described by Mattoli and coauthors. [174] Using this method, 99% of the final cell population contains epithelial cells. The latter were then incubated with IL-1α radiolabeled by a modified chloramine-T method. In addition to binding of specific, single class IL-1α receptors, the latter were also identified by internalization of the receptor, affinity cross-linking, and sodium dodecyl sulfate-polyacrylamide gel electrophoresis (SDS-PAGE). Using unlabeled IL-1α and [^{3}H]dihydroalprenolol

(DHA) for measuring beta-adrenergic mRNA with the guanidium thiocyanate method, the IL-1α-induced accumulation of β_2-adrenoceptor mRNA is demonstrable within two hours and an increase in beta-adrenoceptor concentration within four hours. In other words, concentrated IL-1α derived from human T-lymphocytes binds to a specific, single class surface receptor on human bronchial epithelial cells and induces production of β_2-adrenoceptor mRNA via an associated or separate receptor-linked signaling pathway leading to an increase in epithelial β_2-adrenoceptor concentration. [175]

Subsequently, it was shown that lymphocytic IL-1α is a cell- and species-specific factor in increasing beta-adrenoceptor concentration and induction of its gene. In this study, we used A549 human lung epithelial cells, the A431 human epidermoid cells, the DDT$_1$MF$_2$ hamster smooth muscle cells, cultured human bronchial epithelial cells and smooth muscle cells, cultured canine tracheal epithelial and tracheal smooth muscle cells. Concentrated human lymphocytic IL-1α was then co-cultured with these various cell populations for 24 hours and β_2-adrenoceptors measured radioactively. The originally shown synergistic beta-adrenoceptor upregulation between IL-1α and cortisol [176] was present in the A549 and A431 cells, as well as the human bronchial epithelial cells. Northern blot hybridization showed that levels of β_2-adrenoceptors and β_2-adrenoceptor mRNAs increased significantly by both IL-1α and cortisol, whereas Gs-α, Gi-2α, Gi-3α mRNA levels remained unchanged. In all these situations, the increase in β_2-adrenoceptor mRNAs always preceded the enhanced expression of the receptor. When DDT$_1$MF$_2$ cells, human and canine tracheal smooth muscle cells were used IL-1α had no effect either on β_2-adrenoceptors or β_2-adrenoceptor mRNAs whereas cortisol remained active. [177] This extraordinary degree of cell and species specificity of the β_2-adrenoceptor upregulating effect of lymphocytic IL-1α makes these observations highly important both for normal airway physiology as well as for the possible nature of the beta-adrenergic dysregulation in asthma by adding an entirely new dimension to the beta-adrenergic theory of the atopic abnormality in bronchial asthma as postulated twenty-five years ago by Szentivanyi. [164]

THE FUTURE

The immune-neuroendocrine circuitry represents an immensely complex, powerful, and wide-ranging charter of human physiologic and pathologic possibilities, which, among others, is working its way to the creation of a new immunology based on a vastly enlarged vision of immunologic potential in health and disease. The emergence of this new interdisciplinary field will require a critical reexamination of some of our basic current views on the pathophysiologic and immunopharmacologic realities surrounding the problems of human asthma.

REFERENCES

1. D.W. Talmage, Introduction to basic immunology, *in*: "Immunological Diseases," M. Samter, ed., Little, Brown and Company, Boston (1988).
2. G.Filipp, A. Szentivanyi, and B. Mess, Anaphylaxis and nervous system, *Acta. med. hung. Tomus III, Fasciculus* 2:163 (1952).
3. A. Szentivanyi, G. Filipp, and I. Legeza, Investigations on tobacco sensitivity, *Act. med. hung. Tomus III, Fasciculus* 2:175 (1952).
4. G. Filipp and A. Szentivanyi, Frage der Organlokalisation der allergischen Reaktion, *Wierner klin. Wschr.* 65:620 (1953).

5. G. Filipp and A. Szentivanyi, Experimentelle Data zur regulativen Rolle des Neuroendokriniums in experimenteller Anaphylaxie I. Relazioni e Communicazioni. *Rome Il Pansiero Scientifico* 229:1 (1956).

6. A. Szentivanyi, Allergie und Zentralnervensystem. *Acta. Allergologica* 6:27 (1953).

7. A. Szentivanyi and G. Filipp, Experimentelle Data zur regulativen Rolle des Neuroendokriniums in experimenteller Anaphylaxie. II. Relazionie e Communicazioni. *Rome Il Pansiero Scientifico* 237:1 (1956).

8. A. Szentivanyi and J. Szekely, Effect of injury to, and electrical stimulation of hypothalamic areas on the anaphylactic and histamine shock of guinea pig, *Ann. Allergy* 14:259 (1956).

9. G. Filipp and A. Szentivanyi, Die Wirkung von Hypothalamuslasionen auf den anaphylaktischen Schock des Meerschweinchens. *Allergie und Asthmaforschung Bd.* 1:12 (1957).

10. A. Szentivanyi and J. Szekely, Uber den Effekt der Schadigung und der elektrischen Reizung der hypothalamischen Gegenden auf den anaphylaktischen und Histamin-Schock des Meerschweinchens, *Allergie und Asthmaforschung Bd.* 1:28 (1957).

11. A. Szentivanyi and J. Szekely, Wirkung der konstanten Reizung hypothalamischer Strukturen durch Tiefenelektroden auf den histaminbedingten und anaphylaktischen Schock des Meerschweinchens, *Acta. Physiol. Hung. Suppl. V* 11:41 (1957).

12. A. Szentivanyi and G. Filipp, Anaphylaxis and the nervous system. Part II. *Ann. Allergy* 16:143 (1958).

13. G. Filipp and A. Szentivanyi, Anaphylaxis and the nervous system. Part III. *Ann. Allergy* 16:306 (1958).

14. A. Szentivanyi and J. Szekely, Anaphylaxis and the nervous system. Part IV. *Ann. Allergy* 16:389 (1958).

15. A. Szentivanyi, Hypothalamic influences on antibody formation and on bronchial responses to histamine, *in*: "Proceedings of the Fourth Aspen Conference on Research in Emphysema and Asthma," Aspen, Colorado (1961).

16. A. Szentivanyi and C.W. Fishel, Effect of bacterial products on responses to the allergic mediators, *in*: "Immunological Diseases," M. Samter, ed., Little, Brown and Company, Boston (1965).

17. A. Szentivanyi and C.W. Fishel, Die Amin-Mediatorstoffe der allergischen Reaktion und die reaktionsfahiegheit ihrer Erfolgeszellen, *in*: "Pathogenese und Therapie allergischer Reaktionen," G. Filipp and A. Szentivanyi, eds., Ferdinand Enke Verlag, Stuttgart, Germany (1966).

18. A. Szentivanyi, J.J. Krzanowski and J.B.Polson, The autonomic nervous system: structure, function, and altered effector responses, *in*: "Allergy: Principles and Practice," E. Middleton, C.E. Reed, and E.F. Ellis, eds., The CV Mosby Company, St. Louis (1978).

19. A. Szentivanyi, J.B. Polson and J.J. Krzanowski, The altered reactivity of the effector cells to antigenic and pharmacological influences and its relation to cyclic nucleotides. I. Effector reactivities in the efferent loop of the immune response, *in*: "Pathomechanismus und Pathogenese Allergischer Reaktionen," G. Filipp, ed., Werk-Verlag Dr. Edmund Banachewski, Munich (1980).

20. A. Szentivanyi and D.F. Fitzpatrick, The altered reactivity of the effector cells to antigenic and pharmacological influences and its relation to cyclic nucleotides. II. Effector reactivities in the efferent loop of the immune response, *in*: "Pathomechanismus und Pathogenese Allergischer Reaktionen," G. Filipp, ed., Werk-Verlag Dr. Edmund Banachewski, Munich (1980).

21. A. Szentivanyi and J. Szentivanyi, Immunomodulatory effects of central and peripheral autonomic mechanisms mediated by neuroeffector molecules, *in*: "Proceedings of International Symposium on Biological Response Modifiers in Clinical Oncology and Immunology," Plenum Press, New York (1982).

22. A. Szentivanyi and J. Szentivanyi, The emergence of neuroendocrine disorders as a new group of autoimmune diseases, *in*: "Proceedings of Symposium on Clinical Laboratory Immunology," Plenum Press, New York (1982).

23. G. Filipp and A. Szentivanyi, Anaphylaxis and the nervous system. Part III, *in*: "Foundations of Psychoneuroimmunology," S. Locke, R. Ader, H.O. Besedovsky, N.R. Hall, G. Solomon and T. Strom, eds., Aldine Publishing, Hawthorne, NY (1985).

24. A. Szentivanyi and J. Szentivanyi, Immune-neuroendocrine circuits in antibiotic-bacterial interactions, *in*: "Proceedings of Third International Symposiun on the Influence of Antibiotics on the Host-Parasite Relationship," Springer Verlag, Heidelberg (1987).

25. A. Szentivanyi, S. Reiner, G. Filipp and O. Heim, The influence of anterior hypothalamic lesions on the kinetic parameters of ^{125}I-VIP (vasoactive intestinal peptide) binding to murine

mononuclear cells, *in*: "Proceedings of Workshop 12 on Mediators in Asthma, XII World Congress of Asthmology," Editorial Garsi, Madrid (1987).

26. A. Szentivanyi, J.J. Krzanowski and J.B. Polson, The autonomic nervous system and altered effector responses, *in*: "Allergy: Principles and Practice," E. Middleton, C.E. Reed and E.F. Ellis, eds., The CV Mosby Company, St. Louis (1988).

27. A. Szentivanyi, J. Szentivanyi, K. Haberman and O. Heim, Nonantibiotic properties of antibiotics in relationship to immune-neuroendocrine influences, *Clin. Pharmacol. Therap.* 43:166 (1988)

28. A. Szentivanyi, K. Haberman, O. Heim, P. Schultze, G. Filipp and S. Reiner, Hypothalamic and other central influences on antibiosis and host immunity, *in*: "Proceedings of the Fourth International Conference on Immunopharmacology," Pergamon Press, Oxford (1988).

29. A. Szentivanyi, S. Reiner, O. Heim, G. Filipp and C.M. Abarca, Some biochemical and cellular features of adrenergic mechanisms induced by bacterial lipopolysaccharide endotoxin in rats with or without chemical sympathetic ablation achieved by 6-hydroxydopamine hydrobromide (6-OHDA), *in*: "Proceedings of International Symposium on Endotoxin," Jichi Medical School, Tochigi, Japan (1988).

30. A. Szentivanyi, S. Reiner, O. Heim, G. Filipp and C.M. Abarca, The effect of 6-hydroxydopamine hydrobromide on endotoxin-induced adrenergic mechanisms, *in:* "Proceedings of Second International Meeting on Respiratory Allergy," Pythagora Press, Rome (1988)

31. J. Szentivanyi, A. Szentivanyi, P. Schultze, G. Filipp and O. Heim, Influences of hypothalamic and extrahypothalamic brain structures on the immunogenicity of antibiotic-pretreated bacteria, *in*: "Proceedings of Annual Meeting of the International Society for Interferon Research," Japanese Society for Interferon Research, Kanagawa, Japan (1988).

32. M.E. Schwartz, S. Reiner, O. Heim, C.M. Abarca and A. Szentivanyi, Further observations on the cellular and molecular mechanisms involved in the reciprocal histamine-catecholamine counterregulatory interplay in relation to induction of histidine decarboxylase synthesis by interleukin-3 and granulocyte-macrophage colony stimulating factor, *in*: "Proceedings of XIII International Congress of Allergology and Clinical Immunology," Mosby-Yearbook, St. Louis (1988).

33. A. Szentivanyi, The discovery of immune-neuroendocrine circuits and the concepts of prevailing immunologic thought that impeded the timely recognition of their role in immune-homeostasis, *in*: "Proceedings of the International Symposium on Interactions Between the Neuroendocrine and Immune Systems," Pythagora Press, Rome (1988).

34. A. Szentivanyi, Plenary Lecture: Natural neuropeptides in the immunologic inflammation of the airways in asthma, *in*: "Proceedings of XIV World Congress of Natural Medicines, Malaga, Spain (1988).

35. A. Szentivanyi and J. Szentivanyi, Antibiotic-bacterial interactions in relation to immune-neuroendocrine circuits, *in*: "Proceedings of XIII International Congress of Allergology and Clinical Immunology," Mosby-Yearbook, St. Louis (1988).

36. J. Szentivanyi, A. Szentivanyi, P. Schultze, G. Filipp and O. Heim, Changes in the immune parameters of antibiotic-bacterial interactions induced by hypothalamic and other electrolytic brain lesions produced through stereotaxically implanted depth electrodes, *in*: "The Influence of Antibiotics on the Host-Parasite Relationship," G. Gillissen, W. Opferkuch, G. Peters and G. Pulverer, eds., Springer-Verlag, Heidelberg, Germany (1989).

37. A. Szentivanyi, S. Reiner, O. Heim, G. Filipp and C.M. Abarca, The effect of sympathetic ablation [6-hydroxydopamine hydrobromide (6-OHDA); axotomy] on endotoxin induced adrenergic mechanisms, *The Pharmacologist* 31:118 (1989).

38. J. Szentivanyi, P. Schultz, O. Heim, C. Abarca and A. Szentivanyi, Hypothalamic and other central influences on antibiotic modulated bacterial immunogenicity, *The Pharmacologist* 31:193 (1989).

39. J. Szentivanyi, P. Schultze, O. Heim, S. Reiner, S. Robicsek, C. Abarca and A. Szentivanyi, The effect of hypothalamic and extrahypothalamic nuclear groupings on the antibiotic modulated bacterial immunogenicity and production of IL-1, IFN and TNF, *Cytokine* 1:364 (1989).

40. A. Szentivanyi, S. Reiner, M.E. Schwartz, O. Heim, J. Szentivanyi and S. Robicsek, Restoration of normal beta adrenoceptor concentrations in A549 lung adenocarcinoma cells by leukocyte protein factors and recombinant interleukin-1a, *Cytokine* 1:118 (1989).

41. A. Szentivanyi, J.J. Krzanowski, J.B. Polson and C.M. Abarca, The pharmacology of microbial modulation in the induction and expression of immune reactivities. I. The pharmacologically active effector molecules of immunologic inflammation, immunity, and hypersensitivity. *Immunopharmacology Rev.* 1:159 (1990).

42. A. Szentivanyi, The discovery of immune-neuroendocrine circuits in the fall of 1951, *in*: "Interactions Among the Central Nervous, Neuroendocrine and Immune Systems," J.W. Hadden, G. Nistico and K. Masek, eds., Pythagora Press, Rome (1989).

43. G. Schechter, A good relationship: sensitive, synchronized and synergistic, *Prog. Neuro-Endocrine Immunol.* 2:35 (1989).

44. J.W. Hadden and A. Szentivanyi, eds. "The Pharmacology of the Reticuloendothelial System," Plenum Press, New York (1985).

45. A. Szentivanyi, J. Szentivanyi, E. Middleton, Jr., H. Friedman, L.D. Prockop and C.M. Abarca, The pharmacology of microbial modulation in the induction and expression of immune reactivities. II. Effector mechanisms in the afferent and efferent limbs of the immune response, *Immunopharmacology Rev.* 5 (in press, 1994).

46. C.S. Goodman and K.G. Pearson, Neuronal development: cellular approaches in invertebrates, *Neurosci. Res. Program Bull.* 20:777 (1982).

47. X. He and M.G. Rosenfeld, Mechanisms of complex transcriptional regulation: implication for brain development, *Neuron* 7:183 (1991).

48. D.G. Willkinson and R. Krumlauf, Molecular apporaches to the segmentation of the hindbrain, *Trends Neurosci.* 13:335 (1990).

49. N. LeDouarin. "The Neural Crest," Cambridge University Press, Cambridge (1982).

50. B.A. Cunningham, J.J. Hemperley, B.A. Murray, E.A. Prediger, R. Brackenbury, and G.M. Edelman, Neural cell adhesion molecule: structure, immunoglobulin-like domains, cell surface modulation, and alternative RNA splicing, *Science* 236:799 (1987).

51. R.J. Milner, C. Lai, J.G. Sutcliffe and F.E. Bloom, Expression of immunoglobulin-like proteins in the nervous system: properties of the neural protein 1B236/MAG, *in*: "Neuroimmune Networks: Physiology and Diseases," E.J. Goetzl and N.H. Spector, eds., Alan R. Liss, Inc., New York (1989).

52. A.F. Williams and A.N. Barclay, The immunoglobulin superfamily - domains for cell surface recognition, *Ann. Rev. Immunol.* 6:381 (1988).

53. G.M. Edelman. "Neural Darwinism," Basic Book, Inc., New York (1987).

54. J.F. Bazan, Structural design and molecular evolution of a cytokine receptor superfamily, *Proc. Natl. Acad. Sci. USA* 87:6934 (1990).

55. J.F. Bazan, Neuropoietic cytokines in the hematopoietic fold, *Neuron* 7:197 (1991).

56. T. Yamamori and A. Sarai, Coevolution of cytokine receptor families in the immune and nervous systems, *Neuroscience Res.* 151-161 (1992).

57. A. Miyajima, T. Kitamura, N. Harada, T. Yokota and K. Arai, Cytokine receptors and signal transduction, *Ann. Rev. Immunol.* 10:295 (1992).

58. T. Yamamori, K. Fukada, R. Aebersold, S. Korsching, M.J. Fann and P.H. Patterson, The cholinergic neuronal differentiation factor from heart cells is identical to leukemia inhibitory factor, *Science* 246:1412 (1989).

59. S. Saadat, M. Sendtner and H. Rohrer, Ciliary neurotrophic factor induces cholinergic differentiation of rat sympathetic neurons in culture, *J. Cell Biol.* 108:1807 (1989).

60. T. Hama, Y. Kushima, M. Miyamoto, M. Kubota, N. Takei and H. Hatanaka, Interleukin-6 improves the survival of mesencephalic catecholaminergic and septal cholinergic neurons from postnatal cholinergic neurons from postnatal, two-week-old rats in cultures, *Neuroscience* 40:445 (1991).

61. T. Taga, M. Higi, Y. Hirata, K. Yamasaki, K. Yasukawa, T. Matsuda, T. Hirano and T. Kishimoto, Interleukin-6 triggers the association of its receptor with a possible signal transducer, gp130, *Cell* 58:573 (1989).

62. M. Murphy, K. Reid, D.J. Hilton and P.F. Bartlett, Generation of sensory neurons is stimulated by leukemia inhibitory factor, *Proc. Natl. Acad. Sci. USA* 88:3498 (1991).

63. D.P. Gearing, C.J. Thut, T. VandenBos, S.D. Gimpel, P.B. Delaney, J. King, V. Price, D. Cosman and M.P. Beckmann, Leukemia inhibitory factor receptor is structurally related to the IL-6 signal transducer, gp130, *EMBO J.* 10:2839 (1991).

64. N.Y. Ip, S.H. Nye, T.G. Boulton, S. Davis, T. Taga, Y. Li, S.J. Birren, K. Yasukawa, T. Kishimoto, D.J. Anderson, N. Stahl and G.D. Yancopoulos, CNTF and LIF act on neuronal cells via shared signal pathway that involve the IL-6 signal transducing receptor component, gp130, *Cell* 69:1121 (1992).

65. T. Yamamori, Molecular mechanisms for generation of neural diversity and specificity: roles of polypeptide factors in development of postmitotic neurons, *Neurosci. Res.* 12:545 (1992).

66. S. Davis, T.H. Aldrich, D.M. Valenzuela, V. Wong, M.E. Furth, S.P. Squinto and G.D. Yancopoulos, The receptor for ciliary neurotrophic factor, *Science* 253:59 (1991).

67. D.P. Gearing and D. Cosman, Homology of the p40 subunit of natural killer cell stimulatory factor with the extracellular domain of the interleukin-6 receptor, *Cell* 66:8 (1991).

68. A.K. Hall and M.S. Rao, Cytokines and neurokines: related ligands and related receptors, *Trends Neurosci.* 15:35 (1992).

69. T.M. Jessell and D. A. Melton, Diffusible factors in vertebrate embryonic induction, *Cell* 68:257 (1992).

70. P.H. Patterson, The emerging neuropoietic cytokine family: frist CDF/LIF, CNTF and IL-6; next ONC, MGF, GCSF? *Curr. Opinion Neurobiol.* 2:94 (1992).

71. G.M. Edelman, Topobiology, *Sci. Am.* 260:44 (1989).

72. W. Lanoutte. "Genius in the Shadows: A Biography of Leo Szilard," MacMillan, New York (1992).

73. A. Szentivanyi and H. Friedman, eds. "The Immunologic Revolution: Facts and Witnesses, CRC Press, Boca Raton, FL (1994).

74. N. Geschwind and A.M. Galaburda, Cerebral lateralization: biological mechanisms, associations, and pathology. Parts I-III, *Arch Neurol.* 42:428 (1985).

75. A. Szentivanyi, The immune-neuroendocrine circuitry and its relation to asthma, *in*: "Bronchial Asthma - Mechanisms and Therapeutics," E.B. Weiss and M. Stein, eds., Little, Brown and Company, Boston (1993).

76. A. Szentivanyi, Beta-adrenergic subsensitivity in asthma and atopic dermatitis: A status report, *Acta. Biomed. Hung. Amer.* 1:1 (1991).

77. G. Renoux, K. Biziere, M. Renoux, P. Bardos and D. Degenne, Consequences of bilateral brain neocortical ablation on imuthiol-induced immunostimulation in mice, *in*: "Neuroimmune Interactions: Proceedings of the Second International Workshop on Neuroimmuno-modulation," B.D. Jankovic, B.M. Markovic and N.H. Spector, eds., *Ann. NY Acad. Sci.* 496:346 (1987).

78. R. Malinow and R.W. Tsien, Presynaptic enhancement shown by whole-cell recordings of long-term potentiation in hippocampal slices, *Nature* 346(6290):177 (1990).

79. J.M. Bekkers and C.F. Stevens, Presynaptic mechanism for long-term potentiation in the hippocampus, *Nature* 346(6286):724 (1990).

80. P.N. McFadden and D.E. Koshland, Jr., Habituation in the single cell: diminished secretion of norepinephrine with repetitive depolarization in PC12 cells, *Proc. Natl. Acad. Sci. USA* 87:2031 (1990).

81. P.N. McFadden and D.E. Koshland, Jr., Parallel pathways for habituation in repetitively stimulated P12 cells, *Neuron* 4:615 (1990).

82. B.H. Morimoto and D.E. Koshland, Jr., Excitatory amino acid uptake and N-methyl-D-aspartate-mediated secretion in a neural cell line, *Proc. Natl. Acad. Sci. USA* 87:3518 (1990).

83. B.H. Morimoto and D.E. Koshland, Jr., Induction and expression of long- and short-term neurosecretory potentiation in a neural cell line, *Neuron* 5:875 (1990).

84. Y. Dudai, Neurogenetic dissection of learning and short-term memory in Drosophila, *Ann. Rev. Neurosci.* 11:537 (1988).

85. A. Szentivanyi, P. Maurer and B.W. Janicki, eds. "Antibodies: Structure, Synthesis, Function, and Immunologic Intervention in Disease," Plenum Press, New York (1987).

86. A. Szentivanyi and G. Filipp. "Propriétés Immuno-Chimiques et Physico-Chimiques des Anticorps," Editions Médicales Flammarion, Paris, France (1962).

87. J. Szentivanyi, A. Szentivanyi, J.F. Williams and H. Friedman, Virus associated immune and pharmacologic mechanisms in disorders of respiratory and cutaneous atopy, *in*: "Viruses, Immunity and Immunodeficiency," A. Szentivanyi and H. Friedman, eds., Plenum Press, New York (1986).

88. H. Friedman, T.W. Klein and A. Szentivanyi, eds. "Immunomodulation by Bacteria and Their Products," Plenum Press, New York (1981).

89. A. Szentivanyi, E. Middleton, J.F. Williams and H. Friedman, Effect of microbial agents on the immune network and associated pharmacologic reactivities, *in*: "Allergy: Principles and Practice," E. Middleton, C.E. Reed and E.F. Ellis, eds., The CV Mosby Company, St. Louis (1983).

90. T.W. Klein, S. Specter, H. Friedman and A. Szentivanyi, eds. "Biological Response Modifiers in Human Oncology and Immunology," Plenum Press, New York (1983).

91. J.M. Krueger, F. Obal, Jr., L. Johannsen, A.B. Cady and L. Toth, Endogenous slow-wave sleep substances: a review, *in*: "Current Trends in Slow-Wave Sleep Research," C. Dugsovic and A. Wauquier, eds., Raven Press, New York (1988).

92.   J.M. Krueger, F. Obal, Jr., M. Opp, L. Johannsen, A.B. Cady and L. Toth, Immune response modifiers and sleep, in: "Interactions Among Central Nervous System, Neuroendocrine and Immune Systems," J.W. Hadden, K. Masek and G. Nistico, eds., Pythagora Press, Rome, Italy (1989).

93.   M.P. Fillion, N. Prudhomme, F. Haour, G. Fillion, M. Bonnet, G. Lespinats, K. Masek, M. Flegel, N. Corvaia and J.M. Launay, Hypothetical role of the serotonergic system in neuroimmunomodulation: preliminary molecular studies, in: "Interactions Among Central Nervous System, Neuroendocrine and Immune Systems," J.W. Hadden, K. Masek and G. Nistico, eds., Pythagora Press, Rome, Italy (1989).

94.   P.M. Dougherty and N. Dafny, Central opioid systems are differentially affected by products of the immune response, Soc. Neurosci. Abstr. 13:1437 (1987).

95.   O.F. Eremina and L.V. Devoino, Production of humoral antibodies in rabbits following destruction of the nucleus of the midbrain raphe, Byull. Eksp. Biol. Med. 74:258 (1973).

96.   K. Masek, P. Horak, O. Kadlec and M. Flegel, The interactions between neuroendocrine and immune systems at the receptor level. The possible role of serotonergic system, in: "Interactions Among Central Nervous System, Neuroendocrine and Immune Systems," J.W. Hadden, K. Masek and G. Nistico, eds., Pythagora Press, Rome, Italy (1989).

97.   N. Vekshina and S.V. Magaeva, Changes in the serotonin concentration in the limbic structures of the brain during immunization, Bull. Exp. Biol. Med. 77:625 (1974).

98.   A. Szentivanyi, L.D. Prockop and S.M. Brooks, Immune-neuroendocrine circuitry: component parts, biochemical control mechanisms and implications for atopic diseases, Immunopharmacol. Rev., in preparation (1994).

99.   A. Szentivanyi, M.E. Schwartz, S. Reiner, O. Heim, E. Calderon, C. Abarca and L.D. Prockop, The nature of the central and peripheral adrenergic mechanisms involved in the induction of the de novo synthesis of histidine decarboxylase in hemopoietic progenitor cells of bone marrow, in preparation (1994).

100. A. Szentivanyi, M.E. Schwartz, O. Heim, S. Reiner, E. Calderon, K. Ali, C. Abarca and L.D. Prockop, Dissociation in the time development of adrenergically active beta-arrestin and IL-1a receptor antagonist versus IL-1a in the phenotypical change in T-cell subsets, in preparation (1994).

101. A. Szentivanyi, A. Engel, O. Heim, H. Wagner, E. Calderon, L.D. Prockop and C. Abarca, The regulatory effect of interleukin-1a derived from T-memory cells on neuropeptide (substance P, neuropeptide Y) expression in the ganglion Schwann cell of the rat, in preparation (1994).

102. A. Szentivanyi, J.F. Hackney, O. Heim, S. Robicsek, E. Calderon, L.D. Prockop, K. Ali and C. Abarca, Cultured human cell lines which do or do not respond to lymphocyte conditioned medium of human CD4$^+$ CD45RO$^+$ induction of beta$_2$-adrenoceptor and beta$_2$-adrenergic receptor mRNA synthesis with a parallel dissociation in the induction of tyrosin hydroxylase mRNA, in preparation (1994).

103. K. Ali, E. Calderon, S.M. Brooks, R.G. Coffey, R.F. Lockey and A. Szentivanyi, Modulation of beta-adrenergic responsiveness of A549 human pulmonary epithelial cells by IgE, in preparation (1994).

104. A. Szentivanyi, Adrenergic regulation, in: "Bronchial Asthma - Mechanisms and Therapeutics," E.B. Weiss and M. Stein, eds., Little, Brown and Company, Boston (1993).

105. M.A. Lochrie and M.I. Simon, G protein multiplicity in eukaryotic signal transduction systems, Biochemistry 17:4957 (1988).

106. L. Birnbaumer and A.M. Brown, G proteins and the mechanism of action of hormones, neurotransmitters, and autocrine and paracrine regulatory factors, Am. Rev. Respir Dis. 141:S106 (1990).

107. G.M. Morris, J.R. Hadcock and C.C. Malbon, Cross-regulation between G-protein-coupled receptors. Activation of ß$_2$-adrenergic receptors increases a1-adrenergic receptor mRNA levels, J. Biol. Chem. 266(4):2233 (1991).

108. J.R. Hadcock, J.D. Port and C.C. Malbon, Cross-regulation between G-protein mediated pathways. Activation of the inhibitory pathway of adenylyl/cyclase increases the expression of ß$_2$-adrenergic receptors, J. Biol Chem. 266(18):11915 (1991).

109. E.A. Korneva and V.M. Klimenko, Neuronale hypothalamusaktivitt und homoostatische rektionen, Ergebn. Exp. Med. 23:373 (1976).

110. H.O. Besedovsky, E. Sorkin, D. Felix and H. Haas, Hypothalamic changes during the immune response, *Eur. J. Immunol.* 7:325 (1977).

111. Z. Srebro, I. Spisak-Plonka and E. Szirmai, Neurosecretion in mice during skin allograft rejection, *Agressologie* 15:125 (1974).

112. G.C. Cotzias and L.C. Tang, Adenylate cyclase of brain reflects propensity for breast cancer in mice, *Science* 197:1094 (1977).

113. A.J. Dunn, M.L. Powell, W.V. Moreshead, J.M. Gaskin and N.R. Hall, N.R., Effects of Newcastle disease virus administration to mice on the metabolism of cerebral biogenic amines, plasma corticosterone, and lymphocyte proliferation, *Brain Behav. Evol.* 1:216 (1987).

114. B.M.R.N.J. Woloski, E.M. Smith, W.J. Meyer, G.M. Fuller and J.E. Blalock, Corticotropin-releasing activity of monokines, *Science* 230:1035 (1985).

115. E.W. Bernton, J.E. Beach, J.W. Holaday, R.C. Smallridge and H.G. Fein, Release of multiple hormones by direct action of interleukin-1 on pituitary cells, *Science* 238:519 (1987),

116. P. Kehrer, D. Turnill, J.-M. Dayer, A.F. Muller and R.C. Gaillard, Human recombinant interleukin-1ß and -a, but not recombinant tumor necrosis factor-a stimulate ACTH release from rat anterior pituitary cells in vitro in a prostaglandin E_2 and cAMP independent manner, *Neuroendocrin.* 48:160 (1988).

117. V. Rettori, J. Jurcovicova and S.M. McCann, Central action of interleukin-1 in altering the release of TSH, growth hormone and prolactin in the male rat, *J. Neurosci. Res.* 18:179 (1987).

118. D. Giulian and L.B. Lachman, Interleukin-1 stimulation of astroglial proliferation after brain injury, *Science* 228:497 (1985).

119. D.E. Scarborough, S.L. Leo, C.A. Dinarello and S. Reichlin, Interleukin-1ß stimulates somatostatin biosynthesis in primary cultures of fetal rat brain, *Endocrinology* 124:549 (1989).

120. M. Fukuoka, K. Yasuda, S. Taii, K. Takakura and T. Mori, Interleukin-1 stimulates growth and inhibits progesterone secretion in cultures of porcine granulosa cells, *Endocrinology* 124:884 (1989).

121. R. Sapolsky, C. Rivier, G. Yamamoto, P. Plotsky and W. Vale, Corticotropin-releasing factor-producing neurons in the rat activated by interleukin-1, *Science* 238:522 (1987).

122. F. Berkenbosch, J. van Oers, A. Del Rey, F. Tilders and H. Besedovsky, Corticotropin-releasing factor producing neurons in the rat activated by interleukin-1, *Science* 238:524 (1987).

123. W.L. Farrar, P.L. Kilian, M.R. Ruff, J.M. Hill and C.B. Pert, Visualization and characterization of interleukin-1 receptors in brain, *J. Immunol.* 139:459 (1987).

124. W.S. Zawalich, K.C. Zawalich and H. Rasmussen, Interleukin-1a exerts glucose-dependent stimulatory and inhibitory effects on islet cell phosphoinositide hydrolysis and insulin secretion, *Endocrinology* 124:2350 (1989).

125. R.P. Cornell, Central interleukin-1 elicited hyperinsulinemia is mediated by prostaglandin but not autonomics, *Am. J. Physiol.* 257:R839 (1989).

126. A. Del Rey and H. Besedovsky, Antidiabetic effects of interleukin-1, *Proc. Natl. Acad. Sci. USA* 86:5943 (1989).

127. E.N. Benveniste and J.E. Merrill, Stimulation of oligodendroglial proliferation and maturation by interleukin-2, *Nature* 321:610 (1986).

128. L.R. Smith, S.L. Brown and J.E. Blalock, Interleukin-2 induction of ACTH secretion: presence of an interleukin-2 receptor a-chain-like molecule on pituitary cells, *J. Neuroimmunol.* 21:249 (1989).

129. B. Sherry and A. Cerami, Cachectin/tumor necrosis factor exerts endocrine, paracrine, and autocrine control of inflammatory responses, *J. Cell Biol.* 107:1269 (1988).

130. Y. Naitoh, J. Fukata, T. Tominaga, Y. Nakai, S. Tami, K. Mori and H. Imura, Interleukin-6 stimulates the secretion of adrenocorticotropic hormone in conscious, freely-moving rats, *Biochem. Biophys. Res. Commun.* 155:1459 (1988).

131. E.M. Sternberg, Monokines, lymphokines and the brain, *in*: "The Year in Immunology," J.M. Cruse and J.E. Lewis, eds., Karger, Basel (1989).

132. K. Mealy, B.G. Robinson, J.A. Majzoub and D.W. Wilmore, Hypothalamic-pituitary-adrenal (HPL) axis regulation by tumor necrosis factor, *Prog. Leukocyte Biol.* 10B:225 (1990).

133. B.L. Spangelo, A.M. Judd, P.C. Ross, I.S. Login, W.D. Jarvis, M. Badamchian, A.L. Goldstein and R.M. MacLeod, Thymosin fraction 5 stimulates prolactin and growth hormone release from anterior pituitary cells in vitro, *Endocrinology* 121:2035 (1987).

134. D.Y.M. Leung and R.S. Geha, Regulation of the human IgE antibody response, *Int. Rev. Immunol.* 2:75 (1987)/

135. A. Miyajima, S. Miyatake and J. Schreurs, Coordinate regulation of immune and inflammatory responses by T-cell-derived lymphokines, *FASEB J.* 2:2462 (1988).

136. A.G. Lopez, C.J. Sanderson, J.R. Gamble, H.D. Campbell, I.G. Young and M.A. Vadas, Recombinant human interleukin 5 is a selective activator of human eosinophil function, *J. Exp. Med.* 167:219 (1988).

137. M.E. Rothenberg, W.F. Owen, D.S. Silberstein, R.J. Soberman, K.F. Austen and R.L. Stevens, Human eosinophils have prolonged survival, enhanced functional properties and become hypodense when exposed to human interleukin-3, *J. Clin. Invest.* 81:1986 (1988).

138. W.F. Owen, M.E. Rothenberg and D.S. Silberstein, Regulation of human eosinophil viability, density, and function by granulocyte/macrophage colony-stimulating factor in the presence of 3T3 fibroblasts, *J. Exp. Med.* 166:129 (1987).

139. A.B. Kay, Leucocytes in asthma, *Immunol. Invest.* 17:679 (1988).

140. F. Lee, T. Yokota, T. Otsuka, P. Meyerson, D. Villaret, R. Coffman, T. Mosmann, D. Rennick, N. Roehm, C. Smith, A. Zlotnik and K. Arai, Isolation and characterization of a mouse interleukin cDNA clone that expresses B cell stimulatory factor-1 activities and T-cell and mast cell-stimulating activities, *Proc. Natl. Acad. Sci. USA* 83:2061 (1986).

141. A.B. Kay, T-lymphocytes and their products in atopic allergy and asthma, *Int. Arch. Allergy Appl. Immunol.* 94:189 (1991).

142. J.W. Crump, R.J. Pueringer and G.W. Hunninghake, Bronchoalveolar lavage and lymphocytes in asthma, *Eur. Respir. J.* 4(Suppl 13):39s (1991).

143. A.J. Frew, C.J. Corrigan, P. Maestrelli, J.J. Tsai, K. Kurihara, R.E. O'Hehir, A. Hartnell, O. Cromwell and A.B. Kay, T lymphocytes in allergen-induced late-phase reactions and asthma, *Int. Arch. Allergy Appl. Immunol.* 88:63 (1989).

144. G. Delespesse, M. Sarfati, and R. Peleman, Influence of recombinant IL-4, IFN-a and IFN-c on the production of human IgE binding factor (soluble CD23), *J. Immunol.* 142:134 (1989).

145. Y. Yanagihara, K. Kajiwara, M. Kiniwa, Y. Yui, T. Shida and G. Delespesse, Enhancement of IgE synthesis and histamine release by T-cell factors derived from atopic patients with bronchial asthma, *J. Allergy Clin. Immunol.* 79:448 (1987).

146. R. Alam, J. Rozniecki and K. Selmaj, A mononuclear cell-derived histamine-releasing factor (HRF) in asthmatic patients. I. Histamine release from basophils in vitro, *Ann. Allergy* 53:66 (1984).

147. R. Alam and J. Rozniecki, A mononuclear cell-derived histamine-releasing factor (HRF) in asthmatic patients. II. Activity in vivo, *Allergy* 40:124 (1985).

148. R. Alam, J.A. Grant and M.A. Lett-Brown, Identification of a histamine-release inhibitory factor produced by human mononuclear cells in vitro, *J. Clin. Invest.* 82:2056 (1988).

149. A.J.M. Van Oosterhout and F.P. Nijkamp, Lymphocytes and bronchial hyperresponsiveness, *Life Sci.* 46:1255 (1990).

150. F.P. Nijkamp and P.A.J. Henricks, Beta-adrenoceptors in lung inflammation, *Am. Rev. Respir. Dis.* 141:145s (1990).

151. C. Walker, M.K. Kaegi, P. Braun and K. Blaser, Activated T cells and eosinophilia in bronchoalveolar lavages from subjects with asthma correlated with disease severity, *J. Allergy Clin. Immunol.* 88:935 (1991).

152. B.L. Bradley, M. Azzawi, M. Jacobson, B. Assoufi, J.V. Collins, A.-M. Irani, L.B. Schwartz, S.R. Durham, P.K. Jeffery and A.B. Kay, Eosinophils, T-lymphocytes, mast cells, neutrophils, and macrophages in bronchial biopsy specimens from atopic subjects with asthma: comparison with biopsy specimens from atopic subjects without asthma and normal control subjects and relationship to bronchial hyperresponsiveness, *J. Allergy Clin. Immunol.* 88:661 (1991).

153. S.E. Frigas, D.A. Loegering and G.J. Gleich, Cytotoxic effects of the guinea pig eosinophil major basic protein on tracheal epithelium, *Lab. Invest.* 43:35 (1980).

154. G.J. Gleich, E. Frigas, D.A. Loegering, D.L. Wassom and D. Steinmuller, Cytoxic properties of the eosinophil major basic protein, *J. Immunol.* 123:2925 (1979).

155. Q.A. Hamid, J.C.W. Mak, M.N. Sheppard, B. Corrin, J.C. Venter and P.J. Barnes, Localization of beta$_2$-adrenoceptor messenger RNA in human and rat lung using in situ hybridization: correlation with receptor autoradiography, *Eur. J. Pharmacol.* 206:133 (1991).

156. I. Enander, S. Ahlstedt, H. Nygren and B. Bjorksten, Sensitizing ability of derivatives of picryl chloride after exposure of mice on the skin and in the lung, *Int. Arch. Allergy Appl. Immunol.* 72:59 (1983).

157. J. Garssen, F.P. Nijkamp, H. Van der Vliet and H. Van Loveren, T-cell mediated induction of airway hyperreactivity in mice, *Am. Rev. Respir. Dis.* 144:931 (1991).

158. G. Folkerts, P.A.J. Henricks, P.J. Slootweg and F.P. Nijkamp, Endotoxin-induced inflammation and injury of the guinea pig respiratory airways cause bronchial hyporeactivity, *Am. Rev. Respir. Dis.* 137:1441 (1988).

159. R. Pauwels, R. Peleman and M. Van Der Straeten, Airway inflammation and non-allergic bronchial responsiveness, *Eur. J. Respir. Dis.* 68:137 (1986).

160. C. Murlas and J.H. Roum, Bronchial hyperactivity occurs in steroid-treated guinea pigs depleted of leukocytes by cyclophosphamide, *J. Appl. Physiol.* 58:1630 (1985).

161. J.E. Thompson, L.A. Scypinski, T. Gordon and D. Sheppard, Hydroxyurea inhibits airway hyperresponsiveness in guine pigs by a granulocyte-independent mechanism, *Am. Rev. Respir. Dis.* 134:1213 (1986).

162. W. Cibulas, S.M. Brooks, C.G. Murlas, M.L. Miller and R.T. McKay, Toluene diisocyanate-induced airway hyperreactivity in guinea pigs depleted of granulocytes, *J. Appl. Physiol.* 64:1773 (1988).

163. C.L. Rochester and J.A. Rankin, Is asthma T-cell mediated? *Am. Rev. Respir. Dis.* 144:1005 (1991).

164. A. Szentivanyi, The beta-adrenergic theory of the atopic abnormality in bronchial asthma, *J. Allergy* 42:203 (1968).

165. A.F. Barker, C.A. Hirshman, R. D'Silva and J.M. Hanifin, Airway responsiveness in atopic dermatitis, *J. Allergy Clin. Immunol.* 87:780 (1991).

166. R.J. Hopp, R.G. Townley, R.E. Biven, A.K. Bewtra and N.M. Nair, The presence of airway reactivity before the development of asthma, *Am. Rev. Respir. Dis.* 141:2 (1990).

167. A.J.M. Van Oosterhout and F.P. Nijkamp, Effect of lymphokines on beta-adrenoceptor function of human peripheral blood mononuclear cells, *Br. J. Clin. Pharmacol.* 30:150S (1990).

168. R.J. Hopp, A.K. Bewtra, N.M. Nair and R.G. Townley, Specificity and sensitivity of methacholine inhalation challenge in normal and asthmatic children, *J. Allergy Clin. Immunol.* 74:154 (1984).

169. R.A. Pauwels, Genetic factors controlling airway responsiveness, *Clin. Rev. Allergy* 7:235 (1989).

170. T. Chonmaitree, M.A. Lett-Brown and J.A. Grant, Respiratory viruses induce production of histamine-releasing factor by mononuclear leukocytes: a possible role in the mechanism of virus-induced asthma, *J. Infect. Dis.* 164:592 (1991).

171. F.A. Ennis, A.S. Beare, D. Riley, G.C. Schild, A. Meager, Q. Yi-Hua, G. Schwarz and A.H. Rook, Interferon induction and increased natural killer cell activity - influenza infections in man, *Lancet* ii:891 (1981).

172. F.H. Valone and L.B. Epstein, Biphasic platelet activating factor synthesis by human monocytes stimulated with IL-1ß, tumor necrosis factor or interferon-c, *J. Immunol.* 141:3945 (1988).

173. M.G. O'Sullivan, N.J. MacLachlan, L.N. Fieischer, N.C. Olson and T.T. Brown, Modulation of arachidonic acid metabolism by bovine alveolar macrophages exposed to interferons, *J. Leukocyte Biol.* 44:116 (1988).

174. S. Mattoli, S. Miante, F. Calabro, M. Mezzetti, A. Fasoli and L. Allegra, Bronchial epithelial cells exposed to isocyanates potentiate activation and proliferation of T cells, *Am. J. Physiol.* 259:L320 (1990).

175. S. Robicsek, A. Szentivanyi, E.G. Calderon, O. Heim, P. Schultze, H. Wagner, R.F. Lockey and J.J. Dwornik, Concentrated IL-1a derived from human T-lymphocytes binds to a specific single class surface receptor on human bronchial epithelial cells and induces the production of beta-adrenoceptor mRNA via an associated or separate receptor-linked signalling pathway, *J. Allergy Clin. Immunol.* 89:212 (1992).

176. T. Szentendrei, T. Nakane, E. Lazarj-Wesley, M. Virmani and G. Kunos, Regulation of beta-adrenergic receptor gene expression by interleukin-1, *The Pharmacologist* 33:225 (1991).

177. A. Szentivanyi, E.G. Calderon, O. Heim, P. Schultze, H. Wagner, J. Zority, R.F. Lockey, J.J. Dwornik and S. Robicsek, Cell- and species-specific dissociation in the beta-

adrenoceptor upregulating effects of IL-1a derived from lymphocyte conditioned medium and cortisol, *J. Allergy Clin. Immunol.* 89:274 (1992).

NEUROENDOCRINE CONTROL OF THE THYMIC MICROENVIRONMENT: ROLE OF PITUITARY HORMONES

Wilson Savino, Valeria Mello-Coelho[1] and Mireille Dardenne[2]

[1]Department of Immunology, The Oswaldo Cruz Foundation,
Av. Brasil 4365, Manguinhos, 21045 Rio de Janeiro, Brazil
[2]Hopital Necker, CNRS URA 1471 Paris, France

INTRODUCTION

Within the well demonstrated immunoneuroendocrine network, the nervous, endocrine and immune systems communicate through common mediators and respective receptors, working in fine harmony, they contribute to homeostasis.[1]

A wide variety of findings show that both neurotransmitters and hormones are potent immunomodulators, and that even psychological factors such as typical Pavlovian conditioning can affect the immune response.[2] To better understand such complex interactions, it was important not only to study mature lymphocytes in blood and peripheral lymphoid organs, but also to evaluate primary organs, where precursors of B and T cells are generated.

During recent years, we have investigated the influence of hormones and neuropeptides on the thymus.[3,4] In the present review, we shall focus our attention on the effects of two pituitary hormones, prolactin (PRL) and growth hormone (GH), on the thymic microenvironment. Before presenting the recent data on this subject, some general aspects of the non-lymphoid compartment of the thymus and the thymic microenvironment is briefly discussed in the context of intrathymic T cell differentiation.

INTRATHYMIC T CELL DIFFERENTIATION: GENERAL COMMENTS

The thymus is a central lymphoid organ in which bone marrow-derived T cell precursors undergo a complex process of maturation, eventually leading to the migration of positively selected thymocytes to the T-dependent areas of peripheral lymphoid organs.[5] This differentitation occurs after the rearrangement of T cell antigen receptor genes and involves positive selection of some thymocytes and negative selection of many others (the latter are eliminated by apoptosis). Positively selected

Advances in Psychoneuroimmunology, Edited by
I. Berczi and J. Szélenyi, Plenum Press, New York, 1994

T lymphocytes are released to the periphery and constitute the so-called T cell repertoire. These cells are able to distinguish self and non-self antigens.

It should be pointed out that key events of intrathymic T cell differentiation are influenced by the thymic microenvironment. This environment is a tridimentional network composed of distinct cell types, such as epithelial cells, macrophages and dendritic cells, as well as extracellular matrix elements.

The thymic epithelium is the major component of the thymic microenvironment and has an important and multifaceted influence on early events of T cell differentiation, in at least two distinct ways: a) secretion of a variety of polypeptides such as thymic hormones,[6] interleukins 1, 3 and 6[7,8] and granulocyte-macrophage colony stimulating factor,[9] and b) cell-to-cell contacts, including those occurring through classical adhesion molecules[10] and, most importantly, interactions with differentiating thymocytes, mediated by the major histocompatibility complex (MHC) products, which are highly expressed on thymic epithelial cell membranes.[11-14] Thus, MHC class I proteins interact with the CD8 molecule, whereas MHC class II binds to the CD4 complex. These interactions are important in defining positive versus negative selection of thymocytes bearing the distinct T cell receptor rearrangements. Lastly, thymic epithelial cells (TEC) secrete a number of extracellular matrix proteins, such as fibronectin, laminin and type IV collagen, that apparently also play a role in the general process of intrathymic T cell differentiation.[15]

THE THYMIC EPITHELIUM: A TARGET FOR NEUROENDOCRINE CONTROL

A further concept regarding thymus physiology is that microenvironmental effects upon thymocytes can be modulated by extrinsic factors. In particular, numerous recent studies point to neuroendocrine control of thymus physiology. Different hormones and neuropeptides can modulate distinct aspects of TEC physiology such as growth, hormone secretion and cytokeratin expression.[3,4] For example, we have shown that in vivo treatment of mice with triiodothyronine (T3) induces an increase in the production of thymulin, a chemically-defined thymic hormone,[16] which is produced only by TEC.[17] These findings were later confirmed by Fabris et al.,[18] who further demonstrated that in humans thyroid hormone status modulates thymulin serum levels.[19] Additionally, injection of glucocorticoid hormones modulates not only thymulin production, but also cytokeratin expression and extracellular matrix production.[20,21]

Both thyroid and steroid hormones seem to have a direct effect on TEC, since they were observed in vitro using cultures of murine TEC lines.[22-24] This notion is further supported by the demonstration of specific receptors for each of these hormones in the thymic epithelium.[25,26]

IN VIVO AND IN VITRO MODULATION OF THE THYMIC EPITHELIUM BY PITUITARY HORMONES

In addition to thyroid and steroid hormones, a large piece of work now clearly demonstrates that pituitary hormones, such as PRL and GH, can affect some aspects of TEC physiology.

Effects of Growth Hormone Upon the Thymic Epithelium

Initial data concerning the role of GH in the thymus came from studies on the

Snell-Bagg dwarf mouse. This mutant has a congenital deficiency of GH and thyrotropin production and has a rather immature immune system that parallels thymic atrophy.[26,28] Furthermore, these animals exhibit a precocious age-dependent decline in thymulin serum levels.[29] These data were extended by Goff et al.,[30] who showed that treatment with bovine GH partially restored the low thymulin levels observed in aged dogs. More recently, we observed that GH treatment increased thymulin serum levels in both young and aging mice.[31]

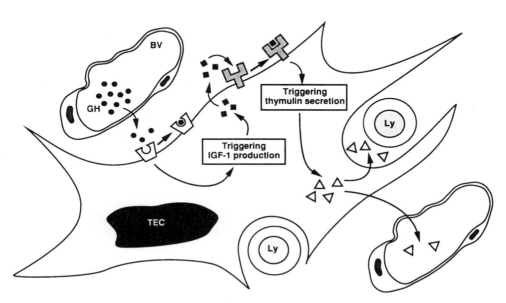

Figure 1. Hypothetical pathway for the action of growth hormone (GH) upon the thymic epithelium. Endogenous GH, once bound to its receptors on the thymic epithelial cell (TEC) membrane, would stimulate production and secretion of IGF-1, which in turn would enhance thymulin secretion via an autocrine circuitry using specific IGF-1 receptors. BV: blood vessel; Ly: thymic lymphocyte.

A second line of evidence concerning the stimulatory role of GH on the thymus appeared in in vivo experiments showing that grafting GH3 pituitary tumor cells (that secrete GH and PRL) into old rats restored thymus structure as well as T cell proliferation and interleukin-2 (IL-2) synthesis.[32] Additionally, this procedure reversed the accumulation of CD4⁻CD8⁻ thymocytes that occurs in these animals.

These findings were in agreement with previous data showing that treatment of dwarf mice with GH and thyroxine (T4) reconstituted their thymic function and markedly prolonged the lifespan of these naturally short-lived animals.[28]

Nonetheless, it should be noted that, at least in normal aging mice, reconstitution of the general microarchitecture of the thymus seems not to be exclusively GH dependent, since such treatment alone, although effective in enhancing thymulin production, did not alter structural parameters of the thymus.[31] In this respect, synergic effects with thyroid hormones and prolactin should be considered potential candidates for future immunoendocrine intervention.

The role of GH in TEC physiology apparently involves direct action on the epithelial cells since: a) growth hormone can stimulate thymulin production by pure cultures of TEC lines,[33,34] and b) both human and murine TEC have GH receptors.[35]

It should be noted, however, that the effects of GH upon the thymic epithelium are likely to be mediated by the production of insulin-like growth factor 1 (IGF-1). First, it was shown that the increased thymulin levels observed in acromegalic patients are positively correlated to IGF-1 levels, but not necessarily to circulating GH values.[34] Moreover, IGF-1 alone increased thymic hormone production and TEC proliferation. Lastly, we observed that the in vitro enhancing effects of GH upon thymulin production could be prevented by anti-IGF-1 or anti-IGF-1 receptor antibodies.[34] Together, these data lead to the hypothesis that TEC constitutively produce IGF-1 and express IGF-1 receptors, both involved in an autocrine IGF-1 dependent circuit (linked to the GH-related endocrine pathway) modulating TEC physiology, as depicted in Figure 1.

Prolactin: A Pleiotropic Modulator of the Thymic Epithelium

It is well established that PRL is a potent immunomodulator with a multifaceted enhancing effect on the immune response, in particular in conditions of stress. Regarding its effect on thymic cells, it was initially shown that in vivo injection of anti-

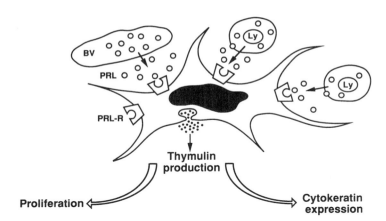

Figure 2. Pleiotropic action of prolactin (PRL) upon thymic epithelial cells. Using an endocrine or paracrine pathway, PRL can bind to its receptor (PRL-R), enhancing not only thymulin production, but also high molecular weight cytokeratin expression as well as cell growth. BV: blood vessel; Ly: thymic lymphocyte.

PRL serum caused changes in thymocyte subpopulations, essentially expressed by increased percentages of CD4$^+$ single positive cells.[36] Moreover, injection of bromocriptine which blocks endogenous PRL production promoted a similar effect.

We recently carried out a number of studies investigating the possible influence of PRL on the thymic epithelium. In vivo injections of PRL into young normal mice consistently increased circulating thymulin. Conversely, hypoprolactinemia induced by bromocriptine had the opposite effect. PRL treatment of old individuals or dwarf mice (both having low thymulin serum levels), significantly augmented thymic hormone production.[33] This was a direct effect of PRL on epithelial cells since it could be reproduced in vitro using human and murine TEC cultures. Importantly, we clearly demonstrated that PRL acts pleiotropically upon TEC since it also enhanced the expression of high molecular weight cytokeratins, and increased in vitro TEC proliferation (see Figure 2).

CAN TEC THYMOCYTE CELL-TO-CELL INTERACTIONS BE INFLUENCED BY PITUITARY HORMONES?

As mentioned above, one aspect of intrathymic T cell differentiation is direct cell-to-cell interactions of thymocytes with thymic epithelial cells. To study the influence of a variety of agents upon TEC/thymocyte adhesion, we recently developed an ELISA

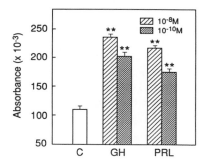

Figure 3. Enhancing effects of prolactin (PRL) and growth hormone (GH) upon the adhesion of thymocytes onto cultures of a mouse thymic epithelial cell line. In this ELISA assay, adhesion degree of thymocytes was estimated using an anti-Thy 1.2 monoclonal antibody. Untreated control cultures (C) were compared to those treated either by GH or PRL at the doses of 10^{-8}M or 10^{-10}M. Results are expressed as the mean of optical absorbance. ** indicates statistical significance by the Student's T test ($p < 0.05$).

system using an anti-Thy 1 monoclonal antibody (a pan T cell marker). With this test we can evaluate the amount of thymocytes adhering to a mouse TEC culture under a given experimental condition. Using this approach, we clearly demonstrated that GH, IGF-1 and PRL were all able to enhance thymocyte adhesion, as can be seen in Figure 3.

The molecular mechanisms involved in this phenomenon remain to be determined. Yet, since this TEC line does not express MHC class II antigens in culture, unless stimulated by interferon-γ,[24] we can postulate that extracellular matrix (and respective receptors) and/or LFA-1/ICAM-1 mediated interactions are involved. In any case, if such events also occur in vivo, they may intervene in the general process of intrathymic T cell migration.

EXPRESSION OF GH AND PRL RECEPTORS IN THE THYMIC EPITHELIUM

To bring further support to the various experimental and human data showing immunomodulatory effects of GH and PRL on the thymic epithelium, it was necessary to demonstrate specific receptors for these pituitary hormones in thymic epithelial cells.

A series of binding experiments using radiolabeled GH allowed us to evidence a GH receptor in both human and murine TEC. Cultured human TEC, for example, bear a range of 210-240 sites per cell, with a Kd of 0.14-0.27 nM.[35]

Since at least some of the GH effects on the thymic epithelium are apparently mediated by IGF-1, it is quite likely that TEC express IGF-1 receptors as well, although the presence of such receptors has not been directly demonstrated. More recently, using immunochemistry and molecular biology approaches, we demonstrated that the thymic epithelium also expresses a PRL receptor (PRL-R), both in situ and in vitro. Interestingly, appropriate amounts of anti-PRL-R monoclonal antibodies, in contact with cultured TEC, revealed agonist or antagonist effect toward PRL in terms of modulating thymulin production and TEC proliferation.[37]

It is now apparent that thymocytes also express PRL-R. Importantly, triple labeling flow cytometry studies revealed that PRL-R is present in both mature and immature stages of thymocyte differentiation.[38,39] These data, together with the recent findings showing that human thymocytes (but not TEC) can express mRNA for PRL,[40] lead to the possibility that PRL can exert endocrine, paracrine and autocrine effects upon the thymus, as postulated in Figure 4.

CONCLUDING REMARKS

The present review brings evidence that pituitary hormones, namely prolactin and growth hormone, should be regarded as modulators of the thymic microenvironment, and more particularly its epithelial component. It is noteworthy that both GH and PRL augment at least one thymocyte differentiation factor, such as thymulin, and enhance in vitro TEC/thymocyte adhesion. Taken together, these data raise the possibility that intrathymic T cell differentiation events may be under pituitary hormone control. In this respect, it is attractive to speculate that even the frequencies of thymocytes bearing distinct T cell receptor gene rearrangements may be influenced by GH and/or PRL, similar to what has been recently reported in relation to estradiol.[41]

Another aspect deserving further investigation concerns the intrathymic production of PRL. Given the fact that thymocytes can produce PRL, and that TEC produce neurohypophyseal hormones such as oxytocin and vasopressin,[42] it can be postulated that, besides being under extrinsic hormonal control, the thymus gland possesses neuroendocrine circuitry that may be physiologically relevant for its function.

ACKNOWLEDGEMENTS

This work was partially funded by grants from CNPq (Brazil) and INSERM (France).

REFERENCES

1. J.E. Blalock, Neuroimmunoendocrinology, *Chem. Immunol.* 52:1 (1992).
2. R. Ader, D.L. Felten and N. Cohen, "Psychoneuroimmunology," Academic Press, San Diego, CA (1991).
3. M. Dardenne and W. Savino, Neuroendocrine control of thymic epithelium: Modulation of thymic endocrine function, cytokeratin expression and cell proliferation by hormones and neuropeptides, *Prog. NeuroEndocrinImmunol.* 3:18 (1990).
4. M. Dardenne and W. Savino, Neuroendocrine circuits controlling the physiology of the thymic epithelium, *Ann. N.Y. Acad. Sci.* 650:85 (1992).
5. W. Van Ewijk, T-cell differentiation is influenced by thymic microenvironments, *Ann. Rev. Immunol.* 9:591 (1991).
6. J.F. Bach, Thymic hormones, *Clin. Immunol. Allergy* 3:133 (1983).
7. P.T. Le, D.T. Tuck, C.A. Dinarello, B.F. Haynes and K.H. Singer, Thymic epithelial cells produce interleukin 1, *J. Immunol.* 138:2520 (1988).
8. P.T. Le, S. Lazorich, L.P. Whichard, Y.C. Yang, S.C. Clarck, B.F. Haynes and K.H. Singer, Human thymic epithelial cells produce IL-6, granulocyte-monocyte CSF and leukemia inhibitory factor, *J. Immunol.* 145:3310 (1990).
9. P.T. Le, J. Kurtzberg, S.L. Brant, J.E. Nieldel, B.H. Haynes and K.H. Singer, Human thymic epithelial cells produce granulocyte and macrophage colony-stimulating factors, *J. Immunol.* 141:1211 (1988).
10. S. Nonayama, M. Nakayama, T. Shiohara and J. Yata, Only dull $CD3^+$ thymocytes bind to thymic epithelial cells. The binding is elicited by both CD2/LFA-3 and LFA-1/ICAM-1 interactions, *Eur. J. Immunol.* 19:1631 (1989).
11. G. Janossy, J.A. Thomas, F.L. Bollum, G. Granzer, G. Pizzolo, K.F. Bradstock, L. Wong, K. Ganeshagun and A.B. Hoffbrand, The human thymic microenvironment: an immunohistologic study, *J. Immunol.* 125:202 (1980).
12. E.J. Jenkinson, W. Van Ewijk and J.J. Owen, Major histocompatibility complex antigen expression on the epithelium of developing thymus in normal and nude mice, *J. Exp. Med.* 153:280 (1981).
13. W. Savino, G. Manganella, J.M. Verley, A. Wolff, S. Berrih, P. Levasseur, J.P.Binet, M. Dardenne and J.F. Bach, Thymoma epithelial cells secrete thymic hormone but do not express class II antigens of the major histocompatibility complex, *J. Clin. Invest.* 76:1140 (1985).
14. W. Van Ewijk, Y. Ron, J. Monaco, J. Kapplier, P. Marrack, H. Le Meur, P. Gerlinger, B. Durand, C. Benoist and D. Mattis, Compartimentalization of MHC class II gene expression in transgenic mice, *Cell* 53:357 (1988).
15. W. Savino and J. Lannes-Vieira, Is there a function for extracellular matrix in thymus physiology and pathology? *Mem. Inst. Oswaldo Cruz* 85:90 (1991).
16. J.F. Bach, M. Dardenne, J.M. Pleau and J. Rosa, Biochemical characterization of a serum thymic hormone, *Nature* 266:55 (1977).
17. W. Savino, M. Dardenne, M. Papiernik and J.F. Bach, Thymic hormone containing cells. Characterization of serum thymic factor in young mouse thymus studied by monoclonal antibodies, *J. Exp. Med.* 156:628 (1982).
18. N. Fabris and E. Mocchegiani, Endocrine control of serum thymic factor in young-adult and old mice, *Cell. Immunol.* 91:325 (1985).
19. N. Fabris, E. Mocchegiani, S. Mariotti, F. Pacini and A. Pinchera, Thyroid function modulates thymic endocrine activity, *J. Clin. Endocrin. Metab.* 62:474 (1986).
20. W. Savino, E. Bartoccioni, F. Homo-Delarche, M.C. Gagnerault, T. Itoh and M. Dardenne, Thymic hormone containing cells - IX. Steroids in vitro modulate thymulin secretion by human and murine thymic epithelial cells, *J. Steroid Biochem.* 19:135 (1988).
21. J. Lannes-Vieira, M. Dardenne and W. Savino, Extracellular matrix components of the mouse thymus microenvironment: ontogenetic studies and modulation by glucocorticoid hormones, *J. Histochem. Cytochem.* 39:113 (1991).

22. W. Savino, E.O. Cirne-Lima, J.F.T. Soares, M.C. Leite de Moraes, I.P.C. Ono and M. Dardenne, Hydrocortisone increases the number of KL1+ cells, a discrete thymic epithelial cell subset characterized by high molecular weight cytokeratin expression, *Endocrinology* 123:2557 (1988).

23. W. Savino, E. Ban, D.M. Villa-Verde and M. Dardenne, Modulation of thymic endocrine function, cytokeratin expression and cell proliferation, by hormones and neuropeptides, *Int. J. Neurosci.* 51:201 (1990).

24. J. Lannes-Vieira, P.H. Van Der Meide and W. Savino, Extracellular matrix components of the mouse thymus microenvironment. II. Gamma-interferon modulates thymic epithelial cell proliferation and extracellular matrix production, *Cell. Immunol.* 137:329 (1991).

25. M. Dardenne, T. Itoh and F. Homo-Delarche, Presence of glucocorticoid receptors in cultured thymic epithelial cells, *Cell. Immunol.* 100:112 (1986).

26. D.M. Villa-Verde, M.P. Defresne, M.A. Vannier-dos-Santos, J.H. Dussault, J. Boniver and W. Savino, Identification of nuclear trioiodothyronine receptors in the thymic epithelium, *Endocrinology* 131:1313 (1992).

27. C. Baroni, Thymus, peripheral lymphoid tissues, and immunological responsiveness of the pituitary dwarf mouse, *Experientia* 23:282 (1967).

28. N. Fabris, W. Pierpaoli and E. Sorkin, Hormones and the immunological activity. IV. Restorative effects of developmental hormones on lymphocytes of the immunodeficiency syndrome of the dwarf mouse, *Clin. Exp. Immunol.* 9:227 (1971).

29. M. Pelletier, S. Montplaisir, M. Dardenne and J.F. Bach, Thymic hormone activity and spontaneous autoimmunity in dwarf mice and their littermates, *Immunology* 30:783 (1976).

30. B.L. Goff, J.A. Roth, L.H. Arp and G.S. Incefy, Growth hormone treatment stimulates thymulin production in aged dogs, *Clin. Exp. Immunol.* 68:580 (1987).

31. R.G. Goya, M.C. Gagnerault, M.C. Leite de Moraes, W. Savino and M. Dardenne, In vivo effects of growth hormone on thymus function in aging mice, *Brain Behav. Immun.* 6:341 (1992).

32. K.W. Kelley, S. Brief, H.J. Weatly, J. Novakofski, P.J. Bechtel, J. Simon and E.B. Walker, GH3 pituitary adenoma cells can reverse thymic aging in rats, *Proc. Natl. Acad. Sci. USA* 83:5663 (1986).

33. M. Dardenne, W. Savino, M.C. Gagnerault, T. Itoh and J.F. Bach, Neuroendocrine control of thymic hormonal production. I. Prolactin stimulates in vivo and in vitro the production of thymulin by human and murine thymic epithelial cells, *Endocrinology* 125:1251 (1989).

34. J. Timsit, W. Savino, W. Safieh, P. Chanson, M.C. Gagnerault, J.F. Bach and M. Dardenne, Growth hormone and insulin-like growth factor-I stimulate hormonal function and proliferation of thymic epithelial cells, *J. Clin. Endocrin. Metab.* 75:183 (1992).

35. E. Ban, M.C. Gagnerault, H. James, M.C. Postel-Vinay, F. Haour and M. Dardenne, Specific binding sites for growth hormone in cultured mouse thymic epithelial cells, *Life Sci.* 48:2141 (1991).

36. D.H. Russell, R. Kibler, L. Matrisian, D.F. Larson, B. Poulos and B.E. Magun, Prolactin receptors on human T and B lymphocytes: antagonism of prolactin binding by cyclosporine, *J. Immunol.* 134:3027 (1985).

37. M. Dardenne, P.A. Kelly, J.F. Bach and W. Savino, Identification and functional activity of prolactin receptors in thymic epithelial cells, *Proc. Natl. Acad. Sci. USA* 88:9700 (1991).

38. M.C. Gagnerault, P. Touraine, W. Savino, P.A. Kelly and M. Dardenne, Expression of prolactin receptors in murine lymphoid cells in normal and autoimmune situations, *J. Immunol.* 150:5673 (1993).

39. M. Dardenne, M.C. Leite de Moraes, P.A. Kelly and M.C. Gagnerault, Prolactin receptor expression in human hematopoietic tissues analysed by flow cytofluorometry, *Endocrinology* (1994) (in press).

40. I. Pellegrini, J.J. Lebrun, S. Ali and P.A. Kelly, Expression of prolactin and its receptor in human lymphoid cells, *Mol. Endocrinol.* 6:1023 (1992).

41. I. Screpanti, D. Meco, S. Morrone, A. Gulino, B.J. Mathielson and L. Frati, In vivo modulation of the distribution of thymocyte subsets: effects of estrogen on the expression of different T cell receptor Vβ gene families in CD4⁻CD8⁻ thymocytes, *Cell Immunol.* 134:414 (1991).

42. V. Geenen, H. Martens, F. Robert, J.J. Legros, M.P. Defresne, J. Boniver, J. Martial, P.J. Lefebvre and P. Franchimont, Thymic cryptocrine signalling and the immune recognition of self neuroendocrine function, *Prog. NeuroEndocrineImmunol.* 4:135 (1991).

IMMUNOMODULATION BY GROWTH HORMONE IN HUMANS

Robert Rapaport

Department of Pediatrics
Division of Pediatric Endocrinology
and Metabolism
Children's Hospital of New Jersey
UMD-New Jersey Medical School
15 South 9th Street
Newark, NJ 07107 USA

INTRODUCTION

The role of growth hormone in the establishment and maintenance of the immune system of animals has been well documented. Beginning with the demonstration of involution of the thymus gland following hypophysectomy,[1] many studies have shown diminished immune functions in hypophysectomized animals that were reversible by growth hormone administration.[2-8] Growth hormone has also been reported to stimulate DNA synthesis and c-myc proto-oncogene expression[9] as well as superoxide anion production[10,11] in hypophysectomized rats.

Anti-growth hormone antiserum administration[12,13] or congenital absence of growth hormone producing cells in certain strains of mice, the Snell-Bagg (dw) and Ames (df) mice, has been reported to result in a wasting syndrome accompanied by immune deficiency that is reversible by growth hormone administration.[14-21] Treatment of the Snell-Bagg mouse with human growth hormone was reported to increase thymic size and improve peripheral blood cell counts.[22,23] Growth hormone treatment was proven to promote engraftment of murine and human T cells in severe combined immunodeficiency mice[24] and reverse thymic aging in mice.[25,26]

The role of growth hormone in immunomodulation in animals is discussed in greater detail elsewhere in this chapter.

CURRENT USE OF GROWTH HORMONE IN MAN

In humans, the interaction between the immune system and the hypothalamic-pituitary-growth hormone axis has been studied less extensively. Hans Selye was one of the very first pioneers in the field of neuroendocrine-immune interactions. He postulated a significant role for growth hormone, which he named STH for somatotrophic hormone, in the pathophysiology of the stress reaction.[27] One could approach the interaction between growth hormone and the immune system from several perspectives: a) examine immune functions in individuals with defects in growth hormone secretion, b) investigate growth hormone dynamics in subjects with immunologic disorders or c) document the occurrence of defects in both systems in the same individual.

Growth hormone, a 191 amino acid peptide, is secreted by the anterior pituitary gland in secretory bursts in response to the net effect of the hypothalamic stimulatory factor (growth hormone releasing hormone or factor - GH-RF)) and inhibitory factor (somatostatin). Growth hormone acts on most tissues to produce small molecular weight peptides (Insulin Like Growth Factors i.e. IGF-I) which in turn act locally and at end organs (bones) to promote growth. An active feedback mechanism exists between IGF-I, GH and the hypothalamic-pituitary unit.[28]

The clinical diagnosis of growth hormone deficiency in poorly growing children is defined as the failure of serum growth hormone to rise above a certain, rather arbitrary, value of between 7 - 10 ng/ml in response to two different pharmacological stimulation tests (i.e. glucagon, L-dopa, clonidine or insulin-induced hypoglycemia). The treatment of human growth hormone deficiency consisted of pituitary-derived human growth hormone (pit-hGH) until the spring of 1985. The quantities of pit-hGH were limited due to the lack of availability of sufficient cadaver pituitary glands. Most children received treatment every other day with standard doses, usually 2 units. Often, treatment regimens consisted of one month without treatment following every two months of therapy. Because of reports of children previously treated with pit-hGH dying of Creutzfeldt-Jacob disease, the use of pituitary derived GH was halted in the spring of 1985 in most countries. To date, 50 cases of Creutzfeldt-Jacob disease have been reported worldwide. The cause seems to be inadvertent contamination of growth hormone derived from cadaver pituitary glands harvested from patients who had clinical or subclinical Creutzfeldt-Jacob disease.[29]

In the fall of 1985 recombinant DNA-derived biosynthetic hGH became available. Since then, deficient children have been receiving continuous treatment with larger doses of hGH, until they reach final stature. Because of the theoretically unlimited availability of biosynthetic hGH, children not previously thought to be candidates for therapy have been receiving treatment, hopefully as part of approved, controlled trials. Human GH has been used to improve the stature of children with Turner Syndrome, non-GH deficient short stature, renal failure and various forms of bone dysgenesis.[30,31]

Growth hormone has started to be used for its anabolic properties in conditions of malnutrition, critical illness, burns. The prospects for the future are that even more conditions will be subject to at least trials of hGH treatment. Growth hormone has also been used extensively in veterinary medicine. These developments clearly make the entire subject of GH-immune interactions much more important and relevant. A detailed discussion of the questions and controversies surrounding treatment with growth hormone is beyond the scope of this review.

IMMUNE FUNCTION IN CHILDREN WITH GH DEFICIENCY

In children with GH deficiency immune parameters have, with few exceptions, been normal. Gupta et al.[32] studied 4 growth hormone deficient children and found increased proportions of suppressor cells, resulting in lower helper/suppressor cell ratio, and increased B cells. They found normal proliferative responses to phytohemagglutinin but decreased T cell mixed lymphocyte reactions in 3/4 patients. However, most investigators have reported that total numbers of white and red blood cells, serum immunoglobulin levels, lymphocyte subsets (cell surface markers) as well as cell proliferation in response to stimuli such as phytohemagglutinin (PHA), poke weed mitogens (PWM) and concanavalin A (con A) have been normal.[33-38]

GH TREATMENT AND IMMUNE FUNCTION IN MAN

Immunoglobulin Levels

Serum immunoglobulin levels during GH treatment do not seem to change significantly.[35,38,39] We measured serum levels of IgA, IgM and IgG in 8 GH deficient children before and every three months during the course of pituitary-derived human growth hormone treatment for up to 16 months and found no statistically significant changes. Bozzola et al.[39] found that growth hormone treatment resulted in decreased in vitro IgM production by unstimulated and also PWM-stimulated lymphocytes obtained from growth hormone deficient subjects.

Cell Surface Markers

Lymphocyte subsets, normal at baseline, have been found by Abbassi and Bellanti[33] to be unchanged by GH treatment. They measured percent T and B cells before and at 9 to 12 months of treatment and found no changes. We measured cell surface markers (by monoclonal antibody markers) at 1, 3, 6, 9, 12, 14 and 16 months of treatment with pituitary-derived hGH in 8 children. Percent total T, helper or suppressor cells did not change significantly. Helper/suppressor (H/S) ratio did transiently decrease due to both a slight decrease in helper cells and an increase in suppressor cells.[35] Bozzola et al.[36] also reported a decrease in T helper, increase in T suppressor and decrease in T helper/suppressor ratio. Church et al.[40] found no changes in cell surface marker expression during growth hormone treatment.

We found a transient but significant decrease in % B cells during pituitary-derived hGH treatment.[35] When immune functions were tested in children who had interruptions of treatment for 1-2 months, reinstitution of treatment also resulted in a decrease in % B cells.[41] When treatment of GH deficiency began with biosynthetic human growth hormone we demonstrated a similar decrease in % B cells.[42] Of interest, 3 out of 4 children had similar decreases in % B cells during biosynthetic human growth hormone treatment as they had during pituitary-derived human growth hormone treatment. In an expanded, multicenter study of the effects of biosynthetic human growth hormone treatment involving 19 subjects, the decrease of % B cells was again demonstrated.[43] This decrease in % B cells was confirmed by others.[38]

In vitro, we exposed peripheral blood lymphocytes derived from growth hormone deficient and normal children to various concentrations of growth hormone. In accordance with our in vivo findings, the expression of surface markers by B cells decreased upon exposure to GH, as detected by monoclonal antibodies using flow cytometry.[44] We suggest that the B cell may play a pivotal role in the interaction between growth hormone and the immune system.

Yoshida et al.[45] studied the effects of growth hormone in vitro on four different cultured B cell lines. Growth hormone failed to enhance immunoglobulin synthesis and thymidine uptake by unstimulated freshly separated B cells but it did stimulate IgG synthesis and thymidine uptake by IM-9 cells, IgA synthesis by GM-1056, and IgM synthesis by CBL cells. GM-1056 and CBL are EBV transformed lymphoblastoid B cell lines that secrete IgA and IgM respectively.

Natural Killer Cells

Natural killer (NK) cell activity has been examined relatively infrequently in growth hormone deficient subjects. Kiess et al.[46,47] reported decreased natural killer cell activity in growth hormone deficient children that was not improved by a short course of growth hormone or growth hormone releasing hormone administration. Growth hormone deficient women (mean age 56.3 ± 8.7 years) did have an increase in natural killer cell activity following growth hormone administration.[48] Bozzola et al.[49] reported impaired natural killer cell function but normal lymphokine-activated killer cell activity in growth hormone deficient children. Long term treatment with growth hormone resulted in normalization of natural killer cell function. Matsuura et al.[37] reported improved natural killer cell activity after a short course of growth hormone treatment. Cells expressing NK cell markers decreased moderately during treatment. In vitro, exposure to growth hormone resulted in no change in NK activity.

We reported normal percentage of cells bearing natural killer cell markers in growth hormone deficient children that did not change during the course of biosynthetic human growth hormone treatment.[42]

Cell Proliferation

Controversy exists regarding the effects of growth hormone administration on mitogen responses. Some have reported no effect of growth hormone on lymphoproliferative function.[50] Astaldi et al.[51] have shown that growth hormone induced lymphocyte blastogenesis of human normal peripheral blood lymphocytes (PBL) in vitro. Kiess et al.[34] showed increased lymphoproliferation of PBL upon exposure to growth hormone.

During growth hormone treatment, when proliferative responses to PHA were measured at 9-12 months of treatment, Abbassi and Bellanti[33] showed an increase stimulation. Using more frequent sampling every 2-3 months, we demonstrated a transient decrease in PHA induced lymphoproliferation.[35] Exposure of cells to growth hormone, at least at supraphysiologic doses, was also shown by others to decrease the proliferative responses of PBL to PHA.[34,52]

We have suggested that the proliferative responses of lymphocytes may depend on the growth hormone status of the subjects tested.[44] In vitro, growth hormone depressed the proliferation of the PBL of normal and untreated growth hormone deficient children. The responses of most normal and treated growth hormone deficient children to PHA were decreased by the addition of growth hormone, confirming previous reports. PBL derived from growth hormone deficient children during treatment with growth hormone exhibited greater spontaneous and PHA stimulated proliferation then did cells of normal children. The addition of growth hormone to the culture medium further increased these children's spontaneous cell proliferation while it seemed to decrease the PHA induced proliferation in 3/4 patients.

Hemopoietic Cells

In humans it has been shown by some that growth hormone treatment leads to an increase in hemopoiesis. Jepson and McGarry[53] reported that prepubertal "panhypopituitary dwarfs" were anemic and had decreased levels of erythropoietin-stimulating factors (ESF). Treatment with growth hormone increased erythropoiesis, red cell mass, transferrin levels, ESF and expanded plasma volume. Others found normal red and white blood cell levels before and during growth hormone treatment of deficient children.[35,37]

In vitro, Blatt et al.[54] showed that GH at physiologic concentrations had no effect on human leukemic lymphoblast colony formation. Supraphysiologic doses did result in increased colony formation. Most other authors demonstrated increased leukemic cell proliferation in response to exposure to growth hormone.[55-57] Human marrow granulopoiesis in vitro was enhanced by hGH but inhibited by anti IGF-I receptor antibodies, suggesting paracrine effects of IGF-I as mediator of the hGH effect.[58] Similarly, IGF-I and IGF-II receptors were implicated as mediators of insulin's effect on the stimulation of certain acute lymphoblastic leukemia cell lines.[59]

Zadik et al.[60] studied the effects of growth hormone and IGF-I on bone marrow cells derived from 5 children with acute lymphocytic leukemia (ALL), 4 children with acute myelogenous leukemic (AML) and on peripheral blood cells of 3 patients with chronic myelogenous leukemia (CML) during remission by using a blast colony formation assay. Both growth hormone and IGF-I increased blast colony numbers in a dose dependent fashion in 8 of the 9 patients with ALL and AML. Growth hormone and IGF-I did not increase peripheral blood blast colony formation in the patients with CML.

Since 1987, reports have appeared documenting the occurrence of leukemia in children undergoing growth hormone therapy. A total of 31 cases of leukemia, preleukemia or myelodysplastic syndrome and one of malignant histiocytosis have been reported.[61] Twelve cases occurred in Japan.[62] Twenty-one of the 31 patients had idiopathic growth hormone deficiency. In addition we collected information in 5 patients[63] and are aware of 3 others[62] who developed leukemia and had growth hormone deficiency but had <u>not</u> received any growth hormone treatment. A careful analysis of the currently existing data suggests a slight increase in leukemia incidence with growth hormone treatment.

Tedeschi et al.[64] recently reported an increase in bleomycin-induced chromosomal aberrations and in spontaneous chromosomal rearrangements in 10 short, non-growth hormone deficient children, during therapy with exogenous recombinant hGH. This report, along with the previous discussion, should serve as a caution against the indiscriminate use of growth hormone in non-growth hormone deficient children. Extremely careful consideration needs to be given to potential GH treatment of children with conditions known to be associated with increased risk of tumor formation or chromosomal aberrations, such as Down, Fanconi or Bloom syndrome.

Phagocytic Cells

The metabolic activity of phagocytes estimated by nitrozolium reductase activity was found to be low in patients with hypopituitarism. Growth hormone administration increased both resting and starch stimulated reductase activity.[65] Beta-D-glucuronidase, myeloperoxidase and lysozyme activities of polymorphonuclear leukocytes (PMN) of normal and hypopituitary subjects were increased by growth hormone. In vitro, incubation with growth hormone inhibited the release of lysosomal enzymes from PMN's.[66]

In growth hormone deficient children we found normal PMN function as measured by chemotaxis under agarose and chemiluminscense.[35] Treatment with hGH did not alter PMN function. In vitro, it has been shown that growth hormone may act to prime human neutrophils by the demonstration of improved respiratory burst production.[67,68] Nitroblue tetrazolium reduction of aged individuals improved to levels seen in younger subjects after the in vitro addition of growth hormone.[69] Native or recombinantly produced growth hormone, increased superoxide anion production by alveolar macrophages stimulated in vitro by opsonized zymosan.[70] The improved superoxide anion secretion of human neutrophils was postulated to be due to the binding of growth hormone to the prolactin receptor.[71] Growth hormone and IGF-I have been shown to increase murine macrophage uptake and degradation of low density lipoprotein (LDL). It was postulated that this effect may provide a possible mechanism for the observed reduction of plasma LDL by growth hormone treatment.[72]

Thymic Cells

Considerable information suggesting a significant role for growth hormone in the development and maintenance of normal thymic function in animals has existed for decades.[73] Mocchegiani et al.[74] have described low levels of a thymic hormone, thymulin, in growth hormone deficient children. Growth hormone injection was able to increase thymulin levels. Timsit et al.[75] have shown that patients with acromegaly (growth hormone excess), had increased levels of thymulin secretion measured both by radioimmunoassay and bioassay. Thymulin levels were correlated with IGF-I levels. In vitro, both growth hormone and IGF-I were able to stimulate thymulin production by human (and rat) thymic epithelial cell lines.

The hormonal regulation of thymic function is discussed in more detail by Savino et al. in this volume.

GROWTH HORMONE STATUS IN IMMUNE DEFICIENCY STATES AND IMMUNE DEFECTS IN SHORT STATURE CONDITIONS

The literature concerning reports of the evaluation of growth hormone secretory dynamics in immune deficiency states is not very extensive. In 1970, Amman et al.[76] described defects in both humoral and cellular immunity in 5 patients with ataxia-telangiectasia, 2 patients with sex linked lymphopenic and one with non-lymphopenic hypogammaglobulinemia. One of the patients with ataxia-telangiectasia had normal growth hormone response to arginine but not to insulin-induced hypoglycemia. That patient would not be considered growth hormone deficient, inasmuch as there was a normal growth hormone response to one stimulus. Amman et al.[77] documented deficient antibody mediated immunity in 2 siblings with short-limbed dwarfism suggesting a classification of short-limbed dwarfism based on the presence of antibody-mediated or cell-mediated immunodeficiency or both. Patients with cartilage-hair hypoplasia have been described as having cell-mediated immune defects and would therefore be considered to have Type 2 short-limbed dwarfism.

T cell defects were described in 5 patients with Schimke immuno-osseous dysplasia, a syndrome of skeletal dysplasia, rapidly progressive nephropathy, lymphopenia and skin changes.[78] Growth retardation, IgA deficiency, low T lymphocyte counts and diminished proliferative responses to mitogens were described in a patient with mutations on the DNA ligase-I gene who had features similar to patients with Bloom syndrome.[79]

An association between X-linked agammaglobulinemia (XLA) and isolated growth hormone deficiency has been described in several families.[80-82] Molecular genetic analysis in several members of two unrelated families suggested the likelihood of this association to be due to a small, contiguous gene deletion syndrome involving the gene for XLA or an allelic variant of that gene.[83]

Isolated growth hormone deficiency in association with IgG2 and IgG4 subclass deficiency[84] and with X-linked combined humoral and cellular immunodeficiency[85] were also described.

Occasional reports have described growth hormone deficiency in patients with various immune or autoimmune related disorders. Attention has recently been focused on the growth of children with human immunodeficiency virus (HIV) infections. While on rare occasions growth hormone deficiency has been described in such children, the growth failure in the majority of children with HIV infections is due to multiple, complex, and as yet unknown causes and not classical growth hormone deficiency. This subject was recently reviewed.[86] Poor nutrition can result in both growth and immunological defects which can be at least in part, reversed by improving the nutritional status.[87-90] A detailed summary of the effects of nutrition on growth and immune function is beyond the intention of this section.

GROWTH HORMONE RECEPTORS ON CELLS OF THE IMMUNE SYSTEM

In view of the overwhelming evidence that growth hormone interacts with the immune system many investigators have tried to elucidate the mechanism by which such an interaction would occur. For the last 20 years cells of the immune system have been examined for the presence of receptors for growth hormone and more recently for IGF-I. Lesniak et al.[91] were the first to describe high-affinity human growth hormone receptors on cultured human lymphocytes. Subsequent reports have demonstrated growth hormone receptors on thymocytes,[92] lymphocytes,[93-95] cultured human fibroblasts,[96] growth plate chondrocytes[97] and human fetal mesenchymal tissue.[98] While the demonstration of GH receptors on cultured lymphocytes, such as the IM-9 cell line, has been reproduced, the isolation of the receptors on circulating lymphocytes has been extremely difficult. IGF-I receptors have been described on human monocytes.[99,100] IGF-I was noted by Stuart et al.[101] to bind selectively to peripheral blood monocytes and B-lymphocytes. Using two-color flow cytometric analysis of human peripheral blood non-activated mononuclear cells stained with monoclonal antibodies they demonstrated IGF-I (and insulin) receptors on only 2% of T-lymphocytes but on nearly all B-lymphocytes and monocytes.

In our previous reports we have suggested that growth hormone treatment affected mostly B cells. We have presented preliminary results demonstrating receptors for fluorescein-conjugated human growth hormone on peripheral blood mononuclear cells, using flow cytometry. Using two-color analysis, receptors were readily detected on B lymphocytes and monocytes. We estimated that there were approximately 6,000 growth hormone receptors/cell, a finding comparable to previous reports. Growth hormone deficient children had normal expression of growth hormone receptors on B cells, as expected.[102]

GH-RF TREATMENT AND IMMUNE FUNCTION

Growth hormone releasing hormone (GH-RF) has been evaluated as potential treatment for short, growth hormone deficient children in whom hypothalamic GH-RF deficiency was the likely etiology. Initial reports suggested that growth hormone releasing hormone was both safe and effective for such therapy.[103]

The effects of growth hormone releasing hormone on immune functions have been inconsistent. Pawlikowski et al.[104] have shown that human peripheral blood lymphocyte natural killer cell activity, estimated by radioactive chromium assay, was suppressed by growth hormone releasing hormone at an effector:target cell ratio of 40:1 but increased at ratios of 20:1 and 10:1. The same authors[105] reported a negative correlation between growth hormone releasing hormone concentration and chemotactic response of PBL but not in the migration inhibition assay. Kiess et al.[47], found no effect of growth hormone releasing hormone administration acutely (intravenous bolus) or chronically (3 - 6 weeks) on either NK activity or lymphocyte subset distribution.

In vitro, Valtora et al.[106] found no effect of growth hormone releasing hormone 1-44, on PHA induced lymphoproliferation, interleukin-2 (IL-2) production or IL-2 receptor expression of PBL of healthy adults. Growth hormone releasing hormone 1-29 however increased PHA responses at low concentrations but inhibited lymphocyte responses, IL-2 secretion and receptor expression at high concentrations.

CYTOKINES AS FEEDBACK SIGNALS REGULATING GH SECRETION

In vitro and animal studies have indicated a strong relationship between cytokines and growth hormone. Interleukin-1 (IL-1) induced growth hormone secretion by dispersed cultures of rat anterior pituitary cells[107,108] and in intact rats.[109] In humans, IL-2 treatment was shown to stimulate growth hormone.[110] Conversely growth hormone may have an effect on interleukin production. Schimpff and Repellin[111] have shown that in the absence (but not in the presence) of PHA, growth hormone can increase human adult mononuclear cell production of both IL-1 and IL-2. In a preliminary study, we found diminished IL-2 production in response to PHA in growth hormone deficient children both before and during growth hormone treatment.[112] We also showed a transient decrease in interleukin-2 receptor levels in most subjects during growth hormone treatment.[42] Interleukin-6 (IL-6) may have a similar bidirectional relationship with growth hormone. Interleukin-6 has been shown to increase growth hormone secretion and GH-RF induced growth hormone secretion.[113,114] Anterior pituitary cells can produce IL-6 under stimulation by various factors including IL-1.[115-118] Tumor necrosis factor (TNF) also has been shown to inhibit growth hormone production by rat anterior pituitary cells.[119,120] Growth hormone can prime rat macrophages to produce TNF-alpha.[121] The effects of various cytokines, including tumor necrosis factor on the hypothalamic-pituitary unit have recently been reviewed.[122] Chapter IV in this volume discusses in more detail the metabolic and neuroendocrine effects of cytokines.

PRODUCTION OF GH/IGF-I BY CELLS OF THE IMMUNE SYSTEM

It has been demonstrated that human alveolar macrophages as well as Epstein-Barr virus transformed human B lymphocytes produce IGF-I - like substances.[99,123] Cells derived from pygmies individuals with non GH-deficient extreme short stature, produced less IGF-I when stimulated with growth hormone then cells from subjects with normal stature.[123] Locally generated IGF-I in conditioned media of transformed T-lymphoblast cell lines was implicated in the mediation of growth hormone's action on T-lymphocytes.[124] Growth hormone and growth hormone releasing hormone have also been shown to be produced by lymphocytes.[125-129] Hattori et al.[130] showed that human lymphocyte secretion of growth hormone was not affected by co-incubation with IGF-I (0-1,000 mcg/L) but that it was up-regulated by co-incubation with growth hormone (0-100 ng/L) in a dose dependent fashion.

The role of lymphocyte derived "classical" hormones, which may have important paracine or autocrine effects is yet to be elucidated.[128]

FUTURE TRENDS

With the increased availability of biosynthetic growth hormone many patients with conditions not previously thought appropriate for growth hormone treatment are now subject to trials of growth hormone treatment. Improvement in stature by growth hormone is sought in patients with Turner syndrome, bone dysplasia, non-growth hormone deficient short stature, renal failure and others. Growth hormone releasing hormone treatment has been tested instead of growth hormone in many European and U.S. trials with good response.[103] Growth hormone is used not only for its growth promoting properties but also for its metabolic effects especially in aging, malnutrition, wound healing and burns.[131]

The interaction of hormones and immune functions in aging has been recognized for more then two decades.[132-134] Studies of the effects of aging have demonstrated a diminution in immune functions and hormonal output, inclusive of growth hormone secretion. The restorative effects of growth hormone treatment in animals have been amply documented.[8] Because of the metabolic similarities between states of growth hormone deficiency and normal aging, e.g. both are associated with decreased levels of growth hormone and IGF-I, it was natural to attempt to reverse some of these effects by growth hormone treatment.[48,135] The relatively limited, short term studies of the effects of growth hormone treatment in human aging have recently been reviewed.[136]

Future studies will evaluate the effects of growth hormone, growth hormone releasing hormone and IGF-I treatment on immune parameters.

PROLACTIN

While the topic of my review is the effect of growth hormone on the immune system I must acknowledge that the role of other stress hormones, and in particular that of prolactin may be crucial for and at times indistinguishable from the effects of growth hormone. This subject has been reviewed in this chapter by Nagy and Berczi.

CONCLUSION

All lines of evidence point to an active communication network between growth hormone and the immune system. While in animal models most of the information supports an immune stimulatory role for growth hormone, data in humans is more diverse with growth hormone having both stimulatory and seemingly inhibitory effects. Measurements of serum levels of hormones cannot adequately detect changes that may be occurring at the cellular level. Autocrine and paracrine effects of hormones also need to be investigated. Alterations in hormonal or immune parameters can result in disruption of the finely tuned balance that exists between the two systems. We believe that B cells may be pivotal for the maintenance of hormonal-immune equilibrium.

Recently a point mutation in the POU-specific portion encoding for the transcription factor Pit-1 was described in 2 unrelated Dutch families.[137] The affected members were deficient in growth hormone, prolactin and thyroid stimulating hormone. Two subjects had normal appearing anterior pituitary glands by magnetic resonance imaging. The mutant protein produced was thought to be able to direct embryonic differentiation and proliferation of somatotropes and lactotropes. In contrast, Snell-Bagg dwarf mice had deficiencies of the same hormones caused by a point mutation in the POU-HD region that resulted in pituitary gland hypoplasia with the absence of somatotropes and lactotropes.

These studies, along with the new information on the growth hormone/prolactin receptor gene family, recently reviewed,[138] and the exciting discoveries of the production of GH/IGF-I by immune cells, have revolutionized our ideas about the role of GH in health and disease, and opened new areas of scientific inquiry.

A great deal of additional studies are needed to elucidate the exact mechanisms of the influences that the hormonal and immune systems exert on each other. Understanding these complex interactions may have important consequences for basic physiology and clinical medicine. Results of basic investigators have to be coupled with clinical observation to provide a unified and useful concept of neuroendocrine-immune interactions.

REFERENCES

1. P.E. Smith, Effect of hypohysectomy upon the involution of the thymus in the rat, *Anat. Rec.* 47:119 (193).
2. P.M. Lundin, Action of hypophysectomy on antibody formation in the rat, *Acta. Pathol. Microbiol. Scand.* 48:351 (1960).
3. L. Enerback, P.M. Lundin and J. Mellgren, Pituitary hormones elaborated during stress. Action on lymphoid tissues, serum proteins and antibody titres, *Acta. Pathol. Microbiol. Scand.* (Suppl.) 144:141 (1961).
4. R.H. Gisler and L. Schenkel-Hulliger, Hormonal regulation of the immune response II. Influence of pituitary and adrenal activity on immune responsiveness in vitro, *Cell. Immunol.* 2:646 (1971).
5. J. Comsa, J.A. Schwarz and H. Neu, Interaction between thymic hormone and hypophyseal growth hormone on production of precipitating antibodies in the rat, *Immunol. Commun.* 3:11 (1974).
6. E. Nagy and I. Berczi, Immunodeficiency in hypophysectomized rats, *Acta Endocrinol.* 89:530 (1978).
7. E. Nagy, I. Berczi and H.G. Friesen, Regulation of immunity in rats by lactogenic and growth hormones, *Acta Endocrinol.* 102:351 (1983).
8. I. Berczi, "Pituitary Function and Immunity," CRC Press, Boca Raton, FL (1986).
9. I. Berczi, E. Nagy, S.M. DeToledo, R.J. Matusik and H.G. Friesen, Pituitary hormones regulate c-myc and DNA synthesis in lymphoid tissue, *J. Immunol.* 146:2201 (1991).
10. C.K. Edwards, III, L.M. Yunger, R.M. Lorence, R. Dantzer and K.W. Kelley, The pituitary gland is required for protection against lethal effects of Salmonella typhimurium, *Proc. Natl. Acad. Sci.* (USA) 88:2274 (1991).
11. C.K. Edwards, III, S.M. Ghiasuddin, L.B. Yunger, R.M. Lorence, S. Arkins, R. Dantzer and K.W. Kelley, In Vivo administration of recombinant growth hormone or gamma interferon activates macrophages: Enhanced resistance to experimental Salmonella typhimurium infection is correlated with generation of reactive oxygen intermediates, *Infect. Immun.* 60:2514 (1992).
12. W. Pierpaoli and E. Sorkin, Relationship between thymus and hypophysis, *Nature* 215:834 (1967).
13. W. Pierpaoli and E. Sorkin, Hormones and immunologic capacity I. Effect of heterologous anti-growth hormone(ASTH)antiserum on thymus and peripheral lymphatic tissue in mice. Induction of a wasting syndrome, *J. Immunol.* 101:1036 (1968).
14. C. Baroni, Thymus, peripheral lymphoid tissues and immunological responses of the pituitary dwarf mouse, *Experientia* 23:282 (1967).
15. C.D. Baroni, N. Fabris and G. Bertoli, Effects of hormones on development and function of lymphoid tissues. Synergistic action of thyroxin and somatotropic hormone in pituitary dwarf mice, *Immunology* 17:303 (1969).
16. C.D. Baroni, P.C. Pesando and G. Bertoli, Effects of hormones on development of lymphoid tissues. II. Delayed development of immunological capacity in pituitary dwarf mice, *Immunology* 21:455 (1971).
17. F. Dumont, F. Roert and P. Bischoff, T and B lymphocytes in pituitary dwarf Snell-Bagg mice, *Immunology* 38:23 (1979).

18. N. Fabris, W. Pierpaoli and E. Sorkin, Hormones and the immunological capacity. III. The immunodeficiency disease of the hypopituitary Snell-Bagg dwarf mouse, *Clin. Exp. Immunol.* 9:209 (1971).

19. N. Fabris, W. Pierpaoli and E. Sorkin, Hormones and the immunological capacity. IV. Restorative effects of developmental hormones or of lymphocytes on the immunodeficiency syndrome of the dwarf mouse, *Clin. Exp. Immunol.* 9:227 (1971).

20. R.J. Duquesnoy, P.K. Kalpaktsoglou and R.A. Good, Immunological studies on the Snell-Bagg pituitary dwarf mouse, *Proc. Soc. Exp. Biol. Med.* 133:201 (1970).

21. R.J. Duquesnoy, Immunodeficiency of the thymus-dependent system of the Ames dwarf mouse, *J. Immunol.* 108:1578 (1972).

22. W.J. Murphy, S.K. Durum and D.L. Longo, Role of neuroendocrine hormones in murine T cell development: growth hormone exerts thymopoietic effects in vivo, *J. Immunol.* 149:3851 (1992).

23. W.J. Murphy, S.K. Durum, M.R. Anver and D.L. Longo, Immunologic and hematologic effects of neuroendocrine hormones, *J. Immunol.* 148:3799 (1992).

24. W.J. Murphy, S.K. Durum and D.L. Longo, Human growth hormone promotes engraftment of murine or human T cells in severe combined immunodeficient mice, *Proc. Natl. Acad. Sci. USA* 89:4481 (1992).

25. K.W. Kelley, S. Brief, H.J. Westly, J. Novakofski, P.J. Bechtel, J. Simon and E.B. Walker, GH3 pituitary adenoma cells can reverse thymic aging in rats, *Proc. Natl. Acad. Sci. USA* 83:5663 (1986).

26. M.Y. Li, D.L. Drunke, R. Dantzer and K.W. Kelley, Pituitary epithelial cell implants reverse the accumulation of CD4-CD8-lymphocytes in thymic glands of aged rats, *Endocrinology* 30:2703 (1992)

27. H. Selye, "The Stress of Life, McGraw Hill Book Company, New York (1986).

28. S.A. Kaplan, "Clinical Pediatric Endocrinology," W.B. Saunders Company, USA (1990).

29. M. Preece, Human pituitary growth hormone and Creutzfeldt-Jacob Disease, *Horm. Res.* 39:95 (1993).

30. J.O.L Jorgensen, Human growth hormoen replacement therapy: Pharmacological and clinical aspects, *Endocr. Rev.* 12:189 (1991).

31. O. Westphal, Non-conventional growth hormone treatment in short children, *Acta Endocrinol.* 128:10 (1993).

32. S. Gupta, S.M. Fikrig and M.S. Noval, Immunological studies in patients with isolated growth hormone deficiency, *Clin. Exp. Immunol.* 54:87 (1983).

33. V. Abbassi and J.A. Bellanti, Humoral and cell-mediated immunity in growth hormone-deficient children: Effect of therapy with human growth hormone, *Pediatr. Res.* 19:299 (1985).

34. W. Kiess, H. Holtmann, O. Butenandt and R. Eife, Modulation of lymphoproliferation by human growth hormone, *Eur. J. Pediatr.* 140:47 (1983).

35. R. Rapaport, J. Oleske, H. Ahdieh, S. Solomon, C. Delfaus and T. Denny, Suppression of immune function in growth hormone-deficient children during treatment with human growth hormone, *J. Pediatr.* 109:434 (1986).

36. M. Bozzola, M. Cisternino, A. Valtorta, A. Moretta, I. Biscaldi, M. Maghnie, M. De Amici and R.M. Schimpff, Effect of biosynthetic methionyl growth hormone (GH) therapy on the immune function in GH-deficient children, *Horm. Res.* 31:153 (1989).

37. M. Matsuura, Y. Kikkawa, T. Kitagawa and S. Tanaka, Modulation of immunological abnormalities of growth hormone-deficient children by growth hormone treatment, *Acta Paediatr. Jpn.* 31:53 (1989).

38. G.I. Spadoni, P. Rossi, W. Ragno, E. Galli, S. Cianfarani, C. Galasso and B. Boscherini, Immune function in growth hormone-deficient children treated with biosynthetic growth hormone, *Acta Paediatr. Scand.* 80:75 (1991).

39. M. Bozzola, R. Maccario, M. Cisternino, D. De Amici, A. Valtorta, A. Moretta, I. Biscaldi and R.M. Schimpff, Immunological and endocrinological response to growth hormone therapy in short children, *Acta Paediatr. Scand.* 77:675 (1988).

40. J.A. Church, G. Costin and J. Brooks, Immune functions in children treated with biosynthetic growth hormone, *J. Pediatr.* 115:420 (1989).

41. R. Rapaport and J. Oleske, Effect of growth hormone therapy on immune functions, *J. Pediatr.* 110:663 (1987).

42. R. Rapaport, B. Petersen, K.A. Skuza, M. Heim and S. Goldstein, Immune functions during treatment of growth hormone-deficient children with biosynthetic human growth hormone, *Clin. Pediatr.* 30:22 (1991).

43. B.H.Petersen, R. Rapaport, D.P Henry, C. Huseman and W.V. Moore, Effect of treatment with biosynthetic human growth hormone (GH) on peripheral blood lymphocyte populations and function in growth hormone-deficient children, *J. Clin. Endocrinol. Metab.* 70:1756 (1990).

44. R. Rapaport, J. Oleske, H. Ahdieh, K. Skuza, B.K. Holland, M.R. Passannante and T. Denny, Effects of human growth hormone on immune functions: In vitro studies on cells of normal and growth hormone-deficient children, *Life Sci.* 41, 2319 (1987).

45. A. Yoshida, C. Ishioka, H. Kimata and H. Mikawa, Recombinant human growth hormone stimulates B cell immunoglobulin synthesis and proliferation in serum-free medium, *Acta Endocrinol.* 126, 524 (1992).

46. W. Kiess, H. Doerr, E. Eisl, O. Butenandt and B.H. Belohradsy, Lymphocyte subsets and natural-killer activity in growth hormone deficiency, *N. Engl. J. Med.* 314:321 (1986).

47. W. Kiess, S. Malozowski, M. Gelato, O. Butenandt, H. Doerr, B. Crisp, E. Eisl, A. Maluish and B.H. Belohradsky, Lymphocyte subset distribution and natural killer activity in growth hormone deficiency before and during short-term treatment with growth hormone releasing hormone, *Clin. Immunol. Immunopathol.* 48:85 (1988).

48. D.M. Crist, G.T. Peake, L.T. MacKinnon, W.L. Sibbitt, Jr. and J.C. Kraner, Exogenous growth hormone treatment alters body composition and increases natural killer cell activity in women with impaired endogenous growth hormone secretion, *Metabolism* 36:1115 (1987).

49. M. Bozzola, A. Valtoria, A. Moretta, M. Cisternino, I. Biscaldi and M.R. Schimpff, In vitro and in vivo effect of growth hormone on cytotoxic activity, *J. Pediatr.* 117:596 (1990).

50. M. Vanderschueren-Lodeweyckx, B. Staf, H. Van Den Berghe, E. Eggermont and R. Eeckels, Growth hormone and lymphocyte transformation, *Lancet* 1:441 (1973).

51. A. Astaldi, Jr., B. Yalcin, G. Meardi, G.R. Burgio, R. Merolla and G. Astaldi, Effects of growth hormone on lymphocyte transformation in cell culture, *Blut* 26:74 (1973).

52. M. Bozzola, A. Valtorta, A. Moretta, D. Montagna, R. Maccario and G.R. Burgio, Modulating effect of growth hormone (GH) on PHA-induced lymphocyte proliferation, *Thymus* 12:157 (1988).

53. J.H. Jepson and E.E. McGarry, Hemopoiesis in pituitary dwarfs treated with human growth hormone and testosterone, *Blood* 39:238 (1972).

54. J. Blatt, S. Wenger, S. Stitely and P.A. Lee, Lack of mitogenic effects of growth hormone on human leukemic lymphoblasts, *Eur. J. Pediatr.* 146:257 (1987).

55. Z. Estrov, R. Meir, Y. Barak, R. Zaizov and Z. Zadik, Human growth hormone and insulin-like growth factor-1 enhance the proliferation of human leukemic blasts, *J. Clin. Oncol.* 9:394 (1991).

56. D.W. Golde, N. Bersch and C.H. Li, Growth hormone modulation of murine erythroleukemia cell growth in vitro, *Proc. Natl. Acad. Sci. USA* 75:3437 (1978).

57. K.E. Mercola, M.J. Cline and D.W. Golde, Growth hormone stimulation of normal and leukemic human T-lymphocyte proliferation in vitro, *Blood* 58:337 (1981).

58. S. Merchav, J. Tatarsky and Z. Hochberg, Enhancement of human granulopoiesis in vitro by biosynthetic insulin-like growth factor 1/somatomedin C and human growth hormone, *J. Clin. Invest.* 81:791 (1988).

59. T.G. Baier, E.W. Jenne, W. Blum, D. Schonberg and K.K.P. Hartmann, Influence of antibodies against IGF-I, insulin or their receptors on proliferation of human acute lymphoblastic leukemia cell lines, *Leuk. Res.* 16:807 (1992).

60. Z. Zadik, Z. Estrov, Y. Karov, T. Hahn and Y. Barak, The effect of growth hormone and IGF-I on clonogenic growth of hematopoietic cells in leukemic patients during active disease and during remission - A preliminary report, *J. Pediatr. Endocrinol.* 6:79 (1993).

61. S.M. Shalet, Leukaemia in children treated with growth hormone, *J. Pediatr. Endocrinol.* 6:109 (1993).

62. S. Watanabe, S. Mizuno, L.H. Oshima, Y. Tsunematsu, J. Fujimoto and A. Komiyama, Leukemia and other malignancies among GH users, *J. Pediatr. Endocrinol.* 6:99 (1993).

63. G.P. Redmond, R. Rapaport, S. Salisbury, R. David, J. Rao and S. Oberfield, Leukemia in growth hormone deficient (GHD) children who did not receive growth hormone (GH), *Pediatr. Res.* 31:83A (1992).

64. B. Tedeschi, G.L. Spadoni, M.L. Sanna, P. Vernole, D. Caporossi, S. Cianfarani, B. Nicoletti and B. Boscherini, Increased chromosome fragility in lymphocytes of short normal children treated with recombinant human growth hormone, *Hum. Genet.* 91:459 (1993).

65. J. Rovensky, M. Vigas, J. Lokai, P. Cuncik, P. Lukac and A. Takac, Effect of growth hormone on the metabolic activity of phagocytes of peripheral blood in pituitary dwarfs and acromegaly, *Endocrinol. Exp.* (Bratisl) 16:128 (1982).

66. J. Rovensky, J. Ferencikova, M. Vigas and P. Lukac, Effect of growth hormone on the activity of some lysosomal enzymes in neutrophilic polymorphonuclear leukocytes of hypopituitary dwarfs, *Int. J. Tissue React.* 7:153 (1985).

67. C.J. Wiedermann, M. Niedermuhlbichler, D. Geissler, H. Beimpold and H. Braunsteiner, Priming of normal human neutrophils by recombinant human growth hormone, *Brit. J. Haematol.* 78:19 (1991).

68. G.L. Spadoni, A. Spagnoli, S. Cianfarani, D. Del Principe, A. Menichelli, S. Di Giulio and B. Boscherini, Enhancement by growth hormone of phorbol diester-stimulated respiratory burst in human polymorphonuclear leukocytes, *Acta Endocrinol. (Copenh)* 124:589 (1991).

69. C.J. Wiedermann, M. Niedermuhlbichler, H. Beimpold and H. Braunsteiner, In vitro activation of neutrophils of the aged by recombinant human growth hormone, *J. Infect. Dis.* 164:1017 (1991).

70. C.K. Edwards, III, S.M. Ghiasuddin, J.M. Schepper, L.M. Yunger and K.W. Kelley, A newly defined property of somatotropin: Priming of macrophages for production of superoxide anion, *Science* 239:769 (1988).

71. Y.K. Fu, S. Arkins, G. Fuh, B.C. Cunningham, J.A. Wells, S. Fong, M.J. Cronin, R. Dantzer and K.W. Kelley, Growth hormone augments superoxide anion secretion of human neutrophils by binding to the prolactin receptor, *J. Clin. Invest.* 89:451 (1992).

72. Z. Hocherg, P. Hertz, G. Maor, J. Oiknine and M. Aviram, Growth hormone and insulin-like growth factor-I increase macrophage uptake and degradation of low density lipoprotein, *Endocrinology* 18:430 (199).

73. M.R. Pandian and G.P. Talwar, Effect of growth hormone on the metabolism of thymus and on the immune response against sheep erythrocytes, *J. Exp. Med.* 134:1095 (1971).

74. E. Mocchegiani, P. Paolucci, A. Balsamo, E. Cacciari and N. Fabris, Influence of growth hormone on thymic endocrine activity in humans, *Horm. Res.* 33:248 (1990).

75. J. Timsit, W. Savino, B. Safieh, P. Chanson, M.C. Gagnerault, J.F. Bach and M. Dardenne, Growth hormone and insulin-like growth factor-I stimulate hormonal function and proliferation of thymic epithelial cells, *J. Clin. Endocrinol. Metab.* 75:183 (1992).

76. A.J. Ammann, R.J. Duquesnoy and R.A. Good, Endocrinological studies in ataxia-telangiectasia and other immunological deficiency disease, *Clin. Exp. Immunol.* 6:587 (1970).

77. A.J. Ammann, W. Sutliff and E. Millinchick, Antibody-mediated immunodefiency in short-limbed dwarfism, *J. Pediatr.* 84:200 (1974).

78. J. Spranger, G.K. Hinkel, H. Stoss, W. Thoenes, D. Eargowski and F. Zepp, Schimke immuno-osseous dysplasia: A newly recognized multisystem disease, *J. Pediatr.* 119:64 (1991).

79. A.D.B. Webster, D. Barnes, C.F. Arlett, A.R. Lehmann and T. Lindahl, Growth retardation and immunodeficiency in a patient with mutations in the DNA ligase l gene, *Lancet* 339:1508 (1992).

80. T.A. Fleisher, R.M. White, S. Broder, S.P. Nissley, R.M. Blaese, J.J. Mulvihill, G. Olive and T.A. Waldmann, X-linked hypogammaglobulinemia and isolated growth hormone deficiency, *N. Engl. J. Med.* 302:1429 (1980).

81. K.W. Sitz, A.W. Burks, L.W. Williams, S.F. Kemp and R.W. Steele, Confirmation of X-linked hypogammaglobulinemia with isolated growth hormone deficiency as a disease entity, *J. Pediatr.* 116:292 (1990).

82. V. Monafo, M. Maghnie, L. Terracciano, A. Valtorta, M. Massa and F. Severi, X-linked agammaglobulinemia and isolated growth hormone deficiency, (Case report) *Acta Paediatr. Scand.* 80:563 (1991).

83. M.E. Conley, A.W. Burks, H.G. Herrod and J.A. Puck, Molecular analysis of X-linked agammaglobulinemia with growth hormone deficiency, *J. Pediatr.* 119:392 (1991).

84. N.W. Wilson, J. Daaboul and J.F. Bastian, Association of autoimmunity with IgG2 and IgG4 subclass deficiency in a growth hormone-deficient child, *J. Clin. Immunol.* 10:330 (1990).

85. M.L. Tang and A.S. Kemp, Growth hormone deficiency and combined immunodeficiency, *Arch. Dis. Child.* 68:231 (1993).

86. R. Rapaport, Endocrine abnormalities in children with HIV infections, *in:* "Management of HIV infections in infants and children," R. Yogev and E. Connor, eds., Mosby-Year Book Inc. (1992)

87. L.C. Alvarez, C.O. Dimas, A. Castro, L.G. Rossman, E.F. Vanderlaan and W.P. Vanderlann, Growth hormone in malnutrition, *J. Clin. Endocr.* 34:400 (1972).

88. A.C. Ferguson, Prolonged impairment of cellular immunity in children with intrauterine growth retardation, *J. Pediatr.* 93:52 (1978).

89. D.N. McMurray, R.R. Watson and M.A. Reyes, Effect of renutrition on humoral and cell-mediated immunity in severely malnourished children, *Am. J. Clin. Nutr.* 34:2117 (1981).
90. A.T. Soliman, A.I. Hassan, M.K. Aref, R.L. Hintz, R.G. Rosenfeld and A.R. Rogol, Serum Insulin-like growth factors I and II concentrations and growth hormone and insulin responses to arginine infusion in children with protein-energy malnutrition before and after nutritional rehabilitation, *Pediatr. Res.* 20:112 (1986).
91. M.A. Lesniak, J. Roth, P. Gorden and J.R. Gavin, III, Human growth hormone radioreceptor assay using cultured human lymphocytes, *Nat. New. Biol.* 241:20 (1973).
92. S. Arrenbrecht, Specific binding of growth hormone to thymocytes, *Nature* 25:255 (1974).
93. R. Eshet, S. Manheimer, P. Chobsieng and Z. Laron, Human growth hormone receptors in human circulating lymphocytes, *Horm. Metab. Res.* 7, 352 (1975).
94. W. Kiess and O. Butenandt, Specific growth hormone receptors on human peripheral mononuclear cells. Reexpression, identification and characterization, *J. Clin. Endocrinol. Metab.* 60:740 (1985).
95. C. Stewart, S. Clejan, L. Fugler, T. Cheruvanky and P.J. Collipp, Growth hormone receptors in lymphocytes of growth hormone-deficient children, *Arch. Biochem. Biophys.* 220:309 (1983).
96. L.J. Murphy, F. Vrhovsek and L. Lazarus, Identification and characterization of specific growth hormone receptors in cultured human fibroblasts, *J. Clin. Endocrinol. Metab.* 57:1117 (1983).
97. G.A. Werther, K.M. Haynes, D.R. Barnar and M.J. Waters, Visual demonstration of growth hormone rectptors on human growth plate chondrocytes, *J. Clin. Endocrinol. Metab.* 70:1725 (1990).
98. G.A. Werther, K. Haynes and M.J. Waters, Growth hormone (GH) receptors are expressed on human fetal mesenchymal tissues - Indentification of messenger ribonucleir acid and GH binding protein, *J. Clin. Endocrinol. Metab.* 76:1638 (1993).
99. W.N. Rom, P. Basset, G.A. Fells, T. Nukiwa, B.C. Trapnell and R.G. Crystal, Alveolar macrophages release an insulin-like growth factor I-type molecule, *J. Clin. Invest.* 82:1685 (1988).
100. R.G. Rosenfeld and L.A. Dollar, Characterization of the somatomedin-C/insulin like growth factor I (SMC/IGF-I) receptor on cultured human fibroblast monolyers: Regulation of receptor concentrations by SM-C/IGFI and insulin, *J. Clin. Endocrinol. Metab.* 55:434 (1982).
101. C.A. Stuart, R.T. Meehan, L.S. Neale, N.M. Cintron and R.W. Furlanetto, Insulin-like growth factor-I binds selectively to human peripheral blood monocytes and B-lymphocytes, *J. Clin. Endocrinol. Metab.* 72:1117 (1991).
102. B.H. Petersen, P. Barrett, L. Green and R. Rapaport, Flow cytometric detection of growth hormone receptors on human peripheral blood cells, (Presented at the 75th Annual Meeting of the Endocrine Society Annual Meeting, Abstract Book), (1993).
103. S.C. Duck, H.P. Schwartz, G. Costin, R. Rapaport, S. Arslanian, A. Hayek, M. Connors and J. Jarmillo, Subcutaneous growth hormone-releasing hormone therapy in growth hormone deficient children: first year therapy, *J. Clin. Endocrinol. Metab.* 75:1115 (1992).
104. M. Pawlikowski, P. Zelazowski, K. Dohler and H. Stepien, Effects of two neuropeptides, somatoliberin (GRF) and corticoliberin (CRF), on human lymphocyte natural killer activity, *Brain. Behav. Immun.* 2:50 (1988).
105. P. Zelazowski, K.D. Dohler, H. Stepien and M. Pawlikowski, Effect of growth hormone-releasing hormone on human peripheral blood leukocyte chemotaxis and migration in normal subjects, *Neuroendocrinology* 50:236 (1989).
106. A. Valtorta, A. Moretta, M. Maccario, M. Bozzola and F. Severi, Influence of growth hormone-releasing hormone (GHRH) on phytoemagglutinin-induced lymphocyte activation: Comparison of two synthetic forms, *Thymus* 18:51 (1991).
107. J.E. Beach, R.C. Smallridge, C.A. Kinzer, E.W. Bernton, J.W. Holaday and H.G. Fein, Rapid release of multiple hormones from rat pituitaries perfused with recombinant interleukin-1, *Life Sci.* 44:1 (1989).
108. E.W. Bernton, J.E. Beach, J.W. Holaday, R.C. Smallridge and H.G. Fein, Release of multiple hormones by a direct action of interleukin-1 on pituitary cells, *Science* 238:519 (1987).
109. V. Rettori, J. Jurcovicova and S.M. McCann, Central action of interleukin-1 in altering the release of TSH, growth hormone, and prolactin in the male rat, *J. Neurosci. Res.* 18:179 (1987).
110. M.B. Atkins, J.A. Gould, M. Allegretta, J.J. Li, R.A. Dempsey, R.A. Rudders, D.R. Parkinson, S. Reichlin and J.W. Mier, Phase I evaluation of recombinant interleukin-2 in patients with advanced malignant disease, *J. Clin. Oncol.* 4:1380 (1986).
111. R.M. Schimpff and A.M. Repellin, Production of interleukin-1-alpha and interleukin-2 by

mononuclear cells in healthy adults in relation to different experimental conditions and to the presence of growth hormone, *Horm. Res.* 33:171 (1990).

112. R. Rapaport, J. Oleske, S. Schenkman, J. Churchill and C. Kirkpatrick, Growth hormone deficiency: Interleukin-2 (IL2) and immune function, *Pediatr. Res.* 19:613 (1985).

113. K. Lyson and S.M. McCann, The effect of interleukin-6 on pituitary hormone release in vivo and in vitro, *Neuroendocrinology* 54:262 (1991).

114. N.L. Spangelo, A.M. Judd, P.C. Isakson and R.M. MacLeod, Interleukin-6 stimulates anterior pituitary hormone release in vitro, *Endocrinology* 125:575 (1989).

115. P. Carmeliet, W. Vankelecom, J. Van Damme, A. Billiau and C. Denef, Release of interleukin-6 from anterior pituitary cell aggregates: developmental pattern and modulation by glucocorticoids and forskolin, *Neuroendorcinology* 53:29 (1991).

116. B.L. Spangelo, R.M. MacLeod and P.C. Isakson, Production of interleukin-6 by anterior pituitary cells in vitro, *Endocrinology* 126:582 (1990).

117. B.L. Spangelo, P.C. Isakson and R.M. MacLeod, Production of interleukin-6 by anterior pituitary cells is stimulated by increased intracellular adenosine 3', 5'-monophosphate and vasoactive intestinal peptide, *Endocrinology* 127:403 (1990).

118. B.L. Spangelo, A.M. Judd, P.C. Isakson and R.M. MacLeod, Interleukin-1 stimulates interleukin-6 release from rat anterior pituitary cells in vitro, *Endocrinology* 128:2685 (1991).

119. R.C. Gaillard, D. Turnill, P. Sappino and A.F. Muller, Tumor necrosis factor alpha inhibits the hormonal response of the pituitary gland to hypothalamic releasing factors, *Endocrinology* 127:101 (1990).

120. P.E. Walton, M.J. Cronin, Tumor necrosis factor-alpha inhibits growth hormone secretion from cultured anterior pituitary cells, *Endocrinology* 125:925 (1989).

121. K.W. Kelley, The role of growth hormone in modulation of the immune response, *Ann. NY Acad. Sci.* 594:95 (1990).

122. S.M. McCann, M.C. Gonzalez, L. Milenkovic, S. Karanth, M.C. Aguila, W.L. Dees, K. Lyson and V. Rettori, The effect of stress and infection on pituitary hormone secretion, *Neuroendocrinol. Lett.* 15:33 (1993).

123. T.J. Merimee, M.B. Grant, C.M. Broder and L.L. Cavalli-Sforza, Insulin-like growth factor secretion by human B-lymphocytes: a comparison of cells from normal and pygmy subjects, *J. Clin. Endocrin. Metab.* 69:978 (1989).

124. M.E. Geffner, N. Bersch, B.M. Lippe, R.G. Rosenfeld, R.L. Hintz and D.W. Golde, Growth hormone mediates the growth of T-lymphoblast cell lines via locally generated insulin-like growth factor I, *J. Clin. Endocrin. Metab.* 71:464 (1990).

125. N. Hattori, A. Shimatsu, M. Sugita, S. Kumagai and H. Imura, Immunoreactive growth hormone (GH) secretion by human lymphocytes: augmented release by exogenous GH, *Biochem. Biophys. Res. Commun.* 168:936 (1990).

126. D.A. Weigent and J.E. Blalock, Expression of growth hormone by lymphocytes, *Intern. Rev. Immunol.* 4:193 (1989).

127. D.A. Weigent and J.E. Blalock, Growth hormone releasing hormone production by rat leukocytes, *J. Neuroimmunol.* 29:1 (1990).

128. D.A. Weigent and J.E. Blalock, Growth hormone and the immune system, *PNEI Review* 3:231 (1990).

129. D.A. Weigent and J.E. Blalock, The production of growth hormone by subpopulations of rat mononuclear leukocytes, *Cell. Immunol.* 134:001 (1991).

130. N. Hattori, K. Shimomura, T. Ishihara, K. Moridera, M. Hino, K. Ikekubo and H. Kurahachi, Growth hormone (GH) secretion from human lymphocytes is up-regulated by GH, but not affected by insulin-like growth factor-I, *J. Clin. Endocrinol. Metab.* 76:937 (1993).

131. Symposium on GH-RH, GH and IGF-I: Basic and clinical advances. Final Program and Abstract Book, Serono Symposia, San Diego, CA (USA), December 9-12, 1993.

132. W.D. Denckla, Interactions between age and the neuro-endocrine and immune systems, *Fed. Proc.* 37:1263 (1978).

133. N. Fabris, W. Pierpaoli and E. Sorkin, Lymphocytes, hormones and aging, *Nature* 240:557 (1972).

134. S. Gillis, R. Kozak, M. Durante and M.E. Weksler, Decreased production of and response to T cell growth factor by lymphocytes from aged humans, *J. Clin. Invest.* 67:937 (1981).

135. D. Rudman, A.G. Feller, H.S. Nagaraj, G.A. Gergans, P.Y. Lalitha, A.F. Goldberg, R.A. Schlenker, L. Cohn, I.W. Rudman and D.E. Mattson, Effects of human growth hormone in men over 60 years old, *N. Engl. J. Med.* 323:1 (1990).

136. E. Corpas, S.M. Harman and M.R. Blackman, Human growth hormone and human aging, *Endo. Rev.* 14:20 (1993).

137. R.W. Pfaffle, G.E. DiMattia, J.S. Parks, M.R. Brown, J.M. Wit, M. Jansen, H. Van der Nat, J.L. Van Den Brande, M.G. Rosenfeld and H. A. Ingraham, Mutation of the POU-specific domain of Pit-1 and hypoituitarism without pituitary hypoplasia, *Science* 257:1118 (1992).
138. P.A. Kelly, J. Djiane, M.C. Postel-Vinay and M. Edery, The prolactin/growth hormoen receptor family, *Endocr. Rev.* 12:235 (1991).

INSULIN-LIKE GROWTH FACTORS AND IMMUNITY: AN OVERVIEW

Sean Arkins and Keith W. Kelley

Laboratory of Immunophysiology
Department of Animal Sciences
207 Plant & Animal Biotech. Lab.
1201 West Gregory Drive
University of Illinois
Urbana IL 61801

INTRODUCTION

The immune and neuroendocrine systems are now known to be linked in an active and bidirectional communication network. One of the most frequently advanced arguments for the existence of this network has been the demonstration of a molecular basis for this circuitry; i.e. the existence of common signal molecules and their receptors in both systems.[1,2] Our laboratory has been investigating the effects of growth hormone and other members of the somatolactogenic family on the immune system. In this area of investigation there is a variety of clinical and experimental data which provide dramatic and direct evidence of neuroendocrine-immune interactions. Genetically hypopituitary species have reduced thymic and splenic weights, decreased cellularity of lymphoid tissues and associated immune dysfunction. Animals which have been hypophysectomized demonstrate similar lymphoid atrophy and have reductions in a number of immune responses.[3-8] These observations have led to the hypothesis that pituitary hormones play an integral role in maintenance of reticuloendothelial tissues and regulation of immune responses. Several workers have demonstrated that growth hormone (GH) has a profound influence on a number of immune events.[5,9,10] The molecular basis of this particular neuroendocrine-immune circuit has been demonstrated by the findings that lymphocytes have receptors for GH,[11] that interleukins regulate pituitary GH production,[12-14] and that lymphocytes produce a molecule similar to pituitary GH.[9,15]

In 1957, Salmon and Daughaday[16] proposed that GH acts on skeletal tissues by inducing the formation of a direct acting intermediary growth factor, or somatomedin. This "somatomedin" hypothesis spawned an active area of research leading to definition of the somatomedins, or insulin-like growth factors (IGF), as anabolic mediators of GH actions *in vivo*.[17] The finding of IGF activity in a wide variety of conditioned media and tissue extracts led to the suggestion that the IGF's are not

Advances in Psychoneuroimmunology, Edited by
I. Berczi and J. Szélenyi, Plenum Press, New York, 1994

simply endocrine hormones but that they also function as autocrine or paracrine hormones.[18]

Structure of IGFs

Insulin-like growth factors I and II are the two major forms of IGF. IGF-I is the major intermediary through which GH exerts its biological effects on postnatal growth whereas IGF-II may perform comparable growth-promoting functions during fetal life.[19] Both IGF-I and IGF-II are single chain molecules with three intrachain disulfide bridges consisting of 70 (7.6 kd) and 67 (7.4 kd) amino acids, respectively. They have identical amino acids in 45 positions, giving a sequence homology of 64%, which suggests a common evolutionary precursor. These peptides also share 50% structural homology to proinsulin. This structural relationship to insulin and the similar *in vitro* actions of IGF's and insulin were the foundations for the terminology "insulin-like growth factors".[17] These three peptides and the more distant relaxin molecule comprise the insulin gene family.

The primary structures of seven mammalian IGF-I species have now been characterized from either the protein or the cDNA sequence (reviewed by Daughaday and Rotwein,[17] Sara and Hall,[18] Rotwein[20]). All mature mammalian IGF-I molecules characterized, to-date, are 70 amino acid residues and each contains an amino terminal B domain (29 aa), a C region (12 aa), an A domain (21 aa) and a carboxy terminal D domain (8 aa). Human, bovine, porcine, ovine and guinea pig IGF-I are identical and rat and mouse IGF-I differ from these by three and four amino acids, respectively. In all mammalian IGF-Is, there exist only five amino acid substitutions of which 3 are relatively conservative.[20] This degree of conservation is truly remarkable, exceeding even that of the insulin protein. Since IGF-I is a secreted peptide, it is synthesized as a precursor molecule, containing a signal peptide which, presumably, directs the nascent chain into the endoplasmic reticulum. The structures of the IGF-I signal peptides determined to date also demonstrate a high degree of amino acid homology, although in a number of species signal peptides of varying length are associated with the mature peptide sequence. A variant form of IGF-I, lacking an amino terminal tripeptide, has been isolated from human brain[21] and bovine colostrum.[22] This variant, termed des-IGF-I, which is thought to arise from post-transcriptional modification of proIGF-I, shows a weak binding affinity for IGF binding proteins (IGFBP) and may represent the locally acting paracrine or autocrine form of IGF-I.

IGF Receptors

The biological actions of the insulin-like growth factors are mediated through their interaction with cell surface receptors. The IGF-I receptor, also known as the type I receptor, binds IGF-I and IGF-II with relatively equal affinity, and also binds insulin but at 100-fold higher concentrations of this ligand. Many of the effects of IGF-II and insulin may be mediated through interaction with this receptor.[23] The IGF-II receptor, or type II receptor, is a single-chain (260 Kd) cation independent mannose-6-phosphate receptor and, unlike the insulin and IGF-I receptors, it lacks tyrosine kinase activity and mediates signal transduction by interaction with G proteins. The type II receptor binds IGF-II with higher affinity than IGF-I and does not bind insulin.[24] In turn, the insulin receptor has an affinity for IGF-I and IGF-II which is 100 times lower than that for insulin.[25]

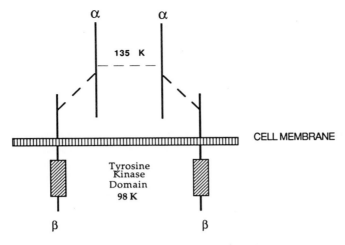

Figure 1. IGF-I receptor. Adapted from Sara and Hall.[17]

The IGF-I receptor is a tetrameric glycoprotein with a molecular weight between 250 to 300 kd and consists of two extracellular α subunits (~130 kd) and two transmembrane β subunits (~98 kd; Fig. 1) with 3 interchain disulfide bonds. This β-α-α-β tetramer receptor complex has both structural and functional similarities to the insulin receptor. Like the insulin receptor, the Type I receptor is synthesized as a single precursor polypeptide which is then glycosylated and cleaved into α and β subunits. The putative hormone binding site is located in the extracellular α subunits and possibly also in the extracellular domain of the β subunits. Intracellular signal transduction is mediated by the auto-phosphorylation of tyrosine residues within the intracellular β subunits.[25] Autophosphorylation of the receptor, an early event, enhances the kinase activity resulting in enhanced phosphorylation of the cellular proteins which are the presumed physiological intermediates in the actions of insulin and IGF-I.[23]

IGF-I Receptors on Leukocytes

The biological activity of any hormone depends on the ability of the target cell to respond to this hormone. This ability is dependent on the existence of both cellular receptors and on intracellular post-receptor signalling pathways. As shown in Table 1, there is a large body of evidence suggesting the existence of IGF-I receptors on leukocytes. Receptors for IGF-I been demonstrated on non-fractionated mononuclear cells, T and B cell lines, pre T cells and granulocytes. However, the presence of receptors, per se, does not confirm function. The kinetics and function of IGF receptor expression on purified populations of human T lymphocytes activated by sepharose-anti-CD3 and IL-2 were recently investigated by Johnson et al.[26] Both CD4$^+$ and CD8$^+$ T cells increased their expression of IGF-I and IGF-II receptors in a sequential manner after activation via the TCR/CD3 complex. Expression of the type I receptor increased early during the activation process, reaching a maximum level at or before the peak of cellular proliferation, whereas type II receptor expression increased later and more transiently in the proliferation cycle. Expression of the insulin receptor remained low throughout the activation period, a result that conflicts with a previous report employing lectin-activated lymphocytes.[27] These data are, however, in broad agreement with those of Kozak et al.[28] and Tapson et al.[29] who assessed IGF-I binding on CD3[+] human T cells and found that upon activation with phytohemagglutinin (PHA), receptor number increased from 45 to 330 sites per cell, with no apparent change in affinity. Thus, IGF receptors appear to belong to the family of "early" antigens which appear on the cell surface on or before the peak of DNA synthesis, and which includes the IL-2 and transferrin receptors.

The intracellular activation signals mediated by IGF-I in leukocytes are largely unknown, although IGF-I and IGF-II, and insulin at supra-physiological concentrations can induce amino acid transport and tyrosine kinase activity in rat thymocytes[30] and IGF-I receptors on B cell line are susceptible to down regulation by IGF-I.[31]

Table 1. IGF-I receptors on leukocytes.

Cell type	Reference
Human mononuclear cells	Thorsson and Hintz[32], Kooijman et al.[33]
Newborn mononuclear cells	Rosenfeld et al.[34]
Rat thymocytes	Verland and Gammeltoft[30]
Mouse thymoma cell lines	Verland and Gammeltoft[30]
Human B lymphoblastoid IM9 line	Rosenfeld and Hintz[31]
Human promyelocyte cell line (HL-60)	Pepe et al.[27]
Human myeloblast line KG1	Pepe et al.[27]
Human leukemic lymphoblasts	Lee et al.[35]
Human T cells	Tapson et al.[29]
Lectin activated human lymphocytes	Kozak et al.[28], Shimpff et al.[36]
Anti-CD3 activated human T lymphocytes	Johnson et al.[26]
Murine X-ray induced thymic lymphoma	Gjerset et al.[37]
Human monocytes	Pepe et al.[27]
Human granulocytes	Pepe et al.[27]

IGF Binding Proteins

Despite its relatively low molecular weight, all of the IGF activity in serum is associated with large molecular weight proteins and little or no activity can be identified in molecular weight fractions corresponding to the free IGF peptides.[38] At least six different categories of IGF-binding proteins (IGFBPs) are now known to be present in mammalian serum.[39] The IGFBPs seem to comprise a new family of proteins, distinct from that of the IGF receptor family, with a good degree of conserved similarity in structure and function. The amino acid homologies for each IGFBP are greater than 70% across species and within a species the IGFBPs show greater than 50% homology, suggesting a common evolutionary precursor gene.[40] The potential role of IGF binding proteins in immunoregulation is only beginning to be explored.

IGF-I and Proliferation of Leukocytes

In both *in vivo* and *in vitro* situations the long term effects of IGF-I are anabolic and result in the promotion of growth.[18,41,42] Not surprisingly then, the primary *in vitro* effect of IGF-I is its ability to stimulate DNA synthesis. The control of cell proliferation is determined in large part by a series of regulatory events which occur in the pre-synthesis or G_1 phase of the cell cycle. In order for cells to be able to respond to IGF-I and move from G_0 to enter the G_1 phase, they must first acquire competence by exposure to PDGF, FGF or other growth factors.[43,44] Following induction of competence, approximately 12 h are required for BALB/c-3T3 cells to progress through G_1 to the S phase. IGF-I was identified by Stiles et al.[43] to be a progression factor in the cell cycle, allowing cells to progress through G_1 to the S phase of the cell cycle. While these studies were performed on density arrested BALB/c-3T3 cells, IGF-I also appears to be necessary in the G_1 phase of exponentially growing cells[44,45] since IGF-I in physiological concentrations can replace the serum requirement for commitment and progression to the S phase. The nature of the cellular signal triggered by IGF-I, acting as a progression factor, is largely unknown. Treatment of cells arrested at the late stage or, V point, of G_1 with an inhibitor of RNA synthesis does not inhibit IGF-I mediated progression to the S phase, suggesting that IGF-I does not need to induce the transcription of new mRNAs for its progression activity.[45] The former study showed that IGF-I-stimulated progression was, however, associated with post-translational modification of some unidentified proteins. These data have led to the current consensus that IGF-I acts as a progression factor by stimulating post-translational modification of proteins rather than inducing new protein synthesis.[44]

Since cellular proliferation is one of the more accessible measurement parameters, there is now a large body of data showing an effect of IGF-I on the basal and mitogen-induced proliferation of a range of leukocytes. These *in vitro* results were heralded by the observation that the stimulating effect of serum on lymphocyte mitogenesis is increased in acromegalics and decreased in pituitary dwarfs.[36] IGF-I has anabolic effects, such as the induction of macromolecular synthesis, and enhancement of intermediary metabolism[30,42] and DNA synthesis,[36] in mitogen-activated lymphocytes cultured in sub-optimal serum levels. The latter results have since been confirmed in a variety of systems; e.g. purified human T cells,[26,29] human leukemic blasts[27,46,47] and human and murine thymic lymphoma cells.[37,48]

These findings have not been universal, however, since Rao et al.[49] found no differences in either resting or PHA-stimulated proliferation in lymphocytes cultured in either normal serum or sera from GH-deficient children and, further, could not detect any differences in somatomedin content of the various cultures. Similarly, while Verland and Gammeltoft[30] demonstrated functional IGF-I receptors on murine thymocytes, they failed to demonstrate a proliferative response to IGF-I either before or after mitogen treatment. The latter result may be a reflection of the cellular developmental stage since Gjerset et al.[37] recently found that a murine pre-thymic lymphoma cell proliferated to IGF-I but was unresponsive to a number of interleukins and hematopoietic growth factors and Smith et al.[48] found that CD3- and CD25- human malignant lymphoblasts responded to IGF-I, under hypoxic conditions, resulting in the long term-establishment of cell lines from patients with acute lymphoblastic leukemias. Tapson et al.[29] have also found IGF-I to be mitogenic for both unstimulated and PHA-activated human CD4+ and CD8+ cells whereas Herzberg and Smith[50] found no effect of IGF-I on human lymphocyte IL-2 production or antigen-driven proliferation. Such divergent observations are difficult to reconcile but at least some of the discrepancies can be explained on the basis of culture method[46] where the inclusion of serum with its high content of IGF-I and IGFBPs is a confounding factor. Indeed, a complete range of responses, including suppression of proliferation and *in vitro* antibody formation,[51] have been reported for IGF-I, but, at least in this study, these results may be ascribed to the use of IGF-I concentrations far in excess of physiological concentrations.

In some of the above instances demonstrating enhancement of proliferation by IGF-I, the specificity of the signaling pathway has been demonstrated by the use of an antibody directed against the human type I receptor which abrogates the IGF-I mediated enhancement.[26,27,46,52]

Differentiation of Hematopoietic Cells

The earliest evidence of a role of IGF-I in modulation of immune events comes from studies demonstrating an effect of relatively impure somatomedin preparations on hematopoietic cells. It was generally reported that colony formation of erythroid cells from embryonic mouse liver and adult bone marrow was stimulated by physiological IGF-I concentrations.[53] More recently, these findings have been corroborated in a number of laboratories with recombinant IGF-I. Phillips et al.[54] treated neonatal rats with IGF-I *in vivo* and found a significant increase in bone marrow erythropoietic cell precursors. Kurtz et al.[55] found that IGF-I stimulated erythropoiesis in hypophysectomized rats both directly and indirectly through the stimulation of increased erythropoietin. Werther et al.[56] recently demonstrated that these growth promoting effects were seen *in vitro* in the absence of either serum or erythropoietin. The effects of GH on erythropoiesis,[4] which include the promotion of both primitive and mature erythroid precursors, also appear to be mediated by macrophage-derived paracrine IGF-I since the GH-mediated enhancement of erythropoiesis is abrogated by an antibody directed to the IGF-I receptor.[57] This group also demonstrated an enhancement of granulopoiesis *in vitro* by both GH and IGF-I.[58] GH enhancement in this instance was dependent on the presence of bone marrow adherent cells, presumably as a source of paracrine IGF-I, and was abrogated by an antibody directed against the IGF-I receptor.

A role for IGF-I in macrophage differentiation has also been suggested by the results of Scheven and Hamilton[59] and, more recently, by Rodriguez-Tarduchy et al.[60] who found that the addition of IGF-I prevented apoptosis in IL-3-driven bone marrow cell lineages upon withdrawal of this cytokine. Our laboratory has recently demonstrated that CSF-1 differentiated macrophages undergo a 50- to 100-fold increase in the level of IGF-I mRNA transcripts during differentiation from bone marrow pluripotent progenitors,[61] suggesting an autocrine-paracrine role for IGF-I in the development of myeloid cells. Indeed, a generalized role for IGF-I in maintaining commitment to proliferation by preventing apoptosis in differentiating cell types is suggested by these results and those of Barres et al.[62]

Our group has also recently described a novel effect of IGF-I on phagocyte effector functions. We found that both IGF-I and GH were as potent as was IFN-γ in eliciting granulocyte superoxide anion production.[63,64] These results indicate a striking range of action for these growth promoting peptides on granulocytes. GH and IGF-I thus seem to be somewhat comparable to the colony stimulating factors (e.g. G-CSF and GM-CSF) which can influence both the differentiation and the effector functions of phagocytic cells.

A generalized effect of IGF-I on lymphoid organ size and cellularity was reported by Guler et al.[65] who found that subcutaneous infusions of IGF-I into hypophysectomized rats were more effective in restoring both spleen and thymic size than GH treatment. IGF-I has also been shown to have significant positive effects on Thy-1 antigen expression and thymocyte maturation in the atrophied thymus of diabetic rats.[66] These results strongly suggest that many of the effects of GH on lymphoid tissue[5] may be mediated by the paracrine synthesis of IGF-I.

Synthesis of IGF-I by Leukocytes

The enhancement of paracrine IGF-I biosynthesis in response to local tissue injury suggests that IGF-I is involved in wound repair by stimulating anabolism and growth of new cells.[18] Animals deficient in IGF-I have been shown to be deficient in wound repair.[67] A possible immunological basis for these phenomena is suggested by the findings of Tapson et al.[29] who showed that IGF-I has both chemoattractant and chemokinetic effects on resting and mitogen-activated human T cells. Recently, Mueller et al.,[67] but not Lynch et al.,[68] showed that local infusions of IGF-I into the wound chambers of hypophysectomized rats increased the overall number of wound leukocytes and macrophages and DNA content of the wound exudate. Since macrophages secrete a range of cytokines which promote fibroblast growth and neovascularization,[69] such a chemoattractant action might obviously be beneficial. In addition, there is now strong evidence, reviewed below, that leukocytes and macrophages in particular, can produce and secrete IGF-I.

As outlined above, there is now ample evidence that IGF-I may influence immune events through a paracrine circuit. Merchav et al.[57,58] and Werther et al.[56] found that GH enhancement of erythropoiesis was mediated through local synthesis of IGF-I. Geffner et al.[52] and Timsit et al.[70] found a similar paracrine circuit in T lymphoblast cell lines and thymic epithelial cells, respectively. Support for such an autocrine/paracrine circuit is also provided by the demonstration by Neely et al.[71] that some human leukemic T and B lymphoblasts produce insulin-like binding proteins 2 and 4. IGFBP production by non-transformed immune cells has not, to our knowledge, been investigated.

Paracrine synthesis of IGF-I in response to GH treatment of thymic epithelial cultures (TEC) is also suggested by the results of Timsit et al.[70] This group found that the enhancement of thymulin secretion from TEC mediated by GH or IGF-I was abrogated by an antibody to either IGF-I or the IGF-I receptor. Such a local synthesis is also supported by the immunohistochemical demonstration by D'Ercole et al.[72] of IGF-I in the thymus of adult rats where its concentration was decreased by 60% following hypophysectomy. Geffner et al.[52] also provided convincing data that the enhancement of T-lymphoblast cell lines by GH was actually mediated by local synthesis of IGF-I and could be abrogated by antibodies directed against either IGF-I or the IGF-I receptor.

The cell type which actually produces IGF-I at local immune foci has, however, received very little attention, until recently. The emerging picture is that a number of immune cell types are capable of producing IGF-I. Merimee et al.[73] found that transformed, but not non-transformed, B cells released IGF-I into the culture medium in response to stimulation with GH. Geffner et al.[52] also provided convincing data that the enhancement of T-lymphoblast cell lines by GH was actually mediated by local synthesis of IGF-I since this could be abrogated by antibodies directed against either IGF-I or the IGF-I receptor. This group also found low but measurable levels of IGF-I in the medium conditioned by these cells. Neely et al.,[71] however, found low to negligible levels of IGF-I and IGF-II in the conditioned medium of human leukemic T and B lymphoblasts but did not investigate the response to GH.

One group has pioneered the study of IGF-I production by non-transformed human cells. In 1988, Rom et al.[74] first characterized a growth factor for fibroblasts from alveolar macrophages of humans with asbestosis. This alveolar macrophage-derived growth factor (AMDGF) had an apparent molecular mass of 26 kd but displaced ^{125}I-IGF-I from its receptor in a binding assay and stimulated IGF-I receptors to phosphorylate a tyrosine-containing artificial substrate. AMDGF stimulated fibroblast proliferation in a serum-free complementation assay and this activity was inhibited in a dose-responsive manner with an anti-IGF-I monoclonal antibody. An antisense RNA transcript, corresponding to exons 1, 2 and 3 of the human IGF-I gene, hybridized to alveolar macrophage mRNA but not to monocyte mRNA, suggesting that the ability to produce IGF-I message might be a reflection of differentiation status. More recently, this group[75] examined the regulation of IGF-I mRNA expression in the human macrophage-like cell line U937. Stimulation with either phorbol myristate acetate (PMA) or a calcium ionophore (A23187) increased the transcription rate of the IGF-I gene four- to five-fold but, surprisingly, resulted in a dramatic decline in cytoplasmic IGF-I mRNA levels. Another surprising aspect of this study was that surface activation of U937 cells by either PMA or the calcium ionophore caused a rapid release of IGF-I, even in the presence of a protein synthesis inhibitor. These results make the unprecedented suggestion that mononuclear phagocytes have a pre-formed, stored, releasable pool of IGF-I.

In the murine system, a number of authors have demonstrated IGF-I mRNA in spleen and thymus by Northern blotting, but in most instances these experiments were performed on crudely homogenized tissues with no effort being made to study IGF-I transcripts in purified cell populations. To-date, there has been only one report[69] demonstrating IGF-I mRNA production by defined leukocyte populations and this also concerned activated wound macrophages where the transcripts were reverse transcribed and amplified by the polymerase chain reaction (PCR). There also exists only one report to-date[76] on the actual production of IGF-I by non-transformed murine cells. This group found IGF-I peptide by direct immunofluorescence, HPLC analysis and a fibroblast bioassay in rat splenocytes. In addition, GH treatment for 24 h induced a 2-fold increase in IGF-I. A recent description of the *in vivo* and *in vitro* down-regulatory effects of the pleiotropic inflammatory mediator, IL-1β, on Leydig cell IGF-I gene expression and steroidogenesis[77] also argues for the existence of a shared regulatory pathway involving cytokines and the autocrine/paracrine production of IGF-I.

We recently provided the first characterization of IGF-I transcripts in cells of the immune system.[61] We used reverse transcription and polymerase chain reaction amplification (RT-PCR) with a variety of leukocytes and purified cell populations to demonstrate that all cells of the immune system can produce IGF-I mRNA transcripts. Both IGF-I Ea and IGF-I Eb cDNAs were cloned and sequenced from purified macrophage populations, a murine macrophage cell line and from mouse liver. This is the first time that IGF-I Eb has been cloned from any cell type other than the liver.

Our results also showed that macrophage IGF-I transcripts arise primarily within exon 1, characteristic of extra-hepatic transcripts. We also defined relative transcript level in a number of leukocytes using an exon 4 specific probe in a solution hybridization assay. Macrophage and macrophage cell lines consistently produced 20 to 40 fold higher levels of IGF-I message than spleen, thymus, T cells or B cells. IGF-II transcripts could not be detected by the same protocol, suggesting that IGF-I is the predominant somatomedin produced by leukocytes. Finally, we used an affinity purified antibody and an enhanced chemiluminescence system in a Western blotting protocol to identify IGF-I protein in lysates of purified macrophages and macrophage cell lines. We identified a protein of approximately 25 Kd, which is larger than the mature protein but is in agreement with previously published reports.

CONCLUSIONS AND OVERVIEW

IGF-I is a peptide growth factor that enhances cellular replication and is, in addition, capable of stimulating cellular differentiation. IGF-I also exerts insulin-like effects on the intermediary metabolism of a variety of tissues. Originally postulated to be an endocrine mediator of GH stimulated anabolism, the significance of IGF-I in paracrine-mediated homeostasis is becoming increasingly appreciated. IGF-I has been shown to have significant *in vivo* and *in vitro* effects on leukocytes. In addition, many of the hitherto described effects of GH on immune function may be mediated by endocrine or paracrine IGF-I. There is strong evidence for such a paracrine circuit in the thymus,[69] bone marrow,[57,58,61,78] transformed B and T cell lines,[52,73] and in splenocytes.[76] In addition, there are now substantial data indicating that leukocytes produce IGF-I.[61,69,74,79] Since cells of the immune system have been shown to produce a variety of neuroendocrine hormones[1,2] including members of the somatolactogenic family[9,10] a short autocrine/paracrine loop is a strong possibility.

Of all of the classical neuroendocrine hormones, IGF-I is perhaps the most likely candidate for involvement in autocrine/paracrine regulation of immune cell functions. It has been shown to be produced by a wide variety of tissues, is involved in regulation of growth and metabolic processes and is involved in cellular differentiation and proliferation, all of which are key features of immune homeostasis. Indeed, it could be argued that immunologists have been inadvertently employing IGF-I for years by using insulin at supra-physiological concentrations in serum-free media, thus ensuring occupancy and stimulation of the IGF-I receptor. The consensus of present results strongly suggest that autocrine or paracrine IGF-I may be involved in the growth and differentiation of myeloid cells and could provide an excellent model for the study of the role of insulin-like growth factors in differentiation events.

ACKNOWLEDGMENTS

Supported by NIH Grant AG-06246 and USDA Grant 92-37206-7777.

REFERENCES

1. D.A. Weigent and J.E. Blalock, Interactions between the neuroendocrine and immune systems: Common hormones and receptors, *Immunol. Rev.* 100:79 (1987).
2. J.E. Blalock, A molecular basis for bidirectional communication between the immune and neuroendocrine systems, *Physiol. Rev.* 69:1 (1989).
3. P.E. Smith, Effect of hypophysectomy upon the involution of the thymus in the rat, *Anat. Rec.* 47:119 (1930).
4. Berczi and E. Nagy, Effects of hypophysectomy on immune function, *in:* "Functional Endocrine Pathology," vol. 2, K. Kovacs, S. Asa, eds., Blackwell Scientific Pub. Inc. (1991).
5. K.W. Kelley, Growth hormone, lymphocytes and macrophages, *Biochem. Pharmacol.* 38:705 (1989).
6. J.H. Exon, J.L. Bussiere and J.R. Williams, Hypophysectomy and growth hormone replacement effects on multiple immune responses in rats, *Brain, Behav. Immun.* 4:118 (1990).
7. R.R. Gala, Prolactin and growth hormone in the regulation of the immune system, *Proc. Soc. Exp. Biol. Med.* 198:513 (1991).

8. W.J.Murphy, S.K. Durum, M.R. Anver and D.L. Longo, Immunologic and hematologic effects of neuroendocrine hormones. Studies on DW/J dwarf mice, *J. Immunol.* 148:3799 (1992).

9. D.A. Weigent and J.E. Blalock, 1990, Growth hormone and the immune system, *PNEI* 3:231 (1990).

10. K.W. Kelley, S. Arkins and Y.M. Li, Growth hormone, prolactin and insulin-like growth factor: new jobs for old players, *Brain Behav. Immun.* 6:317 (1992).

11. S. Arrenbrecht, Specific binding of growth hormone to thymocytes, *Nature* 252:255 (1974).

12. V. Rettori, J. Jurcovicova and S.M. McCann, Central action of interleukin-I in altering the release of TSH, growth hormone and prolactin in the male rat, *J. Neurosci. Res.* 18:179 (1987).

13. P.E. Walton and M.J. Cronin, Tumor necrosis factor-a inhibits growth hormone secretion from cultured anterior pituitary cells, *Endocrinology* 125:925 (1989).

14. P.E. Walton and M.J. Cronin, Tumor necrosis factor-a and interferon-c reduce prolactin release *in vitro, Am. J. Physiol.* 259:(Endocrinol. Metab.22):E672 (1990).

15. D.A. Weigent, J.B. Baxter, W.E. Wear, L.R. Smith, K.L. Bost and J.E. Blalock, Production of immunoreactive growth hormone by mononuclear leukocytes, *FASEB J.* 2:2812 (1988).

16. D. Salmon and W.H. Daughaday, A hormonally controlled serum factor which stimulates sulphate incorporation by cartilage *in vitro. J. Lab. Clin. Med.* 49:825 (1957).

17. W.H. Daughaday and P. Rotwein, Insulin-like growth factors I and II. Peptide, messenger ribonucleic acid and gene structures, serum, and tissue concentrations, *Endocr. Rev.* 10:68 (1989).

18. V.R. Sara and K. Hall, Insulin-like growth factors and their binding proteins, *Physiol. Rev.* 70:591 (1990).

19. P. Rotwein, K.M. Pollock, M. Watson and J.D. Milbrandt, Insulin-like growth factor gene expression during rat embryonic development, *Endocrinology* 121:2141 (1987).

20. P. Rotwein, Structure, evolution and expression and regulation of insulin-like growth factors I and II, *Growth Factors* 5:3 (1991).

21. C. Carlsson-Skwirut, H. Jornvall, A. Holmgren, C. Andersson, T. Bergman, G. Lundquist, B. Sjoren and V.R. Sara, Isolation and characterization of variant IGF-I as well as IGF-II from adult human brain, *FEBS Lett.* 201:46 (1986).

22. G.L. Francis, L.C. Read, F.J. Ballard, C.J. Bagley, F.M. Upton, P.M. Gravestock and J.C. Wallace, Purification and partial sequence analysis of insulin-like growth factor-I from bovine colostrum, *Biochem. J.* 233:207 (1986).

23. H. Werner, M. Woloschak, B. Stannard, Z. Shen-Orr, C.T. Roberts, Jr. and D. LeRoith, The insulin-like growth factor I receptor: molecular biology, heterogeneity and regulation, *in* "Insulin-Like Growth Factors: Molecular and Cellular Aspects," D. LeRoith, ed., CRC Press, Boca Raton, FL (1991).

24. S.P. Nissley, W. Kiess and M.M. Sklar, The insulin-like growth factor-II/mannose 6-phosphate receptor, *in* "Insulin-Like Growth Factors: Molecular and Cellular Aspects," D. LeRoith, ed., CRC Press, Boca Raton FL (1991).

25. M.P. Czech, Signal transmission by the insulin-like growth factors, *Cell* 59:235 (1989).

26. E.W. Johnson, L.A. Jones and R.W. Kozak, Expression and function of insulin-like growth factor receptors on anti-CD3-activated human lymphocytes, *J. Immunol.* 148:63 (1992).

27. M.G. Pepe, N.H. Ginztion, P.D.K. Lee, R.L. Hintz and P.L. Greenberg, Receptor binding and mitogenic effects of insulin and insulin-like growth factors I and II for human myeloid leukemia cells, *J. Cell. Physiol.* 133:219 (1987).

28. R.W. Kozak, J.F. Haskell, L.A. Greenstein, M.M. Rechler, T.A. Waldmann and S.P. Nissley, Type I and II insulin-like growth factor receptors on human phytohemagglutinin-activated T lymphocytes, *Cell. Immunol.* 109:318 (1987).

29. V.F. Tapson, M. Boni-Schnetzler, P.F. Pilch, D.M. Center and J.S. Berman, Structural and functional characterization of the human T lymphocyte receptor for insulin-like growth factor I *in vitro, J. Clin. Invest.* 82:950 (1988).

30. S. Verland and S. Gammeltoft, Functional receptors for insulin-like growth factors I and II in rat thymocytes and mouse thymoma cells, *Mol. Cell. Endocrinol.* 67:207 (1989).

31. R.G. Rosenfeld and R.L. Hintz, Characterization of a specific receptor for somatomedin C (SM-C) on cultured human lymphocytes: Evidence that SM-C modulates homologous receptor concentration, *Endocrinology* 107:1841 (1980).

32. A.V. Thorsson and R.L. Hintz, Specific ^{125}I-somatomedin receptor on circulating human mononuclear cells, *Biochem. Biophys. Res. Comm.* 74:1566 (1977).

33. R. Kooijman, M. Willems, C.J.C. De Haas, G.T. Rijkers, A.L.G. Schuurmans, S.C. Van Buul-Offers, C.J. Heijnen and B.J.M. Zegers, Expression of type I insulin-like growth factor receptors on human peripheral blood mononuclear cells, *Endocrinology* 131:2244 (1992).

34. R.G. Rosenfeld, A.V. Thorsson and R.L. Hintz, Increased somatomedin receptor sites in newborn circulating human mononuclear cells, *J. Clin.Endocrinol. Metab.* 48:456 (1979).

35. P.D.K. Lee, R.G. Rosenfeld, R.L. Hintz and S.D. Smith, Characterization of insulin, insulin-like growth factors I and II and growth hormone receptors on human leukemic lymphoblasts, *J. Endocrinol. Metab.* 62:28 (1986).

36. R.-M. Schimpff, A.-M. Repellin, A. Salvatoni, G. Thieriot-Prevost and P. Chatelain, Effect of purified somatomedins on thymidine incorporation into lectin-activated human lymphocytes, *Acta Endocrinol.* 102:21 (1983).

37. R.A. Gjerset, J. Yeatgin, S.K. Volkman, V. Vila, J. Arya and M. Haas, Insulin-like growth factor-I supports proliferation of autocrine thymic lymphoma cells with a pre-T cell phenotype, *J. Immunol.* 145:3497 (1990).

38. E.R. Froesch, C. Schmid, J. Schwander and J. Zapf, Actions of insulin-like growth factors, *Ann. Rev. Physiol.* 47:443 (1985).

39. S. Shimasaki and N. Ling, Insulin-like growth factor binding proteins, *Growth Factors* 5:243 (1991).

40. G. Lamson, L.C. Giudice and R.G. Rosenfeld, Insulin-like growth factor binding proteins: structural and molecular relationships, *Growth Factors* 5:19 (1991).

41. J.J. Van Wyck, The somatomedins: Biological actions and physiologic control mechanisms, *in:* "Hormonal Proteins and Peptides," E.H. Li, ed., Academic Press, New York (1984).

42. J. Zapf, C. Schmid and E.R. Froesch, Biological and immunological properties of insulin-like growth factors (IGF) I and II, *Clin. Endocrinol. Metab.* 13:3 (1984).

43. C.D. Stiles, G.T. Capone, C.D. Scher, H.N. Antoniades, J.J. Van Wyk and W.J. Pledger, Dual control of cell growth by somatomedins and platelet-derived growth factor, *Proc. Natl. Acad. Sci. USA.* 76:1279 (1979).

44. W.L. Lowe, Jr., Biological actions of the insulin-like growth factors, *in* "Insulin-Like Growth Factors: Molecular and Cellular Aspects," D. LeRoith, ed., CRC Press, Boca Raton, FL (1991).

45. J. Campisi and A.B. Pardee, Post-transcriptional control of the onset of DNA synthesis by an insulin-like growth factor, *Mol. Cell. Biol.* 4:1807 (1984).

46. J. Sinclair, D. McClain and R. Taetle, Effects of insulin and insulin-like growth factor I on growth of human leukemia cells in serum-free and protein-free medium, *Blood* 72:66 (1988).

47. Z. Estrov, R. Meir, Y. Barak, R. Zaizov and Z. Zadik, Human growth hormone and insulin-like growth factor-I enhance the proliferation of human leukemic blasts, *J. Clin. Oncol.* 9:394 (1991).

48. S.D. Smith, P. McFall, R. Morgan, M. Link, F. Hecht, M. Cleary and J. Sklar, Long-term growth of malignant thymocytes *in vitro*, *Blood* 73:2182 (1989).

49. J.K. Rao, B.M. Gebhardt and S.L. Blethen, Somatomedin production and response to mitogen by lymphocytes in children with growth hormone deficiency, *Growth* 50:456 (1986).

50. V.L. Herzberg and K.A. Smith, T cell growth without serum, *J. Immunol.* 139:998 (1987).

51. P. Hunt and D.D. Eardley, Suppressive effects of insulin and insulin-like growth factor-I (IGFI) on immune responses, *J. Immunol.* 136:3994 (1991).

52. M.E.Geffner, N. Bersch, B.M. Lippe, R.G. Rosenfeld, R.L. Hintz and D.W. Golde, Growth hormone mediates the growth of T-lymphoblast cell lines via locally generated insulin-like growth factor-I, *J. Clin. Endocrinol. Metab.* 71:464 (1980).

53. A. Kurtz, W. Jelkmann and C. Bauer, A new candidate for the regulation of erythropoiesis: insulin-like growth factor I, *FEBS Lett.* 149:105 (1982),

54. A.F. Phillips, B. Persson, K. Hall, M. Lake, A. Skottner, T. Sanengen and V.R. Sara, The effects of biosynthetic insulin-like growth factor-I supplementation on somatic growth, maturation, and erythropoiesis in the neonatal rat, *Pediat. Res.* 23:298 (1988).

55. A. Kurtz, J. Zapf, K.-U. Eckardt, G. Clemons, E.R. Froesch and C. Bauer, Insulin-like growth factor I stimulates erythropoiesis in hypophysectomized rats, *Proc. Natl. Acad. Sci. USA.* 85:7825 (1988).

56. G.A. Werther, K. Haynes and G.R. Johnson, Insulin-like growth factors promote DNA synthesis and support cell viability in fetal hemopoietic tissue by paracrine mechanisms, *Growth Factors* 3:171 (1990).

57. S. Merchav, I. Tatarsky and Z. Hochberg, Enhancement of erythropoiesis *in vitro* by human growth hormone is mediated by insulin-like growth factor I, *Brit. J. Hematol.* 70:267 (1988).

58. S. Merchav, I. Tatarsky and Z. Hochberg, Enhancement of human granulopoiesis *in vitro* by biosynthetic insulin-like growth factor-I/somatomedin C and human growth hormone, *J. Clin. Invest.* 81:791 (1988).

59. B.A.A. Scheven and N.J. Hamilton, Stimulation of macrophage growth and multinucleated cell formation in rat bone marrow cultures by insulin-like growth factor I, *Biochem. Biophys. Res. Commun.* 174:647 (1991).

60. G. Rodriguez-Tarduchy, M.K.L. Collins, I. Garcia and A. Lopez-Rivas, A., Insulin-like growth factor-I inhibits apoptosis in IL-3 dependent hemopoietic cells, *J. Immunol.* 149:535 (1992).

61. S. Arkins, N. Rebeiz, A. Biragyn, D.L. Reese and K.W. Kelley, Murine macrophages express abundant insulin-like growth factor-I Ea and Eb transcripts, *Endocrinology* 133:2334 (1993).

62. B.A. Barres, I.K. Hart, H.S.R. Coles, J.F. Burne, J.T. Voyvodic, W.D. Richardson and M.C. Raff, Cell death and control of cell survival in the oligodendrocyte lineage, *Cell* 70:31 (1992).

63. Y.-K. Fu, S. Arkins, B.S. Wang and K.W. Kelley, A novel role of growth hormone and insulin-like growth factor: priming neutrophils for superoxide anion secretion, *J. Immunol.* 146:1602 (1991).

64. Y.-K. Fu, S. Arkins, G. Fuh, B.C. Cunningham, J.A. Wells, S. Fong, M.J. Cronin, R. Dantzer and K.W. Kelley, Growth hormone augments superoxide anion secretion of human neutrophils by binding to the prolactin receptor, *J. Clin. Invest.* 89:451 (1992).

65. H.P. Guler, J. Zapf, E. Scheiwiller and E.R. Froesch, Recombinant human insulin-like growth factor I stimulates growth and has distinct effects on organ size in hypophysectomized rats, *Proc. Natl. Acad. Sci. USA.* 85:4889 (1988).

66. K. Binz, P. Joller, P. Froesch, H. Binz, J. Zapf and E.R. Froesch, Repopulation of atrophied thymus in diabetic rats by insulin-like growth factor I, *Proc. Natl. Acad. Sci. USA.* 87:3690 (1990).

67. R.V. Mueller, E.M. Spencer, A. Sommer, C.A. Maack, D. Suh and T.K. Hunt, The role of IGF-I and IGFBP-3 in wound healing, *in* "Modern Concepts of Insulin-Like Growth Factors," E.M. Spencer, ed., Elsevier Science Publishers, New York (1991).

68. S.E. Lynch, R.B. Colvin and H.N. Antoniades, Growth factors in wound healing. Single and synergistic effects on partial thickness porcine skin wounds, *J. Clin. Invest.* 84:640 (1989).

69. D.A. Rappolee, D. Mark, M.J. Banda and Z. Werb, Wound macrophages express TGF-a and other growth factors *in vivo:* Analysis by mRNA phenotyping, *Science* 241:708 (1988).

70. J. Timsit, W. Savino, B. Safieh, P. Chanson, J.-F. Bach and M. Dardenne, Effect of growth hormone and insulin-like growth factor I on thymic hormonal function in man, *J. Clin. Endocrinol. Metab.* 75:183 (1992).

71. E.K. Neely, S.S. Smith and R.G. Rosenfeld, Human leukemic T and B lymphoblasts produce insulin-like growth factor binding proteins 2 and 4, *Acta Endocrinologica (Copen)* 124:707 (1991).

72. A.J. D'Ercole, A.D. Stiles and L.E. Underwood, Tissue concentrations of somatomedin-C: further evidence for multiple sites of synthesis and paracrine/ autocrine mechanisms of action, *Proc. Natl. Acad. Sci. USA.* 81:935 (1984).

73. T.J. Merimee, M.B. Grant, C.M. Broder and L.L. Cavalli-Sforza, Insulin-like growth factor secretion by human B-lymphocytes: A comparison of cells from normal and pygmy subjects, *J. Clin. Endocrinol. Metab.* 69:978 (1989).

74. W.N. Rom, P. Basset, G.A. Fells, T. Nukiwa, B.C. Trapnell and R.G. Crystal, Alveolar macrophages release an insulin-like growth factor-I type molecule, *J. Clin. Invest.* 82:1685 (1988).

75. I. Nagaoka, B.C. Trapnell and R.G. Crystal, Regulation of insulin-like growth factor I gene expression in the human macrophage-like cell line U937, *J. Clin. Invest.* 85:448 (1990).

76. J.R. Baxter, J.E. Blalock and D.A. Weigent, Characterization of immunoreactive insulin-like growth factor from leukocytes and its regulation by growth hormone, *Endocrinology* 129:1727 (1991).

77. T. Lin, D. Wang, M.L. Nagpal, W. Chang and J.H. Calkins, Down-regulation of Leydig cell insulin-like growth factor-I gene expression by interleukin-I, *Endocrinology* 130:1217 (1992).

78. S. Merchav, I. Silvian-Drachsler, I. Tatarsky, M. Lake and A. Skottner, Comparative studies of the erythroid-potentiating effects of biosynthetic human insulin-like growth factors-I and II, *J. Clin. Endocrinol. Metab.* 73:447 (1992).

79. L.J. Murphy, G.I. Bell, and H.G. Friesen, Tissue distribution of insulin-like growth factor I and II messenger ribonucleic acid in the adult rat, *Endocrinology* 120:1279 (1987).

PROLACTIN AS AN IMMUNOMODULATORY HORMONE

Eva Nagy and Istvan Berczi

Department of Immunology
Faculty of Medicine
University of Manitoba
Winnipeg, Manitoba, Canada

The first experimental observation indicating the effect of prolactin (PRL) on the thymus was made by Smith in 1930. He observed that the thymus gland of hypophysectomized (Hypox) rats ceased to grow and regressed in weight to less than half of controls in long surviving animals. On the other hand, partially hypophysectomized rats showed an absolute weight loss no greater than the controls. These and other observations triggered many investigators to study the influence of hormones on lymphoid tissues. However, the lack of purified hormone preparations and proper assays at the time posed insurmountable difficulties which led to contradictions and confusion.[1] Studies on the regulatory effect of the pituitary gland on bone marrow function have a similar history.[2] During the past decade neuroendocrine immunoregulatory mechanisms have been investigated with increasing interest and the influence of PRL on lymphoid tissue and immune function has been reviewed in recent years.[3-8]

THE REGULATION OF PRIMARY LYMPHOID ORGANS BY PROLACTIN

Prolactin and Bone Marrow Function

The hemopoietic effect of PRL was pointed out first by Jepson and Lowenstein who demonstrated in a series of papers[9] that PRL has an erythropoietic effect in polycythemic mice, is capable of stimulating erythropoiesis in normal or orchidectomized male mice, and it prevents the decrease of red cells and reticulocytes in lactating mice or in mice exposed to hyperoxia.

We observed normochromic-normocytic anemia, leukopenia and thrombocytopenia coupled with impaired DNA and RNA synthesis in the bone marrow of Hypox rats. All these deficiencies could be normalized by syngeneic pituitary grafts (SPG) placed under the kidney capsule or by treatment with ovine or bovine PRL or GH or human placental lactogen.[10-12] Moreover, Hypox rats were found to have 10-20% lactogenic activity in their serum when compared to controls by the Nb2 lymphoma proliferation assay. Such animals were treated daily with a rabbit anti-rat PRL serum. Their serum lactogenic activity diminished further, severe anemia developed and death occurred within 8 weeks. In contrast, untreated Hypox animals gradually increased their serum lactogenic activity starting on the 7th week after pituitary removal which rose up to 50% of control levels by week 9. At this point the anemia of Hypox animals was stabilized. The incorporation of ^{3}H-thymidine by rat bone marrow cells was stimulated in vitro by rat and ovine PRL and GH and by human placental lactogen.[13] PRL also has an influence on the bursa of Fabricius in birds[14,15] which is the site of B lymphocyte maturation.

The observations that the hyperprolactinemia of kidney patients on dialysis is decreased significantly after treatment with recombinant human erythropoietin[16] and that hemorrhage induces pituitary PRL release in swine and rats[17,18] suggest that a feedback regulatory relationship exists between hemopoiesis and PRL secretion.

Prolactin and the Thymus

In dwarf mice ectopic pituitary grafts increase thymus weight and the number of thymocytes.[20] Treatment of dwarf mice with placental lactogen stimulates selectively the growth of thymus.[21] In Hypox animals the thymus involutes rapidly as originally observed by Smith. This involution can be reversed by SPG without stimulating significantly body growth. Moreover, the thymuses of normal adult animals also grow if given SPG. DNA and RNA synthesis in thymocytes of Hypox rats is grossly impaired, which can be restored by SPG or by treatment with PRL, PL or GH.[10,22] These hormones have a direct growth stimulatory effect on thymocytes.[23] There is also evidence to indicate that PRL has an influence on the expression of Thy-1, CD4, and TL antigens by thymoctyes.[24,25]

Syngeneic pituitary grafts did not restore thymic structure after 10 or 21 days in aged mice although immune responsiveness was enhanced at 21 days.[26]

Prolactin affects thymic epithelial cells and regulates the secretion of the thymic hormone thymulin,[27] and promotes the growth of chicken thymocytes in vitro.[28]

Pierpaoli and coworkers[29] observed first that newborn and adult athymic nude mice have markedly diminished levels of serum PRL and abnormally high levels of LH. Thymus grafts normalized the blood levels of both hormones. More recently, Healy et al.[30] reported that fetal thymectomy led to abnormal ovarian differentiation and inhibited oogenesis in Rhesus monkeys. In utero thymectomy elevated plasma FSH and decreased plasma PRL concentrations when compared with intact controls. The weight of the ovaries and adrenal glands were also reduced.

Spangelo and coworkers[31] demonstrated that there is a feedback regulatory interaction between the thymus and PRL secretion by the pituitary gland and they have identified a peptide consisting of 35 amino acids, called MB-35, which is capable of inducing PRL release at 160 ng/ml concentration and GH release at 600 ng/ml concentration from rat anterior pituitary cells in vitro.

PROLACTIN AND IMMUNE FUNCTION

Prolactin Receptors in Lymphoid Tissue

Shiu et al.[32] demonstrated that the Nb2 rat lymphoma cell line has approximately 12,000 high affinity (K_d=75 bM) receptor sites per cell. Thus the affinity of this receptor was approximately 20-fold higher than the receptors found in other tissues. Recent studies revealed that the Nb2 cells express a novel type of the short form of PRL receptor, having a molecular weight of 62 K_d. This receptor has 3.3 higher affinity to PRL when compared to the long form.[33] Russell and coworkers[34] reported first the presence of PRL receptors on human and rat T and B lymphocytes and on monocytes with a calculated receptor number of 360/cell and K_d of 1.66 nM. It was also found that cyclosporin has an effect on PRL binding to its receptor, which could be either enhancing or inhibitory depending on the concentration.[34] These observations have been confirmed by other investigators.[35,36] Natural killer (NK) cells express binding sites for PRL with a calculated receptor number of 660/cell, and K_d of 30 nM. Again, a concentration dependent inhibition or increase of PRL binding to its receptor by cyclosporin A was observed.[37]

PRL receptors in various tissues show three different forms (e.g. short, intermediate and long). Growth hormone receptors have been identified in short and long forms in various species. A number of cytokine receptors have also been shown to belong to the same receptor family which is now called the cytokine/GH/PRL family. Receptors for granulocyte colony stimulating factor (G-CSF), granulocyte-macrophage colony stimulating factor (GM-CSF), erythropoietin (EPO), for the β chain of interleukin-2 (IL-2), for IL-3, IL-4, IL-5, IL-6, and IL-7, all belong to the cytokine/GH/PRL family. Moreover, gp130, which is associated with the IL-6 receptor and plays a role in signal transduction, is also a member of this receptor family.[38-40] A homology has also been found between the receptor for type 3 molecules of fibronectin and the cytokine/GH/PRL family. Receptors for interferon (IFN)-α, β, and γ have been proposed to form a subset of the cytokine/GH/PRL family.[39]

Cell signalling by members of the growth and lactogenic hormone family is not fully resolved and it is possible that more than one signalling pathway plays a role. In the case of PRL, the involvement of G proteins, of thyrosine kinase, and of protein kinase C have been implicated.[39,41] In the case of thymus derived (T) lymphocytes, a direct nuclear signalling by PRL has also been proposed.[42]

The Effect of Prolactin on Lymphoid Cells

In the spleens of Hypox rats DNA and RNA synthesis is deficient, which can be restored to normal levels by SPG or by treatment with PRL, PL or GH. The ability to synthesize nucleic acids in the spleen correlates directly with immunocompetence. PRL and GH show a direct mitogenic effect on rat spleen cells *in vitro*.[23] PRL treated rats show a major increase in ornithine decarboxylase in the thymus and spleen, which is a growth regulatory enzyme.[43] The impaired production of TNFα and O_2^- by peritoneal macrophages of aged rats after activation with IFNγ could be partially or fully restored by SPG from young donors.[44] PRL was shown to induce the expression of IL-2 surface receptors in splenocytes of ovariectomized rats or rats in diestrus, but not in those from estrogen treated OVX rats or rats in estrus.[45] Treatment of mice with PRL led to a biphasic potentiation of lectin induced T cell mitogenesis, low doses being stimulatory whereas high doses were inhibitory.[46] PRL has a growth stimulatory effect on spleen cells of chicken, potentiates the mitogenic response and alters antibody production.[28]

In patients with hyperprolactinemia abnormalities of peripheral blood lymphocyte responses to mitogens, of IL-2 production, an increase in the number of $CD4^+$ cells and the presence of phenotypically immature T cells in the peripheral blood were observed. In some untreated patients enhanced T suppressor cell activity was also found in an *in vitro* pokeweed mitogen driven B cell transformation assay. All these abnormalities could be corrected by treatment with the dopamine agonist drug, bromocriptine.[47,48] Altered chemotaxis of leukocytes was also observed in two of three patients with hyperprolactinemia.[49]

Several laboratories produced evidence that activated lymphocytes produce PRL or GH[36,50-58] and it has been proposed that lymphocyte derived PRL may have an important immunoregulatory function through autocrine and paracrine mechanisms. Others, however, could not find evidence of PRL production by normal lymphocytes using sophisticated techniques of molecular biology.[3,42] It is clear, however, that some human tumor cells of the B cell lineage do, in fact, secrete bioactive PRL, which in some cases, actually serves as an autocrine growth factor.[59-63] Significantly elevated serum PRL levels were observed in 16 of 28 patients with acute myeloid leukemia.[64]

Prolactin and the Immune Response

Humoral Immunity. Hypox rats show a grossly impaired antibody response to the T cell dependent antigen, sheep red blood cells (SRBC), and to bacterial lipopolysaccharide (LPS), which is a T-independent antigen. The antibody response of Hypox rats can be restored by SPG or by treatment with PRL, PL, or GH.[65-67] Treatment of rats with BRC also suppressed the antibody response to SRBC or LPS, which could be restored by additional treatment with either PRL or GH. Treatment with ACTH antagonized the restoring effect of these hormones.[68] PRL treatment was also shown to stimulate, and BRC treatment to inhibit, the anti-SRBC titres in mice.[46] The primary antibody response to SRBC was also enhanced in young female mice given SPG, mice with two grafts and higher PRL levels having an increased antibody responsiveness. Treatment of mice with mouse PRL at the time of immunization also enhanced the antibody response, but was ineffective if delayed by 24 hours.[69] Aged female mice given SPG for 10 days did not exhibit any enhancement in their primary antibody response to SRBC, but in those bearing SPG for 28 days both the IgM and IgG primary antibody response was enhanced, though immunocompetence was not restored to the level of young mice.[26] PRL was also shown to stimulate the antibody response in chicken.[70]

Cell Mediated Immunity. Cell mediated immune responses are deficient in Hypox rats, which can be restored by SPG or treatment with PRL, PL or GH. Treatment with bromocriptine also suppresses cell mediated immunity which may be reversed by PRL or GH treatment and antagonized by additional treatment with ACTH.[65,67,68,71-73] Prolactin deficiency induced in rats by BRC treatment severely reduced the reactivity of lymphocytes in mixed lymphocyte cultures and the severity of local graft versus host reaction. On the other hand, the stimulation of PRL secretion reversed the immunosuppression induced by cyclosporin A. When CSA was administered to rats jointly with BRC, permanent survival of kidney grafts could be achieved at CSA doses which permitted graft rejection if administered alone.[36,74] Similar observations were made in Lewis rats bearing BN cardiac allografts, where treatment with a new dopamine agonist agent, CQP201-403, enhanced greatly the immunosuppressive effect of CSA on graft rejection, and also on the *in vitro* response of lymphocytes to mitogens.[75] Bromocriptine treatment of mice prevented the T cell dependent induction of macrophage tumoricidal activity which could be reversed by additional treatment with PRL. The production of IFN-γ by T lymphocytes and lymphocyte proliferation in response to mitogens was also depressed in spleen cells of BRC-treated mice which were reversed after co-administration of PRL.[76]

Serum PRL levels are increased in most patients with cardiac allografts prior to the primary rejection episode. Such increase was not always observed, however, during recurrent rejections.[77] In patients with kidney grafts and without any evidence of graft rejection or infection, there was a linear correlation between serum levels of CSA and PRL. At the time of immunologic activity this correlation was lost and a significant decrease in serum PRL levels was observed at the time of histologically proven rejection.[78] In BALB/c mice receiving C57BL skin allografts, pituitary PRL mRNA was significantly increased and serum PRL bioactivity was elevated.[79]

Natural killer cell activity of spleen lymphocytes from pituitary grafted male and female Ames dwarf mice was greatly enhanced.[20] There are contrasting observations with regards to the effect of hyperprolactinemia on NK cell activity in man.[80,81] Matera et al.[82] showed that PRL stimulates the proliferation and cytotoxic activity of purified NK cells, but it has no effect on NK activity by unseparated peripheral blood lymphocytes. The different susceptibility of purified NK cells and peripheral blood lymphocytes to PRL was due to the presence in the unseparated cell population of PRL sensitive T cells capable of suppressing NK activity.

PRL enhanced the recovery of the receptor for SRBC on human T lymphocytes after treatment with trypsin. Elevated serum PRL levels induced in males by chlorpromazine and in females by lactation raised the number of large granular lymphocytes in the peripheral blood which exert NK activity. Moreover, human sera containing elevated PRL levels stimulated the metabolic activity of peripheral neutrophilic leukocytes.[83]

Secretory Immune Function. The immune function of the mammary gland is hormonally regulated. Weisz-Carrington and coworkers[84] showed that there is a significant increase in the localization of immunoglobulin forming plasma cells and in epithelial immunoglobulins in the mammary gland of mice during pregnancy and lactation. Cultured mammary epithelial cells from lactating mice bound specifically and internalized dimeric serum IgA if PRL was present in the culture medium.[84] Treatment of virgin guinea pigs with estradiol, progesterone and ovine PRL induced the expression of major histocompatibility antigen class II by the mammary gland epithelium.[85]

The submandibular gland (SMG) of rodents produce substances that have a profound effect on immune and inflammatory reactions. Sabbadini and coworkers have isolated an immunosuppressive cytokine (ISC) of 40 K_d molecular size from the SMG of male rats.[86] Several of the biologically active factors of SMG are under endocrine control, androgens and thyroid hormone having major regulatory role.[87] In Hypox rats a marked atrophy of the SMG and the disappearance of ISC were observed. Treatment of Hypox rats with various pituitary hormones revealed that PRL, TSH and LH induced a significant restoration of ISC production. Additional experiments revealed that treatment with testosterone and thyroid hormone (T3) also induced significant recovery of ISC production. The combination of these two hormones with PRL produced the most effective restoration, whereas estradiol and progesterone had no significant effect. These results indicate that PRL, which has not been considered previously to stimulate the activity of the SMG, plays a role in the production of immunoregulatory factors in this gland.[88]

Prolactin in Inflammation and Infection. The inflammatory response represents the first line of defence against physical, chemical, and biological agents (mostly infectious) causing tissue injury. Irritants acting on sensory nerves may also induce inflammation through the neurogenic pathway as reviewed elsewhere in this volume. Mild injury or the injection of endotoxin elicits a sharp, brief elevation of PRL in the serum. In severe trauma, sepsis, and shock PRL is suppressed, whereas glucocorticoids and catecholamines are elevated. Under these conditions an acute phase response is initiated mainly by IL-1, IL-6, TNFα (several others may be involved), which induce the neuroendocrine response and initiate metabolic alterations in response to severe trauma and/or sepsis.[89]

Prolactin has been shown to potentiate the inflammatory response in rats to various irritants with the exception of dextran and serotonin-induced paw edema and cotton pellet granuloma, which could not be influenced.[90,91]

Mice which were immunosuppressed by bromocriptine showed an increased mortality after infection with *Listeria monocytogenes*. Resistance was restored to normal if additional PRL treatment was given.[76] Prolactin treatment reduced the mortality rate in mice after infection with *Salmonella typhimurium* in a dose dependent manner. Such treatment also increased phagocytosis and intracellular killing by peritoneal macrophages and enhanced chemotaxis by peritoneal granulocytes. On the other hand, the activity of blood leukocytes was not modified by PRL treatment.[92] The treatment of Hypox rats with either GH or PRL elevated superoxide production by macrophages and enhanced phagocytic activity toward opsonized *Listeria monocytogenes in vitro*. GH also enhanced the survival of Hypox rats infected with *Salmonella typhimurium*. Insulin-like growth factor I also primed macrophages *in vitro* in a similar manner.[93,94]

By now there is compelling evidence that a number of autoimmune diseases in animals and man are associated with abnormal PRL secretion. This subject is discussed in detail by Walker et al. in this volume and also in recent reviews.[4,95-99] The abnormalities of PRL in autoimmune disease described to date include (i) abnormal serum levels (hyperprolactinemia); (ii) decreased bioactivity; (iii) abnormal circadian rhythm; (iv) altered responsiveness to a stimulus such as TRH. It appears that hyperprolactinemia may predispose for both antibody mediated (e.g. systemic lupus erythematosus) and cell mediated (e.g. encephalytis, arthritis, hypophysitis) autoimmune disease.[99]

IMMUNE DERIVED AND INFLAMMATORY FEEDBACK SIGNALS REGULATING PRL SECRETION

This subject is discussed repeatedly in this volume, especially by Asa et al. and Gaillard. Here we only provide a summary table of feedback signals.

DISCUSSION

It seems clear that PRL plays an essential role in bone marrow and thymus function, and also in the maintenance of immunocompetence. Cell proliferation and hemopoiesis are stimulated by PRL in the bone marrow. PRL also has a growth promoting effect on the thymus and spleen and is capable of maintaining immunocompetence in hormonally deprived animals. PRL plays an important role in mammary and salivary gland immune function and promotes the inflammatory response and host defence against infection. Abnormalities of PRL secretion were found to be associated repeatedly with autoimmune disease.

We suggested earlier that growth and lactogenic hormones influence the imune system through their general growth regulatory function.[10,104] Cell proliferation is fundamental to the immune response as pointed out first by Burnet in his clonal selection theory.[105] A two signal model for the immune response has been proposed

Table 1. Lymphoid and inflammatory feedback signals regulating PRL secretion.

Feedback signal	Species	Target	Effect on PRL
Hemorrhage	Rat, swine	MBH, ?	↑
Thymosin-α 1	Rat	HYP	↓
MB-35	"	PIT	↑
IL-1α , β	"	HYP	↑
		PIT	↑↓
IL-2	Rat, mouse, man	HYP	↑
		PIT, ?	↑↓
IL-6	"	HYP	0
		PIT	0↑
IFN-γ	"	HYP	0
		PIT	0↓↑
TNF-α	Rat, calf	HYP	↑
		PIT	↑↓ TRH
TGF-β , β 1	Rat	PIT, PITA	↓
PAF	"	PIT	↑
PDGF	Rat	PITA	↓
Endothelin-1	Man	?	↓ TRH
Bradykinin	"	PIT	↑
Histamine	"	HYP	↑

HYP = hypothalamus, IL = interleukin, IFN = interferon, MB-35 = thymic peptide, MBH = median basal hypothalamus, PAF = platelet activating factor, PDGF = platelet derived growth factor, PIT = pituitary, PITA = pituitary adenoma, TGF = transforming growth factor, TNF = tumor necrosis factor, ↑ = increased PRL secretion, ↓ = decreased PRL secretion, 0 = no change in PRL secretion, ↓ TRH = thyroid releasing hormone induced increase is inhibited, ? = site of action, or nature of effect is uncertain.
Based on references[100-103]

by Bretscher and Cohn in 1970[106] on the basis that immunogenic substances must possess at least two antigenic determinants (epitopes) and that the cooperation of bone marrow derived and thymus derived lymphocytes is required to mount an immune response. This theory states that in addition to the antigen signal a second signal that could come from an antibody or T lymphocytes is also necessary to initiate antibody formation by B cells. Today it is firmly established that immunogens initiate the adherence interaction (bridging) of immunocytes, initially the antigen presenting cells with helper T cells and later on the helper T cells with B cells or immature antigen sensitive T cells. The second signal is delivered to the responding cells in the form of a cytokine which varies according to the stage of the immune response.[107]

The model of Bretscher and Cohn, which was constructed on the basis of *in vivo* immunization experiments, ignores the first signal which has to be delivered to lymphocytes and antigen presenting cells by growth and lactogenic hormones as this signal was always available in the animal. Later on during the *in vitro* studies on immune reactions and cell collaboration it was well recognized that fetal calf serum, which is abundant in growth and lactogenic hormones, was necessary for the initiation of immune reactions.

On the basis of current evidence we suggest that immune responses are regulated by three categories of signals, as follows: (i) By growth and lactogenic hormones that maintain immunocompetence. This competence signal enables the immune system to respond to antigen and also to nonspecific stimuli that initiate an inflammatory response. (ii) Antigen or other activating signals. This category of signals is always the result of cell to cell interaction mediated by adhesion molecules. During antigen presentation not only the epitope of the antigen, but also the major histocompatibility (MHC) type of the antigen presenting cell is recognized by receptors of T lymphocytes. Additional adhesion molecules (e.g. CD4, CD8, LFA1) also play a role in lymphocyte activation, proliferation and contribute to, or regulate functional activities, such as immunoglobulin secretion, cytotoxicity, suppression, etc.[108] (iii) The third group of signals is delivered by interleukins or, in general terms, cytokines. During the antigen driven differentiation of mature T and B lymphocytes, the cytokine requirement is subject to change according to the type and/or phase of the immune response.[107,109] Bone marrow cells also change their hemopoietic growth factor requirement as they go through the various differentiation pathways.[110-112] The interaction between stromal cells and B cell precursors and IL-7 play an essential role in B lymphocyte development in the bone marrow.[113] The thymic microenvironment has long been considered a prerequisite for normal T cell development. Cytokines are also produced in the thymus (e.g. IL-1, IL-2, etc.) that contribute to T cell maturation.[114,115] Therefore, cell to cell and cytokine signalling appear to be fundamental also for the development of T and B lymphocytes. The evidence reviewed here indicates that during postnatal life the function of bone marrow, thymus and immune function are all dependent on pituitary PRL and/or GH. These observations imply that the local production of growth factors and cytokines are dependent on pituitary PRL and/or GH and that the rate of their synthesis is regulated by these hormones jointly with stromal/adherence signals. A similar regulatory role of GH/PRL is envisioned for the production of cytokines in other tissues and organs.

In a series of experiments Clevenger and coworkers produced *in vitro* evidence for the three signal model for lymphocyte activation. These workers showed that antigen, PRL and IL-2 were all necessary for the proliferation of a helper T lymphocyte line in culture. They suggested that PRL functioned as a progression factor in this system.[116]

It is suggested that the three signal model outlined here applies in general to growth control in higher animals. There is evidence to indicate that growth and lactogenic hormones play a fundamental role in the development and growth of higher animals, both by providing the competence signal and also by the stimulation of insulin-like growth factors and other cytokines that function as third signals. Morphogenesis is determined by cell to cell signaling and by the local production or activation of growth factors and other cytokines that regulate cell proliferation, differentiation and function. Even programmed cell death is determined by locally delivered (positional) signalling. These stromal/adherence signals are capable of regulating cell growth and function according to their position in the body and are fundamental for species specific, organ specific morphogenesis and the regulation of cell function according to their position in the body. A detailed account of the three signal model for growth control in higher animals is given elsewhere.[117]

It is clear from the evidence reviewed in this chapter that growth and lactogenic hormones play a fundamental role in immune function by regulating both the development and competence of the immune system. Given the heterogeneity of growth and lactogenic hormones and of their receptors, it is conceivable that they participate in immunoregulation at other, as yet unrecognized, levels. These new insights have enhanced already our understanding of immune function significantly and advanced our knowledge regarding the pathogenesis of diseases with underlying immune abnormalities.

REFERENCES

1. T.F.Dougherty, Effects of hormones on lymphatic tissue, *Physiol. Rev.* 32:397 (1952).
2. R.C. Crafts and H.A. Meineke, The anemia of hypophysectomized animals, *Ann. NY Acad. Sci.* 77:501 (1959).
3. H.G. Friesen, G.D. DiMattia, and C.K.L. Too, Lymphoid tumor cells as models for studies of prolactin gene regulation and action, *Prog. Neuroendocrinimmunol.* 4:1 (1991).
4. I.C. Chikanza and G.S. Panayi, Hypothalamic-pituitary mediated modulation of immune function: prolactin as a neuroimmune peptide, *Brit. J. Rheumatol.* 30:203 (1991).
5. L. Matera, G. Bellone and A. Cesano, Prolactin and the immune network, *Adv. Neuroimmunol.* 1:158 (1991).
6. K. Skwarlo-Sonta, Prolactin as an immunoregulatory hormone in mammals and birds, *Immunol. Lett.* 33:105 (1992).
7. K.W. Kelley, S. Arkins and Y.M. Li, Growth hormone, prolactin, and insulin-like growth factors: new jobs for old players, *Brain Behav. Immun.* 6:317 (1992).
8. R.R. Gala, Prolactin and growth hormone in the regulation of the immune system, *Proc. Soc. Exp. Biol. Med.* 198:513 (1991).
9. J.H. Jepson and L. Lowenstein, Hormonal control of erythropoiesis during pregnancy in the mouse, *Br. J. Haematol.* 14:555 (1968).
10. I. Berczi and E. Nagy, The effect of prolactin and growth hormone on hemolymphopoietic tissue and immune function, *in:* "Hormones and Immunity," I. Berczi and K. Kovacs, eds., MTP Press, Lancaster, UK (1987).
11. I. Berczi and E. Nagy, Human placental lactogen is a hemopoietic hormone, *Br. J. Haematol.* 79:355 (1991).
12. E. Nagy and I. Berczi, Pituitary dependence of bone marrow function, *Br. J. Haematol.* 71:457 (1989).
13. E. Nagy and I. Berczi, Hypophysectomized rats depend on residual prolactin for survival, *Endocrinology* 146:2776 (1991).
14. G. Bhat, S.K. Gupta and B.R. Maiti, Influence of prolactin on mitogenic activity of the bursa of Fabricius of the chick, *Gen. Comp. Endocrinol.* 52:452 (1983).
15. K. Skwarlo-Sonta, D. Rosolowska-Huszc, J. Sotowska-Brochocka and A. Gajewska, Daily variations in response of certain immunity indices to proalctin in white leghorn chickens, *Exp. Clin. Endocrinol.* 87:195 (1986).
16. R.M. Schaefer, F. Kokot, B. Kurner, M. Zech and A. Heidland, Normalization of elevated prolactin levels in hemodialysis patients on erythropoietin, *Nephron* 50:400 (1988).
17. D.E. Carlson, H.G. Klemcke and D.S. Gann, Response of prolactin to hemorrhage is similar to that of adrenocorticotropin in swine, *Am. J. Physiol.* 258:R645 (1990).
18. J. Jurcovicová, R. Kvetnansky, M. Dobrakovová, D. Jezová, A. Kiss and G.B. Makara, Prolactin response to immobilization stress and hemorrhage: The effect of hypothalamic deafferentations and posterior pituitary denervation, *Endocrinology* 126:2527 (1990).
20. A.I. Esquifino, M.A. Villanua, A. Szary, J. Yau and A. Bartke, Ectopic pituitary transplants restore immunocompetence in Ames dwarf mice, *Acta Endocrinol.* 125:67 (1991).
21. C. Arezzini, V. De Gori, P. Tarli and P. Neri, Weight increase of body and lymphatic tissues in dwarf mice treated with human chorionic somatomammotropin (HCS), *Proc. Soc. Exp. Biol. Med.* 141:98 (1972).
22. I. Berczi and E. Nagy, Prolactin and other lactogenic hormones, *in:* "Pituitary Function and Immunity," I. Berczi, ed., CRC Press, Boca Raton, FL (1986).

23. I. Berczi, E. Nagy, S.M. De Toledo, R.J. Matusik and H.G. Friesen, Pituitary hormones regulate c-myc and DNA synthesis in lymphoid tissue, *J. Immunol.* 146:2201 (1991).
24. U. Singh and J.J.T. Owen, Studies on the maturation of thymus stem cells. The effects of catecholamines, histamine and peptide hormones on the expression of T cell alloantigens, *Eur. J. Immunol.* 6:59 (1976).
25. D.H. Russel, K.T. Mills, F.J. Talamantes and H.A. Bern, Neonatal administration of prolactin antiserum alters the developmental pattern of T- and B-lymphocytes in the thymus and spleen of BALB/c female mice, *Proc. Nat. Acad. Sci. USA* 85:7404 (1988).
26. R.J. Cross, J.L. Campbell, W.R. Markesbery and T.L. Roszman, Transplantation of pituitary grafts fail to restore immune function and to reconstitute the thymus glands of aged mice, *Mech. Age. Dev.* 56:11 (1990).
27. M. Dardenne, W. Savino, M.C. GagneraultC, T. Itoh and J.F. Bach, Neuroendocrine control of thymic hormonal production. I. Prolactin stimulates in vivo and in vitro the production of thymulin by human and murine thymic epithelial cells, *Endocrinology* 125:3 (1989).
28. K. Skwarlo-Sonta, Mitogenic effect of prolactin on chicken lymphocytes in vitro, *Immunol. Lett.* 24:171 (1990).
29. W. Pierpaoli, H.G. Kopp and E. Bianchi, Interdependence of thymic and neuroendocrine functions in ontogeny, *Clin. Exp. Immunol.* 24:501 (1976).
30. D.L. Healy, J. Bacher and G.D. Hodgen, Thymic regulation of primate fetal ovarian-adrenal differentiation, *Biol. Reprod.* 32:1127 (1985).
31. M. Badamchian, B.L. Spangelo, T. Damavandy, R.M. MacLeod and A.L. Goldstein, Complete amino acid sequence analysis of a peptide isolated from the thymus that enhances release of growth hormone and prolactin, *Endocrinology* 128:1580 (1991).
32. R.P.C. Shiu, H.P. Elsholtz, T. Tanaka, H.G. Friesen, P.W. Gout, C.T. Beer and R.L. Noble, Receptor-mediated mitogenic action of prolactin in a rat lymphoma cell line, *Endocrinology* 113:159 (1983).
33. S. Ali, I. Pellegrini, and P.A. Kelly, A prolactin-dependent immune cell line (Nb2) expresses a mutant form of prolactin receptor, *J. Biol. Chem.* 266:20110 (1991).
34. D.H. Russell, R. Kibler, L. Matrisian, D.G. Larson, B. Poulos and B.E. Magun, Prolactin receptors on human T and B lymphocytes: antagonism of prolactin binding by cyclosporine, *J. Immunol.* 134:3027 (1985).
35. G. Bellussi, G. Muccioli, C. Ghe and R. Dicarlo, Prolactin binding sites in human erythrocytes and lymphocytes, *Life Sci.* 41:951 (1987).
36. P.C. Hiestand, P. Melker, R. Nordmann, A. Grieder and C. Permongkol, Prolactin as a modulator of lymphocyte responsiveness provides a possible mechanism of action for cyclosporine, *Proc. Natl. Acad. Sci. USA* 83:2599 (1986).
37. L. Matera, G. Muccioli, A. Cesano, G. Bellussi and E. Genazzani, Prolactin receptors on large granular lymphocytes: dual regulation by cyclosporin A, *Brain Behav. Immun.* 2:1 (1988).
38. S.A. Aaronson, Growth factors and cancer, *Science* 254:1146 (1991).
39. P.A. Kelly, J. Djiane, M.C. Postel-Vinay and M. Edery, The prolactin/growth hormone receptor family, *Endocrine Rev.* 12:235 (1991).
40. J.A. Wells and A.M. de Vos, Structure and function of human growth hormone: implications for the hematopoietins, *Annu. Rev. Biophys. Biomol. Struct.* 22:329 (1993).
41. H. Rui, J.Y. Djeu, G.A. Evans, P.A. Kelly and W.L. Farrar, Prolactin receptor triggering. Evidence for rapid tyrosine kinase activation, *J. Biol. Chem.* 267:24076 (1992).
42. C.V. Clevenger, A.L. Sillman and M.B. Prystowsky, Interleukin-2 driven nuclear translocation of prolactin in cloned lymphocytes-T, *Endocrinology* 127:3151 (1990).
43. J.F. Richards, Ornithine decarboxylase activity in tissues of prolactin treated rats, *Biochem. Biophys. Res. Commun.* 63:292 (1975).
44. D.R. Davila, C.K. Edwards III, S. Arkins, J. Simon and K.W. Kelley, Interferon-c-induced priming for secretion of superoxide anion and tumor necrosis factor-a declines in macrophages from aged rats, *FASEB J.* 4:2906 (1990).
45. P. Mukherjee, A.M. Mastro and W.C. Hymer, Prolactin induction of interleukin-2 receptors on rat splenic lymphocytes, *Endocrinology* 126:88 (1990).
46. B.L. Spangelo, N.R.S. Hall, P.C. Ross and A.L. Goldstein, Stimulation of in vivo antibody production and concanavalin-A-induced mouse spleen cell mitogenesis by prolactin, *Immunopharmacology* 14:11 (1987).
47. A. Vidaller, L. Llorente, F. Larrea, J.P. Mendez, J. Alcocer-Varela and D. Alarcon-Segovia, T-cell dysregulation in patients with hyperprolactinemia: effect of bromocriptine treatment, *Clin. Immunol. Immunopathol.* 38:337 (1986).

48. R. Gerli, C. Riccardi, I. Nicoletti, S. Orlandi, C. Cernetti, F. Spinozzi and P. Rambotti P, Phenotypic and functional abnormalities of T lymphocytes in pathological hyperprolactinemia, *J. Clin. Immunol.* 7:463 (1987).

49. R.D. Harris, N.E. Kay, E.L. Seljeskog, K.J. Murray and S.D. Douglas, Prolactin suppression of leukocyte chemotaxis in vitro, *J. Neurosurg.* 50:462 (1979).

50. D.W. Montgomery, C.F. Zukoski, G.N. Shah GN, A.R. Buckley, T. Pacholczyk and D.H. Russell, Concanavalin A-stimulated murine splenocytes produce a factor with prolactin-like bioactivity and immunoreactivity, *Biochem. Biophys. Res. Commun.* 145:692 (1987).

51. D.W. Montgomery, J.A. LeFevre, E.D. Ulrich, C.R. Adamson and C.F. Zukoski, Identification of prolactin-like proteins synthesized by normal murine lymphocytes, *Endocrinology* 127:2601 (1990).

52. D.P. Hartmann, E.W. Bernton, T.K. Shakarijian and J.W. Holaday, Antibodies to prolactin inhibit murine and human lymphocyte proliferation in vitro by inhibiting lymphocyte response to T- and B-cell growth factors, *FASEB J.* 2:A1642 (1988) (Abstract).

53. D.P. Hartmann, J.W. Holadayand D.W. Bernton, Inhibition of lymphocyte proliferation by antibodies to prolactin, *FASEB J.* 3:2194 (1989).

54. N. Hattori, A. Shimatsu, M. Sugita, S. Kumagai and H. Imura, Immunoreactive growth hormne (GH) secretion by human lymphocytes: augmented release by exogenous GH, *Biochem. Biophys. Res. Commun.* 168:396 (1990).

55. D.A. Weigent, J.E. Riley, F.S. Galin, R.D. Leboeuf and J.E. Blalock, Detection of growth hormone and growth hormone-releasing hormone-related messenger RNA in rat leukocytes by the polymerase chain reaction, *Proc. Soc. Exp. Biol. Med.* 198:643 (1991).

56. K.D. O'Neal, D.W. Montgomery, T.M. Truong and L.Y. Yulee, Prolactin gene expression in human thymocytes, *Mol. Cell. Endocrinol.* 87:R19 (1992).

57. I. Pellegrini, J.J. Lebrun, S. Ali and P.A. Kelly, Expression of prolactin and its receptor in human lymphoid cells, *Mol. Endocrinol.* 6:1023 (1992).

58. P. Sabharwal, R. Glaser, W. Lafuse, S. Varma, Q. Liu, S. Arkins, R. Kooijman, L. Kutz, K.W. Kelley and W.B. Malarkey, Prolactin synthesized and secreted by human peripheral blood mononuclear cells - an autocrine growth factor for lymphoproliferation, *Proc. Natl. Acad. Sci. USA* 89:7713 (1992).

59. G.E. DiMattia, B. Gellersen, H.G. Bohnet and H.G. Friesen, A human B-lymphoblastoid cell line produces prolactin, *Endocrinology* 122:2508 (1988).

60. B. Gellersen, G.E. DiMattia, H.G. Friesen and H.G. Bohnet, Regulations of prolactin secretion in the human B-lymphoblastoid cell line IM-9-P3 by dexamethasone but not other regulators of pituitary prolactin secretion, *Endocrinology* 125:2853 (1989).

61. B. Gellersen, G.E. DiMattia, H.G. Friesen and H.G. Bohnet, Prolactin (PRL) mRNA from human decidua differs from pituitary PRL mRNA but resembles the IM-9-P3 lymphoblast PRL transcript, *Mol. Cell. Endocrinol.* 64:127 (1989).

62. G.E. DiMattia, B. Gellersen, M.L. Duckworth and H.G. Friesen, Human prolactin gene expression: the use of an alternative noncoding exon in decidua and the IM-9-P3 lymphoblastoid cell line, *J. Biol. Chem.* 265:16412 (1990).

63. L.A. Baglia, D. Cruz and J.E. Shaw, An Epstein-Barr virus-negative Burkitt lymphoma cell line (sfRamos) secretes a prolactin-like protein during growth in serum free medium, *Endocrinology* 128:2266 (1991).

64. S.J. Hatfill, R. Kirby, M. Hanley, E. Rybicki and L. Bohm, Hyperprolactinemia in acute myeloid leukemia and indication of ectopic expression of human prolactin in blast cells of a patient of subtype M4, *Leukemia Res.* 14:57 (1990).

65. E. Nagy E and I. Berczi, Immunodeficiency in hypophysectomized rats, *Acta Endocrinol.* 89:530 (1978).

66. I. Berczi, E. Nagy, K. Kovacs and E. Horvath, Regulation of humoral immunity in rats by pituitary hormones, *Acta Endocrinol.* 98:506 (1981).

67. E. Nagy, I. Berczi and H.G. Friesen, Regulation of immunity in rats by lactogenic and growth hormones, *Acta Endocrinol.* 102:351 (1983).

68. E. Nagy, I. Berczi, G.E. Wren, S.L. Asa and K. Kovacs, Immunomodulation by bromocriptine, *Immunopharmacology* 6:231 (1983).

69. R.J. Cross, J.L. Campbell and T.L. Roszman, Potentiation of antibody responsiveness after the transplantation of a syngeneic pituitary gland, *J. Neuroimmunol.* 25:29 (1989).

70. J. Sotowska-Brochocka, D. Rosolowska-Huszcz, K. Skwarlo-Sonta and A. Gajewska, Effect of exogenous prolactin on immunity in chickens, *Res. Vet. Sci.* 37:123 (1984).

71. E. Nagy and I. Berczi, Prolactin and contact sensitivity, *Allergy* 36:429 (1981).

72. E. Nagy and I. Berczi, Immunomodulation by tamoxifen and pergolide, *Immunopharmacology* 12:145 (1986).

73. I. Berczi, E. Nagy, S.L. Asa and K. Kovacs, Pituitary hormones and contact sensitivity in rats, *Allergy* 38:325 (1983).

74. P.C. Hiestand, J.M. Gale and P. Mekler, Soft immunosuppression by inhibition of prolactin release: synergism with cyclosporine in kidney allograft survival and in the localized graft-versus-host reaction, *Transplant Proc.* 18:870 (1986).

75. M.L. Wilner, R.B. Ettenger, M.A. Koyle and J.T. Rosenthal, The effect of hypoprolactinemia alone and in combination with cyclosporine on allograft rejection, *Transplantation* 49:264 (1990).

76. E.W. Bernton, M.T. Meltzer and J.W. Holaday, Suppression of macrophage activation and T-lymphocyte function in hypoprolactinemic mice, *Science* 239:401 (1988).

77. C. Cosson, I. Myara, R. Guillemain, C. Amrein, G. Dreyfus and N. Moatti, Serum prolactin as a rejection marker in heart transplantation, *Clin. Chem.* 35:492 (1989).

78. G. Mayer, J. Kovarik, E. Pohanka, M. Schwarz, H. Graf and W. Wolosczuk, Serum prolactin levels after kidney transplantation, *Transplant Proc.* 19:3724 (1987).

79. G.K. Shen, D.W. Montgomery, E.D. Ulrich, K.R. Mahoney and C.F. Zukoski, Upregulation of prolactin gene expression and feedback modulation of lymphocyte proliferation during acute allograft rejection, *Surgery* 112:387 (1992).

80. R. Gerli, P. Rambotti, I. Nicoletti, S. Orlandi, G. Migliorati and C. Riccardi, Reduced number of natural killer cells in patients with pathological hyperprolactinemia, *Clin. Exp. Immunol.* 64:399 (1986).

81. L. Matera, E. Ciccarelli, A. Cesano, F. Veglia, C. Miola and F. Camanni, Natural killer activity in hyperprolactinemic patients, *Immunopharmacology* 18:143 (1989).

82. L. Matera, A. Cesano, G. Muccioli and F. Veglia, Modulatory effect of prolactin on the DNA synthesis rate and NK activity of large granular lymphocytes, *Int. J. Neurosci.* 51:265 (1990).

83. J. Rovensky, M. Vigas, J. Marek, S. Blazickova, L. Vydetelkova and A. Takac, Evidence for immunomodulatory properties of prolactin in selected invitro and invivo situations, *Int. J. Immunopharmacol.* 13:267 (1991).

84. P. Weisz-Carrington, Secretory immunobiology of the mamary gland, *in:* "Hormones and Immunity," I. Berczi and K. Kovacs, eds., MTP Press, Lancaster, UK (1987).

85. L. Klareskog, U. Forsum and P.A. Peterson, Hormonal regulation of the expression of Ia antigens on mammary gland epithelium, *Eur. J. Immunol.* 10:958 (1980).

86. M. Abdelhaleem and E. Sabbadini, Identification of immunosuppressive fractions of the rat submandibular gland, *Immunology* (1992) (submitted)

87. P. Walker, The mouse submaxillary gland: a model for the study of hormonally dependent growth factors, *J. Endocrinol. Invest.* 4:183 (1982).

88. E. Nagy, I. Berczi and E. Sabbadini, Endocrine control of the immunosuppressive activity of the submandibular gland, *Brain Behav. Immun.* 6:418 (1992).

89. I. Berczi and E. Nagy, Neurohormonal control of cytokines during injury, *in:* "Brain Control of the Responses to Trauma," N.J. Rothwell and F. Berkenbosch, eds., Cambridge University Press (1994) (in press).

90. R. DiCarlo, R. Meli and G. Muccioli, Effects of prolactin on rat paw oedema induced by different irritants, *Agents Action* 36:87 (1992).

91. R. Meli, O. Gualillo, G.M. Raso and R. DiCarlo, Further evidence for the involvement of prolactin in the inflammatory response, *Life Sci.* 53:PL105 (1993).

92. R. DiCarlo, R. Meli, M. Galdiero, I. Nuzzo, C. Bentivoglio and R. Carratelli, Prolactin protection against lethal effects of Salmonella typhimurium, *Life Sci.* 53:981 (1993).

93. C.K. Edwards, S.M. Ghiasuddin, L.M. Yunger, R.M. Lorence, S. Arkins, R. Dantzer and K.W. Kelley, Invivo administration of recombinant growth hormone or gamma-interfron activates macrophages. Enhanced resistance to experimental Salmonella-typhimurium infection is correlated with generation of reactive oxygen intermediates, *Infect. Immun.* 60:2514 (1992).

94. C.K. Edwards, S. Arkins, L.M. Yunger, A. Blum, R. Dantzer and K.W. Kelley, The macrophage-activating properties of growth hormone, *Cell. Mol. Neurobiol.* 12:499 (1992).

95. I. Berczi, F.D. Baragar, I.M. Chalmers, E.C. Keystone, E. Nagy and R.J. Warrington, Hormones in self tolerance and autoimmunity: a role in the pathogenesis of rheumatoid arthritis? *Autoimmunity* 16:45 (1993).

96. D. Buskila, D. Sukenik and Y. Shoenfeld, The possible role of prolactin in autoimmunity - a review, *Am. J. Reprod. Immunol.* 26:118 (1991).

97. L.J. Jara, C. Lavalle and L.R. Espinoza, Does prolactin have a role in the pathogenesis of systemic lupus erythematosus, *J. Rheumatol.* 19:1333 (1992).

98. S.E. Walker, S.A. Allen and R.W. McMurray, Prolactin and autoimmune disease, *Trends Endocrinol. Metab.* 4:147 (1993).

99. I. Berczi, The role of prolactin in the pathogenesis of autoimmune disease, *Endocr. Pathol.* 4:178 (1993).

100. I. Berczi, The immunology of prolactin, *Sem. Reprod. Endocrinol.* 10:196 (1992).

101. S. McCann, S. Karanth, A. Kamat, W.L. Dees, K. Lyson, M. Gimeno and V. Rettori, Induction by cytokines of the pattern of pituitary hormone secretion in infection, *Neuroimmunomodulation* 1:2 (1994).

102. I. Berczi, Hormonal interactions between the pituitary and immune system, *in:* "Hormones and Immunity," D.J. Grossman, ed., Springer-Verlag (1994) (in press).

103. H. Vierhapper, U. Hollenstein, M. Roden and P. Nowotny, Effect of endothelin-1 in man. Impact on basal and stimulated concentrations of luteinizing hormone, follicle-stimulating hormone, thyrotropin, growth hormone, corticotropin, and prolactin, *Metabolism* 42:902 (1993).

104. I. Berczi, Neurohormonal immunorgulation, *Endocr. Pathol.* 1:197 (1990).

105. F.M. Burnet. "The Clonal Selection Theory of Acquired Immunity," Cambridge University Press (1959).

106. P. Bretscher and M. Cohn, A theory of self-nonself discrimination, *Science* 169:1042 (1970).

107. N.A. Mitchison, Unique features of the immune system: their logical ordering and likely evolution, *in:* "Cell to Cell Interaction," M.M. Burger, B. Sordat and R.M. Zinkernagel, eds., Karger, Basel (1990).

108. B.E. Bierer and S.J. Burakoff, T cell receptors: adhesion and signaling, *Adv. Cancer Res.* 56:49 (1991).

109. W.E. Paul, The immune system, an introduction, *in:* "Fundamental Immunology," 2nd edition, W.E. Paul, ed., Raven Press, New York (1989).

110. L. Sachs, The molecular control of blood cell development, *Science* 238:1374 (1987).

111. S.C. Clark and R. Kamen, The human hematopoietic colony-stimulating factors, *Science* 236:1229 (1987).

112. D. Metcalf, The granulocyte-macrophage colony-stimulating factors, *Science* 229:16 (1985).

113. F. Melchers, Cell to cell cooperations in B-cell development and B-cell responses, *in:* "Cell to Cell Interaction," M.M. Burger, B. Sordat and R.M. Zinkernagel, eds., Karger, Basel (1990).

114. J.J.T. Owen, Lymphocyte interaction in the thymus, *in:* "Cell to Cell Interaction," M.M. Burger, B. Sordat and R.M. Zinkernagel, eds., Karger, Basel (1990).

115. J.W. Hadden, P.H. Malec, J. Coto and E.M. Hadden, Thymic involution in aging. Prospects for correction, *Ann. NY Acad. Sci.* 673:231 (1992).

116. C.V. Clevenger, A.L. Sillman, J. Hanleyhyde and M.B. Prystowsky, Requirement for prolactin during cell cycle regulated gene expression in cloned lymphocytes-T, *Endocrinology* 130:3216 (1992).

117. I. Berczi, The role of the growth and lactogenic hormone family in immune function, *Neuroimmunomodulation* (1994) (in press).

PROLACTIN AND AUTOIMMUNITY: INFLUENCES OF PROLACTIN IN SYSTEMIC LUPUS ERYTHEMATOSUS

Sara E. Walker[1], Duane H. Keisler[3], Susan H. Allen[1], Cynthia L. Besch-Williford[4], Robert W. Hoffman[1,2], and Robert W. McMurray[1]

Departments of [1]Internal Medicine and [2]Pathology, School of Medicine, Department of [3]Animal Sciences, and the [4]College of Veterinary Medicine, University of Missouri-Columbia
Columbia, MO 65201-5297

INTRODUCTION

The autoimmune disease, systemic lupus erythematosus (SLE), is incurable, and individuals with this illness are at risk of severe and potentially fatal involvement of the central nervous system and kidneys.[1,2] SLE occurs most commonly in women of childbearing age, and evidence has suggested that this disease is influenced by reproductive hormones.[3] Recent studies in this laboratory have employed a hormone-sensitive murine model of SLE, the NZB X NZW (B/W) mouse,[4] to investigate the role of the lactotrope, prolactin,[5,6] in spontaneously occurring autoimmune disease. This report will summarize our findings in groups of female B/W mice made hyperprolactinemic or, conversely, treated with bromocriptine, an inhibitor of prolactin secretion.[7]

The impression that prolactin may contribute to disease activity in SLE was strengthened recently, when we identified four women with longstanding hyperprolactinemia who had developed relatively mild SLE. In two of the patients, stopping bromocriptine treatment was followed by active disease.[8,9] Our cases are described with the anticipation that other clinicians will be able to identify additional individuals in whom elevated prolactin concentrations are associated with autoimmune disease. We are currently applying our observations to a selected group of SLE patients, in whom we are studying suppresssive effects of bromocriptine. This relatively nontoxic drug may provide a novel means of suppressing autoimmunity.

PROLACTIN

The lactotropic polypeptide hormone, prolactin, belongs to the growth hormone family and is secreted by acidophilic cells of the anterior pituitary gland.[5] In the

Advances in Psychoneuroimmunology, Edited by
I. Berczi and J. Szélenyi, Plenum Press, New York, 1994

absence of stimuli such as suckling or stress, secretion of prolactin is restricted by the hypothalamus. Dopamine has been identified as the inhibiting substance,[10,11] and treatment with the dopamine agonist, bromocriptine, inhibits secretion of prolactin selectively with little effect on other hormones of the anterior pituitary gland.[7]

Prolactin and the Immune System

Prolactin regulates a number of biochemical processes in tissues and cells, including induction and stimulation of ornithine decarboxylase, an enzyme which controls biosynthesis of alipathic polyamine compounds required for cell growth and division[12] and activation of protein kinase C.[13] Prolactin binding sites have been identified in a variety of mammalian tissues, including T cells and B cells.[13] Prolactin stimulates growth related gene mRNA[14,15] and interferon (IFN) regulatory factor 1 mRNA in mouse lymphocytes,[16] and serves as an important growth factor for lymphocytes. Cloned mouse T helper lymphocytes have been identified which require prolactin as a co-factor in IL-2 driven proliferation,[17] and anti-prolactin antibody has been reported to inhibit proliferation in an interleukin (IL)-2 dependent cell line.[18]

Prolactin may also affect macrophage activity. Bromocriptine treatment of mice led to impaired T lymphocyte production of IFN-γ and prevented T cell dependent induction of macrophage tumoricidal activity. These effects were reversed by treatment with bovine prolactin.[19]

The immune stimulating properties of prolactin have been demonstrated in experimental animals *in vivo*. Production of antibodies to sheep red blood cells (SRBC) was diminished in lymphocytes from hypophysectomized rats, and response was restored by giving daily injections of prolactin or by implanting an autonomously functioning pituitary gland.[20] In pituitary-implanted mice, serum prolactin concentrations were correlated directly with primary plaque forming cell responses to SRBC.[21] Cell mediated immune responses are also affected by prolactin. Dinitrochlorobenzene-induced contact dermatitis was suppressed in hypophysectomized rats, and this suppression was restored by injections of bovine prolactin.[22]

Prolactin in Animal Models of Inflammatory Disease

Prolactin has been implicated in stimulating activity of a number of immune-mediated diseases in experimental animals. In Fischer rats, induction of adjuvant arthritis was inhibited by hypophysectomy or bromocriptine therapy, but arthritis developed if the animals were treated subsequently with prolactin or growth hormone. Giving hypophysectomized rats follicle stimulating hormone, luteinizing hormone, or thyroid stimulating hormone had no effect on the arthritis.[23] Bromocriptine has also been found to suppress clinical signs in rats with experimental allergic encephalitis.[24] In a recent study, immunization of Lewis rats with spinal cord homogenate was associated with a three-fold increase in basal prolactin levels. Implantation of a continuous-release bromocriptine pellet two days before immunization ablated the increase in prolactin but did not cause remarkable suppression of the serum prolactin concentration. Nevertheless, bromocriptine significantly suppressed severity of neurological signs; therapy administered one week after immunization, and in the setting of chronic disease, was also beneficial.[25]

Cyclosporine inhibits the binding of prolactin to receptor sites on T lymphocytes,[26] and the question has been raised that cyclosporine exerts immunosuppressive effects by blocking the prolactin receptor. This finding served as the basis for studies designed to compare therapeutic efficacy of bromocriptine, cyclosporine, and both drugs given in combination. In the diabetes-prone BB rat model, intense inflammation of pancreatic islets is followed by β cell destruction, diminished insulin levels, and glycosuria.[27] Treatment with bromocriptine, 10 mg/kg every 2 weeks, resulted in glycosuria in 75% of rats; occurrence of glycosuria was also 75% in controls. Cyclosporine therapy, 5 mg/kg/day, and bromocriptine combined with cyclosporine, were associated with glycosuria in 25% and 17% of rats, respectively.[28] Results of combination therapy were more promising in Lewis rats with experimental autoimmune uveitis. In this model, high dose cyclosporine (10 mg/kg/day, for two weeks) ablated uveitis, and a similar beneficial results was found in animals receiving both low dose bromocriptine and cyclosporine, 2 mg/kg.[29]

The NZB/W Model of Autoimmunity

Matings between inbred New Zealand Black (NZB) and New Zealand White (NZW) strains produce F_1 NZB/NZW (NZB/W) hybrid mice, which spontaneously develop anti-DNA antibodies and die with immune complex glomerulonephritis.[4,30,31] NZB/W mice are accepted as models of lupus. The autoimmune disorder in females is triggered by sexual maturity[32] and leads to early death at 10 to 12 months of age, whereas males have late onset disease.[4]

Influences of Sex Hormones in NZB/W Disease

The lupus-like disease of NZB/W mice responds to gonadal hormones. Treatment with androgens protected female mice from premature death,[33] whereas administration of exogenous 17 β -estradiol (E2) in crystalline implants containing 6-7 mg hormone stimulated autoantibodies and led to premature death.[32,34-36] We have recently reexamined the effects of exogenous estrogens in NZB/W females, employing mice implanted at the age of 6 weeks with 6 mg 17 β -estradiol or 0.5 mg ethinyl estradiol (EE). Mice treated with estrogen died early, with mean age at death 19 weeks ± 1.4 SEM vs control mice (33 weeks ± 3; p <0.005). Detailed necropsies showed that estrogen-treated mice developed severe urinary tract infections and endometritis. Estrogen stimulated production of very high concentrations of circulating prolactin, 10X to 90X greater than controls (Table 1), and pituitary adenomas were identified in 50% of treated animals.[37,38]

In retrospect, these findings should not be surprising. Hans Selye reported in 1939 that estrogens caused toxic genitourinary complications in rodents,[39] and the prolactin stimulating properties of exogenous estrogens are well established.[40] Although it has been quoted widely that "estrogen accelerates lupus", this supposition must now be qualified in NZB/W mice receiving traditional experimental doses of estrogens. It is plausible to assume that high levels of prolactin contribute to stimulation of the immune system and autoantibody production in estrogen-treated mice. This assumption is supported by our recent investigations of the effects of hyperprolactinemia on autoimmune disease in NZB/W mice.

Prolactin in Murine Autoimmunity

We tested the hypothesis that prolactin is capable of stimulating SLE in NZB/W mice by making young female mice hyperprolactinemic; each of these animals received two transplanted syngeneic pituitary glands under the renal capsule. In this situation the implanted glands, removed from the inhibitory influence of hypothalamic dopamine, autonomously secrete large amounts of prolactin.[41] A second group of females was treated with daily injections of bromocriptine, 300μ g per day, and control mice received sham operations plus daily injections of vehicle.

Table 1. Serum prolactin concentrations.*

	Weeks of Treatment		
Implant	3	8	12
0 (control)	184 ± 53	53 ± 14	18 ± 7
	(6)	(12)	(17)
EE	2206 ± 300	1574 ± 147	1638 ± 282
	(6)	(6)	(8)
E2	2055 ± 275	509 ± 32	1363 ± 399
	(5)	(3)	(3)

*ng/ml, mean SEM. Parentheses enclose numbers of mice tested at each point. Numbers of mice tested were limited by early mortality in the ethinyl estradiol (EE) and 17β estradiol (E2) treatment groups. Table adapted from and reproduced with permission of *Arthritis and Rheumatism*.[38]

In hyperprolactinemic mice, serum concentrations of prolactin were maintained 3 to 18-fold greater than control mice through 30 weeks of age. Mean longevity was 28.7 weeks (Figure 1), and acceleration of disease was reflected in premature appearance of albuminuria, elevated levels of circulating immune complexes (gp70-anti-gp70), and high serum IgG (Figure 2). Bromocriptine injections produced inconsistent suppression of prolactin in serum obtained from anesthetized study mice. Nevertheless, bromocriptine recipients had delayed elevation of anti-DNA antibodies and serum IgG, and longevity was prolonged to 38 weeks. Analyses of thymic and splenic lymphocytes revealed no differences in lymphocyte subpopulations in mice with altered prolactin levels. We therefore considered that prolactin accelerated autoimmune disease in NZB/W mice and we postulated that lymphocyte function was affected, possibly through stimulation of T helper (Th) cell cytokine production.[6]

Roles of Cytokines in Murine SLE

The NZB/W hybrid mouse, as well as two additional established murine models of SLE -- the MRL/Mp/lpr/lpr (MRL-lpr) substrain, and BXSB mice -- are characterized by generalized B cell hyperactivity.[31] Th cells, which have surface CD4 antigen, are also important in expression of autoimmune disease in these mice; treatment with anti-L3T4/CD4 antibodies was beneficial in all three models.[42-44] These cells produce cytokines, which are soluble regulators of immune function that control lymphocyte proliferation and differentiation.[45]

It is assumed that cytokines participate in immune-mediated disease in lupus mice. IL-2, which is produced by cloned Th1 cells, provides important stimulus to proliferation of antigen-activated T cells.[45] Under specific culture conditions, concanavalin-A (Con-A)-stimulated spleen cells from autoimmune mice were capable of producing IL-2.[46] Th1 cells (as well as natural killer cells) also produce IFN-γ, a cytokine which up-regulates expression of class II antigens, activates macrophages, and acts as a B cell differentiation factor.[47] IFN-γ clearly has a role in lupus disease of NZB/W mice. Injecting these animals with IFN-γ shortened lifespans significantly, whereas treatment with DB-1 anti-rat IFN-γ antibodies prolonged longevity.[48] IL-4, which is produced preferentially by cloned mouse Th2 cells,[49] is one of a number of factors that initiate clonal expansion of B cells and promote B cell growth and differentiation.[50] Another Th2 product, IL-6, is of special interest in the context of autoimmune disease. IL-6 exerts its effects primarily at the final maturation stage of B cells and is an important factor for production of IgM and IgG.[51] Autoimmune MRL/lpr mice have been reported to produce excessive amounts of this cytokine.[52]

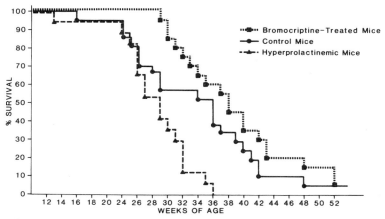

Figure 1. Survival curves for female NZB/W mice. Bromocriptine-treated mice had daily subcutaneous injections of drug plus sham surgery, control mice received vehicle plus sham surgery, and hyperprolactinemic mice were injected with vehicle and implanted with two syngeneic pituitary glands. Treatment began at 5 weeks of age and continued until spontaneous death. Breslow's test[53] indicated significant differences between survival curves for bromocriptine-treated mice and controls (p = 0.05), and survival curves for controls and hyperprolactinemic mice (p = 0.05).[6] Figure reproduced with permission, J. Immunol.

More recent studies have investigated expression of cytokine genes in lupus mice. Con-A stimulated spleen cells from MRL/lpr and BXSB mice produced IL-2, IL-4, and IFN-γ message and protein, but levels of IL-2 mRNA in 6-7 month old MRL/lpr mice with active autoimmune disease were 10-fold lower than other mice.[54] The expanded subset of abnormal B220$^+$, Thy-1$^+$, CD4$^-$, CD8$^-$ T cells in diseased MRL/lpr mice expressed a different group of cytokine genes. These cells, examined without mitogenic stimulation, contained genes encoding IFN-γ, tumor necrosis factor-α and -β, and IL-6.[55]

Cytokine Gene Expression in Hyperprolactinemic NZB/W Mice

Our investigations showed that Con-A stimulated splenocytes from hyperprolactinemic NZB/W females had augmented IL-4 mRNA and IL-6 mRNA. Bromocriptine treatment of NZB/W mice resulted in spleen cells with altered expression of mRNA for IL-2, IL-4, and IFN-γ following Con-A stimulation. In another preliminary experiment, fresh lymphocytes were isolated from a NZB/W female made extremely hyperprolactinemic by implantation of four, rather than two, pituitary glands. The splenocytes had prominent expression of IL-4 and IL-6 mRNA.[56] These findings were in accord with a report of increased IL-6 gene expression in unstimulated peripheral blood mononuclear cells from patients with SLE.[57]

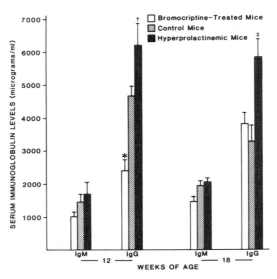

Figure 2. Hypergammaglobulinemia in NZB/W females with experimentally manipulated serum prolactin. Bromocriptine-treated, control, and hyperprolactinemic animals were studied at 12 and 18 weeks of age. Serum IgG concentration was suppressed in bromocriptine-treated mice at the age of 12 weeks. In contrast, hyperprolactinemic females had elevated IgG at both test points. *p = 0.004, †p = 0.03, ‡p = 0.0002.[6] Figure reproduced with permission, J. Immunol.

Prolactin in Human Autoimmune Diseases

A number of reports have suggested that prolactin has a role in stimulating organ specific autoimmune disease in humans. Autoimmune thyroid disease has been reported in individuals with hyperprolactinemia.[58] In three individuals with the syndrome of macroprolactinemia, hyperprolactinemia was associated with autoimmune thyroid disease; one patient had chronic thyroiditis, and two had Graves' disease.[59] Fourteen patients with autoimmune uveitis associated with a variety of disorders, including Behcet's syndrome and sarcoidosis, benefitted from therapy with combination bromocriptine and low dose cyclosporine A. Therapy suppressed serum prolactin below 2 ng/ml, and growth hormone levels did not change. Visual acuity improved in 8 patients.[60] In four patients with anterior uveitis, combination treatment with bromocriptine and cyclosporine resulted in significant decreases in circulating autoantibody levels.[61] Bromocriptine given as the sole treatment was effective in a small series of patients with iridocyclitis or iritis.[62]

Based on our experience with NZB/W mice, it is logical to raise the question that prolactin may stimulate increased disease activity in patients with a generalized autoimmune disorder such as SLE. Secretion of prolactin is stimulated by pregnancy, parturition, and suckling,[5] and exacerbations of lupus have been observed in the postpartum period.[3] Follicle stimulating hormone and luteinizing hormone levels are normal in SLE patients.[63] However, a number of reports have confirmed the association between SLE and hyperprolactinemia. Five pregnant women with SLE have been reported in whom elevated prolactin concentrations were associated with flares of the disease.[64] Jara and associates[65] suggested that hyperprolactinemia was more common in lupus than had been suspected previously. Ten of 45 SLE patients, including two pregnant women, had elevated (>20 ng/ml) serum prolactin; the mean concentration was 47 ng/ml, and values ranged from 20 to 188. Seven of the 10 hyperprolactinemic patients had active lupus. In contrast, only 5 of 35 SLE patients with normal serum prolactin had clinically active disease (P <0.05). Six of the hyperprolactinemic patients were followed serially, and persistently high prolactin occurred with continued clinical activity. In contrast, 3 SLE patients with normal circulating prolactin levels remained in remission. One patient has been observed with central nervous system lupus resistant to conventional therapy, whose clinical status improved remarkably after bromocriptine and intravenous immunoglobulin were added to the treatment regime.[66]

The association between elevated serum prolactin and SLE has also been observed in men. In 8 male patients with SLE, mean serum prolactin was elevated significantly, 7 times greater than controls.[67] Folomeev et al.[68] studied 29 men with SLE. Mean serum prolactin was 358 mU/ml ± 67, a value that significantly exceeded serum prolactin in 10 male controls (136 ± 10). It was of interest to note that serum prolactin concentrations were not increased in men with another inflammatory disease, rheumatoid arthritis. In 14 men with classical rheumatoid arthritis, serum prolactin was 169 ± 20.

The source of excessive serum prolactin in lupus patients has not been established. Lymphocytes have the capacity for production of prolactin: Con-A stimulation of murine lymphocytes resulted in secretion of a molecule resembling prolactin,[69] and immunoreactive prolactin was produced by a line of human lymphoblastoid B-cells[70] and by blast cells from an acute myeloid leukemia patient.[71] Human peripheral blood mononuclear cells, incubated with Con-A or phytohemagglutinin, secreted a high molecular mass immunoreactive prolactin that appeared to function in an autocrine manner as a growth factor for lymphoproliferation.[72]

The possibility exists that cytokines, produced by immunologically active cells in lupus patients, stimulate the pituitary to produce excessive prolactin. IL-6 gene expression was increased tenfold in unstimulated peripheral blood mononuclear cells from SLE patients.[57] In an *in vitro* model, recombinant mouse IL-6 stimulated release of prolactin, growth hormone, and LH from cultured rat pituitary cells.[73] The increased expression of IL-6 in SLE, as well as the reported stimulatory effects of IL-6 on the anterior pituitary, suggest that IL-6 (and possibly other cytokines) may have a role in accelerating secretion of immune-enhancing pituitary hormones in lupus patients.

The role of the pituitary in mediating immune responses may be much broader than we have realized. Folliculo-stellate cells of the anterior pituitary from mice and rats have been found to produce IL-6 *in vitro*,[74] and cells from a variety of human pituitary adenomas secreted IL-6 in monolayer cultures.[75]

SLE in Patients with Longstanding Hyperprolactinemia

We have shown in an animal model that longstanding hyperprolactinemia accelerates SLE. The assumption that prolactin stimulates autoimmunity was strengthened recently when we identified four women, aged 22 to 39, with hyperprolactinemia and SLE. The women were self-referred, or referred from rheumatology practices in this area within a three month period. All had symptomatic hyperprolactinemia, of 27 to 120 months' duration, and three had pituitary microadenomas. The disease we observed was relatively mild; our patients had photosensitivity, malar rashes, arthralgias, and positive fluorescent antinuclear antibody tests. In one patient, the onset of lupus was within one month after discontinuing bromocriptine therapy. In a second instance, a flare of active disease occurred during a 6-month period after bromocriptine was stopped.[8,9]

These initial patients have provided the impetus for a study in progress, designed to examine the effects of suppressing prolactin secretion in active SLE. The current drugs available for therapy of lupus -- corticosteroids, antimalarials, and cytotoxic agents -- are potentially toxic. Bromocriptine may be a novel, relatively safe form of immunosuppression that can be useful in selected patients with SLE.

REFERENCES

1. N.F. Rothfield, Systemic lupus erythematosus, *in:* "Rheumatic Diseases, Diagnosis and Management," W.A. Katz, ed., J.B. Lippincott Co., Philadelphia (1977).
2. W.H. Reeves and R.G. Lahita, Clinical presentation of systemic lupus erythematosus in the adult, *in:* "Systemic Lupus Erythematosus," R.G. Lahita, ed., John Wiley & Sons, New York (1987).
3. J.S. Smolen and A.D. Steinberg, Systemic lupus erythematosus and pregnancy: clinical, immunological, and theoretical aspects, *Prog. Clin. Biol. Res.* 70:283 (1981).
4. J.B. Howie and B.J. Helyer, The immunology and pathology of NZB mice, *Adv. Immunol.* 9:215 (1968).
5. C. Tougard and A. Tixier-Vidal, Lactotropes and gonadotropes, *in:* "The Physiology of Reproduction," E. Knobil and J.D. Neill, eds., Raven Press, New York (1988).
6. R. McMurray, D. Keisler, K. Kanuckel, S. Izui and S.E. Walker, Prolactin influences autoimmune disease activity in the female B/W mouse, *J. Immunol.* 147:3780 (1991).
7. D. Parkes, Bromocriptine, *N. Engl. J. Med.* 301:873 (1979).
8. R. W. McMurray, S.H. Allen, A. Braun and S.E. Walker, Longstanding hyperprolactinemia (HYPERPR) in systemic lupus erythematosus (SLE): possible hormonal stimulation of an autoimmune syndrome, *Arthritis Rheum.* 35(suppl):S168 (1992).
9. R.W. McMurray, S.H. Allen, A. Braun, F. Rodriguez and S.E. Walker, Longstanding hyperprolactinemia associated with systemic lupus erythematosus: possible hormonal stimulation of an autoimmune disease, *J. Rheumatol.* (in press).
10. D.M. Gibbs and J.D. Neill, Dopamine levels in hypophyseal stalk secretions in the rat are sufficient to inhibit prolactin secretion in vivo, *Endocrinology* 102:1895 (1978).
11. D.A. Leong, L.S. Frawley and J.D. Neill, Neuroendocrine controls of prolactin secretion, *Annu. Rev. Physiol.* 45:109 (1983).
12. D.H. Russell, D.F. Larson, S.B. Cardin and J.G. Copeland, Cyclosporine inhibits prolactin induction of ornithidine decarboxylase in rat tissues, *Mol. Cell. Endocrinol.* 35:159 (1984).
13. D.H. Russell, New aspects of prolactin and immunity: a lymphocyte-derived prolactin-like product and nuclear protein kinase C activation, *Trends Pharmacol. Sci.* 10:40 (1989).
14. L.Y. Yu-Lee, A.M. Stevens, J.A. Hrachovy and L.A. Schwartz, Prolactin-mediated regulation of gene transcription in lymphocytes, *Ann. N.Y. Acad. Sci.* 594:146 (1990).
15. I. Berczi, E. Nagy, S.M. de Toledo, R.J. Matusik and H.G. Friesen, Pituitary hormones regulate c-myc and DNA synthesis in lymphoid tissue, *J. Immunol.* 146:2201 (1991).
16. A.M. Stevens and L.Y. Yu-Lee, Regulation of interferon regulatory factor 1 (IRF-1) by prolatin in lymphocytes, *Arthritis Rheum.* 34(Suppl):S48 (1991).
17. C.V. Clevenger, D.H. Russell, P.M. Appasamy and M.B. Prystowsky, Regulation of interleukin 2-driven T-lymphocyte proliferation by prolactin, *Proc. Natl. Acad. Sci. USA* 87:6460 (1990).
18. D.P. Hartmann, J.W. Holaday and E.W. Bernton, Inhibition of lymphocyte proliferation by antibodies to prolactin, *FASEB J.* 3:2194 (1989).
19. E.W. Bernton, M.S. Meltzer and J.W. Holaday, Suppression of macrophage activation and T-lymphocyte function in hypoprolactinemic mice, *Science* 239:401 (1988).
20. E. Nagy and I. Berczi, Immunodeficiency in hypophysectomized rats, *Acta Endocrinol.* 89:530 (1978).
21. R.J. Cross, J.L. Campbell and T.L. Roszman, Potentiation of antibody responsiveness after the transplantation of a syngeneic pituitary gland, *J. Neuroimmunol.* 25:29 (1989).
22. I. Berczi, E. Nagy, S.L. Asa and K. Kovacs, Pituitary hormones and contact sensitivity in rats, *Allergy* 38:325 (1983).
23. I. Berczi, E. Nagy, S.L. Asa and K. Kovacs, The influence of pituitary hormones on adjuvant arthritis, *Arthritis Rheum.* 27:682 (1984).

24. E. Nagy, I. Berczi, G.E. Wren, S.L. Asa and K. Kovacs, Immunomodulation by bromocriptine, *Immunopharmacology* 6:231 (1983).
25. P.N. Riskind, L. Massacesi, R.H. Doolittle and S.L. Hauser, The role of prolactin in autoimmune demyelination: suppression of experimental allergic encephalomyelitis by bromocriptine, *Ann. Neurol.* 29:542 (1991).
26. D.H. Russell, R. Kibler, L. Matrisian, D.F. Larson, B. Poulos and B.E. Magun, Prolactin receptors on human T and B lymphocytes: antagonism of prolactin binding by cyclosporine, *J. Immunol.* 134:3027 (1985).
27. A.F. Nakhooda, A.A. Like, C.I. Chappel, C.N. Wei and E.B. Marliss, The spontaneously diabetic Wistar rat (the "BB" rat). Studies prior to and during development of the overt syndrome, *Diabetologia* 14:199 (1978).
28. J.L. Mahon, H.C. Gunn, K. Stobie, C. Gibson, B. Garcia, J. Dupre and C.R. Stiller, The effect of bromocriptine and cyclosporine on spontaneous diabetes in BB rats, *Transplant. Proc.* 20(suppl):197 (1988).
29. A.G. Palestine, C.G. Muellenberg-Coulombre, M.K. Kim, M.C. Gelato and R.B. Nussenblatt, Bromocriptine and low dose cyclosporine in the treatment of experimental autoimmune uveitis in the rat, *J. Clin. Invest.* 79:1078 (1987).
30. A.D. Steinberg, T. Pincus and N. Talal, DNA-binding assay for detection of anti-DNA antibodies in NZB/NZW F_1 mice, *J. Immunol.* 102:788 (1969).
31. A.N. Theofilopoulos and F.J. Dixon, Experimental murine systemic lupus erythematosus, *in:* "Systemic Lupus Erythematosus," R.G. Lahita, ed., John Wiley & Sons, New York (1987).
32. J. Roubinian, N. Talal, P.K. Siiteri and J.A. Sadakian, Sex hormone modulation of autoimmunity in NZB/NZW mice, *Arthritis Rheum.* 22:1162 (1979).
33. K.A. Melez, W.A. Boegel and A.D. Steinberg, Therapeutic studies in New Zealand mice. VII. Successful androgen treatment of NZB/NZW F_1 females of different ages, *Arthritis Rheum.* 12:41 (1980).
34. J.R. Roubinian, N. Talal, J.S. Greenspan, J.R. Goodman and P.K. Siiteri, Effect of castration and sex hormone treatment on survival, anti-nucleic acid antibodies, and glomerulonephritis in NZB/NZW F_1 mice, *J. Exp. Med.* 147:1568 (1978).
35. J.R. Roubinian, R. Papoian and N. Talal, Androgenic hormones modulate autoantibody responses and improve survival in murine lupus, *J. Clin. Invest.* 59:1066 (1979).
36. P.K. Siiteri, L.A. Jones, J. Roubinian and N. Talal, Sex steroids and the immune system - I. Sex difference in autoimmune disease in NZB/NZW hybrid mice, *J. Steroid Biochem.* 12:425 (1980).
37. S.E. Walker, R.W. McMurray, C.L. Besch-Williford, L.W. Keisler and D.H. Keisler, Exogenous 17-β -estradiol (E2) and ethinyl estradiol (EE) therapy in the NZB/W mouse model of systemic lupus erythematosus (SLE); disease modification, toxicity, and pituitary response, *Lupus* 1(suppl):18 (1992).
38. S.E. Walker, R.W. McMurray, C.L. Besch-Williford and D.H. Keisler, Premature death with bladder outlet obstruction and hyperprolactinemia in New Zealand Black X New Zealand White mice treated with ethinyl estradiol and 17-β -estradiol, *Arthritis Rheum.* 35:1387 (1992).
39. H. Selye, On the toxicity of oestrogens with special reference to diethylstilbestrol, *Can. Med. Assoc. J.* 41:48 (1939).
40. S. Franks, Regulation of prolactin secretion by oestrogens: physiological and pathological significance, *Clin. Sci.* 65:457 (1983).
41. R.A. Adler, The anterior pituitary-grafted rat: a valid model of chronic hyperprolactinemia, *Endocr. Rev.* 7:302 (1986).
42. D. Wofsy, Administration of monoclonal anti-T cell antibodies retards murine lupus in BXSB mice, *J. Immunol.* 136:4554 (1986).
43. D. Wofsy and W.E. Seaman, Reversal of advanced murine lupus in NZB/NZW F_1 mice by treatment with monoclonal antibody to L3T4, *J. Immunol.* 138:3247 (1987).
44. T.J. Santoro, J.P. Portanova and B.L. Kotzin, The contribution of L3T4[+] cells to lymphoproliferation and autoantibody production in MRL-lpr/lpr mice, *J. Exp. Med.* 167:1713 (1988).
45. N.E. Street and T.R. Mosmann, Functional diversity of T lymphocytes due to secretion of different cytokine patterns, *FASEB J.* 5:171 (1991).
46. T.J. Santoro, T.A. Luger, E.S. Raveche, J.S. Smolen, J.J. Oppenheim and A.D. Steinberg, In vitro correction of the interleukin 2 defect of autoimmune mice, *Eur. J. Immunol.* 13:601 (1983).

48. C.O. Jacob, P.H. van der Meide and H.O. McDevitt, In vivo treatment of (NZB X NZW) F_1 lupus-like nephritis with monoclonal antibody to c interferon, *J. Exp. Med.* 166:798 (1987).

49. T.R. Mosmann, H. Cherwinsky, M.W. Bond, M.A. Giedlin and R.L. Coffman, Two types of murine helper T cell clones. I. Definition according to profiles of lymphokine activities and secreted proteins, *J. Immunol.* 136:2348 (1986).

50. E.S. Vitetta, J. Ohara, C.D. Myers, J.E. Layton, P.H. Krammer and W.E. Paul, Serological, biochemical and functional identity of B cell-stimulatory factor 1 and B cell differentiation factor for IgG1, *J. Exp. Med.* 162:1726 (1985).

51. T. Kishimoto and T. Hirano, Molecular regulation of B lymphocyte response, *Annu. Rev. Immunol.* 6:485 (1988).

52. G.J. Prud'homme, C.L. Park, T.M. Fieser, R. Kofler, F.J. Dixon and A.N. Theofilopoulos, Identification of a B cell differentiation factor(s) spontaneously produced by proliferating T cells in murine lupus strains of the lpr/lpr genotype, *J. Exp. Med.* 157:730 (1983).

53. N. Breslow, Covariance analysis of censored survival data, *Biometrics* 30:89 (1974).

54. S. Umland, R. Lee, M. Howard and C. Martens, Expression of lymphokine genes in splenic lymphocytes of autoimmune mice, *Mol. Immunol.* 26:649 (1989).

55. L.J. Murray, R. Lee and C. Martens, In vivo cytokine gene expression in T cell subsets of the autoimmune MRL/Mp-lpr/lpr mouse, *Eur. J. Immunol.* 20:163 (1990).

56. R.W. McMurray, R.W. Hoffman and S.E. Walker, In vivo prolactin manipulation alters in vitro IL-2, IL-4, and IFN-c mRNA levels in female B/W mice, *Clin. Res.* 39:734A (1991).

57. M. Linker-Israeli and R. Deant, Dysregulated lymphokine production in systemic lupus erythematosus (SLE), *Ann. NY Acad. Sci.* 557:567 (1989).

58. C. Ferrari, M. Boghen, A. Paracchi, P. Rampini, F. Raiteri, R. Benco, M. Romussi, F. Codecasa, M. Mucci and M. Bianco, Thyroid autoimmunity in hyperprolactinemic disorders, *Acta Endocrinol.* 104:35 (1983).

59. W.B. Malarkey, R. Jackson and J. Wortsman, Long-term assessment of patients with macroprolactinemia, *Fertil. Steril.* 50:413 (1988).

60. A.G. Palestine, R.B. Nussenblatt and M. Gelato, Therapy for autoimmune uveitis with low-dose cyclosporine plus bromocriptine, *Transplant. Proc.* 20:131 (1988).

61. M. Blank, A. Palestine, R. Nussenblatt and Y. Shoenfeld, Down-regulation of autoantibody levels of cyclosporine and bromocriptine treatment in patients with uveitis, *Clin. Immunol. Immunopathol.* 54:87 (1990).

62. L.P. Hedner and G. Bynke, Endogenous iridocyclitis relieved during treatment with bromocriptine, *Am. J. Ophthalmol.* 100:618 (1985).

63. C. Lavalle, D. Gonzalez-Barcena, A. Graef and A. Fraga, Gonadotropins pituitary secretion in systemic lupus erythematosus, *Clin. Exp. Rheumatol.* 2:163 (1984).

64. L. Jara-Quezada, A. Graef and C. Lavalle, Prolactin and gonadal hormones during pregnancy in systemic lupus erythematosus, *J. Rheumatol.* 18:349 (1991).

65. L.J. Jara, C. Gomez-Sanchez, L.H. Silveira, P. Martinez-Osuna, M. Seleznick, F.B. Vasey and L.R. Espinosa, Hyperprolactinemia in systemic lupus erythematosus: association with disease, *Am. J. Med. Sci.* 303:222 (1992).

66. C.E. Rabinovich, L.E. Schanberg and D.W. Kredich, Intravenous immunoglobulin and bromocriptine in the treatment of refractory neuropsychiatric systemic lupus erythematosus, *Arthritis Rheum.* 33(suppl):R22 (1990).

67. C. Lavalle, E. Loyo, R. Paniagua, J.A. Bermudez, J. Herrera, A. Graef, D. Gonzalez-Barcena and A. Fraga, Correlation study between prolactin and androgens in male patients with systemic lupus erythematosus, *J. Rheumatol.* 14:268 (1987).

68. M. Folomeev, T. Prokaeva, V. Nassonova, E. Nassonov, V. Masenko and N. Ovtraht, Prolactin levels in men with SLE and RA, *J. Rheumatol.* 17:1569 (1990).

69. D.W. Montgomery, C.F. Zukoski, G.N. Shah, A.R. Buckley, T. Pacholczyk and D.H. Russell, Concanavalin A-stimulated murine splenocytes produce a factor with prolactin-like bioactivity and immunoreactivity, *Biochem. Biophys. Res. Commun.* 145:692 (1987).

70. G.E. DiMattia, B. Gellerson, H.G. Bohnet and H.G. Friesen, A human B-lymphoblastoid cell line produces prolactin, *Endocrinology* 122:2508 (1988).

71. S.J. Hatfill, R. Kirby, M. Hanley, E. Rybicki and L. Bohm, Hyperprolactinemia in acute myeloid leukemia and indication of ectopic expression of human prolactin in blast cells of a patient of subtype M4, *Leuk. Res.* 14:57 (1990).

72. P. Sabharwal, R. Glaser, W. Lafuse, S. Varma, Q. Liu, S. Arkins, R. Kooijman, L. Kutz, K.W. Kelley and W.B. Malarkey, Prolactin synthesized and secreted by human peripheral blood

mononuclear cells: an autocrine growth factor for lymphoproliferation, *Proc. Natl. Acad. Sci. USA* 89:7713 (1992).

73. B.L. Spangelo, A.M. Judd, P.C. Isakson and R.M. MacLeod, IL-6 stimulates anterior pituitary hormone relase in vitro, *Endocrinology* 125:575 (1989).

74. H. Vankelecom, P. Carmeliet, J. Van Damme and C. Denef, Production of interleukin-6 by follicuo-stellate cells of the anterior pituitary gland in a histiotypic cell aggregate culture system, *Neuroendocrinology* 49:102 (1989).

75. T.H. Jones, A. Price, S. Justice and K. Chapman, Interleukin-6 secretion by human pituitary adenomas in vitro, *J. Endocrinol.* 127(suppl):86 (1990).

EFFECTS OF CYTOKINES ON THE HYPOTHALAMO-PITUITARY-ADRENAL AXIS

Rolf C. Gaillard

Division of Endocrinology and Metabolism
Department of Medicine
University Hospital (CHUV)
CH-1011 Lausanne/Switzerland

INTRODUCTION

The interactions between the immune and endocrine systems have long been recognized, the most well-known effect being the immunosuppressive actions of glucocorticoids. More recently, however, there is growing evidence that peptide hormones have immune effects as well. Most exciting is the evidence that immune-neuroendocrine interactions or communications are bidirectional since it is now quite obvious that in addition to the recognized effects of hormones on the immune system, the latter system exerts a reciprocal effect on the endocrine system by producing substances that may profoundly affect endocrine function. In other words, it has been shown that the immune system is subject to endocrine and neural control, and that it exerts a reciprocal effect on the neuroendocrine system.

HORMONAL MODULATION OF THE IMMUNE FUNCTION

Numerous studies have demonstrated the existence of an immune-endocrine network through which hormones, neuropeptides and neurotransmitters are able to affect immune processes.[1-5] This modulation of the immune response involves the presence of specific receptors for all these subtances on immunocytes. More than 30 different types of hormonal receptors have been detected on immunocytes (for review[6,7]). Receptors for pituitary hormones, for steroid and thyroid hormones, for endogenous opioid peptides, for hypothalamic hypophysiotropic releasing and inhibiting hormones and for many other neuropeptides and neurotransmitters have been characterized in lymphoid and accessory cells.

These substances may regulate the immune function through an endocrine, a paracrine or even an autocrine fashion. Owing to their large numbers, it is impossible to summarize all the effects of these various substances on the different components of the immune response. However, numerous studies have shown that hormone administration can lead to stimulated or depressed immune responses, depending on the kind of hormones, the dose used and the timing of their administration. In general, glucocorticoids, sex hormones and endogenous opioids depress the the immune

Advances in Psychoneuroimmunology, Edited by
I. Berczi and J. Szélenyi, Plenum Press, New York, 1994

response *in vivo*, whereas growth hormone, prolactin, thyroxine and insulin increase the response.[2,8-14]

Glucocorticoids are important and even crucial mediators of the endocrine-immune interactions.[8] They exert a negative control over the secretion of peptides from both the hypothalamo-pituitary axis and the immune system. As immunomodulators, glucocorticoids have an overall inhibitory effect on many immune system functions. They induce a decrease in lymphoid cell growth, antibody formation, cellular cytotoxicity and inflammatory reactions.[8] In contrast, prolactin (PRL) and growth hormone (GH) have both recently been demonstrated to enhance immune responsiveness.[9,10,13,14] Inhibition of PRL by dopamine agonists such as bromocriptine results in a decrease in both humoral and cell mediated immunity,[11,12,15] whereas hyperprolactinemia results in enhanced immune responsiveness.[15] It has been suggested that PRL may affect immune target tissues by antagonizing certain glucocorticoid actions on immune cells. Treatment with PRL or with drugs that stimulate PRL secretion antagonizes immunosuppression due to chronic stress, hypercortisolism or treatment with the immunosuppressive peptide cyclosporin.[16] Sex steroids play important roles on the immune response; physiological, experimental and clinical data substantiate differences between the two sexes in terms of immune response.[17] Evidence for a role of sex steroids in the immune response is the presence of a naturally occurring sexual dimorphism in many immunological parameters. Indeed, females seem to have a more vigourous immune response, a better developed thymus, more resistance to the induction of immunological tolerance and a greater ability to reject tumors and allografts. In addition, most patients suffering from various types of autoimmune diseases are female.[18]

The effects of various stressors have been extensively studied in both humans and animals.[19-22] The generally accepted notion is that stress depresses the immune response, but there are exceptions to the rule. In humans, several observations have shown, a decrease in cellular immune response and in natural killer cell activity in different groups of individuals after acute or chronic stress; for instance, students during exam periods, people with poor marital relationships, etc... In animals, studies concerning the effect of stress on susceptibility to infectious diseases and numerous immune parameters provide contradictory results even if the general tendency is towards decreased responses. Indeed, in a stressful situation, all immune parameters may not vary in the same way: some parameters can be depressed, while others remain unchanged or are enhanced.[6]

MODULATION OF ENDOCRINE RESPONSES BY THE IMMUNE SYSTEM

The reciprocal arm of the bidirectional relationship between the neuroendocrine and immune system is the fact that the immune system may influence the neuroendocrine responses through messengers released by activated immune cells. These messengers are the various **cytokines**. In addition, the immune system can be considered as a neuroendocrine organ since it is able to produce hormones and neuropeptides.

The synthesis of hormones and neuropeptides by immune cells has been first demonstrated by J.E. Blalock et al.[23] Immunocytes are able to synthesize pro-opiomelanocortin (POMC)-related peptides, GH, PRL, TSH, LH, FSH, and pro-enkephalin-derived peptide.[7,24-34] With regard to the modulation of peptide production, immunocytes respond, in most cases, like pituitary cells to hypothalamic stimulatory factors and hormones involved in negative feedback regulation.

The reason for the production of hormones by cells of the immune system may be a paracrine, or even an autocrine regulation. These hormones modulate a number of immune functions such as natural killer cell activity,[35-37] inhibition of the production of interferon[38] and inhibition of the antibody response. In addition, the hormones produced by the immunocytes may represent a link between the immune and neuroendocrine systems in the case of aggression by non-cognitive stimuli (bacteria, viruses, tumors, etc...).

However, the major link between the immune and neuroendocrine systems are the various cytokines. Cytokines are polypeptides produced by the activated immune system and, as recently shown, also in a variety of other tissues.[39] In addition to the regulation of the immune system, they have a broad spectrum of physiological and pathophysiological actions.[40-44] Most interestingly, cytokines have recently been shown to affect both endocrine and metabolic functions (for review[5]), and they are now regarded as important modulators of the secretion of various hormones and neuropeptides. Interleukin-1, interleukin-2, interleukin-6, tumor necrosis factor α, γ-interferon and various thymic factors are the cytokines whose effects have been best studied on the endocrine system (for review[45-48]).

The existence of a physiological interaction between the endocrinological and immune cell functions was suggested some years ago by Besedovsky, among others, since it was observed that animals undergoing responses to various antigens showed, at the same time, increased glucocorticoid blood levels. Similar responses were observed after inoculation of Newcastle disease virus into mice.[49] The same marked increase in plasma ACTH and corticosterone levels was observed when mice were injected with supernatants from cultures of Newcastle disease virus - infected mouse spleen cells or human peripheral leucocytes, indicating that these cells might be stimulated by the virus to produce a factor that influences the hypothalamo-pituitary-adrenal axis. It was therefore quite obvious that substances released by the immune system could influence the neuroendocrine function. Evidence that this factor could be interleukin-1 arose from experiments in which the response was abrogated by injection of anti-interleukin-1 antibodies.[49] Since then, additional evidence has clearly demonstrated that cytokines can affect the hypothalamo-pituitary-adrenal axis.[45,46] In spite of some discrepancies in the literature, interleukin-1 (IL-1) and interleukin-6 (IL-6) seem to stimulate hormone secretion of the hypothalamo-pituitary-adrenal axis by acting at all three levels: the hypothalamus, the pituitary gland and adrenal glands inducing CRH, ACTH and cortisol production.

Effects of Cytokines at the Pituitary Level

Many studies have shown evidence for a direct effect of cytokines at pituitary level. Thymosin fraction 5,[50] IL-1 and IL-6 are reported to stimulate hormonal release directly from cultured[51,52] and perfused[53,54] pituitary cells. In our hands, IL-1α and IL-1β induce significant ACTH release from rat pituitary cells, but only after a prolonged (more than 10 hours) period of exposure to the interleukins.[51,55] In contrast to other ACTH secretagogues such as vasopressin, angiotensin II and catecholamines,[56,57] IL-1 does not potentiate CRF-41 activity on the corticotroph cells and its ACTH releasing ability is only subject to partial glucocorticoid negative feedback mechanism.[51] Local prostaglandin synthesis often mediates part of the physiological effects induced by cytokines in the peripheral target tissue. At the pituitary level, however, despite the concomitant and parallel release of prostaglandins and ACTH, prostaglandins are not the triggering mechanism for the IL1-induced ACTH release, since complete blockade of endogenous PGE_2 by Indomethacin does not affect this ACTH release.[51] In addition to their effect on ACTH secretion, cytokines have been reported to stimulate synthesis of pro-opiomelanocortin; both IL-1 and IL-2 treatment enhance the expression of

POMC mRNA by normal primary pituitary cells as well as by AtT-20 cells.[58,59] Further evidence is the discovery that the pituitary gland possesses specific binding sites of high affinity for IL-1 and IL-6.[60,61] These receptors can therefore bind the peripheral circulating cytokines but are also able to bind the cytokines produced within the pituitary gland itself. Indeed, there is an intrapituitary production of cytokines since a subpopulation of thyreotroph cells and the folliculo-stellate cells have been shown to produce IL-1 and IL-6 respectively.[62,63] It may therefore be speculated that these substances act as intrinsic modulators of secretion, playing a paracrine role at the pituitary level in regulating pituitary function during infectious challenge. Furthermore, IL-1 seems also to be involved in a mechanism which prevents the corticotroph cells to switch from a β-adrenergic ACTH regulation to an α-adrenergic pattern; in other words, IL-1 prevents the loss of the *in vivo* corticotropic responsiveness to β-adrenergic stimulation when the pituitary cells are incubated *in vitro*.[64]

There is however some controversy on a direct pituitary effect of cytokines, since some studies were unable to obtain an IL1-induced ACTH release using cultures of rat pituitary cells. The reason for this discrepancy is unclear. It seems unlikely that this may be due to differences in the sex of the rats employed, since cells cultured from female and male rats have been found to be responsive to IL-1. It is however possible, as recently suggested, that the preparation of the cells may play a critical role. It has been put forward that the presence of folliculo-stellate cells may be necessary to observe a pituitary effect of the cytokines, and that the failure to observe such an effect could be due to the absence of folliculo-stellate cells in the preparation.[65] The folliculo-stellate cells may therefore constitute a kind of interface through which the pituitary gland perceives changes in the activation state of the immune system.

Tumor necrosis factor α (TNFα) appears to be a marker for morbidity and mortality during septic shock. It therefore occupies a crucial role in the pathophysiological responses to infection and may induce lethal shock.[66] Interestingly the effect of TNFα at the pituitary level is different to that of IL-1 and IL-6. In contrast to IL-1 and IL-6, TNFα has no effect on basal pituitary hormone secretion but it significantly inhibits the stimulated levels of ACTH, GH, LH and PRL.[67] This inhibitory effect is not uniform, some stimulations are completely inhibited, others only partially. Thus TNFα fully blocks angiotensin II-induced ACTH release, GHRF-stimulated GH release and TRH-stimulated PRL release, whereas its inhibition is only partial on LHRH-induced LH release and CRH- and vasopressin-stimulated ACTH release. It appears that the corticotroph axis cannot be fully blocked by TNFα even at very high concentrations. This suggests that during septic shock, TNFα may inhibit the body's ability to develop an appropriate ACTH and glucocorticoid response to the stress of endotoxic shock. In other words, the corticotroph axis could still be stimulated, but not adequately.[67] These data indicate that TNFα inhibits the release of pituitary hormones which play important key roles during pathophysiological situations such as septic shock or cancer. It inhibits the corticotroph axis implicated in the stress response, the somatotroph axis which may be implicated in metabolic and anabolic actions and pituitary prolactin, which has been shown to have a critical influence on the maintenance of lymphocyte function and on lymphokine-dependent macrophage activation.[67-69] An inhibitory effect of TNFα does not seem to be specific for the pituitary, since this cytokine has also been shown to inhibit the action of ACTH and angiotensin II at the adrenal level[70,71] as well as to inhibit the hormonal production at the gonadal level.[72]

Effects of Cytokines at the Hypothalamic Level

In contrast to the slow onset of effects of the cytokines at the level of the pituitary *in vitro*, intravenous or intraperitoneal injections of cytokines *in vivo* cause a prompt rise in plasma ACTH with the peak responses appearing already within 15 to 30 minutes.[73-75] There is now convincing evidence that *in vivo* the acute effects of cytokines given peripherally are mainly exerted at the **hypothalamic level** by enhancing the release of CRH.[73,74,76,77] It has been found that IL-1 produces a marked depletion of hypothalamic CRH, due to increased release of CRH and not reduced synthesis.[76,78,79] The CRH-mediation of interleukins' action was further confirmed by an increase of CRH in the hypothalamo-hypophysial portal vessels following the injection of IL-1 in rats,[77] and by the observation that the ACTH-stimulating effect of cytokines was abolished by pre-treatment with anti-CRH antisera.[74,77,80] However, CRH does not seem to be involved in the ACTH-stimulating effect of all cytokines. Indeed, in contrast to IL-1, IL-6 and TNF, IL-2 action on ACTH release is not abolished in presence of an anti-CRH antiserum (Fukata J. and Imura, unpublished observation). Furthermore, we and others have clearly shown that cytokines induce CRH release from hypothalamic fragments *in vitro*.[80-82] Using this paradigm, we have shown that IL-1, IL-6, TNFα and Thymosin fraction 5, but not endotoxins, induce a significant dose-dependent and rapid (within 10-20 minutes) release of CRH.[81] Interestingly, none of the cytokines stimulated vasopressin secretion suggesting that cytokines act only on CRH containing neurons and not on neurons where CRH and vasopressin are co-localized. In addition to its effect on CRH release, IL-1 also stimulates CRH biosynthesis since it increases CRH mRNA in the hypothalamus.[83]

Further evidence for a direct effect of cytokines at the hypothalamic level is the presence of cytokine receptors within the hypothalamus and other brain areas[84-86] as well as the capability of the brain and particularly the hypothalamus to synthesize various cytokines.[39,87-91] *In situ* hybridization techniques reveal the existence of mRNAs for various cytokines in central nervous system cells, not only glial cells (microglia and astrocytes) but also neurons.[39,85,91] Cells of the nervous system appear to be able to produce IL-1, IL-2, IL-6, TNF, interferon and possibly some thymic hormones.[85] Co-localization of IL-1 immunoreactive fibers with CRH in the human hypothalamus, notably in the paraventricular nucleus (PVN) has been demonstrated.[92] The mechanisms by which IL-1 and IL-6 initiate the release of CRH are not well-known, but several possibilities have been put forward. Catecholamines play an important role in the regulation of CRH and ACTH secretion[93,94] and the relevance of catecholaminergic pathways on immune-hypothalamo-pituitary-adrenal axis interactions have been suggested, but there are also some controversies.[5,64,95-98] It has been proposed that the local release of catecholamines, acting on both α - and β -adrenergic receptors in the median eminence, may be important,[98,99] although circulating catecholamines do not appear to be involved.[100] The activation of noradrenergic neurons may not be sufficient to explain the activation of the hypothalamo-pituitary-adrenal axis, because the effect of IL-1 on plasma ACTH and corticosterone is not altered by administration of α -1 or β -antagonists.[101,102] In contrast, another study has shown that i.c.v. prazocin did block the effect of IL-1 administered i.c.v. on the elevation of plasma ACTH and corticosterone.[103]

In contrast to the cytokines effects at the pituitary gland level, those produced at hypothalamic levels are clearly mediated by the prostaglandins. Indeed, the arachidonic acid cascade pathway plays an important role, in so far as IL-1, IL-6 and TNFα stimulated release of CRH is blocked by cyclo-oxygenase (but not lipo-oxygenase) inhibitors.[80,104-106] Alternative pathways involving epoxygenase products via cytochrome P-450 and phospholipase A_2 pathway have also been suggested.[107] All these data indicate that cytokines do indeed regulate pituitary hormone secretion and they suggest that the dominant route of this modulation appears to be via the hypothalamus.

Cytokines and the Blood-Brain Barrier

Since cytokines are relatively large molecular weight proteins and water-soluble, they probably do not cross the blood-brain barrier in significant amounts. Thus a question still unanswered is the interaction between circulating and intracerebral cytokines. In other words, how do these substances reach the brain? How are the blood-borne cytokine messages transduced into neuronal signals? A blood-brain barrier is absent in several small areas of the brain in the so-called circumventricular organs, which are located at various sites within the walls of the cerebral ventricles of the brain. These include the median eminence, the organum vasculosum of the laminae terminalis of the hypothalamus (OVLT), the subfornical organ, the choroid plexus, and the area postrema at the base of the fourth ventricle.[108,109] We studied if the median eminence could represent a target for activation of the hypothalamo-pituitary-adrenal axis by the cytokines. Whereas Sharp et al.[110] reported that instillations of IL-1 near the median eminence stimulate ACTH release, we found that IL-6, but neither IL-1 nor TNFα, could increase CRH release from perifused median eminence *in vitro*.[81] IL-6 may therefore stimulate the hypothalamo-pituitary-adrenal axis through an action on CRH nerve terminals, whereas TNFα and IL-1 may act on the perikarya. There is, however, some controversy in this field, since Navarra et al.[111] found that neither IL-1 nor IL-6 could increase CRH release from the median eminence *in vitro*.

Another possible site of IL-1 action is the OVLT. Indeed, recent studies[112-114] suggest an important role for the OVLT, as a possible site of entry of blood-borne IL-1β into the brain, and for the preoptic area (POA), which may contain the neurons required for the response of the corticotrophs axis. It is suggested that instead of being a portal of entry of cytokines into the brain, the OVLT, alternatively, might be a site where the chemical messages of the blood-borne cytokines are transduced into neuronal signals, such that secondary messengers might be evoked that transmit original signals to the POA.[114] Prostaglandins produced by astrocytes in this area seem to mediate the action of IL-1 and to act on neurons surrounding this area.

But cytokines are also able to act within the central nervous system. It has been shown that intrahypothalamic infusion of IL-1 in the vicinity of the PVN can directly activate CRH containing neurons.[115] In addition, destruction of the PVN significantly inhibits the hypothalamo-pituitary-adrenal response to central IL-1 infusion. The PVN lesions did not completely block the response suggesting the involvement of other brain regions in this regulation.[116] This is consistent with the presence of IL-1 receptors in the rat and mouse hypothalamus and also with the visualisation of IL-1 immunoreactive fibers in the PVN of human brain.

Effects of Cytokines at the Adrenal Level

Briefly, cytokines may also act directly on the adrenal gland.[117-120] Similarly to the pituitary effect, the adrenal effect of the cytokines is of slow onset necessitating several hours of incubation.[120] It has been suggested that the direct secretory effect of IL-1 on the adrenals may involve the activation of an intraglandular CRF/ACTH system.[121] Like other endocrine tissues, the adrenal gland contains cytokines. IL-1 has been found in the chromaffin cells of the adrenal medulla and in the cortex,[122] whereas glomerulosa cells produce IL-6.[123] Interestingly TNFα is a potent inhibitor of angiotensin II-induced aldosterone synthesis[71] and of ACTH-induced corticosterone secretion.[70]

CONCLUSION

In conclusion, cytokines modulate the hypothalamo-pituitary-adrenal axis acting at all three levels, the hypothalamus, the pituitary gland and the adrenals. Acute effects of cytokines are mainly exerted at the hypothalamic level by enhancing the release of CRH, whereas slower effects are exerted at pituitary and adrenal levels. These interactions between the immune and endocrine systems may represent a potent negative feedback mechanism through which the immune system, by stimulating the hypothalamo-pituitary-adrenal axis and therefore stimulating the production of the immunosuppressive glucocorticoids, avoids an overshoot of the inflammatory and febrile effect during the acute-phase response.[8] Because glucocorticoids inhibit virtually all the components of the immune response, the consequence of the activation of the hypothalamo-pituitary-adrenal axis will be the suppression and/or modulation of the inflammatory responses to invading organisms. Therefore any dysfunction of these interactions could lead to an inflammatory disease. This has indeed recently been illustrated by studies in the Lewis rats, a strain of rats which are unable to respond to inflammation with an increase in glucocorticoid secretion. The susceptibility of these rats to the development of arthritis is clearly associated with the inability of their hypothalamo-pituitary-adrenal axis to adequately respond to inflammatory stimuli. The modulation of the hypothalamo-pituitary-adrenal axis by the cytokines is a mechanism which may play important physiological and pathophysiological roles.

ACKNOWLEDGEMENTS

I wish to thank Ms Lynn Trieste for typing the manuscript. This work was supported by the Swiss National Science Foundation (grant no 31.298676.90).

REFERENCES

1. R. Ader, D. Felten and N. Cohen, "Psychoneuroimmunology," Academic Press, New-York (1991).
2. I. Berczi and K. Kovacs, "Hormones and Immunity," MTP Press, Lancaster (1987).
3. J.W. Hadden, K. Masek and G. Nistico, "Interactions among CNS, Neuroendocrine and Immune Systems," Pythagore Press, Rome (1989).
4. M.S. O'Dorisio and A. Panerai, eds., Neuropeptides and immunopeptides: Messengers in a neuroimmune axis, *Ann. NY Acad. Sci.*, vol. 594 (1990).
5. H.O. Besedovsky and A. Del Rey, Immune-neuroendocrine circuits: integrative role of cytokines. (Vol. 13), *in:* "Frontiers in Neuroendocrinology", Raven Press Ltd, New-York (1992).
6. F. Homo-Delarche and M. Dardenne, The neuroendocrine-immune axis, *Springer Semin. Immunopathol.* 14:221 (1993).
7. D.A. Weigent and J.E. Blalock, Interactions between the neuroendocrine and immune systems: common hormones and receptors, *Immunol. Rev.* 100:79 (1987).
8. A. Munck, P.M. Guyre, and N.J. Holbrook, Physiological functions of glucocorticoids in stress and their relation to pharmacological actions, *Endocr. Rev.* 5:25 (1984).
9. R.J. Cross and T.L. Roszman, Neuroendocrine modulation of immune function: The role of prolactin, *Prog. Neuroendocr. Immunol.* 2:17 (1989).
10. B.L. Spangelo, N.R.S Hall, P.C. Ross, and A.L. Goldstein, Stimulation of in vivo antibody mouse spleen cell mitogenesis by prolactin, *Immunopharmacology* 14:11 (1987).
11. P.C. Hiestand, P. Mekler, R. Nordmann, A. Grieder, and C. Permmongkol, Prolactin as a modulator of lymphocyte responsiveness provides a possible mechanism of action for cyclosporin, *Proc. Natl. Acad. Sci. USA* 83:2599 (1986).
12. E.W. Bernton, M.S. Meltzer, and J.W. Holaday, Suppression of macrophage activation and T-lymphocyte function in hypoprolactinemic mice, *Science* 239:401 (1988).
13. K.W. Kelley, S. Brief, H.J. Westly, J. Novakofski, P.J. Betchel, J. Simon, and E.B. Walker, GH3

pituitary adenoma cells can reverse thymic aging rats, *Proc. Natl. Acad. Sci. USA* 83:5663 (1986).

14. D.A. Weigent and J.E. Blalock, Growth hormone and the immune system, *Prog. Neuroendocr. Immunol.* 3:231 (1990).

15. B.L. Spangelo, N.R.S. Hall, P.C. Ross, and A.L. Goldstein, Stimulation of in vivo antibody mouse spleen cell mitogenesis by prolactin, *Immunopharmacology* 14:11 (1987).

16. E.W. Bernton, Prolactin and immune host defenses, *Prog. Neuroendocr. Immunol.* 2:21 (1989).

17. F. Homo-Delarche, F. Fitzpatrick, M. Christeff, E.A. Nunez, J.F. Bach, and M. Dardenne, Sex steroids, glucocorticoids, stress and autoimmunity, *J. Steroid. Biochem.* 40:619 (1991).

18. C.J. Grossman, Are the underlying immune-neuroendocrine interactions responsible for immunological sexual dimorphism ? *Prog. Neuroendocr. Immunol.* 3:75 (1990).

19. S. Cohen and G.M. Williamson, Stress and infectious disease in humans, *Psychol. Bull.* 109:5 (1991).

20. D.N. Khansari, A.J. Murgo, and R.E. Faith, Effects of stress on the immune system, *Immunol. Today* 11:170 (1990).

21. M. Stein, Depression, the immune system and health and illness, *Arch. Gen. Psychiatry* 48:171 (1991).

22. R. Dantzer and K.W. Kelley, Stress and immunity : an integrated view of relationship between brain and the immune system, *Life Sci.* 44:1995 (1989).

23. J.E. Blalock, Production of neuroendocrine peptide hormones by the immune system, *Prog. Allergy* 43:1 (1988).

24. M.J. Ebaugh and E.M. Smith, Human lymphocyte production of immunoreactive luteinizing hormone, *FASEB J.* 2:68 (1988).

25. D.W. Montgomery, J.A. LeFevre, E.D. Ulrich, C.R. Adamson, and C.F. Zukoski, Identification of prolactin like proteins synthetized by normal murine lymphocytes, *Endocrinology* 127:2601 (1990).

26. J.R. Kenner, J.W. Holaday, E.W. Bernton, and P.F. Smith, Prolactin like protein in murine splenocytes: morphological and biochemical evidence, *Prog. Neuroendocr. Immunol.* 3:188 (1990).

27. D.A. Weigent, J.B. Baxter, W.E. Wear, R.L. Smith, K.L. Bost, and J.E. Blalock, Production of immunoreactive growth hormone by mononuclear leucocytes, *FASEB J.* 2:2812 (1988).

28. D.A. Weigent and J.E. Blalock, Immunoreactive growth-hormone-releasing-hormone in rat leucocytes, *J. Neuroimmunol.* 29:1 (1990).

29. J.E. Blalock, Proopiomelanocortin-derived peptides in the immune system, *Clin. Endocrinol.* 22:823 (1985).

30. H.J. Westly, A.J. Kleiss, K.W. Kelley, P.K. Wong, and P.-H. Yuen, Newcastle disease virus-infected splenocytes express the pro-opiomelanocortin gene, *J. Exp. Med.* 163:1589 (1986).

31. E.M. Smith, A.C. Monil, W.J.III Mayer, and J.E. Blalock, Corticotropin releasing factor induction of leucocyte-derived immunoreactive ACTH and endorphins, *Nature* 321:881 (1986).

32. G. Zurawski, M. Benedik, B.J. Kamb, J.S. Abrams, S.M. Zuraswki, and F.D. Lee, Activation of mouse T-helper cells induces abundant preproenkephalin mRNA synthesis, *Science* 232:772 (1986).

33. H.M. Johnson, E.M. Smith, B.A. Torres, and J.E. Blalock, Regulation of the in vitro antibody response by neuroendocrine hormones, *Proc. Natl. Acad. Sci. USA* 79:4171 (1982).

34. J. Wybran, T. Appelboom, J.P. Famaey, and A. Govaerts, Suggestive evidence for receptors for morphine and methionine enkephalin on normal human blood T lymphocytes, *J. Immunol.* 123:1068 (1979).

35. D.J. Carr and G.R. Klimpel, Enhancement of the generation of cytoxic T cells by endogenous opiates, *J. Neuroimmunol.* 12:75 (1986).

36. R.N. Mandler, W.E. Biddison, R. Mandler, and S.A. Serrate, Beta-endorphin augments the cytolytic activity and interferon prediction of natural killer cells, *J. Immunol.* 136:934 (1986).

37. R.E. Faith, H.J. Liang, A.J. Murgo, and N.P. Plotnikoff, Neuroimmunomodulation with enkephalins: enhancement of human natural killer (NK) cell activity in vitro, *Clin. Immunol. Immunopathol.* 31:412 (1984).

38. H.M. Johnson, B.A. Torres, E.M. Smith, L.D. Dion, and J.E. Blalock, Regulation of lymphokine (gamma-interferon) production by corticotropin, *J. Immunol.* 132:246 (1984).

39. J.I. Koenig, Presence of cytokines in the hypothalamo-pituitary axis, *Prog. Neuroendocr. Immunol.* 4:143 (1991).

40. C.A. Dinarello, Inflammatory cytokines : interleukin-1 and tumor necrosis factor as effector molecules in autoimmune diseases, *Curr. Opin. Immunol.* 3:941 (1991).

41. B. Beutler and A. Cerami, Cachectin and tumor necrosis factor as two sides of the same biological coin, *Nature* 320:584 (1988).
42. T. Kishimoto, The biology of interleukin-6, *Blood* 74:1 (1989).
43. C.A. Dinarello, Interleukin-1 and its biologically related cytokines, *Adv. Immunol.* 44:153 (1989).
44. J. Le and J. Vilcek, Biology of disease. Tumor necrosis factor and interleukin-1 : cytokines with multiple overlaping biological activities, *Lab. Invest.* 56:234 (1987).
45. A. Bateman, A. Singh, T. Kral, and S. Salomon, The immune hypothalamic-pituitary-adrenal axis *Endocr. Rev.* 10:92 (1989).
46. H. Imura, J.J. Fukata, and T. Mori, Cytokines and endocrine function : an interaction between the immune and neuroendocrine systems, *Clin. Endocrinol.* 35:107 (1991).
47. E.M. Smith, Hormonal activities of lymphokines, monokines and other cytokines, *Prog. Allergy* 43:121 (1988).
48. N. Azad, L. Agrawal, M.A. Emanuele, M.R. Kelley, N. Mohagheghpour, A.M. Lawrence, and N.V. Emanuele, Neuroimmunoendocrinology, *Am. J. Reprod. Immunol.* 26:160 (1991).
49. H. Besedovsky, A. Del Rey, E. Sorkin, and C.A. Dinarello, Immuno-regulatory feedback between interleukin-1 and glucocorticoid hormones, *Science* 233:652 (1986).
50. B.L. Spangelo, A.M. Judd, P.C. Ross, I.S. Login, W.D. Jarvis, M. Badamchian, A.L. Goldstein, and R.M. MacLeod, Thymosin Fraction 5 stimulates prolactin and growth hormone release from anterior pituitary cells in vitro, *Endocrinology* 121:2035 (1987).
51. P. Kehrer, D. Turnill, J.M. Dayer, A.F. Muller, and R.C. Gaillard, Human recombinant interleukin-1 beta and alpha, but not recombinant tumor necrosis factor alpha stimulate ACTH release from rat anterior pituitary cells in vitro in a prostaglandin E2 and cAMP independent manner, *Neuroendocrinology* 48:160 (1988).
52. E.W. Bernton, J.E. Beach, J.W. Holaday, R.C. Smallridge, and H.G. Fein, Release of multiple hormones by a direct action of interlukin-1 on pituitary cells, *Science* 238:519 (1987).
53. J.E. Beach, R.C. Smallridge, C.A. Kinzer, E.W. Bernton, J.W. Holaday, and U.G. Fein, Rapid release of multiple hormones from rat pituitaries perifused with recombinant Interleukin-1, *Life Sci.* 44:1 (1993).
54. B.L. Spangelo, A.M. Judd, P.C. Isakson, and R.M. MacLeod, Interleukin-6 stimulates anterior pituitary hormone release in vitro, *Endocrinology* 125:575 (1989).
55. P. Kehrer, R.C. Gaillard, J.M. Dayer, and A.F. Muller, International Congress of neuroendocrinology, *Neuroendocrinology Suppl.*:(Abstract 327) (1986).
56. P. Schoenenberg, P. Kehrer, A.F. Muller, and R.C. Gaillard, Angiotensin II potentiates corticotropin-releasing activity of CRF 41 in rat anterior pituitary cells: mechanism of action, *Neuroendocrinology* 45:86 (1987).
57. W. Vale, J. Vaughan, M. Smith, G. Yamamoto, J. Rivier, and C. Rivier, Effects of synthetic ovine corticotropin-releasing factor, glucocorticoids, catecholamines, neurohypophyseal peptides and other substances on cultured corticotropin cells, *Endocrinology* 113:1121 (1983).
58. S.L. Brown, L.R. Smith, and J.E. Blalock, Interleukin-1 and interleukin-2 enhance proopiomelanocortin gene expression in pituitary cells, *J. Immunol.* 139:3181 (1987).
59. J. Fukata, T. Usui, Y. Naitoh, Y. Makai, and H. Imura, Effects of recombinant human interleukin-1 alpha, -1 beta, 2 and 6 on ACTH synthesis and release in the mouse pituitary tumour cell line AtT-20, *J. Endocrinol.* 122:33 (1989).
60. F.G. Haour, E.M. Ban, G.M. Milon, D. Baron, and G.M. Fillion, Brain interleukin-1 receptors : characterization and modulation after lipopolysaccharide injection, *Prog. Neuroendocr. Immunol.* 3:196 (1990).
61. M. Ohmichi, K. Hirota, K. Koike, H. Kurachi, S. Ohtsuka, N. Matsuzaki, M. Yamaguchi, A. Miyaka, and O. Tanizawa, Binding sites for interleukin-6 in the anterior pituitary gland, *Neuroendocrinology* 55:199 (1992).
62. H. Vankelecom, P. Carmeliet, J. Van Damme, A. Billau, and C. Denef, Production of interleukin 6 by folliculo-stellate cells of the anterior pituitary gland in a histiotypic cell aggregate culture system, *Neuroendocrinology* 45:102 (1989).
63. J. Koenig, K. Snow, B.D. Clark, R. Toni, J.G. Cannon, A.R. Show, C.A. Dinarello, S. Reichlin, S.L. Lee, and R.M. Lechan, Intrinsic pituitary interleukin-1 beta is induced by bacterial lipopolysaccharide, *Endocrinology* 126:3053 (1990).
64. M. Boyle, G. Yamamoto, M. Chen, J. Rivier, and W. Vale, Interleukin-1 prevents loss of corticotropic responsiveness to beta-adrenergic stimulation in vitro, *Proc. Natl. Acad. Sci. USA* 85:5556 (1988).
65. H. Vankelecom, M. Andries, A. Billiau, and C. Denef, Evidence that folliculo-stellate cells mediate the inhibitory effect of interferon-gamma on hormone secretion in rat anterior pituitary cell

cultures, *Endocrinology* 130:3537 (1992).

66. K.J. Tracey, S.F. Lowry, T.J. Fahey III, J.D. Albert, Y. Fong, D. Hesse, B. Beutler, K.R. Manogue, S. Calvano, A. Cerami, and G.T. Shires, Cachetin/tumor necrosis factor induces lethal shock and stress hormone responses in the dog, *Surg. Gynecol. Obstet.* 164:415 (1987).

67. R.C. Gaillard, D. Turnill, P. Sappino, and A.F. Muller, Tumor necrosis factor alpha inhibits the hormonal response of the pituitary gland to hypothalamic releasing factors, *Endocrinology* 127:101 (1990).

68. H.G. Friesen, G.E. Mattia, and C.K.L. Too, Lymphoid tumor cells as models for studies of prolactin gene regulation and action, *Prog. Neuroendocr. Immunol.* 4:1 (1991).

69. C.K. Edwards, III, S.M. Giasuddin, J.M. Shepper, L.M. Yunger, and K.W. Kelley, A newly defined property of somatotropin : priming of macrophages for production of superoxide anion, *Science* 239:769 (1988).

70. M.J. Brennan, J.A. Betz, and M. Poth, Tumor necrosis factor inhibits ACTH stimulated corticosterone secretion by rat adrenal cortical cells, *70th Annual Meeting of the Endocrine Society, Seattle, USA* (Abstract 1453) (1989).

71. R. Natarajan, S. Ploszaj, R. Horton, and J. Nadler, Tumor necrosis factor and interleukin-1 are potent inhibitors of angiotensin-II-induced aldosterone synthesis, *Endocrinology* 125:3084 (1989).

72. T. Van der Poll, J.A. Romijn, E. Endert, and H. Sauerwein, Effects of tumor necrosis factor on the hypothamamic-pituitary-testicular axis in healthy men, *Metabolism* 42:303 (1993).

73. A. Uehara, P.E. Gottschall, R.R. Dahl, and A. Arimura, Interleukin-1 stimulates ACTH release by an indirect action which requires endogenous corticotropin releasing factor, *Endocrinology* 121:1580 (1987).

74. Y. Naitoh, J. Fukata, T. Tominaga, Y. Nakai, S. Tamai, K. Mori, and H. Imura, Interleukin-6 stimulates the secretion of adrenocorticotropic hormone in conscious freely-moving rats, *Biochem. Biophys. Res. Commun.* 155:1459 (1988).

75. R.C. Gaillard, Pituitary-Immune systems interactions, *Proceedings of the 3rd International Pituitary Congress* (1993).

76. F. Berkenbosch, J. Van Oers, A. Del Rey, F. Tilders, and H. Besedovsky, Corticotropin-releasing factor producing neurons in the rat activated by interleukin-1, *Science* 238:524 (1987).

77. R. Sapolsky, C. River, G. Yamamoto, P. Plotsky, and W. Vale, Interleukin-1 stimulates the secretion of hypothalamic releasing factor, *Science* 238:522 (1987).

78. F. Berkenbosch, J. Van Oers, A. Del Rey, F. Tilders, and H. Besedovsky, Corticotropin-releasing activity of monokines, *Science* 230:1035 (1987).

79. P. Saudan, J.M. Dubuis, R. Corder, P. Kehrer, A.F. Muller, and R.C. Gaillard, Activation of the corticotroph axis by interleukin-1 beta may involve both direct hypothalamic and pituitary effects, Journées Internationales HP Klotz d'Endocrinologie Clinique, Paris, France. *Hormone Research*. (Abstract), (1988).

80. R. Bernardini, T.C. Kamilaris, A.E. Calogero, E.O. Johnson, M.T. Gomez, P.W. Gold, and G.P. Chrousos, Interactions between tumor necrosis factor alpha, hypothalamic corticotropin releasing hormone, and adrenocorticotropin secretion in the rat, *Endocrinology* 126:2876 (1990).

81. E. Spinedi, R. Hadid, T. Daneva, and R.C. Gaillard, Cytokines stimulate the CRH but not the vasopressin neuronal system : evidence for a median eminence site of interleukin-6 action, *Neuroendocrinology* 56:46 (1992).

82. S. Tsagarakis, G. Gillies, L.H. Rees, M. Besser, and A. Grossman, Interleukin-1 directly stimulates the release of corticotropin releasing factor from rat hypothalamus, *Neuroendocrinology* 49:98 (1989).

83. T. Suda, F. Tozawa, T. Ushiyama, T. Sumitomo, M. Yamada, and H. Demura, Interleukin-1 stimulates corticotropin-releasing factor gene expression in rat hypothalamus, *Endocrinology* 126:1223 (1990).

84. W.L. Farrar, P.L. Kilian, M.R. Ruff, J.M. Hill, and C.B. Part, Visualization and characterization of interleukin-1 receptors in brain, *J. Immunol.* 139:459 (1987).

85. C.R. Plata-Salaman, Immunoregulations in the nervous system, *Neurosci. Biobehav. Rev.* 15:185 (1991).

86. L.J. Cornfield and M.A. Sills, High affinity interleukin-6 binding sites in bovine hypothalamus, *Eur. J. Pharmacol.* 202:113 (1991).

87. E. Hetier, J. Ajela, P. Dentfle, A. Boussean, P. Rouget, M. Mallat, and A. Prochiantz, Brain macrophages synthesize interleukin-1 and interleukin-1 mRNAs in vitro, *J. Neurosci. Res* 21:391 (1988).

88. A.P. Lieberman, P.M. Pitha, H.S. Shin, and M.L. Shin, Production of tumor necrosis factor and

other cytokines by atrocytes stimulated with lipopolysaccharide or neurotrophic virus, *Proc. Natl. Acad. Sci. USA* 86:6348 (1989).

89. K. Frei, U.V. Malipiew, J.P. Leist, R.M. Zinkernagel, M.E. Schwab, and A. Fontana, On the cellular source and function of interleukin-6 produced in the central nervous system in viral diseases, *Eur. J. Immunol.* 19:689 (1989).

90. B.L. Spangelo, A.M. Judd, R.M. MacLeod, D.W. Godman, and P.C. Isackson, Endotoxin-induced release of interleukin-6 from rat medial basal hypothalamus, *Endocrinology* 125:575 (1990).

91. G. Nistico and G. De Sarro, Is interleukin-2 a neuromodulator in the brain ? *Trends neurosci.* 14:146 (1991).

92. C.D. Breder, C.A. Dinarello, and C.B. Japler, Interleukin-1 immunoreactive innervation of the human hypothalamus, *Science* 240:321 (1988).

93. P.M. Plotzky, T. Emmet, Jr. Cunningham, and E.P. Widmaier, Catecholaminergic modulation of corticotropin-releasing factor and adrenocorticotropin secretion, *Endocr. Rev.* 10:437 (1989).

94. A. Szafarczyk, V. Guillaume, B. Conte-Devolx, G. Alonso, F. Malaval, N. Pares-Herbute, C. Oliver, and I. Assenmacher, Central catecholaminergic system stimulates secretion of CRH at different sites, *Am. J. Physiol.* 255:E463 (1988).

95. H. Besedovsky, A. Del Rey, E. Sorkin, M. Da Prada, R. Burri, and C. Honegger, The immune response evokes changes in brain noradrenergic neurons, *Science* 221:564 (1983).

96. S.L. Carlson, D.L. Felten, S. Livnat, and S.Y. Felten, Alterations of monoamines in specific central autonomic nuclei following immunization in mice, *Brain Behav. Immun.* 1:52 (1987).

97. A. Kabiersch, A. Del Rey, C.G. Honegger, and H.O. Besedovsky, Interleukin-1 induces changes in norepinephrine metabolism in the rat brain, *Brain Behav. Immun.* 2:267 (1988).

98. A.J. Dunn, Systemic interleukin-1 administrations stimulates hypothalamic norepinephrine metabolism paralleling the increased plasma corticosterone, *Life Sci.* 43:429 (1988).

99. S.G. Matta, J. Singh, R. Newton, and B.M. Sharp, The adrenocorticotropin response to interleukin-1 beta instilled into the rat median eminence depends on the local release of catecholamines, *Endocrinology* 127:2175 (1990).

100. C. Rivier, W. Vale, and M. Brown, In the rat, interleukin-1 alpha and beta stimulate adrenocorticotropin and catecholamine release, *Endocrinology* 125:3096 (1989).

101. S.B. Mizel, The interleukins, *FASEB J.* 3:2379 (1989).

102. J.D. Adrian, Interleukin-1 as a stimulator of hormone secretion, *Prog. Neuroendocr. Immunol.* 3:26 (1990).

103. J. Weidenfeld, O. Abramsky, and H. Ovadia, Evidence for the involvement of the central adrenergic system in interleukin-1 induced adrenocortical response, *Neuropharmacology* 28:1411 (1989).

104. R. Bernardini, A.E. Calogero, G. Mauceri, and G.P. Chrousos, Rat hypothalamic corticotropin-releasing hormone secretion in vitro is stimulated by interleukin-1 in an eicosanoid-dependent manner, *Life Sci.* 47:1601 (1990).

105. P. Navarra, S. Tsagarakis, M.S. Faria, L.H. Rees, G.M. Besser, and A.B. Grossman, Interleukin-1 and -6 stimulate the release of corticotropin-releasing hormone-41 from rat hypothalamus in vitro via the eicosanoid cyclooxygenase pathway, *Endocrinology* 128:37 (1990).

106. P.W. Gold and G.P. Chrousos, Arachidonic acid metabolites modulate rat hypothalamic corticotropin-releasing hormone secretion in vitro, *Neuroendocrinology* 50:708 (1989).

107. K. Lyson and S.M. McCann, Involvement of arachidonic acid cascade pathways in interleukin-6-stimulated corticotropin-releasing factor release in vitro, *Neuroendocrinology* 55:708 (1992).

108. J.T. Stitt, Evidence for the involvement of the organum vasculosum laminae terminalis in the febril responses of rabbits and rats, *J. Physiol.* 366:501 (1985).

109. J.T. Stitt, Passage of immunomodulators across the blood-brain barrier, *Yale J. Biol. Med.* 63:121 (1990).

110. B.M. Sharp, S.G. Matta, P.K. Peterson, R. Newton, C. Chav, and K. MacAllen, Tumor necrosis factor-alpha is a potent adreno-corticotropin secretagogue : comparison to interleukin-1 beta, *Endocrinology* 124:3131 (1989).

111. P. Navarra, G. Pozzoli, L. Brunetti, E. Ragazzoni, G.M. Besser, and A. Grossman, Interleukin-1 beta and interleukin-6 specifically increase the release of prostaglandin E2 from rat hypothalamic explants in vitro, *Neuroendocrinology* 56:61 (1992).

112. G. Katsuura, A. Arimura, K. Koves, and P.E. Gottschall, Involvement of organum vasculosum of lamina terminalis and preoptic area in interleukin 1 beta-induced ACTH release, *Am. J. Physiol.* 258:E163 (1990).

113. C.M. Blatteis, Neuromodulative actions of cytokines, *Yale J. Biol. Med.* 63:133 (1990).

114. C.M. Blatteis, Role of the OVLT in the febrile response to circulating pyrogens, *Prog. Brain Res.* 91:409 (1992).
115. G. Barbanel, G. Ixart, A. Szafarczyk, F. Malaval, and I. Assenmacher, Intrahypothalamic infusion of interleukin-1beta increases the release of corticotropin-releasing hormone (CRH-41) and adrenocorticotropic hormone (ACTH) in free-moving rats bearing a push-pull cannula in the median eminence, *Brain Res.* 516:31 (1990).
116. S. Rivest and C. Rivier, Influence of the paraventricular nucleus of the hypothalamus in the alteration of neuroendocrine functions induced by intermittent footshock or interleukin, *Endocrinology* 129:2049 (1991).
117. M.S. Roh, K.A. Drazenovich, J.J. Barbose, C.A. Dinarello, and C.F. Cobb, Direct stimulation of the adrenal cortex by interleukin-1, *Surgery* 102:140 (1987).
118. J.S.D. Winter, K.W. Gow, Y.S. Perry, and A.H. Greenberg, A stimulatory effect of interleukin-1 on adrenocortical cortisol secretion mediated by prostaglandins, *Endocrinology* 127:1904 (1990).
119. M.A. Salas, S.W. Evans, M.J. Levell, and J.T. Whicher, Interleukin-6 and ACTH act synergistically to stimulate the release of corticosterone from adrenal gland cells, *Clin. Exp. Immunol.* 79:470 (1990).
120. T. Tominaga, J. Fukata, Y. Naito, T. Usui, N. Murakami, M. Fukushima, Y. Nakai, Y. Hirai, and H. Imura, Prostaglandin-dependent in vitro stimulation of adrenocortical steroidogenesis by interleukins, *Endocrinology* 128:526 (1991).
121. P.G. Andreis, G. Neri, A.S. Belloni, G. Mazzocchi, A. Kasprzak, and G.G. Nussdorfer, Interleukin-1 beta enhances corticosterone secretion by acting directly on the rat adrenal gland, *Endocrinology* 129(1):53 (1991).
122. M. Schultzberg, C. Anderson, A. Unden, M. Troye-Blomberg, S.B. Svenson, and T. Bartfai, Interleukin-1 in adrenal chromaffin cells, *Neuroscience* 30:805 (1989).
123. A.M. Judd, B.L. Spangelo, and R.M. MacLeod, Rat adrenal zona glomerulosa cells produce interleukin-6, *Prog. Neuroendocr. Immunol.* 3(4):282 (1990).

THE ROLE OF CYTOKINES IN IMMUNE NEUROENDOCRINE INTERACTIONS

Hugo O. Besedovsky and Adriana del Rey

Department of Physiology
Philipps-Universitat Marburg
Fachbereich Humanmedizin
Deutschhausstrabe 2
D-35037 Marburg, Germany

In order to consider the interaction of immune and neuroendocrine systems, the following facts should be taken into account. (a) The immune system is capable of processing and responding to an enormous amount of information. Indeed, the immune system is surpassed only by the central nervous system (CNS) in terms of complexity. (b) The immune system participates in both physiological and pathological processes, and (c) both the immune and neuroendocrine systems are constantly in operation. Several soluble mediators released by the immune system have already been identified as messengers carrying afferent signals towards the CNS and there is little doubt that additional messengers will be found. Lymphokines, monokines, certain complement split products, immunoglobulins, histamine, serotonin, mediators of inflammation, thymic hormones, etc. are possible candidates for immune-CNS communication. Thus, it is possible that these messengers would carry information to the CNS about the type of immune response in operation, whereas the site of the response could be signaled through the local stimulation of nerve fibers by immune cell products released in their vicinity. The existence of afferent pathways from the immune system to the CNS implies, as we have previously suggested, that the immune system is a receptor-sensory organ.

The CNS exerts its regulatory influence on the immune system via hormones, neurotransmitters and neuropeptides, all of which have been demonstrated already to affect immunological events. These agents may affect their target cells directly or, alternatively, may modify immune reactions by the alteration of cytokine production, the suppressor-helper cell relationship, or the antibody class and idiotypic feedback regulatory network. Therefore, the classical view that the immune system is a self controlled and self monitored system is no longer tenable. Below we provide selective examples derived from our own work which support the concept of neuroendocrine-immune interactions. These observations show that (a) neuroendocrine mechanisms exert a tonic control on immune responses; (b) activation of immune cells affects neuroendocrine mechanisms; and (c) immune cell products are capable of affecting neuroendocrine functions and metabolic mechanisms.

Advances in Psychoneuroimmunology, Edited by
I. Berczi and J. Szélenyi, Plenum Press, New York, 1994

A. NEUROENDOCRINE EFFECTS ON IMMUNE PROCESSES

By now there is ample evidence to indicate that hormones, neurotransmitters and neuropeptides are able to affect immune processes.[1-6] It is also established that immunological organs are innervated. A direct contact between nerve fibers and immunocytes has also been observed. Lymphocytes and accessory cells express receptors for corticosteroids, insulin, prolactin, growth hormone, estradiol, testostserone, β-adrenergic agents, acetylcholine, endorphins, enkephalins, substance P (SP) and vasointestinal peptide (VIP). Therefore, cells of the immune system are able to interact with neuroendocrine messengers.

Numerous reports indicate that hormones are capable of stimulating or depressing immune reactions. In general, glucocorticoids, androgens, estrogens, progesterone, and endorphins depress the immune response *in vivo*, whereas growth hormone, prolactin, thyroxin and insulin have a stimulatory effect. Sex differences in immune reactivity and in the susceptibility to autoimmune disease are well documented. Endogenous opioid peptides also function as immunomodulators. (For literature review see 1 and 7.)

Although glucocorticoids are widely applied in medicine as immunosuppressive and antiinflammatory agents, the role of endogenous glucocorticoids in immunoregulation has not been investigated until recently. We studied the relationship between the level of endogenous corticosterone and the number of splenic immunoglobulin secreting cells in non-overtly immunized mice.[8] The number of immunoglobulin secreting cells was several fold increased in adrenalectomized animals and markedly reduced in stressed animals when compared to controls. Therefore an inverse relationship could be established between endogenous glucocorticoid levels and plaque forming cells in the spleen.

The effects of mediators of the autonomic nervous system on immune reactions is contradictory, but it seems certain that these neurotransmitters do have an influence, both *in vivo* and *in vitro*. We[9] and others[10] have observed that neonatal sympathectomy with 6-hydroxydopamine (6-OHDA) enhances the immune response to several antigens. Surgical denervation of the spleen had a similar effect in adult rats. However adult chemical sympathectomy had the opposite effect.[11-13] We also denervated animals at birth with 6-OHDA treatment and found that the number of immunoglobulin secreting cells in the spleen of non-overtly immunized animals was increased 2 to 3-fold at 5 months of age, compared to control animals. These results agree with the enhancing effect of sympathectomy on specific antibody forming cells in immunized animals and they demonstrate the role of sympathetic innervation under basal conditions in nonimmunized animals. Because 6-OHDA administered at birth does not only interfere with the sympathetic innervation of peripheral organs, but also with central adrenergic neurons, the studies do not allow a definite conclusion with regards to the final mechanism that affected the function of B cells.

Parasympathetic agents have been shown to increase antibody formation and cell mediated cytotoxicity. There is also compelling evidence showing that the manipulation of the brain by electrical lesions or stimulation of various regions can affect the immune response. Such manipulations may lead to the alteration of endocrine and autonomic control mechanisms. Similar neuroendocrine pathways may be responsible for the immunomodulatory effect of stress, circadian rhythms, and of conditioning of the immune response. (For literature review see 1 and 7.)

Neurotransmitters are capable of altering the level of intracellular second messengers and thereby modulating lymphocyte proliferation and activation. Glucocorticoids inhibit the production of IL-1, -2, -6, TNF, colony stimulating factor, and γ -interferon.[14-19] Other studies indicate that receptors for IL-1, IL-2 and TNF are also affected by glucocorticoids, insulin or prolactin, but a general conclusion cannot be drawn from the data at this time. Neurotransmitters and neuropeptides also influence the production of cytokines. For example, norepinephrine (NE) has been shown to decrease IL-1 production;[20] α -2 adrenergic agonists increase TNF release by lipopolysaccharide stimulated macrophages;[21] somatostatin (SOM) and VIP decrease IFNγ ; SP induces an increase in IL-1, TNF and IL-6 release from human peripheral blood leukocytes.[22,23] Epinephrine and SOM induce an increase in the number of receptors for TNF on monocytes,[24] β -adrenergic agonists induced a reduction of IL-2 receptors in lymphocytes,[25] and SP enhanced IL-2 receptors in gut associated lymphoid tissue.[26] Glucocorticoids affect the cytokine mediated expression of class I and II major histocompatibility (MHC) antigens on macrophages[18,27,28] and NE and VIP inhibited class II antigen expression on astrocytes.[29,30]

The evidence discussed above permits the following conclusions: (a) The immune system functions in an environment of hormones, neurotransmitters and neuropeptides. (b) Receptors for some of these mediators have been identified in immune cells. (c) The manipulation of the endocrine and central nervous systems can affect the immune response. (d) Naturally occurring stimuli mediated by the brain, such as stress and circadian rhythms, influence the immune response, and immune reactions can be behaviourally conditioned. (e) Neuroendocrine and antigenic stimuli share common intracellular pathways. (f) Hormones, neurotransmitters and neuropeptides can influence cytokine production and action and expression of MHC antigens.

B. IMMUNE PROCESSES AFFECT NEUROENDOCRINE MECHANISMS

It has been known for a long time that infection, inflammation and neoplastic disease elicit neuroendocrine alterations.[31] Such derangements could be directly caused by microorganisms and neoplastic cells and/or their products, or by substances released following tissue injury. The activation of the immune system also elicits neuroendocrine responses as illustrated by the examples given below. Increased glucocorticoid blood levels have been observed during the course of specific immune responses to different inocuous antigens.[32] These changes were detected only when the immune response was intense enough to reach a given threshold. In the case of Biozzi high-low responder strain of mice, significant increases in corticosterone blood levels were observed only in the immunologically high responder strain.[33] The elevation of corticosteroids in blood during immunization can interfere with the response to other unrelated antigens. A profound increase in adrenocorticotrophic hormone (ACTH) and corticosterone blood levels was also observed after the inoculation of mice with Newcastle disease virus (NDV), a virus that produces a mild disease in rodents.[34-36] The levels of ACTH and corticosterone in blood were increased several-fold at 2 hours after virus inoculation. Coculture of peripheral blood leukocytes with NDV resulted in the induction of a factor that stimulated the pituitary-adrenal axis in normal noninfected animals. This effect was neutralized when the active material was incubated with anti-IL-1 serum. Thus it was demonstrated for the first time that IL-1 can mediate the increase in glucocorticoid levels following the administration of a natural infective agent.[35]

Bacterial endotoxins also can stimulate the pituitary-adrenal axis and the increase in glucocorticoid levels has a protective effect during septic shock.[37] The depletion of macrophages in rats prior to the injection of LPS abrogates the increase of ACTH and corticosterone in the blood.[38] This endocrine response is predominantly elicited by IL-1 as administration of an IL-1 receptor antagonist also abrogates the increase in adrenocortical hormones following LPS administration.[39]

An early increase in glucocorticoid blood levels was also observed following syngeneic tumor transplantation.[40] A biphasic early and late increase in glucocorticoid levels was observed following the administration of EL-4 lymphoma cells. The presence of T cells and compatibility at MHC loci was required for the manifestation of this endocrine response.[41,42]

Immunization may elicit a reduction in the level of thyroid hormones,[32] an early increase in prolactin (PRL).[43] The inoculation of tumoral, but not of normal cells caused an increase in PRL and corticosterone and a decrease in thyroxin, insulin and testosterone blood levels 24-48 hours later.[40] These endocrine changes preceded the overt appearance of the tumor by several days. The changes in insulin and testosterone levels were biphasic.

The autonomic nervous system also responds to immunization. We have observed a decreased sympathetic activity in the spleen of rats as evaluated by NE content and turnover during the immune response.[9,44] The administration of Freund's complete adjuvant also induces a decrease in the total NE content of spleen that lasts several days[45] and the NE content in lymphoid organs of germ free animals is higher than those of immunologically more active specific pathogen free animals.[46]

The hypothalamus integrates most endocrine and autonomic functions. For this reason, we studied the effect of immune reactions on the rate of firing of individual hypothalamic neurons at various intervals after injection to rats of SRBC or TNP hemocyanin.[42] During the response to TNP hemocyanin the frequency of firing in the ventromedial nucleus of the hypothalamus was increased on day 2 (peak of the IgM response) and in SRBC stimulated animals on day 5. No statistically significant changes were detected at the time of the peak immune response to SRBC in the arcuate, premamillaris dorsalis, paraventricular nuclei, the preoptic, anterior and posterior areas of the hypothalamus or in the reuniens thalamic nucleus (unpublished results). Therefore the hypothalamic response to antigenic challenge does not involve general neural activation. On the contrary, the data suggests that the signals derived from activated immune cells follows specific pathways. Changes in the rate of firing and of neurons of the paraventricular hypothalamic nucleus and in the anterior hypothalamic area were detected after immunization with SRBC of free moving conscious rats bearing chronic recording electrodes.[48]

Catecholamine turnover rates were studied in different regions of the CNS during the immune response to SRBC.[49] In immunologically high responder rats a marked decrease in the hypothalamic NE turnover rate compared to saline injected controls was observed 4 days after antigenic stimulation, whereas the turnover rate of low responder animals was almost equal to those of controls. Others observed that during the immune response to SRBC the NE content is particularly reduced in the paraventricular nucleus and hypothalamus.[50]

C. IMMUNE CELL PRODUCTS CAN EFFECT NEUROENDOCRINE FUNCTIONS

Our prediction was that among other agents lymphokines and monokines could serve as mediators of immune-brain communication. The injection of supernatants from lymphocytes stimulated with mitogens or antigens *in vitro* into normal animals increased ACTH and corticosterone output[51] and decreased NE content of the hypothalamus,[49] thus mimicking two of the neuroendocrine changes observed following *in vivo* immunization. Later we showed that both natural purified and recombinant human IL-1 can stimulate ACTH and corticosterone output in mice and rats and that IL-1 is the most likely mediator of glucocorticoid changes induced by certain viruses.[35] This effect of IL-1 seems to be rather specific because it is not parallelled by changes in other stress hormones such as growth hormone, prolactin and α-melanocyte stimulating hormone. The levels of catecholamines in blood are only modestly increased following IL-1 injection.[52] It was also shown that IL-1 induced increase in corticosterone levels is mediated *in vivo* by corticotropin releasing factor.[53-55] However, other investigators[56,57] observed a direct effect of IL-1 on ACTH release using long term cultures of normal pituitary cells. It was also shown that IL-6, TNF and IFNγ stimulate the pituitary-adrenal axis.[58,59] Thymosin fraction 5 also increased corticosterone blood levels *in vivo*.[60] We found that IL-1 is more potent than TNF or IL-6 in ACTH release.[61] Conversely, a macrophage derived factor that inhibits ACTH steroidogenesis has also been found.[62] By now it is clear that the pituitary-thyroid axis, the pituitary-gonadal axis, and growth hormone secretion can also be affected by IL-1 and/or IL-6 and/or TNF and/or γ-interferon. (For literature review see 7.)

There is evidence that immune cells, most likely by releasing soluble mediators, can influence peripheral nerve activity. For example, we have observed that the presence of T cells in the spleen can modulate the sympathetic innervation of this organ.[63] Others observed that IL-1 induces nerve growth factor (NGF) production *in vitro*.[64] Lymphokines and monokines can also affect CNS mechanisms. For example, certain cytokines inhibit NE metabolism in the brain and affect the electrical activity of brain neurons. They can also influence sleep and termoregulation. Moreover, stimulated glial cells and certain neurons can produce various cytokines and supernatants containing lymphokines can support the growth of neurons in culture in the absence of NGF. Other potential immunologic messengers include histamine, serotonin, pituitary like peptide hormones, vasopressin, oxytocin. (For literature review see 7.)

D. BIOLOGICAL RELEVANCE OF IMMUNE NEUROENDOCRINE NETWORKS

The neuroendocrine response to immunization implies the existence of a regulatory interaction between these two systems.[65] A major difficulty is to explain how nonspecific regulatory agents, such as hormones and neurotransmitters, can contribute to the control of a highly specific mechanism such as the immune response to a given antigen. We postulated earlier that this problem could be circumvented if resting and activated immunological cells would have different sensitivities to neuroendocrine signals.[7,65,66] For instance, antigen presentation and the release of certain lymphokines and monokines that are necessary for the control of immune reactions are sensitive to glucocorticoids. However, the early events of the immune response precede the increase of endogenous glucocorticoids and thus clonal expansion will not be affected. It was also reported that IL-1 protects specific T helper cells, but not suppressor and cytotoxic T cells from inhibition by glucocorticoids.[67] Another as yet unidentified factor derived from mitogen stimulated spleen cells also seems to block selectively the suppressive action of glucocorticoids on helper cells.[68] As the immune response proceeds, the stimulation of the hypothalamus-pituitary-adrenal (HPA) axis will result in inhibition of the production of certain lympho-monokines.[15-19] The inhibitory effects of glucocorticoids during the course of an immune response explains, at least in part, the phenomenon of sequential antigenic competition in which the response to one antigen inhibits the response to a second unrelated antigen administered later. This competition can be abrogated to a large extent by adrenalectomy or hypophysectomy.[66,69]

There is reasonable proof to indicate that normal communication between the HPA axis and the immune system can play a protective role, whereas disruption of this communication can lead to predisposition to, or aggravation of, disease. It is known that glucocorticoids reach a very high level in the blood during sepsis. This increase must be protective for the host as indicated by the high susceptibility of adrenalectomized animals to endotoxins.[37] The stimulatory effect of LPS and glucocorticoid output is mediated by cytokines released by macrophages,[38] most likely IL-1. Another example of the protective action of the HPA axis is observed in Lew-N female rats which are susceptible to experimentally induced arthritis. These rats cannot respond to immunization with an elevation of serum glucocorticoid levels and thus are unable to attenuate autoimmune reactions.[70] A disturbance of the HPA axis has also been observed in patients with rheumatoid arthritis.[71]

Obese (OS) chickens, which develop spontaneously autoimmune thyroiditis, do not respond with an elevation of corticosterone blood levels to immunization with SRBC, whereas normal chickens do.[72] Serum glucocorticoid levels increase in rats before and during the overt clinical development of experimental allergic encephalomyelitis (EAE). That such increased levels help to moderate the clinical course of the disease is illustrated by the observation that adrenalectomized rats with EAE die during the first attack.[73,74] However, immune mediated glucocorticoid release may not always be beneficial to the host. In the case of virus infection where glucocorticoid levels are also elevated, the resulting immunosuppression may in fact be responsible for the appearance of opportunistic superinfections that are so common during viral diseases.

The existence of endogenous pyrogens have been discovered more than 40 years ago. It was also well known that during certain infectious and neoplastic diseases profound metabolic derangements occur. That immune responses can also affect metabolic processes is illustrated by the observation that 5 days after the administration of SRBC to rats a modest but significant (15%) decrease in glucose blood levels occurs. Similar changes can be induced by supernatants of stimulated lymphocytes in mice.[7] Moreover, there is compelling evidence that certain cytokines, like IL-1, TNF and IL-6, can alter intermediate metabolism.[75-79] Although much of the evidence indicating the metabolic effect of cytokines comes from pharmacological studies, there is indication that antigens and cytokines mediate similar metabolic alterations during the course of diseases that involve the immune sytem. We have observed that the administration of nanogram amounts of recombinant IL-1β into C3H/HeJ mice resulted in the reduction by 40-50% of glucose blood levels. This effect is long lasting as the animals remain hypoglycemic for 8-12 hours. At the same time IL-1 produced a transient increase (2-3-fold) in insulin and glucagon blood levels.[76] In other strains of mice IL-1 induced a comparable hypoglycemia but no increase in insulin levels.[80] Glucose tolerance tests indicated that the "set point" of the regulatory mechanisms that stabilize glucose concentrations in the blood is adjusted to a lower level during IL-1 treatment.[76,80] The IL-1 induced increase in glucose levels develops against increased levels of hormones such as glucocorticoids and glucagon and counteregulatory mechanisms mediated by epinephrine and glucocorticoids can only moderate IL-1 induced hypoglycemia.[76]

Treatment of insulin resistant db/db mice with IL-1 resulted in a profound decrease in glucose blood levels. The levels attained were comparable to those of normal animals. As expected, db/db mice did not respond to high doses of insulin. The euglycemic heterozygous db/+ mice responded to IL-1 in the same way as normal animals with long lasting hypoglycemia. No changes in insulin blood levels were observed in either db/db or db/+ mice. Similar results were obtained when IL-1 was injected to the insulin resistant diabetic C57BL/6J ob/ob mice.[81] The injection of IL-1 to the insulin resistant diabetic Zucker fa/fa rats also induced hypoglycemia and reduction of total lipids and triglyceride levels. The triglyceride response seemed to be biphasic because levels were increased 4 hours after IL-1. Free fatty acids in blood were increased (del Rey & Besedovsky unpublished). No increases in insulin levels after IL-1 administration were observed in diabetic animals at any of the time points studied. The hypoglycemic effect of IL-1 could be based on reduction of endogenous glucose production by the direct inhibition of gluconeogenic enzymes, for instance. IL-1 was already shown to partially inhibit the *in vitro* induction of phosphoenolpyruvate-carboxy-kinase (PEPCK) by glucocorticoids which is a rate limiting enzyme in gluconeogenesis.[82] This finding is in line with previous data indicating the existence of a monocyte derived factor that antagonizes the induction of PEPCK by glucocorticoids. However, hypoglycemia is more pronounced in adrenalectomized animals which seems to contradict this hypothesis. Crude supernatants from activated macrophages were shown to stimulate glucose oxidation in fat pads,[83] which could indicate an alternate mechanism. IL-1 is capable of increasing glucose transport in adipocytes though less efficiently than insulin.[84] Other authors found that cytokines can increase glucose clearance rate.[85]

E. CONCLUSIONS

The function of the immune system is to eliminate foreign microorganisms and macromolecules and aberrant cells with altered self antigens. Through this activity the immune system contributes to the maintenance of the *milieu interieur*. Therefore, the immune system, like all other homeostatic systems, is under neuroendocrine control. Evidence is rapidly increasing that the immune system is able to receive neuroendocrine signals and in turn emits feedback signals that affect autonomic and CNS mechanisms. On this basis we have proposed that the degree of activity of such a complex network of functional interactions is changed by stimuli acting at the level of either the immune or neuroendocrine system. Networks between locally produced or released hormones and neurotransmitters and cytokines may operate at organ and tissue level. The degree of activity of the network can be changed by stimuli acting at or generated from any of its components, e.g. antigens at the level of the immune system, or psychosocial stimuli at the level of CNS. In turn, due to the tonic functioning of the network, all components are kept informed and influenced mutually by their functional activity. For instance, the effect of stress on the immune system could be thought of as stress induced changes in the degree of tonic interaction within the network. Behavioural conditioning of the immune response can also be interpreted as a process that affects certain levels of a constantly operating network of brain-immune interactions. Conversely the activation of the immune system may have consequences beyond immunoregulation and affect general homeostasis. Homeostatic mechanisms that operate under basal conditions may differ qualitatively and quantitatively from those that are active during pathological states involving the immune system. In fact the set point for the regulation of essential variables may need to be readjusted during the course of these states and powerful immune derived messengers may mediate these adjustments. However immune-neuroendocrine interactions could contribute to the aggravation of certain diseases, which indicates that the system may become antihomeostatic in some situations.

REFERENCES

1 .R. Ader, D. Felten, and N. Cohen, eds., "Psychoneuroimmunology," 2nd ed., Academic Press, New York (1991).
2. I. Berczi and K. Kovacs, eds., "Hormones and Immunity," MTP Press, Lancaster, U.K. (1987).
3. N. Fabris, E. Garaci, V. Hadden, and N.A. Mitchison, eds., "Immunoregulation," Plenum Press, New York (1983).
4. R. Guillemin, M. Cohn, and T. Melnechuk, eds., "Neural Modulation of Immunity," Raven Press, New York (1985).
5. J.W. Hadden, K. Masek, and G. Nistico, eds., "Interactions Among CNS, Neuroendocrine and Immune Systems," Pythagora Press, Rome (1989).
6. M.S. O'Dorisio and A. Panerai, eds. "Neuropeptides and Immunopeptides: Messengers in a Neuroimmune axis," Ann. N.Y. Acad. Sci., New York (1990).
7. H.O. Besedovsky and A. del Rey, Immune-neuroendocrine circuits: integrative role of cytokines, *in:* "Frontiers in Neuroendocrinology," W.F. Ganong and L. Martini, eds., Raven Press, NY (1992).
8. A. del Rey, H.O. Besedovsky, and E. Sorkin, Endogenous blood levels of corticosterone control the immunological cell mass and B cell activity in mice, *J. Immunol.* 133:572 (1984).
9. H.O. Besedovsky, A. del Rey, E. Sorkin, M. Da Prada, and H.A. Keller, Immunoregulation mediated by the sympathetic nervous system, *Cell. Immunol.* 48:346 (1979).
10. L. Miles, J. Quintans, E. Chelmicka-Schorr, and E.G.W. Arnason, The sympathetic nervous system modulates antibody response to thymus-independent antigens, *J. Neuroimmunol.* 1:101 (1981).
11. N.R. Hall, J.E. McClure, S.K. Hu, N.S. Tare, C.M. Seals, and A.L. Goldstein, Effects of 6-hydroxydopamine upon primary and secondary thymus dependent immune responses, *Immunopharmacol.* 5:39 (1982).
12. K. Kasahara, S. Tanaka, T. Ito, and Y. Hamashima, Suppression of the primary immune response by chemical sympathectomy, *Res. Commun. Chem. Pathol. Pharmacol.* 16:687 (1977).
13. S. Livnat, S.Y. Felten, S.L. Carlson, D.L. Bellinger, and D.L. Felten, Involvement of peripheral and central catecholamine systems in neural-immune interactions, *J. Neuroimmunol.* 10:5 (1985).
14. B. Beutler, N. Krochin, I.W. Milsark, C. Luedke, and A. Cerami, Control of cachectin (tumor necrosis factor) synthesis: mechanisms of endotoxin resistance, *Science* 232:977 (1986).
15. R.A. Daynes and B.A. Aranco, Contrasting effects of glucocorticoids on the capacity of T cells to produce the growth factors interleukin 2 and interleukin 4, *Eur. J. Immunol.* 19:2319 (1989).
16. S. Gillis, G.R. Crabtree, and K. Smith, Glucocorticoid-induced inhibition of T cell growth factor production. I. The effect on mitogen-induced lymphocyte proliferation, *J. Immunol.* 123:1624 (1979).
17. A. Kelso and A. Munck, Glucocorticoid inhibition of lymphokine secretion by alloreactive T lymphocyte clones, *J. Immunol.* 133:784 (1984).
18. D.S. Snyder and E.R. Unanue, Corticosteroids inhibit murine macrophage Ia expression and interleukin-1 production, *J. Immunol.* 129:1803 (1982).
19. S.M. Wahl, L.C. Altman, and D.L. Rosenstreich, Inhibition of in vitro lymphokine synthesis by glucocorticoids, *J. Immunol.* 115:476 (1975).
20. W.C. Koff, A.V. Fann, M.A. Dunegan, and L.B. Lachman, Catecholamine-induced suppression of interleukin-1 production, *Lymphokine Res.* 5:239 (1986).
21. R.N. Spengler, R.M. Allen, D.G. Remick, R.M. Strieter, and S.L. Kunkel, Stimulation of alpha-adrenergic receptor augments the production of macrophage-derived tumor necrosis factor, *J. Immunol.* 145:1430 (1990).
22. M. Lotz, J.H. Vaughan, and D.A. Carson, Effect of neuropeptides on the production of inflammatory cytokines by human monocytes, *Science* 241:1218 (1988).
23. M. Muscettola and G. Grasso, Somatostatin and vasoactive intestinal peptide reduce interferon gamma production by human peripheral blood mononuclear cells, *Immunobiology* 180:419 (1990).
24. L.E. Bermudez, M. Wu, and L.S. Young, Effect of stress-related hormones on macrophage receptors and response to tumor necrosis factor, *Lymphokine Res.* 9:137 (1990).
25. R.D. Feldman, G.W. Hunninghake, and W.L. McArdle, Beta-adrenergic-receptor-mediated suppression of interleukin 2 receptors in human lymphocytes, *J. Immunol.* 139:3355 (1987).
26. R. Hart, H. Dancygier, F. Wagner, H. Niedermeyer, and M. Classen, Substance P modulates

lymphokine activities in supernatants of cultured human duodenal biopsies, *Immunol. Lett.* 19:133 (1988).

27. J. Rhodes, J. Ivanyi, and P. Cozens, Antigen presentation by human monocytes: effects of modifying major histocompatibility complex class II antigen expression and interleukin 1 production by using recombinant interferons and corticosteroids, *Eur. J. Immunol.* 16:370 (1986).

28. L. Shen, P. Guyre, E. Ball, and M. Fanger, Glucocorticoid enhances gamma interferon effects on human monocyte antigen expression and ADCC, *Clin. Exp. Immunol.* 65:387 (1986).

29. E. Frohman, T. Frohman, B. Vayuvegula, S. Gupta, and S. van den Noort, Vasoactive intestinal polypeptide inhibits the expression of the MHC class II antigens on astrocytes, *Neurol. Sci.* 88:339 (1988).

30. E. Frohman, B. Vayuvegula, S. Gupta, and S. van den Noort, Norepinephrine inhibits gamma-interferon-induced major histocompatibility class II (Ia) antigen expression on cultured astrocytes via beta-2-adrenergic signal transduction mechanisms, *Proc. Natl. Acad. Sci.* 85:1292 (1988).

31. W.R. Beisel, Magnitude of the host nutritional response to infection, *Am. J. Clin. Nutrition* 30:1236 (1977).

32. H.O. Besedovsky, E. Sorkin, M. Keller and J. Muller, Changes in blood hormone levels during the immune response, *Proc. Soc. Exp. Biol.* 150:466 (1975).

33. P.N. Shek and B.H. Sabiston, Neuroendocrine regulation of immune processes: changes in circulating corticosterone levels induced by the primary antibody response in mice, *Int. J. Immunopharmacol.* 5:23 (1983).

34. H.O. Besedovsky and A. del Rey, Mechanism of virus-induced stimulation of the hypothalamus-pituitary-adrenal axis, *J. Steroid Biochem.* 34:235 (1989).

35. H.O. Besedovsky, A. del Rey, E. Sorkin, and C.A. Dinarello, Immunoregulatory feedback between interleukin-1 and glucocorticoid hormones, *Science* 233:652 (1986).

36. A.J. Dunn, M.L. Powell, W.V. Moreshead, J.M. Gaskin, and N.R. Hall, Effect of Newcastle Disease Virus administration to mice on the metabolism of cerebral monoamines, plasma corticosterone and lymphocyte proliferation, *Brain Behav. Immun.* 1:216 (1987).

37. K. Nakano, S. Suzuki, and C. Oh, Significance of increased secretion of glucocorticoids in mice and rats injected with bacterial endotoxin, *Brain Behav. Immun.* 1:159 (1987).

38. R. de Rijk, N. van Rooijen, H. Besedovsky, A. del Rey, and F. Berkenbosch, Selective depletion of macrophages prevents pituitary-adrenal activation in response to subpyrogenic but not to pyrogenic doses of bacterial endotoxin in rats, *Endocrinology* 129:330 (1991).

39. C. Rivier, R. Chizzonite, and W. Vale, In the mouse, the activation of the hypothalamic-pituitary-adrenal axis by a lipopolysaccharide (endotoxin) is mediated through interleukin-1, *Endocrinology* 125:2800 (1989).

40. H.O. Besedovsky, A. del Rey, M. Schardt, E. Sorkin, S. Normann, J. Baumann and J. Girard, Changes in plasma hormone profile after tumor transplantation into syngeneic and allogeneic rats, *Int. J. Cancer* 36:209 (1985).

41. S. Normann, H.O. Besedovsky, M. Schardt, and A. del Rey, Hormonal changes following tumor transplantation and the relationship of corticosterone to tumor induced anti-inflammation, *Int. J. Cancer* 41:850 (1988).

42. S. Normann, H.O. Besedovsky, M. Schardt, and A. del Rey, Interactions between endogenous glucocorticoids and inflammatory responses in normal and tumor bearing mice: role of T cells, *J. Leukocyte Biol.* 44:551 (1988).

43. M. Neidhart and D.F. Larson, Freund's complete adjuvant induces ornithine decarboxylase activity in the central nervous system of male rats and triggers the release of pituitary hormones, *J. Neuroimmunol.* 26:97 (1990).

44. H.O. Besedovsky, A. del Rey, and E. Sorkin, Immunnoneuroendocrine interactions, *J. Immunol.* 135S:750s (1985).

45. F.J. Mackenzie, J.P. Leonard, and M.L. Cuzner, Changes in lymphocyte b-adrenergic receptor density and norepinephrine content of the spleen are early indicators of immune reactivity in acute experimental allergic encephalomyelitis in the Lewis rat, *J. Neuroimmunol.* 23:93 (1989).

46. A. del Rey, H.O. Besedovsky, E. Sorkin, M. Da Prada, and S. Arrenbrecht, Immunoregulation mediated by the sympathetic nervous system II, *Cell. Immunol.* 63:329 (1981).

47. H.O. Besedovsky, E. Sorkin, D. Felix, and H. Haas, Hypothalamic changes during the immune response, *Eur. J. Immunol.* 7:323 (1977).

48. D. Saphier, O. Abramsky, G. Mor, and H. Ovadia, Multiunit electrical activity in conscious rats

during an immune response, *Brain Behav. Immun.* 1:40 (1987).

49. H.O. Besedovsky, A. del Rey, E. Sorkin, M. Da Prada, R. Burri, and C.G. Honegger, The immune response evokes changes in brain noradrenergic neurons, *Science* 221:564 (1983).

50. S.L. Carlson, D.L. Felten, S. Livnat, and S.Y. Felten, Alterations of monoamines in specific central autonomic nuclei following immunization in mice, *Brain Behav. Immun.* 1:52 (1987).

51. H.O. Besedovsky, A. del Rey, and E. Sorkin, Lymphokine containing supernatants from con A-stimulated cells increase corticosterone blood levels, *J. Immunol.* 126:385 (1981).

52. F. Berkenbosch, D.E.C. de Goeij, A. del Rey, and H. Besedovsky, Neuroendocrine, sympathetic and metabolic responses induced by interleukin-1, *Neuroendocrinology* 50:570 (1989).

53. F. Berkenbosch, J. Van Oers, A. del Rey, F. Tilders, and H.O. Besedovsky, Corticotropin releasing factor producing neurons in the rat are activated by interleukin-1, *Science* 238:524 (1987).

54. R. Sapolsky, C. Rivier, G. Yamamoto, P. Plotsky, and W. Vale, Interleukin-1 stimulates the secretion of hypothalamic corticotropin-releasing factor, *Science* 238:522 (1987).

55. A. Uehara, P.E. Gottschall, R.R. Dahl, and A. Arimura, Interleukin-1 stimulates ACTH release by an indirect action which requires endogenous corticotropin releasing factor, *Endocrinology* 121:1580 (1987).

56. E.W. Bernton, J.E. Beach, J.W. Holaday, R.C. Smallridge, and H.G. Fein, Release of multiple hormones by a direct action of interleukin-1 on pituitary cells, *Science* 238:519 (1987).

57. P. Kehrer, D. Turnill, J-M. Dayer, A.F. Muller, and R.C. Gaillard, Human recombinant interleukin-1 beta and -alpha, but not recombinant tumor necrosis factor alpha stimulate ACTH release from rat anterior pituitary cells in vitro in a prostaglandin E_2 and cAMP independent manner, *Neuroendocrinology* 48:160 (1988).

58. F. Holsboer, G.K. Stalla, U. von Bardeleben, K. Hamman, H. Muller, and O.A. Muller, Acute adrenocortical stimulation by recombinant gamma interferon in human controls, *Life Sci.* 42:1 (1988).

59. B.M. Sharp, S.G. Matta, P.K. Peterson, R. Newton, C. Chao, and K. McAllen, Tumor necrosis factor-alpha is a potent ACTH secretagogue: comparison to interleukin-1 beta, *Endocrinology* 124:3131 (1989).

60. G.V. Vahouny, E. Kyeyune-Nyombi, J.P. McGillis, N.S. Tare, K.-Y. Huang, R. Tombes, A.L. Goldstein and N.R. Hall, Thymosin peptides and lymphomonokines do not directly stimulate adrenal corticosteroid production in vitro, *J. Immunol.* 130:791 (1983).

61. H.O. Besedovsky, A. del Rey, I. Klusman, H. Furukawa, G. Monge-Arditi, and A. Kabiersch, Cytokines as modulators of the hypothalamus-pituitary-adrenal axis, *Molec. Biol.* 40:613 (1991).

62. J.C. Mathison, R.D. Schreiber, A.C. La Forest, and R.J. Ulevitch, Suppression of ACTH-induced steroidogenesis by supernatant from LPS-treated peritoneal exudate macropahges, *J. Immunol.* 130:2757 (1983).

63. H.O. Besedovsky, A. del Rey, E. Sorkin, R. Burri, C.G. Honegger, M. Schlumpf and W. Lichtensteiger, T lymphocytes affect the development of sympathetic innervation of mouse spleen, *Brain Behav. Immun.* 1:185 (1987).

64. D. Lindholm, R. Heumann, M. Meyer, and H. Thoenen, Interleukin-1 regulates synthesis of nerve growth factor in non-neuronal cells of rat sciatic nerve, *Nature* 330:658 (1987).

65. H.O. Besedovsky and E. Sorkin, Immune neuroendocrine netowrk, *Clin. Exp. Immunol.* 27:1 (1977).

66. H.O. Besedovsky, A. del Rey, and E. Sorkin, Antigenic competition between horse and sheep red blood cells as a hormone-dependent phenomenon, *Clin. Exp. Immunol.* 37:106 (1979).

67. L.M. Bradley and R.I. Mishell, Selective protection of murine thymic helper T cells from glucocorticoid inhibition by macrophage-derived mediators, *Cell. Immunol.* 73:115 (1982).

68. S.S. Fairchild, K. Shannon, E. Kwan, and R.I. Mishell, T cell-derived glucosteroid response-modifying factor ($GRMF_r$): a unique lymphokine made by normal T lymphocytes and a T cell hybridoma, *J. Immunol.* 132:821 (1984).

69. S. Tokuda, L.C. Trujillo, R.A. Nofchissey, Hormonal regulation of the immune response, *in:* "Stress, Immunity and Aging," E.L. Cooper, ed., Marcel Dekker Inc., New York (1984).

70. E.M. Sternberg, J.M. Hill, G.P. Chrousos, T. Kamilaris, S.J. Listwak, P.W. Gold and R.L. Wilder, Inflammatory mediator-induced hypothalamic-pituitary-adrenal axis activation is defective in streptococcal cell wall arthritis-susceptible Lewis rats, *Proc. Natl. Acad. Sci. USA* 86:2374 (1989).

71. G. Neek, K. Federlin, V. Graef, D. Rusch, and K.L. Schmidt, Adrenal secretion of cortisol in patients with rheumatoid arthritis, *J. Rheumatol.* 17:24 (1990).

72. K. Schauenstein, R. Faessler, H. Dietrich, S. Schwarz, G. Kroemer, and G. Wick, Disturbed

immune-endocrine communication in autoimmune disease. Lack of corticosterone response to immune signals in obese strain chickens with spontaneous autoimmune thyroiditis, *J. Immunol.* 139:1830 (1987).

73. S. Levine, R. Sowinski, and B. Steinetz, Effects of experimental allergic encephalomyelitis on thymus and adrenal: relation to remission and relapse, *Proc. Soc. Exp. Biol. Med.* 165:218 (1980).

74. I.A.M. MacPhee, F.A. Antoni, and W.D. Mason, Spontaneous recovery of rats from experimental allergic encephalomyelitis is dependent on regulation of the immune system to endogenous adrenal corticosteroids, *J. Exp. Med.* 169:431 (1989).

75. B. Beutler, I.W. Milsark, and A.C. Cerami, Cachectin tumor necrosis factor: production, distribution and metabolic rate in vivo, *J. Immunol.* 135:3972 (1985).

76. A. del Rey and H.O. Besedovsky, Interleukin-1 affects glucose homeostasis, *Am. J. Physiol.* 253:R794 (1987).

77. S.K. Durum, J.J. Oppenheim, and R. Neta, Immunophysiologic role of interleukin 1, *in*: "Immunophysiology: the Role of Cells and Cytokines in Immunity and Inflammation," J.J. Oppenheim, E.M. Shevach, eds., Oxford University Press Inc., New York (1990).

78. E.A. Flores, B.R. Bistrian, J.J. Pomposelli, C.A. Dinarello, G.L. Blackburn, and N.W. Istfan, Infusion of tumor necrosis factor/cachectin promotes muscle catabolism in the rat. A synergistic effect with interleukin 1, *J. Clin. Invest.* 83:1614 (1989).

79. N.J. Rothwell, CRF is involved in the pyrogenic and thermogenic effects of interleukin 1b in the rat, *Am. J. Physiol.* 256:E111 (1989).

80. H.O. Besedovsky and A. del Rey, Interleukin-1 and glucose homeostasis: an example of the biological relevance of immune-neuroendocrine interactions, *Horm. Res.* 31:94 (1989).

81. A. del Rey and H.O. Besedovsky, Antidiabetic effects of interleukin-1, *Proc. Natl. Acad. Sci. USA* 86:5943 (1989).

82. M.R. Hill, R.D. Stith, and R.E. McCallum, Human recombinant IL-1 alters glucocorticoid receptor function in Reuber hepatoma cells, *J. Immunol.* 141:1522 (1988).

83. J.P. Filkins, Endotoxin-enhanced secretion of macrophage insulin-like activity, *J. Reticuloendothel. Soc.* 27:507 (1980).

84. A. Garcia-Welsh, J.S. Schneiderman, and D.L. Baly, Interleukin-1 stimulates glucose transport in the rat adipose cells. Evidence for receptor discrimination between IL-1 beta and alpha, *FEBS Lett.* 269:421 (1990).

85. E.A. Flores, N. Istfan, J.J. Pomposelli, G.L. Blackburn, and B.R. Bistria, Effect of interleukin-1 and tumor necrosis factor/cachectin on glucose turnover in the rat, *Metabolism* 39:738 (1990).

CYTOKINES AND SLEEP MECHANISMS

James M. Krueger[1] and Linda Toth[2]

University of Tennessee, Memphis
Departments of Physiology & Biophysics[1] and Comparative Medicine[2]
894 Union Avenue
Memphis, TN 38163

INTRODUCTION

The primary thesis of this essay is that infectious disease induces changes in sleep and that these sleep alterations are mediated by cytokines. More specifically, we propose that specific microbial products stimulate cytokine production and that these cytokines, in turn, induce the changes in sleep that occur during infections. Included in this scheme is the proposal that the cytokine interleukin-1 (IL1) is also involved in physiological sleep regulation.

Sleep is classified into two major types, non-rapid eye movement sleep (NREMS), which also called slow-wave sleep (SWS) in animals, and REMS, which is also called paradoxical sleep. Recognition and differentiation of the two states of sleep or of sleep and quiet wakefulness is difficult on the basis of behavioral criteria; thus, physiological measurements including the electroencephalogram (EEG), brain temperature (T_{br}), movement and electromyogram (EMG), are often used. NREMS is associated with high amplitude EEG slow-waves (1/2 - 4 Hz) and decreasing T_{br}, while REMS is characterized by a more rapid EEG, similar to wakefulness (W), a rapid increase in T_{br} and muscle atonia (a flat EMG). In many animals, episodes of NREMS and REMS are relatively short, usually lasting less than 2 min. Typically, an animal will enter NREMS from W, and will then either wake up or enter REMS. Animals seldom enter REMS from W; in fact, this sequence is a sign of sleep pathology. Collectively, NREMS and REMS episodes occupy about 50% of the time of laboratory rats and rabbits. Rats also exhibit a strong circadian organization of sleep; in this species, most sleep occurs during daylight hours.

The study of sleep in some sense lags behind other fields of neurobiology; fever, for example, has been studied for centuries. In contrast, no one systematically described or even measured sleep over the course of infection/disease until our 1988 studies. This neglect is surprising given that almost everyone is subjectively aware that feelings of sleepiness are associated with illness. Part of the reason for the lack of attention to sleep is that the functions of sleep, *i.e.*, what sleep does for the brain at any level, remain unknown.

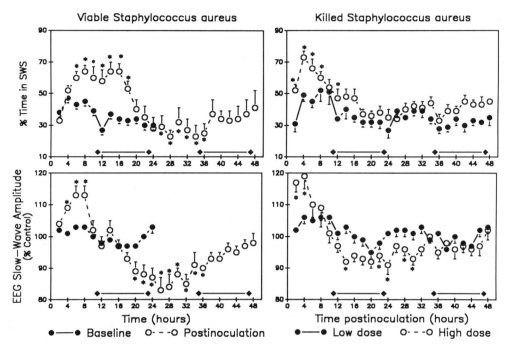

Figure 1. Sleep patterns in rabbits after inoculation with the gram-positive bacterium *Staphylococcus aureus*. Panels on the left indicate the effects of intravenous inoculation of rabbits (n=16) with viable *S. aureus* (10^7 to 10^8 CFU) on the percentage of time spent in SWS (top) and the amplitudes of EEG slow-waves during SWS (bottom) for 48 h after inoculation (o) relative to values obtained prior to inoculation (•). Panels on the right indicate the same measurements taken for 48 h after the intravenous administration of 8 x 10^7 (•; n=8) or 7 x 10^9 (o; n=12) CFU of heat-killed *S. aureus*. Baseline data for these animals are not shown but were not significantly different from data shown in the left panels. For all panels, data points represent the mean   S.E.M. of values obtained from each rabbit during the preceding 2-h period. Horizontal lines along the abscissa indicate the lights-off period. *, p < 0.03 relative to corresponding baseline values.

Sleep During Infections

If rabbits are inoculated with a gram positive bacterium, they increase the amount of time spent in NREMS.[1] This period of NREMS enhancement is followed by a period of inhibition of NREMS; thus, sleep responses are biphasic (*e.g.*, Fig. 1). A similar biphasic response in the amplitudes of EEG slow-waves (delta waves) during NREMS also occurs after bacterial inoculation (Fig. 1). Enhanced amplitudes of EEG slow-waves during sleep are thought to reflect the depth or intensity of NREMS. For example, supranormal EEG slow-waves are observed during the deep sleep that follows sleep deprivation.[2] Thus, the initial response to bacterial inoculation is characterized by a period of deep or intense sleep, and is followed by a period of sleep that is relatively shallow compared to corresponding control periods. The sleep

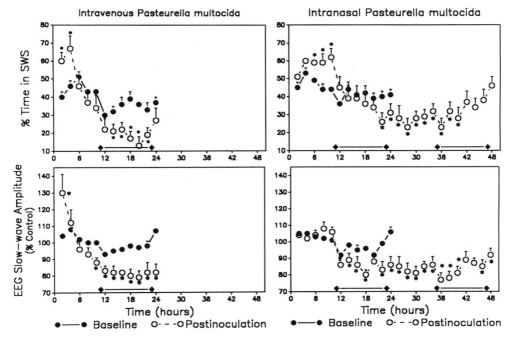

Figure 2. Sleep patterns in rabbits after inoculation with the gram-negative bacterium *Pasteurella multocida*. SWS (top panels) and EEG slow-wave amplitudes during sleep (bottom panels) were monitored for 24 h before (●) and 24-48 h after (o) inoculation of rabbits with viable *P. multocida* via intravenous (n=8; $6.7 \pm 1.2 \times 10^7$ CFU) or intranasal (n=10; $9.6 \pm 1.2 \times 10^8$ CFU) routes. For all panels, data points represent the mean S. E. M. of values obtained from each rabbit during the preceding 2-h period. The horizontal lines along the abscissa indicate the lights-off period. *, p 0.05 relative to corresponding baseline values.

responses seem to be independent, in part, from other physiological pertubations induced by the infection. For example, the time courses of fevers and changes in white blood cell counts are distinct from those of NREMS responses.[3,4]

The specific magnitude and time course of NREMS responses are dependent on both the specific infectious agent used and on the route of administration. The intravenous injection of a fungal organism induces sleep responses that are similar to those observed after gram positive bacteria.[3] In contrast, responses induced by intravenous administration of gram negative bacteria are much more rapid in onset, but the enhancement of NREMS is much shorter in duration (compare Fig. 1 to 2). However, if the gram-negative bacteria are given by a different route, (*e.g.*, intranasally [Fig. 2]), the temporal pattern of responses more closely resemble those induced by intravenous gram-positive bacteria.[4]

In rabbits, REMS is almost always reduced after injection of bacteria or fungal agents. This inhibition of REMS lasts throughout the routine 48 hr recording period in our experiments.[2-4] Similar findings were reported in rats with fungal infections.[5] In contrast to these results, rabbits treated with influenza virus, (an abortive infection in this species), donot exhibit reductions in REMS, although animals do display biphasic NREMS responses after receiving the virus.[6]

An important question is whether the sleep responses to infection are adaptive. Although this question cannot currently be answered with certainty, some evidence suggests that sleep aids in the recovery from infection. After either bacterial or fungal challenge, the specific sleep patterns that develop are related to the clinical responses of the animal. A more favorable prognosis is associated with a robust NREMS response after infectious challenge. In contrast, animals that display short periods of enhanced sleep or an enhanced inhibition phase of the NREMS response have a much higher probability of succumbing to the infections during the experiment[7] (Fig. 3). Indirect evidence supporting the hypothesis that sleep helps in recuperative processes are observations that sleep-deprived mice that were previously immunized to influenza virus failed to clear the virus from their lungs if challenged a second time; control immunized mice completely cleared the virus from their lungs.[8] Finally, Everson showed that prolonged sleep-deprivation of rats resulted in opportunistic infections.[9]

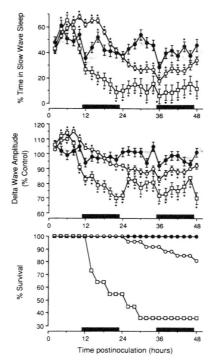

Figure 3. Sleep alterations and survival in rabbits inoculated with *S. aureus*. Upper panel, percentage of time in slow-wave sleep (SWS); middle panel, EEG slow wave amplitudes (delta wave amplitudes) during SWS (expressed as a percentage of baseline values); bottom panel, percentage of rabbits surviving. For all panels, filled circles, open circles and open squares respectively represent rabbits demonstrating sleep patterns with minimal infection-induced enhancement of sleep (n=8), prolonged enhancement of sleep (n=26), and brief enhancement of sleep (n=11). Baseline values for individual groups were not significantly different and were combined for statistical analysis. The shaded area represents the baseline values; these values are repeated during the second 24-h interval to facilitate visual comparisons. The bars along the abscissa indicate the lights-off period. *, p < 0.03 relative to comparable baseline periods.

Some of the bacterial and viral components that elicit sleep responses have been identified. Bacterial replication *per se* is not required for sleep responses; *e.g.*, heat killed gram-positive bacteria, if given in sufficient dosage, elicit sleep responses (Fig. 1). Isolated cell walls from either gram positive,[10] gram negative[10] or even those from archaebacteria[11] elicit sleep responses. Similarly, cell wall components, *e.g.*, muramyl peptides (reviewed[12-14]) or endotoxin[15] or its lipid A moiety[16] also elicit sleep responses. Macrophages can phagocytize and digest bacterial cell walls to release low molecular weight somnogenic muramyl peptides.[17,18] Similarly, the digestion of bacterial cell walls with lysozyme produces biologically active muramyl peptides.[18] In the case of viruses, influenza virus double-stranded RNA has the capacity to elicit sleep and fever responses.[19] Similarly, the synthetic double-stranded RNA, poly (rI:rC) also is somnogenic.[20] All of these microbial components have the capacity to enhance cytokine production (reviewed[21]) (Fig. 4).

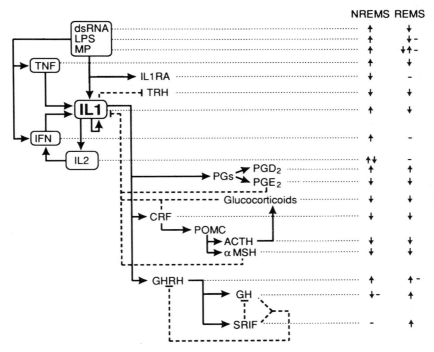

Figure 4. Putative sleep regulatory scheme. Solid lines with arrows (→) indicate stimulation (or potentiation) of production/activity. Broken lines with (− |) indicates inhibition of production/activity. Horizontal dotted lines from individual substances to columns indicate how those substances affect non-rapid eye movement sleep (NREMS) and REMS. Specific references for each of these activities are cited in reference 21. Arrows in the right columns indicate increases (↑), decreases (↓) and no change (←). Abbreviations are as follows: dsRNA = double stranded RNA; LPS = lipopolysaccharide; MP = muramyl peptide; IL1 = interleukin-1; IL1RA = interleukin-1 receptor antagonist; TRH = thyroid releasing hormone; IL2 = interleukin-2; IFN = interferon; TNF = tumor necrosis factor; PG = prostaglandin; CRF = corticotropin releasing factor; POMC = proopiomelanocyte stimulating hormone; ACTH = adrenocorticotropin hormone; aMSH = alpha melanocyte stimulating hormone; GHRH = growth hormone releasing hormone; GH = growth hormone; SRIF = somatostatin.

Cytokines and Sleep

Most data concerning the relationship between cytokines and sleep describe the somnogenic actions of IL1β . Briefly, the evidence implicating IL1 in sleep regulation is as follows. IL1β enhances sleep in rats,[22,23] cats,[24] rabbits[25] and monkeys,[26] and it induces feelings of sleepiness in humans undergoing IL1 therapy.[27] In some species, sleep responses to IL1 are complex. In rats, for example, low doses of IL1 (2.5 ng) enhance sleep whether given during the day or night. In contrast, high doses (10 ng) enhance sleep if given during the night, but inhibit sleep if given during the day.[22] Finally, low doses enhance NREMS without affecting REMS, whereas high doses inhibit NREMS and REMS. Substances that enhance IL1 production (e.g., muramyl peptides) enhance sleep.[12-14] Substances that inhibit IL1 production, (e.g., PGE_2 [28] or glucocorticoids[21]), or the actions of IL1 (e.g., α MSH[29] or CRF[30]) inhibit sleep. Furthermore, an endogenous IL1 receptor antagonist (IL1RA) with a half-life of about 10 min transiently inhibits NREMS. Exogenous administration of the IL1RA also completely blocks sleep responses induced by exogenous IL1.[31] Finally, IL1RA attenuates the sleep rebound that occurs after sleep deprivation.[32] Antibodies to IL1β also reduce normal sleep and attenuate the excess sleep that occurs after sleep deprivation.[33] A fragment of IL1, $IL1_{208-240}$, also has somnogenic actions;[34] interestingly, the IL1RA blocks these actions of $IL1_{208-240}$.[35]

Additional evidence suggests a physiological role for IL1 in sleep regulation. IL1β mRNA is present in normal brain.[36,37] IL1 receptors are widely distributed in normal brain,[38,39] and neurons that stain for IL1-like immunoreactivity are found in human[40] and rat[41] anterior hypothalamus. In humans, plasma levels of IL1 peak at the onset of NREMS.[42] Finally, in cats, cerebrospinal fluid levels of IL1 vary in phase with sleep-wake cycles.[43] Collectively, such data support a role for IL1 in normal sleep regulation.

Other cytokines are also likely to be involved in sleep, but they have not been investigated to the same degree as IL1. In rabbits, human recombinant, TNFα [44] or TNFβ [45] enhances NREMS. Similarly, fragments of TNFα that contain amino acid residues 31-36, which is a region known to interact with the TNF receptor, are somnogenic. Other TNFα fragments were pyrogenic, and one reduced food intake of rats but was not somnogenic.[46] Such data suggest that functional sites on TNFα for various biological activities are different. TNF is produced by astrocytes,[47] and neurons that stain for TNF-like immunoreactivity are present in normal brain.[48] Furthermore, the ability of systemic cells to produce TNFα is enhanced during sleep deprivation; other forms of stress did not have this effect.[49] Finally, anti-TNF or the TNF binding protein inhibit NREMS (Kapás, et al., unpublished).

Another group of cytokines implicated in sleep regulation are the interferons (IFNs). Human recombinant IFNα_2[50] or rabbit IFN (Kimura-Takeuchi et al., unpublished) enhance rabbit NREMS, although the effective dose of rabbit IFN in rabbits is several orders of magnitude less than the dose of human IFNα needed to induce similar sleep responses. IFNs are effective whether given IV or ICV. In monkeys, human IFNs reduce latency to REMS,[51] and humans undergoing IFN therapy report excessive sleepiness.[52] In virally-challenged rabbits, the maximum sleep responses occur at a time when plasma IFN levels are elevated.[53] Further, human IFNα or rat IFN microinjected into the locus coeruleus of rats induces a transient enhancement of sleep.[54] IFNα and β are produced by most nucleated cells and the ability of leukocytes to produce IFN is enhanced during and after sleep deprivation.[55] Thus, IFNα/β and/or TNFα may also be involved in sleep regulation. Furthermore, these substances also have the capacity to induce IL1 production, and their somnogenic actions could be mediated via IL1.

IL2 and IL6 have also been tested for somnogenic activity, although in both cases, more work is needed. Injection of IL2 into either the third ventricle or the locus coeruleus enhances sleep in rats; in contrast, injection into the hippocampus or ventral medial hypothalamus enhances motor activity.[54] Doses of human recombinant IL6 that induce mild fevers in rabbits do not affect sleep in this species.[56]

Pathways Involved in Cytokine-Induced Sleep

IL1 induces growth hormone (GH) release, and this effect is mediated via a hypothalamic mechanism.[57] IL1-induced GH release is blocked if animals are pretreated with anti-GH releasing hormone (GHRH) antibodies.[58] Both GH release and GHRH are associated with sleep (reviewed[59]). Thus, plasma levels of GH vary in phase with sleep-wake cycles; in man, they are highest during the first stage 4 (deep) NREMS episode. GHRH is a hypothalamic releasing factor for GH; neurons containing GHRH are in the arcuate nucleus and project to the median eminence. Another locus of GHRH neurons is found around the ventral medial nucleus; these neurons project to the median eminence and to the basal forebrain area. Those GHRH-containing neurons projecting to the median eminence are thought to be involved in pituitary GH release, while those projecting to the basal forebrain may modulate sleep (reviewed[60]). This latter region is the only area in the brain from which sleep can be induced by rapid stimulation; moreover, removal of this area induces insomnia (reviewed[61]).

Several lines of evidence link GHRH to sleep regulation, and some data suggest that IL1-enhanced sleep involves GHRH release. GHRH enhances sleep in rats[62], rabbits[63] and man[64]. If rats are pretreated with either a peptide GHRH antagonist [(N-Ac-Tyr^1-D Arg^2)-GRF (1-29) NH$_2$]^{65} or with anti-GHRH antibodies,[66] normal sleep is inhibited. Furthermore, anti-GHRH antibodies inhibit sleep rebound after sleep deprivation.[66] Finally, if rats pretreated with anti-GHRH are then given somnogenic doses of IL1, the expected enhancement of NREMS is not observed (Payne *et al.*, unpublished).

Another set of hormones likely to be involved in the modulation of sleep induced by IL1 and possibly by other cytokines is the CRF-ACTH (α MSH)-glucocorticoid axis. IL1 induces hypothalamic CRF release[67,68] and ACTH release directly from the pituitary.[69] CRF, ACTH, ACTH$_{1-13}$ (α MSH) and glucocorticoids all inhibit normal sleep (reviewed[21]). CRF[30] and α MSH[29] also inhibit IL1-enhanced sleep. The inhibitory actions of these substances may represent a multiple-level negative feedback mechanism for regulating the somnogenic actions of IL1 in the brain, the pituitary gland and the adrenal gland. The distribution of CRF in the brain is similar to that of IL1;[36] however, distinctions between brain CRF-IL1 interactions and interactions of IL1 and the CRF-pituitary-adrenal axis are not available with regard to sleep mechanisms.

Another negative feedback mechanism likely to modulate the somnogenic actions of IL1 involves prostaglandin E$_2$. The level at which this negative feedback loop operates with regard to sleep is not clear; it may be at the intracellular level. IL1 induces PGE$_2$ production, and PGE$_2$ inhibits IL1 production (reviewed[27]) and inhibits sleep.[28,70] However, single i.c.v. injections of PGE$_2$ induce a rapid inhibition of sleep (minutes) that lasts less than one hour.[28] Such a rapid time course would suggest that PGE$_2$ exerts direct inhibitory effects on sleep, as opposed to actions exerted via inhibition of IL1 production.

Another second messenger system implicated in cytokine actions is nitric oxide (NO) (reviewed[71]). Inhibition of nitric oxide synthase in rabbits results in substantial reductions in sleep.[72] However, this treatment has multiple effects that could secondarily alter sleep (*e.g.*, hypertension).

In conclusion, we have shown that complex sleep responses occur during infections. Several microbial products that could induce these sleep responses have been identified (*e.g.*, muramyl peptides). These microbial products, in turn, amplify the endogenous production of cytokines, perhaps in brain. The somnogenic actions of cytokines, in turn, are probably mediated via neuroendocrine mechanisms involving, among others, GHRH and CRF. The mechanisms involved in sleep responses to infection are also likely to function in the normal physiological regulation sleep, although at a lower basal level, and are also probably involved in the sleep rebound that occurs after sleep deprivation.

ACKNOWLEDGEMENTS

This work was supported, in part, by the Office of Naval Research (N00014-90-J-1069) and the National Institutes of Health (NS-25378, NS-27250 and NS-26429). We thank Drs. Opp, Kapás, Kimura-Takeuchi, and Payne for allowing us to mention some of their unpublished data.

REFERENCES

1. L.A. Toth and J.M. Krueger, Alteration of sleep in rabbits by *Straphylococcus aureus* infection, *Infect. Immun.* 56:1785 (1988).
2. J.R. Pappenheimer, G. Koski, V. Fencl, M.L. Karnovsky, and J.M. Krueger, Extraction of sleep-promoting factor S from cerebrospinal fluid and from brains of sleep-deprived animals, *J. Neurophysiol.* 38:1299 (1975).
3. L.A. Toth and J.M. Krueger, Effects of microbial challenge on sleep in rabbits, *FASEB J.* 3:2062 (1989).
4. L.A. Toth and J.M. Krueger, Somnogenic, pyrogenic and hematologic effects of experimental pasteurellosis in rabbits, *Am. J. Physiol.* 258:R536 (1990).
5. S. Kent, M. Price, and B. Satinoff, Fevers alters characteristics of sleep in rats, *Physiol. Behav.* 44:709 (1988).
6. M. Kimura-Takeuchi, J.A. Majde, L.A. Toth, and J.M. Krueger, Tolerance to virally-induced changes in sleep of rabbits, *Sleep Res.* 20:268 (1991).
7. L.A. Toth and J.M. Krueger, Sleep as a prognostic indicator during infectious disease in rabbits, *Proc. Soc. Expt. Biol. and Med.* 203:179 (1993).
8. R. Brown, G. Pang, A.J. Husband, and M.G. King, Suppression of immunity to influenza virus infection in the respiratory tract following sleep disturbance, *Regional Immunol.* 2:321 (1989).
9. C.A. Everson, Sustained sleep deprivation impairs host defense, *Am. J. Physiol.* (in press).
10. L. Johannsen, L.A. Toth, R.S. Rosenthal, M.R. Opp, F. Obál Jr., A.B. Cady, and J.M. Krueger, Somnogenic, pyrogenic and hematologic effects of bacterial peptidoglycan, *Am. J. Physiol.* 259:R182 (1990).
11. L. Johannsen, H. Labischinski, and J.M. Krueger, Somnogenic activity of pseudomurein in rabbits, *Infection and Immunity* 59:2502 (1991).
12. J.M. Krueger, Somnogenic activity of immune response modifiers, *TIPS* 11:122 (1990).
13. J.M. Krueger, J.R. Pappenheimer, and M.L. Karnovsky, Sleep-promoting effects of muramyl peptides, *Proc. Nat. Acad. Sci. USA* 79:6102 (1982).
14. J.M. Krueger, J. Walter, M.L. Karnovsky, L. Chedid, J.P. Choay, P. Lefrancier, and E. Lederer, Muramyl peptides: variation of somnogenic activity with structure, *J. Exp. Med.* 159:68 (1984).
15. J.M. Krueger, S. Kubillus, S. Shoham, and D. Davenne, Enhancement of slow-wave sleep by endotoxin and lipid A, *Am. J. Physiol.* 251:R591 (1986).
16. A.B. Cady, G. Riveau, L. Chedid, C.A. Dinarello, L. Johannsen, and J.M. Krueger, Interleukin-1-induced sleep and febrile responses differentially altered by a muramyl dipeptide derivative, *Int. J. Immunopharma.* 11:887 (1989).
17. M.W. Vermeulen and G.R. Grey, Processing of Bacillus subtilis peptidoglycan by a mouse macrophase cell line, *Infect. Immunity* 46:476 (1984).

18. L. Johannsen, J. Wecke, F. Obal Jr., and J.M. Krueger, Macrophages produce somnogenic and pyrogenic muramyl peptides during the digestion of staphylococci, *Am. J. Physiol.* 260:R126 (1991).

19. J.A. Majde, R.K. Brown, M.W. Jones, C.A. Dieffenbach, N. Maitra, J.M. Krueger, A.B. Cady, C.W. Smitka, and H.F. Maassab, Detection of toxic viral-associated double-stranded RNA (dsRNA) in influenza-infected lung, *Microb. Pathogen.* 10:105 (1991).

20. J.M. Krueger, J.A. Majde, C.M. Blatteis, J. Endsley, R.A. Ahokas, and A.B. Cady, Polyriboinosinic: polyriboytidlylic acid (poly I:C) enhances rabbit slow-wave sleep, *Am. J. Physiol.* 255:R748 (1988).

21. J.M. Krueger, F. Obal Jr., M. Opp, L. Toth, L. Johannsen, and A.B. Cady, Somnogenic cytokines and models concerning their effects on sleep, *Yale J. Biol. Med.* 63:157 (1990).

22. M. Opp, F. Obal Jr., and J.M. Krueger, Responsiveness of rats to interleukin-l: Temporal dose-related effects, *Am. J. Physiol.* 260:R52 (1991).

23. I. Tobler, A.A. Borbely, M. Schwyzer, and A. Fontana, Interleukin-1 derived from astrocytes enhances slow-wave activity in sleep EEG of the rat, *Eur. J. Pharmacol.* 104:191 (1984).

24. V. Susic and S. Totic, "Recovery" function of sleep: effects of purified human interleukin-1 on the sleep and febrile response of cats, *Met. Brain Dis.* 4:73 (1989).

25. J.M. Krueger, J. Walter, C.A. Dinarello, S.M. Wolff, and L. Chedid, Sleep-promoting effects of endogenous pyrogen (interleukin-1), *Am. J. Physiol.* 246:R994 (1984).

26. E.M. Friedman, S. Boinski, and C.E. Coe, Interleukin-1 induces sleep-like behavior and alters call structure in juvenile rhesus macaques, *Am. J. Primatol.* (in press).

27. C.A. Dinarello, Interleukin-1 and Interleukin-1 antagonism, *Blood* 8:1627 (1991).

28. J.M. Krueger, L. Kapas, M. Opp, and F. Obal Jr., Prostaglandins E$_2$ and D$_2$ have little effect on rabbit sleep, *Physiol. and Behav.* 51:481 (1992).

29. M.R. Opp, F. Obal Jr., and J.M. Krueger, Effects of alpha-MSH on sleep, behavior, and brain temperature: interactions with IL1, *Am. J. Physiol.* 255:R914 (1988).

30. M. Opp, F. Obal Jr., and J.M. Krueger, Corticotropin-releasing factor attenuates interleukin-1 induced sleep and fever in rabbits, *Am. J. Physiol.* 257:R528 (1989).

31. M.R. Opp and J.M. Krueger, An interleukin-1 receptor antagonist blocks interleukin-1 induced sleep and fever, *Am. J. Physiol.* 260:R453 (1991).

32. M.R. Opp and J.M. Krueger, Effects of an interleukin-1 receptor antagonist on recovery sleep of rabbits after total sleep deprivation, *Sleep Res.* 20:416 (1991).

33. M. Opp and J.M. Krueger, Anti-interleukin-1b reduces sleep and sleep rebound after sleep deprivation in rats, *Am. J. Physiol.* (in press).

34. F. Obal Jr., M.R. Opp, A.B. Cady, L. Johannsen, A.E. Postlethwaite, H.M. Poppleton, J.M. Seyer, and J.M. Krueger, Interleukin-1a and an interleukin-1b fragment are somnogenic, *Am. J. Physiol.* 259:R439 (1990).

35. M.R. Opp, A.E. Postlethwaite, J.M. Seyer, and J.M. Krueger, Interleukin-1 receptor antagonist blocks somnogenic and pyrogenic responses to an IL1 fragment, *Proc. Natl. Acad. Sci. USA* 89:3726 (1992).

36. F. Berkenbosch, N. Robakis, and M. Blum, Interleukin-1 in the central nervous system: a role in the acute phase response and in brain injury, brain development and the pathogenesis of Alzheimer's disease, *in*: Peripheral Signaling of the Brain, R.C.A. Frederickson *et al.*, eds., Hogrefe & Hunter, Toronto (1991).

37. W.L. Farrar, J.M. Hill, A. Harel-Bellan, and M. Vinocour, The immune logical brain, *Immunol. Rev.* 100:361 (1987).

38. W.L. Farrar, P.L. Kilan, M.R. Ruff, J.M. Hill, and C.B. Pert, Visualization and characterization of interleukin-1 receptors in brain, *J. Immunol.* 139:459 (1987).

39. F.G. Haour, E.M. Ban, G.M. Milon, D. Baran, G.M. Fillion, Brain interleukin-1 receptors: characterization and modulation after lipopolysaccharide injection, *Prog. NeuroEndocrinImmunology* 3:196 (1990).

40. C.D. Breder, C.A. Dinarello, C.B. Saper, Interleukin-1 immunoreactive innervation of the human hypothalamus, *Science* 240:321 (1988).

41. R. Lechan, R. Toni, B. Clark, J. Cannon, A. Shaw, C.A. Dinarello, and S. Reichlin, Immunoreactive interleukin-1 beta localization in rat forebrain, *Brain Res.* 514:135 (1990).

42. H. Moldofsky, F.A. Lue, J. Eisen, E. Keystone, and R.M. Gorczynski, The relationship of interleukin-1 and immune functions to sleep in humans, *Psychosom. Med.* 48:309 (1989).

43. F.A. Lue, M. Bail, J. Jephthah-Ocholo, K. Carayanniotis, R. Gorczynski, and H. Moldofsky, Sleep and cerebrospinal fluid interleukin-1 like activity in the cat, *Intern. J. Neurosci.* 42:179 (1988).

44. S. Shoham, D. Davenne, A.B. Cady, C.A. Dinarello, and J.M. Krueger, Recombinant tumor necrosis factor and interleukin-1 enhance slow-wave sleep in rabbits, *Am. J. Physiol.* 253:R142 (1987).

45. L. Kapas and J.M. Krueger, Tumor necrosis factor b induces sleep, fever and and anorexia, *Am. J. Physiol.* 263:R703 (1992).

46. L. Kapas, L. Hong, A.B. Cady, M.R. Opp, A.E. Postlethwaite, J.M. Seyer, and J.M. Krueger, Somnogenic, pyrogenic and anorectic effects of TNF∝ fragments, *Am. J. Physiol.* 263:R708 (1992).

47. J. Vilcek and T.H. Lee, Tumor necrosis factor: New insights into the molecular mechanisms of its multiple actions, *J. Biol. Chem.* 266:7313 (1991).

48. C.D. Breder and C.B. Saper, Tumor necrosis factor immunoreactive innervation in the mouse brain, *Soc. for Neurosci. Absts.* 14:1280 (1988).

49. K. Yamasu, Y. Shimada, M. Sakaizumi, G.I. Soma, and D-I. Mizono, Activation of the systemic production of tumor necrosis factor after exposure to acute physical stress, *Third Internat. Cont. on TNF and Related Cytokines Absts.* 174 (1990).

50. J.M. Krueger, C.A. Dinarello, S. Shoham, D. Davenne, J. Walter, and S. Kubillus, Interferon alpha-2 enhances slow-wave sleep in rabbits, *Int. Immunopharmacol.* 9:23 (1987).

51. M. Reite, M. Landenslager, J. Jones, C. Crnic, and K. Kaening, Interferon decreases REMS latency, *Biol. Psychiatry* 22:104 (1987).

52. H. Smedlley, M. Katrak, K. Sikora, and T. Wheeler, Neurological effects of recombinant human interferon, *Br. J. Med.* 286:262 (1983).

53. M. Kimura-Takeuchi, M.A. Majde, L.A. Toth, and J.M. Krueger, Influenza virus-induces changes in rabbit sleep and other acute phase responses, *Am. J. Physiol.* 263:R1115 (1992).

54. G.B. De Sarro, C. Ascioti, M.G. Audino, U. Rispoli, G. Nistico, Behavioral and ECOG spectrum changes induced by intracerebral microfusion of interferons and interleukin-2 in rats are antagonized by naloxone, *in*: Interactions among central nervous system, neuroendocrine and immune systems, J.W. Hadden *et al.*, eds. Pythagona Press, Rome (1989).

55. J. Palmblad, K. Cantell, H. Strander, J. Froberg, C.G. Karlsson, L. Levi, M. Granstrom, and P. Unger, Stressor exposure and immunological response in man: interferon-producing capacity and phagocytosis, *J. Psychosom. Res.* 29:193 (1976).

56. M. Opp, F. Obal, Jr., A.B. Cady, L. Johannsen, and J.M. Krueger, Interleukin-6 is pyrogenic but not somnogenic, *Physiol. Behav.* 45:1069 (1989).

57. V. Rettori, J. Jurcovicova, and S.M. McCann, Central action of interleukin-1 in altering the release of TSH, growth hormone and prolactin in the male rat, *J. Neurosci. Res.* 18:179 (1987).

58. L. Payne, F. Obal, Jr., M.R. Opp, J.M. Krueger, Stimulation and inhibition of growth hormone secretion by interleukin-1B: the involvement of GHRH, *Neuroendocrinol.* 56:118 (1992).

59. F. Obal, Jr. and J.M. Krueger, Growth hormone-releasing hormone and interleukin-1 in sleep regulation, *FASEB J.* 7:645 (1993).

60. F. Obal, Jr., M. Opp, G. Sary, and J.M. Krueger, Endocrine mechanisms in sleep regulation, *in*: Endogenous Sleep Factors, S. Inoue and J.M. Krueger, eds. The SPB Academic Publishing bv, Hague (1990).

61. D.J. McGinty, Physiological equilibrium and the control of sleep states, *in*: Brain Mechanisms of Sleep, D.J. McGirty *et al.*, eds., Raven Press, New York (1985).

62. F. Obal, Jr., Effects of peptides (DSIP, DSIP analogues, VIP, GRF and CCK) on sleep in the rat, *Clin. Neuropharmacol.* 9:459 (1986).

63. F. Obal, Jr., P. Alfoldi, A.B. Cady, L. Johannsen, G. Sary, and J.M. Krueger, Growth hormone-releasing factor enhances sleep in rats and rabbits, *Am. J. Physiol.* 255:R310 (1988).

64. A. Steiger, J. Guldner, U. Hemmeter, B. Rothe, K. Wiedemann, F. Holsboer, Changes in sleep-EEG and nocturnal hormonal secretion under pulstile application of GHRH or somatostatin, *Sleep Res.* 20A:195 (1991).

65. F. Obal, Jr., L. Payne, L. Kapas, M. Opp, and J.M. Krueger, Inhibition of growth hormone-releasing factor suppresses both sleep and growth hormone secretion in the rat, *Brain Res.* 557:149 (1991).

66. F. Obal, Jr., L. Payne, M. Opp, P. Alfoldi, L. Kapas, and J.M. Krueger, Antibodies to growth hormone-releasing hormone suppress sleep and prevent enhancement of sleep after sleep deprivation in the rat, *Am. J. Physiol.* 263:R1078 (1992).

67. R. Sapolsky, C. Rivier, G. Yamamoto, P. Plotsky, and W. Vale, Interleukin-1 stimulates the secretion of hypothalamic corticotropin-releasing factor, *Science* 238:522 (1987).

68. F. Berkenbosch, J. van Oers, A. Del Rey, F. Tilders, H. Besedovsky, Cortico-tropin-releasing factor-producing neurons in the rat activated by interleukin-1, *Science* 238:524 (1987).

69. E.W. Bernton, J.E. Beach, J.W. Holaday, R.C. Smallridge, and H.G. Fein, Release of multiple hormones by a direct action of interleukin-1 on pituitary cells, *Science* 238:519 (1987).
70. O. Hayaishi, Sleep-wake regulation by prostaglandin D_2 and E_2, *J. Biol. Chem.* 263:14593 (1988).
71. J. Garthwaite, Glutamate, nitric oxide and cell-cell signalling in the nervous system, *Trends in Neuro. Sci.* 14:60 (1991).
72. L. Kapás, M. Shibata, and J.M. Krueger, Inhibition of nitric oxide synthesis suppresses sleep in rabbits, *Am. J. Physiol.* (in press).

INTERLEUKIN INVOLVEMENT IN THE REGULATION OF PITUITARY CELL GROWTH

E. Arzt,[1] J. Sauer, U. Renner, G.K. Stalla

Max-Planck-Institute of Psychiatry, Clinical Institute, Munich, Germany
[1]Present address: Instituto de Investigaciones Medicas, Universidad de Buenos Aires, Donato Alvarez 3150, 1427 Bs. As., Argentina

INTRODUCTION

After the first description by Besedovsky et al.[1] of the stimulatory action of interleukin-1 (IL-1) on the secretion of hormones by the hypothalamic-pituitary-adrenocortical (HPA) axis, the action of this and other cytokines have been extensively studied. IL-1 is the most potent cytokine stimulating the axis. As we will concentrate on other aspects of cytokine action, it is beyond the scope of this chapter to review the action of cytokines on hormone secretion. It is now well accepted that these cytokines not only constitute a lymphocyte message but also that they are autocrine and paracrine factors in the regulation of the HPA axis, especially the pituitary. We will start with a brief description about the intrinsic expression of interleukins and their receptors on neuro-endorine tissues and then concentrate on their effects on the growth of these cells.

INTRINSIC PRODUCTION OF INTERLEUKINS AND INTERLEUKIN RECEPTORS ON NEURO-ENDOCRINE TISSUES

It has been described that the endocrine tissues may not only be a target but also a site of origin of lymphokines. Accordingly, IL-1β immunoreactive fibers were found in the hypothalamus, innervating key endocrine cells groups.[2] Rat medial basal hypothalamic cultures release IL-6 and express its mRNA.[3] Concerning the pituitary, recently IL-1 production by pituitary cells was demonstrated. IL-1β immunoreactive material and mRNA was found in rat pituitaries increasing after bacterial lipopolysaccharide treatment of the animals.[4] Moreover, IL-1 receptors and mRNA were characterized in mouse pituitary cells and AtT-20 corticotrophs.[5,6] Also, IL-6 production by cells obtained from anterior pituitary gland from rats was demonstrated,[7,8] and shown to be stimulated by IL-1.[9,10] As well, the expression of IL-6 mRNA in rat anterior pituitary[3] and corticotrophic adenoma cell cultures[11] and the release of IL-6 from human pituitary adenoma cultures[12] were reported. The presence of IL-6 in different types of pituitary adenomas was shown recently by immunocytochemistry.[13] Furthermore, IL-6 receptors are expressed in rat anterior pituitary cells.[14] Accordingly, we have recently described the expression of IL-2 and its receptor (IL-2R) by pituitary cells of different species.[15] In the mouse AtT-20 pituitary tumor cell line, we detected IL-2 mRNA expression after stimulation with corticotropin-releasing hormone (CRH) or phorbol myristate acetate (PMA). In human corticotrophic adenoma cells, basal IL-2 mRNA expression as well as IL-2 secretion were further stimulated by PMA. In both adenoma and AtT-20 cells we found detectable amounts of IL-2R mRNA and by using immunofluorescence, IL-2R membrane expression. We have also observed[15,16] the expression of IL-2R in the membrane of normal pituitary cells (Figure 1). The IL-2R shows high colocalization with ACTH, GH and PRL producing cells. The presence of the receptor in these cells is in agreement with previous studies showing the action of IL-2 on the secretion of these hormones.[17,18]

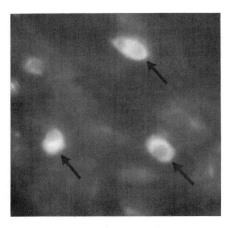

Figure 1. Immunofluorescence photograph (200x) showing the presence of IL-2R positive cells in normal anterior pituitary cells. A specific anti-rat TAC IL-2R (alpha subunit) FITC-conjugated monoclonal antibody (OX39) was used diluted 1:20. Cells were stained after 48 h of culture at basal condition. Positive (arrows) and negative cells can be observed.

INTERLEUKIN ACTION ON CELLULAR GROWTH

The action of interleukins as growth stimulatory and inhibitory factors is not restricted to cells of the immune system[19,20] or tumors,[21,22] but also affects other cells, including those of the neuroendocrine system. IL-2 may control the growth of glial cells of the nervous system, exerting proliferative as well as anti-proliferative effects on oligodendrocytes.[23-25] It stimulates the proliferation and maturation of rat oligodendrocytes at concentrations greater than 10 U/ml.[24] The oligodendrocyte response to IL-2 is developmentally regulated: concentrations of IL-2 below 0.5 U/ml inhibit the proliferation of less mature oligodendrocytes.[25] In addition, IL-2 has a direct effect on neurones stimulating sympathetic neurite outgrowth at concentrations around 2 U/ml.[26] IL-6 as well as TNF-α also influence the proliferation of astrocytes.[27] IL-2 decreased in a dose dependent manner the ^{3}H-thymidine uptake of Graves' thyrocytes but had no effect on the uptake in normal thyroid cells.[28] The biphasic effect of IL-2 on the proliferation of cultured vascular smooth muscle cells was also reported.[29] Recently IL-1 has been shown to stimulate the growth of primary cultures of bovine adrenocortical zona fasciculata/reticularis cells.[30] IL-1 inhibits human thyroid carcinoma cell growth.[31] In addition to their effects on the function of the neuro-endocrine cells, interleukins seem to have growth regulatory properties in this system.

PITUITARY GROWTH

Although the anterior pituitary gland is an organ of low mitosis, autoradiographic studies have shown that ^{3}H-thymidine is incorporated into rat anterior pituitary nuclei *in vivo*.[32] The anterior pituitary gland is considered, according to its proliferative activity, a slow but expanding cell population, as described for other endocrine glands (thyroid and adrenal gland).[33] DNA synthesis also occurs in short-term cultures in cells committed to the synthesis of specific hormones.[34-36] The anterior pituitary can develop adenoma tumors of different sizes, classified as micro and macroadenomas, frequently associated with high levels of hormone secretion and showing a benign behaviour.[37,38] Some of the tumors of adeno-hypophyseal origin grow invasively, representing biologically intermediate forms between the sharply demarcated benign adenomas and the rarely observed metastasizing pituitary carcinomas.[37] Invasive tumor growth is associated with higher proliferation activity.[39] Although recently efforts have been made to characterize the growth factors produced and acting in the pituitary,[40-42] a detailed insight into the mechanism and factors governing the cellular proliferation in this gland is lacking, despite the potentially great practical value of identifying such factors as growth regulators of pituitary adenomas. The presence of growth factors was also described in normal human pituitaries.[43] Several factors influencing hormone secretion have been shown to regulate pituitary cell growth. Although many of these studies have focused on the tumor cell line GH$_3$, demonstrating inhibitory and stimulatory factors,[42,44] other studies with normal pituitaries showed an effect on the proliferative activity of all the different cell types[33] and the existence of paracrine growth regulation between these cells.[36] Cell-to-cell communication influencing the secretion of hormones and other pituitary activities seems to be very important for normal pituitary function.[45] Abnormalities of these reulatory factors may be involved in the development of certain types of pituitary tumors, particularly those with little or no malignant potential (micro-macroadenomas).

REGULATION OF PITUITARY CELL GROWTH BY INTERLEUKIN

The demonstration that interleukins control the growth of the pituitary adds new insights into the functional interactions of the brain with the endocrine and immune systems[16] (Table 1). For instance, IL-2 and IL-6 stimulate the growth of GH$_3$ cells and inhibit the proliferation of normal rat anterior pituitary cells (Table 1). Several reasons, including the tumoral stage of GH$_3$ cells or the release of other inhibitory mediators by IL-2 and IL-6 in the normal pituitary preparations, could account for these differential effects.

There are well known species differences in the immune system, particularly between mouse/human and the rat, in which IL-2 constitutes one of the molecules that shows a high degree of species-specific activity.[46,47] This is particularly true for

IL-2 effects on the endocrine system as rat IL-2 was shown to induce corticosterone production by rat adrenal cells in culture and ACTH release by normal rats *in vivo*, while human IL-2 had no effect in these systems.[48,49] For this reason we used rat IL-2 and species specific anti-IL-2R antibodies to establish the presence of the IL-2R as well as IL-2 action on pituitary cells. The specificity of IL-2 action in concert with its receptor expression on pituitary cells strongly suggests a physiological regulation role for IL-2 in pituitary function.

Table 1. Effect of IL-2 and IL-6 on growth of pituitary cells.

Cell type	Culture condition[2]	Effect on growth[6]	
		IL-2[7]	IL-6[8]
GH$_3$	DMEM with 0.5% FCS[3]	150-400%	210-320%
GH$_3$	DMEM without FCS[4]	120-300%	150-300%
Normal[1]	D-Valine-MEM[5] with 2 to 10% FCS	60-80%	65-75%
Normal[1]	DMEM with 0.5% FCS	70-80%	75-80%
Normal[1]	DMEM with 2 to 10% FCS	No effect	No effect

[1] Normal rat anterior pituitary cells. [2] Cells were cultured for 3 days. [3] Similar results were obtained when cells were cultured for up to 7 days. [4] Containing transferrin, T3 and insulin. [5] This medium specifically inhibits fibroblast cell growth. [6] Growth was measured as ^{3}H-Thymidine incorporation. Results are expresed compared to control without stimulation = 100%. In addition in the GH$_3$ cells cell count was determined showing the same results as ^{3}H-Thymidine incorporation. [7] IL-2 was used at 1-100 IU/ml. [8] IL-6 was used at 10-1000 IU/ml.

The autocrine (and paracrine) secretion of growth factors is a unifying concept in the search for the molecular and cellular basis of malignant transformation.[50] Transforming growth factor-β (TGF-β) production and receptor expression was demonstrated by primate kidney cells. TGF-β inhibited the growth of these cells, thus confirming the hypothesis that cells may produce their own inhibitors of growth.[50] IL-2 and IL-6 are produced and their receptors are expressed by pituitary cells[4,7,14,15]. The stimulatory/inhibitory effects of these lymphokines on pituitary cell proliferation strongly suggest their involvement in the process of tumor formation in the anterior pituitary. Furthermore, both IL-2 and IL-6 are produced by pituitary adenoma cells.[12,15] Therefore, interleukins may act in concert with other growth factors in the regulation of the development and growth of the anterior pituitary cell types.

Recent research has demonstrated that in experimental tumor models immunotherapy with low-dose IL-2 can be highly effective against metastatic cancer, when applied locally at a site of tumor growth.[51] The production of IL-2 by tumor cells transfected with the IL-2 gene causes a vigorous rejection of these tumor cells.[52] Furthermore, insertion of a functional IL-2 gene into plasmocytoma cell lines substantially reduced tumorigenicity of resulting clones.[53] Also, tumor cells induced to secrete IL-2 following transfer and expression of IL-2 cDNA amplify the host's local anti-tumor response in a paracrine fashion resulting in the destruction of malignant metastatic tumor cells.[54]   IL-6 was shown to inhibit the growth of mesangial cells in culture[55] and anti-human IL-6 receptor antibodies inhibit human myeloma growth *in vivo*.[56]   A paracrine role of IL-6 produced by breast fibroblasts in the inhibition of breast cancer cell growth has also been postulated.[57]   IL-6 production by the network of folliculo-stellate cells in the pituitary[7] could play a similar role.   Considering that the uncontrolled division of cells may result either from excessive growth stimulation or deficient growth inhibition, the regulation of pituitary cell growth by IL-2 and IL-6 might imply that these cytokines participate in the process of initiation or maintenance of pituitary tumors.

The study of paracrine and autocrine interactions within the anterior pituitary has been the focus of expanding research efforts in the last 3-4 years.   We provide a new concept of interleukin involvement in pituitary (patho)-physiology as intrinsic factors of the gland.   Acting through specific functional receptors on the same or neighbouring cells, interleukins constitute autologous and/or inter-cellular factors which are involved in the coordinate regulation not only of hormone secretion, but also of cell proliferation.

REFERENCES

1. H. Besedovsky, A. Del Rey, E.C.A. Sorkin and C.D. Dinarello, Immunoregulatory feedback between interleukin-1 and glucocorticoid hormones, *Science* 233:652 (1986).
2. C.D. Breder, C.A. Dinarello and C.B. Saper, Interleukin-1 immunoreactive innervation of the human hypothalamus, *Science* 240:321 (1988).
3. B.L. Spangelo, A.M. Judd, R.M. MacLeod, D.W. Goodman and P.C. Isakson, Endotoxin-induced release of interleukin-6 from rat medial basal hypothalami, *Endocrinology* 127:1779 (1990).
4. J.I. Koenig, K. Snow, B.D. Clark, R. Toni, J.G. Cannon, A.R. Shaw, C.A. Dinarello, S. Reichlin, S.L. Lee and R.M. Lechan, Intrinsic pituitary IL-1b is induced by bacterial LPS, *Endocrinology* 126:3053 (1990).
5. E.B. De Souza, E.L. Webster, D.E. Grigoriadis and D.E. Tracey, Corticotropin-releasing factor (CRF) and interleukin-1 (IL-1) receptors in the brain-pituitary-immune axis, *Psychopharmacol. Bull.* 25:299 (1989).
6. J. Bristulf, A. Simoncsits and T. Bartfai, Characterization of a neuronal interleukin-1 receptor and the corresponding mRNA in the mouse anterior pituitary cell line AtT-20, *Neurosci. Lett.* 128:173 (1991).
7. H. Vankelecom, P. Carmeliet, J. Van Damme, A. Billiau and C. Denef, Production of IL-6 by folliculostellate cells of the anterior pituitary gland in a histiotypic cell aggregate culture system, *Neuroendocrinology* 49:102 (1989).
8. B.L. Spangelo, R.M. MacLeod and P.C. Isakson, Production of interleukin-6 by anterior pituitary cells in vitro, *Endocrinology* 126:582 (1990).
9. B.L. Spangelo, A.M. Judd, P.C. Isakson and R.M. MacLeod, Interleukin-1 stimulates interleukin-6 release from rat anterior pituitary cells in vitro, *Endocrinology* 128:2685 (1991).
10. M. Yamaguchi, N. Matsuzaki, K. Hirota, A. Miyake and O. Tanizawa, Interleukin 6 possibly induced by interleukin 1b in the pituitary gland stimulates the release of gonadotropins and prolactin, *Acta Endocrinol. (Copenh.)* 122:201 (1990).
11. B. Velkeniers, G. D'Haens, G. Smets, P. Vergani, L. Vanhaelst and E.L. Hooghe-Peters, Expression of IL-6 mRNA in corticotroph cell adenomas, *J. Endocrinol. Invest.* 14 (Suppl. 1) 31 (Abstr.) (1991).
12. T.H. Jones, S. Justice, A. Price and K. Chapman, Interleukin-6 secreting human pituitary adenomas in vitro, *J. Clin. Endocrinol. Metab.* 73:207 (1991).
13. S. Tsagarakis, G. Kontogeorgeos, P. Giannou, N. Thalassinos, J. Woolley, G.M. Besser and A. Grossman, Interleukin-6, a growth promoting cytokine, is present in human pituitary adenomas: an immunocytochemical study, *Clin. Endocrinol.* 37:163 (1992).
14. M. Ohmichi, K. Hirota, K. Koike, H. Kurachi, S. Ohtsuka, N. Matsuzaki, M. Yamaguchi, A. Miyake and O. Tanizawa, Binding sites for interleukin-6 in the anterior pituitary gland, *Neuroendocrinology* 55:199 (1992).

15. E. Arzt, G. Stelzer, U. Renner, M. Lange, O.A. Muller and G.K. Stalla, Interleukin-2 and IL-2 receptor expression in human corticotrophic adenoma and murine pituitary cell cultures, *J. Clin. Invest.* 90:1944 (1992).

16. E. Arzt, R. Buric, G. Stelzer, J. Stalla, J. Sauer, U. Renner and G.K. Stalla, Interleukin involvement in anterior pituitary cell growth regulation: effects of interleukin-2 and interleukin-6, *Endocrinology* 132:459 (1993).

17. S. Karanth and S.M. McCann, Anterior pituitary hormone control by interleukin-2, *Proc. Natl. Acad. Sci. USA* 88:2961 (1991).

18. S. Karanth, The influence of dopamine (DA) on interleukin-2 induced release of prolactin (PRL), luteinizing hormone (LH) and follicle stimulating hormone (FSH) by the anterior pituitary, Program of the 73rd Annual Meeting of The Endocrine Society, Washington, D.C., p. 210 (Abstr.) (1991).

19. R.J. Robb, Interleukin 2: the molecule and its function, *Immunol. Today* 5:203 (1984).

20. M. Helle, J.P.J. Brakenhoff, E.R. Degroot and L.A. Aarden, Interleukin-6 is involved in interleukin-1 induced activities, *Eur. J. Immunol.* 18:957 (1988).

21. T. Kishomoto, The biology of interleukin-6, *Blood* 74:1 (1989).

22. L. Chen, L.M. Shulman and M. Revel, IL-6 receptors and sensitivity to growth inhibition by IL-6 in clones of human breast carcinoma cells, *J. Biol. Reg. Homeos. Agents* 5:125 (1991).

23. J. Merrill, Macroglia: neural cells responsive to lymphokines and growth factors, *Immunol. Today* 8:146 (1987).

24. E.N. Benveniste and J.E. Merrill, Stimulation of oligodendroglial proliferation and maturation by interleukin-2, *Nature* 321:610 (1986).

25. R.P. Saneto, A. Altman, R.L. Knobler, H.M. Johnson and J. de Vellis, Interleukin-2 mediates the inhibition of oligodendrocyte progenitor cell proliferation in vitro, *Proc. Natl. Acad. Sci. USA* 83:9221 (1986).

26. P.K. Haugen and P.C. Letourneau, Interleukin-2 enhances chick and rat sympathetic, but not sensory, neurite outgrowth, *J. Neurosci. Res.* 25:443 (1990).

27. K.W. Selmaj, M. Farooq, W.T. Norton, C.S. Raine and C.F. Brosnan, Proliferation of astrocytes in vitro in response to cytokines. A primary role for tumor necrosis factor, *J. Immunol.* 144:219 (1990).

28. M. Itoh, N. Shimomura, T. Okugawa, Y. Murata and H. Seo, Effects of interleukin 2 or 6 on DNA synthesis and expression of c-fos mRNA in Graves' thyroid cells, Program of the 73rd Annual Meeting of The Endocrine Society, Washington, D.C., p. 340 (Abstr.) (1991).

29. T. Nabata, S. Morimoto, K. Fukuo and T. Ogihara, Biphasic effects of interleukin-2 on proliferation of cultured vascular smooth muscle cells, *Jpn. J. Pharmacol.* 59s2:363 (1992).

30. N. Hanley, B.C. Williams, M. Nicol, I.M. Bird and S.W. Walker, Interleukin-1b stimulates growth of adrenocortical cells in primary culture, *J. Mol. Endocrinol.* 8:131 (1992).

31. H. Kimura, S. Yamashita, H. Namba, T. Tominaga, M. Tsuruta, N. Yokoyama, M. Izumi and S. Nagataki, Interleukin-1 inhibits human thyroid carcinoma cell growth, *J. Clin. Endocrinol. Metab.* 75:596 (1992).

32. W.A.J. Crane and R.S. Loomes, Effect of age, sex, and hormonal state on tritiated thymidine uptake by rat pituitary, *Brit. J. Cancer* 21:787 (1967).

33. E. Carbajo-Perez, S. Carbajo, A. Orfao, J.L. Vicente-Villardon and R. Vazquez, Circadian variation in the distribution of cells throughout the different phases of the cell cycle in the anterior pituitary gland of adult male rats as analysed by flow cytometry, *J. Endocrinol.* 219:329 (1991).

34. A. Mastro, E. Shelton and W.C. Hymer, DNA synthesis in the rat anterior pituitary. An electron microscope radioautographic study, *J. Cell Biol.* 43:626 (1969).

35. N. Billestrup, L.W. Swanson and W. Vale, Growth hormone-releasing factor stimulates proliferation of somatotrophs in vitro, *Proc. Natl. Acad. Sci.* 83:6854 (1986).

36. D. Tilemans, M. Andries and C. Denef, Luteinizing hormone-releasing hormone and neuropeptide Y influence deoxyribonucleic acid replication in three anterior pituitary cell types. Evidence for mediation by growth factors released from gonadotrophs, *Endocrinology* 130:882 (1991).

37. B.W. Scheithauer, K.T. Kovacs, E.R. Laws and R.V. Randall, Pathology of invasive pituitary tumors with special reference to functional classification, *J. Neurosurg.* 65:733 (1986).

38. A.M. Landolt, T. Shibata, P.K. Keihues and E. Tuncdogan, Growth of human pituitary adenomas: facts and speculations, *Adv. Biosci.* 69:53 (1988).

39. M. Buchfelder, R. Fahlbusch, E.M. Adams, M. Roth and P. Thierauf, Growth characteristics of

invasive pituitary adenomas, *J. Endocrinol. Invest.* 14(suppl 1):33 (1991).

40. S. Ezzat and S. Melmed, The role of growth factors in the pituitary, *J. Endocrinol. Invest.* 13:691 (1990).

41. H. Houben and C. Denef, Regulatory peptides produced in the anterior pituitary, *Trends Endocrinol. Metab.* 1:398 (1990).

42. J. Webster, J. Ham, J.S. Bevan, C.D.T. Horn and M.F. Scanlon, Preliminary characterization of growth factors secreted by human pituitary tumours, *J. Clin. Endocrinol. Metab.* 72:687 (1991).

43. J. Halper, P.G. Parnell, B.J. Carter, P. Ren and B.W. Scheithauer, Presence of growth factors in human pituitary, *Lab. Invest.* 66:639 (1992).

44. Y. Yajima and T Saito, The effects of epidermal growth factor on cell proliferation and prolactin production by GH_3 rat pituitary cells, *J. Cell. Physiol.* 120:249 (1984).

45. J. Schwartz and R. Cherny, Intercellular communication within the anterior pituitary influencing the secretion of hypophysial hormones, *Endocrine Rev.* 13:453 (1992).

46. A.J. McKnight and B.J. Classon, Biochemical and immunological properties of rat recombinant interleukin-2 and interleukin-4, *Immunology* 75:286 (1992).

47. L.P. Thompson and G. DiSabato, Purification and characterization of two forms of rat interleukin-2, *Cell. Immunol.* 114:12 (1988).

48. T. Tominaga, J. Fukata, Y. Naito, T. Usui, N. Murakami, M. Fukushima, Y. Nakai, Y. Hirai and H. Imura, Prostaglandin-dependent in vitro stimulation of adrenocortical steroidogenesis by interleukins, *Endocrinology* 128:526 (1991).

49. Y. Naito, J. Fukata, T. Tominaga, Y. Masui, Y. Hirai, N. Murakami, S. Tamai, K. Mori and H. Imura, Adrenocorticotropic hormone-releasing activities of interleukins in a homologous in vivo system, *Biochem. Biophys. Res. Commun.* 164:1262 (1989).

50. M.B. Sporn and A.B. Roberts, Autocrine growth factors and cancer, *Nature* 313:745 (1985).

51. R.A. Maas, H.J. Van Weering, H.F.J. Dullens and W. Den Otter, Intratumoral low-dose interleukin-2 induces rejection of distant solid tumour, *Cancer Immunol. Immunother.* 33:389 (1991).

52. C. Roth, L. Mir, M. Cressent, F. Quintin-Colonna, V. Ley, D. Fradelizi and P. Kourilsky, IL-2 gene transduction in malignant cells: applications in cancer containment, *Bone Marrow Transplant.* 9:174 (1992).

53. J. Bubenik, E. Lotzova, M. Indrova, J. Simova, T. Jandlova and D. Buvenikova, Use of IL-2 gene transfer in local immunotherapy of cancer, *Cancer Lett.* 62:257 (1992).

54. S. Eccles, S. Russell and M. Collins, Inhibition of tumour growth and metastasis by IL-2 secreting rodent sarcoma cells, Program of the 4th Int. Cong. on Hormones and Cancer, Amsterdam, p. 328 (Abstr.) (1991).

55. M. Ikeda, U. Ikeda, T. Ohara, E. Kusano and S. Kano, Recombinant interleukin-6 inhibits the growth of rat mesangial cells in culture, *Am. J. Pathol.* 141:327 (1992).

56. H. Suzuki, K. Yasukawa, T. Saito, R. Goitsuka, A. Hasegawa, Y. Ohsugi, T. Taga and T. Kishimoto, Anti-human interleukin-6 receptor antibody inhibits human myeloma growth in vivo, *Eur. J. Immunol.* 22:1989 (1992).

57. E.F. Adams, B. Rafferty and M.C. White, Interleukin 6 is secreted by breast fibroblasts and stimulates 17b-oestradiol oxidoreductase activity of MCF-7 cells: possible paracrine regulation of breast 17b-oestradiol levels, *Int. J. Cancer* 49:118 (1991).

LIPID SATURATION IN THE TARGET CELLS PLASMA MEMBRANE BLOCKS TUMOR NECROSIS FACTOR MEDIATED CELL KILLING

Erno Duda,[1] Sandor Benko,[2] Ibolya Horvath,[1] Erzsebet Galiba,[1] Tibor Pali,[3] Ferenc Joo,[4] and Laszlo Vigh[1]

[1]Institute of Biochemistry and [3]Institute of Biophysics, MTA Biological Center, Szeged
[2]Clinic of Internal Medicine, A. Szent-Gyorgyi Medical School, Szeged
[4]Department of Physical Chemistry, L. Kossuth University of Debrecen, Hungary

SUMMARY

Destruction of sensitive tumor cells by tumor necrosis factor (TNF) is greatly influenced by the composition and the physical state of the lipids within the plasma membrane of the target cells. Experimental conditions that reduce the fluidity of the lipid bilayer of the membrane decrease or completely abolish TNF sensitivity of the target cells. One possible explanation for this phenomenon is the restricted availability of ceramide and arachidonic acid containing phospholipids to enzymes mediating the effects of TNF or decreased activity of the enzymes towards these phospholipids in membranes of decreased fluidity. Since microviscosity and lipid composition of the plasma membrane is known to be altered by the diet and by mediators derived from the neuro-endocrine and the immune systems, the physiological significance of these findings is clear. This phenomenon might also serve as a basis for better treatments for malignant diseases.

INTRODUCTION

TNF-alpha is a cytokine with anti-tumor activity and a variety of other biological activities. It is homotrimer of 17kDa subunits,[1,2] synthesized by activated mononuclear phagocytes and by some other cells. TNF causes hemorrhagic necrosis of certain tumors in experimental animals and in cancer patients and exerts cytotoxic or cytostatic activity on many transformed cell lines in vitro.[2-6] The action of TNF requires specific binding to high-affinity cell surface receptors, which are expressed not only on most malignant cells but also on normal diploid cells of various tissues.[7-10]

Despite high affinity binding, many tumor cells and most normal cells appear to be resistant to the cytotoxic or cytostatic action of TNF, indicating that sensitivity is controlled at post-receptor level.[7-10]

A series of animal experiments proved that TNF has a considerable potential in the elimination of tumors. Subsequent clinical trials revealed that the medical applications of TNF are jeopardized by its systemic toxicity, which is manifested in high fever, hypotension, thrombocytopenia, convulsions, intravascular coagulation, etc. Therefore, TNF cannot be administered safely, unless we can restrict its action to the malignant tissues or protect normal cells against its cytotoxicity.

TNF induces apoptosis, programmed cell death in most susceptible target cells. The exact mechanism of TNF induced cell killing is not known, though in a model system ceramide, an intermediate of the TNF signalling pathway functions as an intracellular mediator of apoptosis induced by TNF.[11] TNF is a powerful pyrogen and the sensitivity of target cells is very much temperature dependent, as is the microviscosity of the cellular membranes. We wanted to explore the role of the physical status of the plasma membrane and of phospholipase(s) in the TNF sensitivity of tumor target cells.

MATERIALS AND METHODS

Recombinant Human TNF-alpha

The human TNF-alpha gene was cloned in *E. coli* cells, TNF was produced and purified to homogeneity in our laboratory. The specific activity of the preparations was more than 20 million U/mg, with no significant endotoxin content (<30 pg/mg protein).

TNF was radioiodinated by chloramine T in a multi-drop system: TNF, Na^{125}I (IZINTA, Budapest) were dissolved in 100 mM K-phosphate buffer, pH 7.0. 50 μ l of this solution, in the form of a drop on siliconized glass surface was exposed to chlorine produced by neighbouring drops of chloramine T solution (SIGMA, St. Louis, MO, 10 mg/ml, 50 μ l each).

Hydrogenation of the Living Cells

The catalyst, Pd(QS) (palladium di[sodium alizarine monosulfonate]), was a product of Molecular Probes, Eugene, Oregon.

Mouse L929 tumor cells were grown in DMEM (SERVA, Heidelberg), containing 5% newborn calf serum (J. Boys SA., Reims), to near confluency. The cells were trypsinized, the action of trypsin was arrested by the addition of minute amounts of serum. The cells were collected by centrifugation and washed free of serum proteins, suspended in 10 mM HEPES buffer (N-2-hydroxy-ehtylpiperazine-N'-2-ethanesulfonic acid, SIGMA, St. Louis, MO), pH 7.4 containing isotonic solution of mannitol (hydrogenation medium). The suspension contained 5×10^5 cells/ml.

A stock solution of catalyst (5 mg/ml) was prepared in degassed HEPES buffer (10 mM, pH 7.4). Part of the catalyst solution was activated by incubating it in hyrogen atmosphere for 5 min before use, in order to convert it to its hydride (B) form.[12] Thirty-forty ml aliquots of L929 cell suspension were transferred into custom designed reaction vessels and placed into a water bath at 37°C. The reaction vessels were connected to a manifold and the gas phase was evacuated and replaced by hydrogen or nitrogen (control), at atmospheric pressure. To initiate the reaction, aliquots of inactive or activated forms of the catalyst were then injected into the reaction vessels through a rubber septum. The final concentration of either form of the complex was 0.1 mg/ml.

During the hydrogenation procedure (max. 5 min), the reaction vessels were rotated at 45 rpm. At the end of the desired incubation time L929 cells were removed, the reaction stopped by aerating the suspension with 5% CO_2 containing air. The cells were repeatedly washed to remove the catalyst, harvested for lipid extraction, ESR measurements or plated for TNF assay.

Viability of hydrogenated cells was estimated by dye exclusion. Long survival of the cells was quantitated by plating aliquotes on Petri dishes in DMEM, 10% FCS and counting the arising clones two weeks later.

Lipid Extraction and Fatty Acid Analysis

Lipids of whole cells were extracted by chloroform/methanol (2:1, v/v), followed by Folch's partition. The individual phospholipid classes were isolated by one dimensional TLC, using pre-coated silica gel plates (Kieselgel 60 F254, Merck). Total or isolated complex lipids were subjected to transesterification in HCl/methanol at 80°C for 2 hours under N_2 protection. Fatty acid methyl esters were separated isothermally (180°C) on a JEOL JGC-20K gas chromatograph, equipped with an SP-2230 capillary column.[12]

ESR Measurements: Estimation of Plasmamembrane Fluidity

To measure the degree of lipid fluidity in the plasmamembrane of control and briefly hydrogenated cells the fatty acid spin label 2-(3-carboxypropyl)-2-trydecyl-4,4-dimethyl-3-oxazolidinyloxyl,5-doxylsterate (Molecular Probes, Eugene, Oregon) was used. Five µl aliquots of solution of the probe (2 mg/ml in ethanol) were added to 400 µl cell suspensions (1.5 x 10^7 cells/ml) and vortexed occasionally for 2 min. After 5 min of incorporation of the spin probe, the cell suspension was centrifuged and the cells were transferred into 1 mm diameter quartz ESR capillaries. Experiments were performed in an X-band spectrometer (JEOL JES-PE-1X) using 100 kHz modulation technique. Spectra were recorded at 37°C. The time dependent changes in the concentration of the spin probe were monitored by setting the magnetic field of the spectrometer at the peak of the midfield line of ESR spectrum.[13] Decrease of the line height was recorded and plotted against incubation time.

Data were collected, stored and manipulated with an IBM PC computer. Orientation order parameter, S, was calculated from the position of the inner and outer splittings and converted for polarity as described by Griffith et al.[14]

Following the ESR experiments, cell viability was routinely tested by dye exclusion. More than 90% of control or briefly (2 min) hydrogenated cells proved to be alive by this test.

Determination of TNF Cytotoxicity

TNF-alpha was quantitated in a modified bioassay using the highly sensitive mouse L929 tumor cells[15] and the metabolic indicator tetrazolium dye, MTT (3-[4,5-dimethylthiazol-2yl]-2,5-diphenyl 2H-tetrazolium bromide, SIGMA, St. Louis, MO).

$3x10^4$ cells in 200 µl medium per well of microtiter plates were plated one day before the assay. MTT was added (10 mg/ml, 10 µl per well) 16 hrs after the exposure to TNF. The color of the dye rapidly changed in the presence of living cells. After 5 hrs of incubation plates were centrifuged, medium removed and 200 µl DMSO was added to dissolve the precipitated dye and measure the absorbance at 540 nm.

Data were calculated as percent survival rate or as units of TNF-alpha, 1 U/ml being defined as the concentration that resulted in the death of 50% of the L929 cells in the presence of 0.1 g/ml actinomycin D at 37°C.

For the characterization of the cytotoxic activity in our TNF preparation samples were incubated either with an excess of rabbit anti-recombinant human TNF-alpha (generously provided by Dr. K. Nielsen) or with control rabbit sera. After 2 hrs of incubation at 37°C, residual cytotoxicity was determined on L929 cells.

TNF Receptor Assay

^{125}I labelled TNF was incubated with the monolayers (grown on microscopic cover slides) at different temperatures for 10, 20, 30 and 60 min in the presence or absence of 10^4 fold excess of non-labelled TNF. The cells were washed at 0°C with PBS. Non-internalized TNF was washed off with 50 mM Na-acetate buffer, pH 3.0. The amounts of receptor bound, non-specifically bound and internalized TNF were calculated on the basis of counts measured by immersing the cover slips into scintillation liquid (dried after consecutive washes of TCA and chloroform:methanol).

RESULTS

Exposure of L929 cells to TNF at different temperatures resulted in very different LD_{50} values. Incubation of tissue cultures with liposomes composed of different phospholipids, free fatty acids and cholesterol modified significantly the TNF sensitivity of the cells (Galiba and Duda, unpublished).

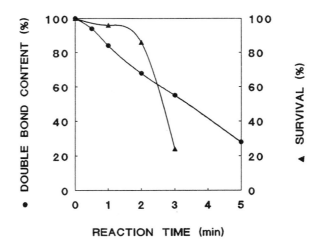

Figure 1. Changes in the viability and the level of fatty acid saturation of L929 cells during catalytic hydrogenation. Membrane lipids of living L929 cells were hydrogenated as described in the Methods. The level of fatty acid saturation (% decrease in total fatty acid double bonds) of L929 cells increased with time during catalytic hydrogenation. Virtually all cells survived, exposed to reduced catalyst up to two minutes long hydrogenation.

To define the exact role of the physical state of the plasmamembrane in the cytotoxic action of TNF we used the method of catalytic hydrogenation.[12,16] Treatment of living L929 mouse tumor cells seemed to be a useful method to alter the lipid composition of the (plasma)membranes in a rapid and reproducible way without seriously interfering with the metabolism of the cells. The hydrogenation procedure can be accomplished by incubation of the cells with the active form of Pd(QS) catalyst under atmospheric presure of H_2 gas. The Pd(QS) complex catalysed hydrogenation was shown to be specific for fatty acids within membrane constituents.[16]

The effects of the presence of the catalyst in its inactive (oxidized) or active (reduced) form, the hydrogenation reaction and the rotation of the reaction vessels were individually tested on living cells. When cells were subjected to the conditions of hydrogenation in the absence of hydrogen, cell viability was not effected adversely. Brief exposures to the active catalyst in the presence of hydrogen did not decrease the number of viable cells signficantly, 90 to 95% of the cells tolerated 2 min hydrogenation and more than 80% survived reaction times up to 3 min.

Further reduction of the fluidity level of cellular membranes during longer periods of hydrogenation decreased the viability of the cultured cells. Fig. 1 shows time dependent saturation of phospholipids and con-comitant change in the viability of L929 cells, respectively. It is worth-while to mention that treated cells adhered to the plates much more slowly than trypsinized, untreated controls. A detailed study was made of lipid hydrogenation rates in living cells. Hydrogenation proceeded as described in previous reports[16] with polyunsaturated fatty acids being reduced to saturates or monosaturates, including some unnatural trans-isomers.

Table 1. Change of fatty acid composition of total lipids and phospholipids in L929 cells after catalytic hydrogenation (mol % of major fatty acids, C: control cells; H: cells hydrogenated for 2 min).

Fatty acids	Total lipids		Phosphatidyl-choline		Phosphatidyl-ethanolamine	
	C	H	C	H	C	H
16:0	17.1	20.9	18.1	22.5	10.7	10.0
16:1	3.0	1.4	4.6	2.8	2.9	1.6
18:0	18.9	26.2	14.2	27.6	29.1	32.2
18:1	53.1	46.8	59.1	45.1	50.4	50.1
18:2	1.2	0.4	0.2	0.0	1.4	0.9
18:3	3.5	2.0	2.6	0.9	1.9	1.7
20:4	2.4	2.3	1.2	1.1	3.6	3.5

Table 1 details data of fatty acid compositions of lipids of the cellular membranes of control cells (treated with inactive catalyst) and for those treated with the active catalyst for 2 min (almost complete survival of treated cells, as shown on Fig. 1). Efficient reduction of the major fatty acid, oleic acid was the most striking difference found between the two samples. Hydrogenation occurred also in the palmitoleic acid and in polyunsaturated C18 fatty acids as well, but these were only minor compounds. It was noteworthy, that no significant hydrogenation of 20:4, arachidonic acid was found. (Due to a variety of hydrogenation products that were distributed among several minor peaks, we could not identify saturated derivatives of arachidonic acid.) A more detailed analysis revealed that certain lipid classes were virtually untouched by 2 min hydrogenation of L929 cells. Focusing on two main classes of complex phospholipids, PE was somewhat more resistant to hydrogenation than PC. Preference of the catalyst for PC was not found in mixed dispersions of these isolated complex lipids. This finding therefore was considered to be an indication of restricted access of the catalyst to some acyl lipids (found primarily in the inner layer of the membranes). Evidence for selective catalytical hydrogenation of a peripherally located membrane of intact cells was reported earlier.[17] In accordance with these findings, we assumed that endomembranes of living L929 cells are not readily accessible to the catalyst during short hydrogenations, whereas lipids of the plasmamembrane were exposed to the effect immediately.

To prove this point, we carried out experiments using nitroxid spin labels incorporated into L929 cells. It was reported[13] that nitroxid spin labels used to probe membrane dynamics are reduced by mammalian cells to the corresponding hydroxylamines. It appeared that reduction of doxylstearates takes place at the ubiquinon level in the respiratory chain in mitochondria. Therefore, these results suggested that the rate of the reduction of the doxyl moiety is partially controlled by the internalization of the spin probe. In view of the above considerations the kinetics of the reduction of 5-doxylstearate was determined in both control and 2 min hydrogenated cells (Fig. 2).

Time-dependent changes in the concentration of the spin label run closely parallel in both samples as shown on Fig. 2. Our results also indicated that for a while (up to 10 min) the reduction rate is non-detectable.

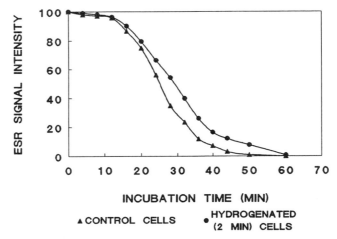

Figure 2. Time dependent decrease of electron spin resonance signal in hydrogenated and control cells. Membrane lipids of living L929 cells were hydrogenated and the ESR signal was measured as described in the Methods. All data were recorded at 37°C. Values shown for non-hydrogenated cells were obtained using N_2 gas instead of H_2 with the catalyst.

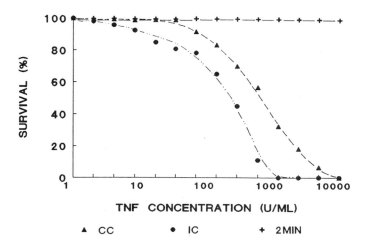

Figure 3. The effect of the reaction time of hydrogenation on the order parameter (S'A) of fatty acid chains by using 5-doxyl-stearate spin probe. Membrane lipids of living L929 cells were hydrogenated and order paprameter was measured as described in the Methods. Control represents order parameter obtained from plasma membrane of cells grown at 37°C without hydrogenation, in the presence of inactive catalyst.

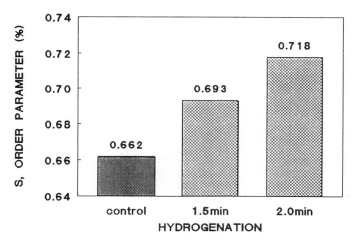

Figure 4. The effect of hydrogenation of membrane lipids of L929 cells on TNF mediated cytotoxicity. Membrane lipids of living L929 cells were hydrogenated and their TNF sensitivity was measured as detailed in the Methods. Saturation of the membrane lipids decreased the sensitivity of L929 cells to TNF by at least 100-fold. Non-hydrogenated cells, containing trace amounts of catalyst proved to be more sensitive to TNF than control cells. CC: control cells; IC: inactive catalyst; 2 min: cells hydrogenated for 2 min.

During this time the fatty acid spin label is located in the membranes of the cell surface. In accordance with this finding, the orientation order parameters were calculated from the ESR spectra of those cells prelabelled for 5 min with 5-doxylstearate (Fig. 3). The structural order parameter of membrane lipids was not significantly affected by the presence of inactive catalyst (data not shown). Parallel with the progress of hydrogenation at conditions resulting in no reduction of the viability of the cells, ordering state of the plasmamembrane increased markedly.

The above described experiments underlined the usefulness of catalytic hydrogenation in changing the acyl chain composition of the membranes of living L929 cells and also proved that the effect of brief treatments is confined to the plasma membrane.

The effect of hydrogenation on the susceptibility of treated and control tumor cells to the cytotoxic effect of TNF was then investigated. As it is shown in Fig. 4, a brief 2 min hydrogenation of L929 cells resulted in a dramatic change in their TNF sensitivity. TNF practically lost its cytotoxicity toward treated cells. In order to rule out that the effect is caused by traces of Pd(QS), in one control experiment we did not wash out the catalyst completely. The compound did not protect the cells against TNF; actually, it slightly enhanced the killing effect of the cytokine (Fig. 4).

Receptor binding experiments with ^{125}I labelled TNF molecules showed almost no difference between TNF binding of hydrogenated and control cells. Internalization of TNF-receptor complexes slowed down in hydrogenated cells, not unlike cells exposed to TNF at 4°C (data not shown).

DISCUSSION

Exposure of tumor cells to TNF at different temperatures resulted in very different LD_{50} values. Incubation of tissue cultures with liposomes composed of different phospholipids, free fatty acids and cholesterol modified significantly the TNF sensitivity of the cells (Galiba and Duda, unpublished).

In this paper we report the establishment of experimental conditions that allowed almost complete survival of cells while greatly reduced their sensitivity to TNF.

Following brief treatments in the hydrogenation condition, where viability of the hydrogenated cells remain over 90%, quantitative analyses detected a remarkably low change in arachidonic acid content. Obviously, this alteration was hardly enough to explain a dramatic change in TNF sensitivity. The order parameter (S'A) of 5-doxyl-stearic acid rose steadily throughout the range of hydrogenation.

Receptor-binding experiments revealed that TNF-receptor interaction was not impaired by catalytic hydrogenation. Internalization of TNF-receptor complexes, however, slowed down significantly, but the effect was completely reversible. This finding was not surprising, since living cells rapidly restore the fluidity of their membranes following hydrogenation.[18]

In light of the above findings, we assumed that the explanation of the decreased sensitivity should be found elsewhere. It was shown that membrane fluidity controls PLA_2 activity. A relatively low degree of hydrogenation markedly influenced PLA catalysed arachidonic acid release, as demonstrated by using a reconstituted system,[19] ciliary membranes of Tetrahymena,[20] or rat liver mitochondria.[21]

Arachidonic acid pathway metabolites play a definitive role in TNF mediated cell killing, inhibitors of the cyclooxygenase and lipoxygenase pathways protect sensitive cells against the cytotoxic effect of TNF.[22] It was shown that in certain cells TNF activates the sphingomyelin signal transduction pathway liberating ceramide as the second messenger molecule.[23,24] TNF resistance of the hydrogenated cells could therefore stem from the fact that liberation of arachidonic acid or the release of ceramide is blocked by the elevation of the membrane microviscosity. Our previous data suggest that intramembraneous hydrogenation of lipids reduces the substrate availability, but not the enzyme activity of endogenous PLA_2.[19]

In conclusion, we can assume that temporary changes in the physical state of the plasmamembrane caused by the environment, the diet or the actions of the neuro-endocrine and/or the immune systems can influence the sensitivity of the tumor cells.

Since membrane viscosity can be modified by hypo- or hyperthermia, pharmacological agents, local treatment of tumors might render malignant cells more sensitive to TNF action, or, alternatively, sensitive healthy tissues could be protected against TNF toxicity. By increasing the sensitivity of the target cells (e.g. vascular endothelium of the tumor) it would be possible to lower the therapeutic doses of TNF, eliminating or significantly decreasing the systemic toxicity of this powerful cytokine.

ACKNOWLEDGEMENTS

This work was supported by grants from the Hungarian Scientific Research Foundation (OTKA, #543 to L.V. and #547 to E.D.).

REFERENCES

1. E.A. Carswell, L.J. Old, R.L. Kassel, S. Green, N. Fiore and B. Williamson, An endotoxin-induced serum factor that causes necrosis of tumors, *Proc. Natl. Acad. Sci. USA* 72:3666 (1975).
2. B.B. Aggarwal, W.J. Kohr, P.E. Hass, B. Moffat, S.A. Spencer, W.J. Henzel, T.S. Bringman, G.E. Nedwin, D.V. Goeddel and R.N. Harkins, Human tumor necrosis factor, *J. Biol. Chem.* 260:2345 (1985).
3. B.J. Sugarman, G.G. Aggarwal, P.E. Hass, I.S. Figari, M.A. Palladino, M.M. Shephard, Recombinant human TNF alpha: effects on proliferation of normal and transformed cells *in vitro*. *Science* 230:943 (1985).
4. C. Peetre, U. Gullbert, E. Nilson and J. Olson, Effects of recombinant TNF on proliferation and differentiation of leukemic and normal hemopoietic cells *in vitro.*, *J. Clin. Invest.* 78:1694 (1986).
5. B. Williamson, E.A. Carswell, B.Y. Rubin, J.S. Prendergast and L.J. Old, Human TNF produced by human B cell lines: Synergistic cytotoxic interaction with human interferon gamma, *Proc. Natl. Acad. Sci. USA* 80:5397 (1983).
6. C.S. Johnson, M. Chang and P. Furmanski, *In vivo* hematopoietic effects of TNF alpha in normal and erythroleukemic mice: characterization and therapeutic applications, *Blood* 72:1875 (1988).
7. F.C. Kull, S. Jacobs and P. Cuatrecasas, Cellular receptor for ^{125}I labeled TNF: Specific binding, affinity labeling and relationship to sensitivity, *Proc. Natl. Acad. Sci. USA* 82:5756 (1985).
8. B.Y. Rubin, S.L. Anderson, S.A. Sulliman, B. Williamson, E.A. Carswell and L.J. Old, High affinity binding of ^{125}I human TNF (LuKII) to specific cell surface receptors, *J. Exp. Med.* 162:1099 (1985).
9. M. Tsujimoto, Y.K. Yip and J. Vilcek, TNF: Specific binding and internalization in sensitive and resistant cells, *Proc. Natl. Acad. Sci. USA* 82:7626 (1985).
10. H. Loetscher, E.J. Schlaeger, H.W. Lahm, Y.C.E. Pan, W. Lesslauer and M. Brockhaus, Purification and partial amino acid sequence analysis of two distinct TNF receptors from HL60 cells, *J. Biol. Chem.* 265:20131 (1990).
11. L.M. Obeid, D.M. Linardic, L.A. Karolak and Y.A. Hannun, Programmed cell death induced by ceramid, *Science* 259:1769 (1983).
12. F. Jóo, N. Balogh, L.I. Horváth, G. Filep, I. Horváth and L. Vigh, L., Complex hydrogenation/oxidation reactions of the water-soluble hydrogenation catalyst palladium di(sodium alizarinmonosufonate) and details of homogenous hydrogenation of lipids in isolated biomem-branes and living cells, *Anal. Biochem.* 194:34 (1991).
13. K. Chen, P.D. Morse, II and H.M. Swarts, Kinetics of enzyme-mediated reduction of lipid soluble nitroxide spin labels by living cells, *Biochem. Biophys. Acta* 943:477 (1988).

14. O.H. Griffith and P.C. Jost, Lipid spin labels in biological membranes, *in:* "Spin Labeling: Theory and Applications," L.J. Berliner, ed., Acad. Press, New York (1976).

15. K. Frei, C. Siepl, P. Groscurth, S. Bodmer, C. Schwerdel and A. Fontana, Interleukin-HP-1 related hybridoma and plasmacytoma growth factors induced by lipopolysaccharide *in vivo, Eur. J. Immunol.* 17:1217 (1987).

16. M. Schlame, L. Horváth and L. Vigh, Relation between lipid saturation and lipid-protein interaction in liver mitochondria modified by catalytic hydrogenation with reference to cardiolipin molecular species, *Biochem. J.* 265:79 (1990).

17. S. Benko, H.J. Hilkmann, L. Vigh and W.J. van Blitterswijk, Catalytic hydrogenation of fatty acid chains in plasma membranes: effect of membrane lipid fluidity and expression of cell surface antigens, *Biochem. Biophys. Acta* 896:129 (1987).

18. L. Vigh, I. Horváth and G.A. Thompson, Jr., Recovery of *Dunaliella salina* cells following hydrogenation of lipids in specific membranes by a homogenous palladium catalyst, *Biochem. Biophys. Acta* 937:42 (1988).

19. R. Kannagi and K. Koisumi, Effect of different physical states of phospholipid substrates on partially purified platelet phospholipase A_2 activity, *Biochem. Biophys. Acta* 556:423 (1979).

20. I. Horváth, L. Vigh, T. Pali and G.A. Thompson, Jr., Effect of catalytic hydrogenation of *Tetrahymena* ciliary phospholipid fatty acids on ciliary phospholipase A activity, *Biochem. Biophys. Acta* 1002:409 (1989).

21. M. Schlame, I. Horváth, Zs. Török, L.I. Horváth and L. Vigh, Intramembraneous hydrogenation of mitochondrial lipids reduces the substrate availability but not the enzyme activity of endogenous phospholipase A. The role of polyunsaturated phospholipid species, *Biochem. Biophys. Acta* 1045:1 (1990).

22. P. Suffys, R. Beyaert, F. Van Roy and W. Fiers, Reduced TNF-induced cytotoxicity by inhibitors of arachidonic acid metabolism, *Biochem. Biophys, Res. Comm.* 149:735 (1987).

23. M-Y. Kim, C. Linardic, L. Obeid and Y. Hannun, Identification of sphingomyelin turnover as an effector mechanism for the action of TNF alpha and gamma IFN, *J. Biol. Chem.* 266:484 (1991).

24. K.A. Dressler, S. Mathias and R.N. Kolesnick, TNF alpha activates the sphingomyelin signal transduction pathway in a cell free system, *Science* 255:1715 (1992).

BIDIRECTIONAL COMMUNICATION BETWEEN THE IMMUNE AND ENDOCRINE SYSTEMS - MEDIATION BY HORMONES FROM THE GONADS

C.J. Grossman and M.A. Neinaber

V.A. Medical Center, Research Service
3200 Vine Street, Cincinnati, Ohio 45229
and the Department of Biology, Xavier University
Cincinnati, Ohio, U.S.A.

SUMMARY

Communication between immune and endocrine systems depend on steroid hormones. Both gonadal steroids (GS) and adrenal steroids (AS) are involved because they target diverse cell types within the immune system. Lymphoid cells sequestered within the cortical thymic compartment are responsive to modulation by both GS and AS and, in addition, reticuloendothelial (RE) cells of thymic matrix are also targeted. During maturation lymphoblasts within the cortex possess steroid receptors for both GS and AS but mature thymocytes only retain their AS receptors. This suggests that one or more stages in early development of lymphocytes may be regulated by both GS and AS within the thymic microenvironment. Also, maturation depends on regulatory substances such as thymic hormones secreted by the RE cells under the control of GS and AS. Thus a concert of hormones present during early maturation may program developing lymphocyte subpopulations and lead to immunological sexual dimorphism.

The question of hormonal feedback from lymphocytes to regulate endocrine events during maturation remains unanswered. However, such feedback pathways are present for mature effector cells undergoing an active immune response. Included in this scheme are cytokines that impact the hypothalamus thereby regulating the pituitary axes controlling both the gonads and adrenals. This results in down regulation of immune effector cells via activation of the adrenal axis accompanied by inhibition of the gonadal axis.

COMMUNICATION BETWEEN THE IMMUNE SYSTEM AND OTHER BODY SYSTEMS

In complex multilevel systems, communication between individual elements is absolutely necessary. To facilitate this communication, multicellular organisms have developed both nevous and endocrine systems. These communication pathways provide exquisite regulation between the various systems within the multicellular organism. Classically, negative feedbacks are central to the regulation provided by both the nervous and endocrine systems.

Additional levels of regulation take place between the nervous and endocrine systems themselves mediated by neurohormones secreted by cells in such diverse locations as the hypothalamus and the adrenal medulla. Thus the immune system regulates, and in turn is regulated by, the other systems of the body. This linkage is accomplished via direct and indirect connections between the nervous system and immune tissues, by endocrine hormones that interact with tissues and cells of the immune system and by immunohormones (such as cytokines) elaborated by cells of the immune system that impact upon various levels and elements of the nervous and endocrine systems.

While many endocrine hormones have been implicated in immune system regulation, the steroid hormones from the gonadal and adrenal axes are of particular interest because of their possible roles in immunological stress responses, autoimmunity and immunological sexual dimorphism. Effects elicited by these steroids may be mediated directly through receptors within the effector lymphocyte subpopulations. In addition, these steroids may function indirectly by acting to effect the release of other endocrine substances which then impact the effector lymphocytes.

LYMPHOCYTE STEROID RECEPTORS

The action of both sex and adrenal steroids directly on effector lymphocytes has been documted by studies on lymphocyte steroid receptors. Ahmed[1] proposed that sex steroids function at the level of the stem cell, pre-T and pre-B lymphocytes and can also regulate adult T-cells and monocytes; Gulino found that fetal thymus tissue[2] and large immature thymocytes[3] both contain receptors for estrogen. Estrogen receptors have also been demonstrated in human thymocytes by others,[4] but in addition, they have been identified in human peripheral blood mononuclear cells,[4,5] which would imply that certain mature lymphocytes continue to possess receptors for estrogen.

With regard to androgen receptors in lymphocytes, the results reported by Pearce[6] suggest that those lymphocyte populations within the thymus that are glucocorticoid resistant may also possess androgen receptors. Support for the presence of androgen receptors within thymocytes can be found in the study by Kovacs and Olsen[7] who also reported that androgen receptors were not present in mature blood lymphocytes or thoracic duct T-cells, (a finding similar to that reported by Gulino for thymocyte estrogen receptors[3]). Furthermore, Kovacs[8] has also reported that the function of the thymocyte androgen receptors may be to reduce IL-2 in cultures of thymocytes exposed to dihydrotestosterone,[7] although Dauphinee et al.[8] have reported that treatment of NZB mice with androgen enhances IL-2 production *in vivo*. In addition to estrogen and androgen receptors, glucocorticoid receptors have been found in many different populations of lymphocytes both during early development and within the mature forms.[9-13]

The presence of sex steroid receptors and glucocorticoid receptors both in developing and mature lympohcytes would imply that both sex steroids and glucocorticoids are capable of regulating lymphocyte development and effector cell function. In addition, sex steroid receptors have been identified within the reticuloepithelial matrix of the thymus, suggesting that both estrogens, androgens and progestins play a role in defining the thymic microenvironment that programs developing T-lymphocytes.

THYMIC RETICULOEPITHELIAL (RE) STEROID RECEPTORS

Estrogen,[14,15] androgen,[16,17] progestin[18,19] and glucocorticoid[13] receptors have all been identified in reticuloepithelial (RE) cells of the thymus. One of the functions of the RE cell receptors may be to regulate the release of thymic hormones. This conclusion is supported by our early studies[20,21] where we demonstrated that serum prepared from castrate rats stimulated blastogenic transformation of thymocytes in culture.

One of the functions of the RE cell receptors may be to regulate the release of thymic hormones.[20,21] This conclusion is supported by our early studies[20,21] where we demonstrated that serum prepared from castrate rats stimulated blastogenic transformation of thymocytes in culture. However, serum prepared from castrate rats that had been treated with physiological levels of estradiol (E2) lost this stimulatory property. In addition, serum prepared from castrate thymectomized rats lost its stimulatory properties, suggesting that the source of the stimulatory factor in the serum was the thymus. The function of these thymic hormones remains under investigation, but it would seem logical that they might be involved both in thymocyte development within the thymic microenvironment and regulation of mature T-lymphocytes in location distant from the thymus.

These older studies have been extended by the work of a Japanese group[22,23] who were able to demonstrate histochemically that the RE cells within the thymus, that possess receptors for estrogen and progestin, were identical to the cells that secreted the thymic hormone thymulin. This would support the view that the sex steroid hormones are involved in regulating the secretion of thymulin.

In addition to the presence of sex steroid hormone receptors within the thymic reticuloepithelial matrix, receptors for glucocorticoid have also been identified immunohistochemically,[13] but the function as related to regulation of thymic hormone secretion remains hypothetical. However, both sex steroid hormones and glucocorticoids can cause massive structural changes within the thymic tissue[24-26] which can be reversed at least temporarily by castration.[15]

EFFECTS OF CASTRATION ON LYMPHOID TISSUES

After castration and/or adrenalectomy, there is a significant increase in the mass of lymphoid tissues such as thymus, spleen and lymph nodes. This would suggest that these tissues are all targets for sex steroids and glucocorticoids, although sex steroid receptors to date have only been identified in thymus. After castration, the thymic cortex and medulla are both increased in mass and lymphocyte cellularity.[25] Furthermore, estrogen treatment has been shown to reduce thymic mass, disorganize thymic architecture and promote thymocyte destruction especially of cortical thymocytes.[25] There may also be an increase in mass of the spleen after castration, but the magnitude of the response is significantly less than for the thymus.[27]

In studies by Greenstein and Fitzpatrick[28-30] castration of female rats retarded thymic regression. Replacement with testosterone (T) and estrogen increased thymic atrophy while dihydrotestosterone appeared to retard the atrophic effect.[28,29] In addition, these authors found that thymic regeneration in old male rats could be produced if they were treated with an analogue of LHRH.[30] The authors suggest that the mechanism responsible may be the desensitization of the pituitary gland to hypothalamic LHRH, however "a direct action on the thymus, or an action through other pituitary hormones cannot be ruled out."

THE IMPACT OF STEROID HORMONES ON T-LYMPHOCYTE SUBPOPULATIONS

Subpopulations of T-lymphocytes appear to be specifically effected by steroids. For example, Ahmed et al.[31] reported that treatment of 2-3 month old female C57BL/6 mice with E2 and T for 2 weeks reduced total lymphocyte recovery from thymus and spleen, but not from bone marrow or lymph nodes. This would appear to support the assumption that lymphocytes in these locations are targets for sex steroids.

Furthermore, Screpanti et al.[32] reported that the CD4$^+$/CD8$^+$ double positive subpopulations were signfiicantly depleted by short-term *in vivo* treatment with estradiol and the CD4$^+$ (T-helper) and to a lesser extent the CD8$^+$ (T-suppressor) was increased. The authors suggest that estrogen treatment in mice created a functional imbalance between these subpopulations leading to an increased CD4/CD8 ratio. In addition, Ho et al.[33] have demonstrated that human patients with estrogen deficiencies have significantly lower CD4/CD8 ratios (1.15, n=19) vs normal patients with normal levels of estrogen (1.80, n=24), and with reduced estrogen there was a significantly increased number of CD8 (T-suppressor/cytotoxic) cells.

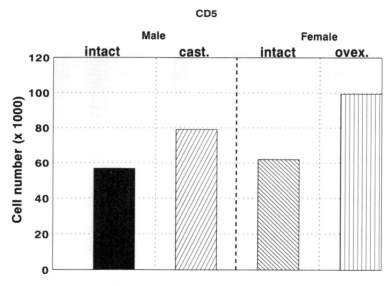

Figure 1. Lymphocytes were purified from the spleens of male and female Sprague Dawley rats 3 months of age. The cells were washed repeatedly in Hanks Balanced Salt Solution and centrifuged over Histopaque 1077. Cells were then incubated in the presence of fluorescently labelled monoclonal antibodies (Caltag Labs, South San Francisco, CA) specific for T-lymphocyte subpopulations. The concentration of the monoclonal antibodies used in these studies was in the range of 10 1g/10^6 cells/ml. Incubation was carried out in Hanks Balanced Salt Solution (including Calcium/Magnesium) + 4% Horse Serum for 18 hours in the cold. Labelled cells were then counted in a fluorescent cell sorter (FACS). The units of Cell #/Body Weight (i.e. corrected cell number) presented in this figure were derived from the percent of a particular labelled cell population as measured in the FACS x total cell number/spleen and this was then divided by body weight (BW) of the animal. In this figure the corrected cell number of total T-cells (CD5) is presented for normal (intact) male and female rats and for castrate male and castrate (ovariectomized) female rats.

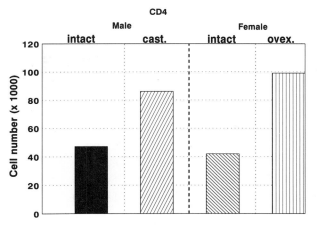

Figure 2. Lymphocytes were prepared, fluorescently labelled and counted in the FACS as described in Figure 1. In this figure the corrected cell number of T-helper cells (CD4) is presented.

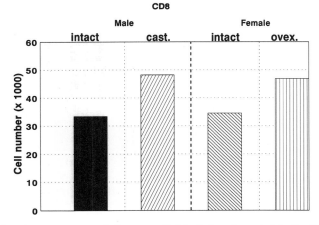

Figure 3. Lymphocytes were prepared, fluorescently labelled and counted in the FACS as described in Figure 1. In this figure the corrected cell number of T-suppressor/cytotoxic cells (CD8) is presented.

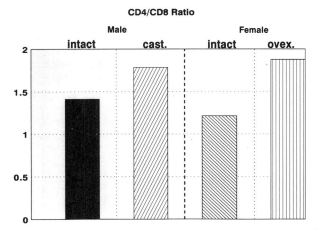

Figure 4. In this figure the CD4/CD8 ratio derived from previous data in Figures 2 and 3 is presented for intact male and female rats and for castrate male and female rats.

Generally, these findings are in agreement with preliminary studies currently in progress in our own laboratory. In the initial experiments we have castrated male and female rats, separated splenocytes by Histopaque centrifugation and labelled the individual lymphocyte subpopulations with fluorescent marker indirectly conjugated to specific monoclonal antibodies. The labelled cells were then counted on a FACS. As can be seen in Figure 1, castration increased the number of total T-cells (CD5) in both male and female rats. In addition, castration in males and females also increased the CD4 (T-helper) (Figure 2), and to a lesser extent CD8 (T-suppressor) cells (Figure 3). As expected, the CD4/CD8 ratio was also elevated after castration for both male and female rats (Figure 4).

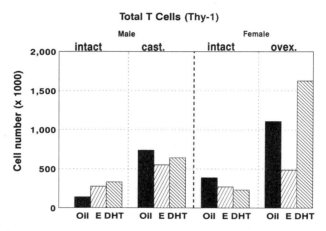

Figure 5. Animals received replacement steroid injections of estrogen (E) or dihydrotestosterone (DHT) dissolved in corn oil. Corn oil alone was also injected as a control. These steroids were injected at a concentration of 50 1g/kg BW/day for 3 days prior to sacrifice. After sacrifice lymphocytes were prepared, fluorescently labelled and counted in the FACS as described in Figure 1. In this figure the corrected cell number of total T-cells as measured by Thy-1 monoclonal antibody is presented. Thy-1 is known to label both mature and certain immature forms of T-cells. CD5, on the other hand, (as utilized in Figure 1) labelled more mature forms of T-cells.

To learn if the effects of castration can be reversed by replacement with sex steroids we then proceeded to castrate male and female rats and treat them with steroid oil injections for three days. The concentration of steroids used was 50μ g/kg BW. From previous studies we know that this concentration results in physiological blood levels. Splenocytes were separated by Histopaque centrifugation and lymphocyte subpopulations were again immunofluorescently labelled using monoclonal antibodies directly conjugated to the fluorgen. Labelled cells were then counted in the FACS. It is important to point out that these results are still in the preliminary stages and the number of animals utilized too small to allow for the use of accurate statistical calculations. Castration increased in both males and females the populations of total T-cells (Thy-1, Figure 5), of CD4 T-helper cells (Figure 6) and CD8 T-suppressor/ cytotoxic cells (Figure 7). These effects were more pronounced in females. Moreover, while treatment of intact males with estrogen and DHT mildly stimulated total T-cells (Figure 5), CD4 cells (Figure 6) and CD8 cells (Figure 7), treatment of intact females with E or DHT seemed to reduce the numbers of these different cell subpopulations,

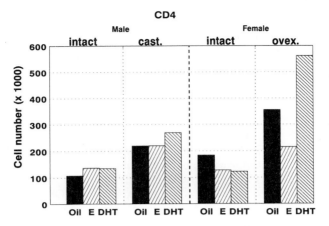

Figure 6. Animals were steroid replaced as described in Figure 5 and lymphocytes were prepared, fluorescently labelled and counted in the FACS as described in Figure 1. In this figure the corrected cell numbers for T-helper cells (CD4) are presented.

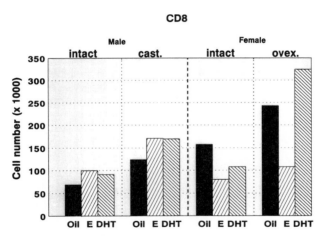

Figure 7. Animals were steroid replaced as described in Fig. 5 and lymphocytes were prepared, fluorescently labelled and counted in the FACS as described in Fig. 1. In this figure the corrected cell numbers for T-suppressor/cytotoxic cells (CD8) are presented.

especially for CD8 cells (Figures 5-7). In addition, while treatment of castrate males with E or DHT slightly reduced total T cells (Figure 5), treatment of castrate females with E drastically reduced total T-cells while DHT strongly stimulated them (Figure 5). The pattern for CD4 cells (Figure 6) and CD8 cells (Figure 7) in castrate females appeared similar to that for total T-cells, where E2 reduced the CD4 and CD8 cells and DHT stimulated them. However, for castrate males E2 and DHT appeared mildly stimulatory on CD8 cells (Figure 7), but for CD4 cells (Figure 6) E2 had no apparent effect, while DHT might have been mildly stimulatory. These results can be summarized in the CD4/CD8 ratio (Figure 8) where E2 appeared to increase this ratio in the intact and castrate females, castration stimulated this ratio in the castrate males, and DHT reduced this ratio in the castrate male. Clearly, sex steroids impact differently on the T-lymphocyte subpopulations in males and females. These results would appear to support immunological sexual dimorphism (ISD), and, in addition, they imply that one of the underlying causes of ISD may be due to differences in response of these T-cell subpopulations to sex steroids.

Our results are corroborated by Aboudkhil et al.[34] who found that treatment of male mice with the androgen Depo-testosterone increased the CD4 and CD8 cells in the thymus and CD8 cells in the spleen. However, in these studies they utilized very high levels of their androgen (5 mg/kg) and in addition they reported that three weeks after castration there was a decrease in both the CD4 and CD8 cells in the spleen, a finding our data does not support. The reasons for this difference remains unclear but the length of time after castration, or the levels of steroid utilized, may be a factor.

Olsen et al.[35] also studied the effects of androgen treatment on thymocyte and splenocyte subpopulations in male mice after castration. They reported an increase in both thymic and splenic mass (as we observed) and also that treatment with testosterone (T) reduced thymic (but not spleen) mass. In addition, they reported that removal of androgens by castration shifted the balance within the subpopulations towards CD4 cells. This supports our finding that castration increases the CD4/CD8 ratio. They also found that treatment with T increased the CD8 population while we observed that treatment with DHT reduced the CD4/CD8 ratio. However, we never studied the effects of T in our system because it can be metabolically converted into both E and DHT *in vivo*, resulting in problems with interpretation of the results.

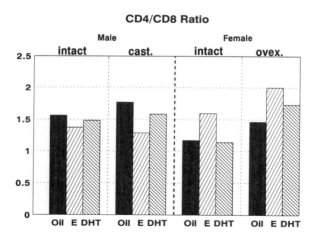

Figure 8. In this figure the CD4/CD8 ratio derived from previous data in Figs. 6 and 7 is presented for intact male and female rats and for castrate male and female rats.

Our findings are also supported by the work of Screpanti et al.[36] who reported that in male mice treatment with estrogen *in vivo* increased CD4 and CD8 thymocytes. However, these authors did not study the effects of estrogen on subpopulations of thymocytes from female mice.

POSSIBLE IMPACT OF SEX STEROIDS ON LYMPHOCYTE SUBPOPULATIONS DURING DEVELOPMENT

Recently[37] we suggested that one of the underlying causes of autoimmune disease was due to inappropriate development of T-suppressor cells. According to Weinstein and Berkovich,[38] androgens target subsets of T-suppressor cells resulting in functional immunosuppression of other effector lymphocyte subclasses by the T-suppressor cells. In the event that this immunosuppression is lost, autoimmune disorders may be one result. Perhaps T-suppressor lesions (which may be genetically predetermined) are brought to fruition by events within the microenvironment during development and programming of lymphocyte subclasses. Certainly multiple effects can be elicited in the microenvironment by steroid hormones and thymic hormones; and sex steroid hormones target effector lymphocytes at various points in the functional immune networks. However, sex hormones that function in the microenvironment to effect development are of special interest because immature lymphocytes have been reported to possess sex steroid receptors[2,3] and estrogen targets developing thymocytes.[32,36]

While the action of suppressor T-cells as regulators of immune function must be underscored, the importance of the thymic processes during development of T-cells is clearly paramount. Considering that over 90% of the pre-T lymphocytes entering the thymus from the bone marrow undergo autolysis, what does this suggest? Clearly this implies that potentially autoreactive T-cells are being removed from the lymphocyte pool prior to their maturation, thus limiting the development of autoimmune diseases. While the steps involved in such cyto-destruction have not been fully elucidated, it has been suggested that the process is based on the presence/absence of certain cell surface markers, that it involves interactions with reticulo-epithelial elements of the thymic matrix, and that autocrine/cytokine hormonal messengers play a role.[39,40,41] Additionally, it is not unreasonable to conclude that because sex hormones and glucocorticoids impact on the various thymic elements, they are also involved in regulation of T-cell development, and ultimately in the adult immune responder cell populations.[37,42]

In conclusion, a variety of factors impact upon T-lymphocytes which can then fucntion to either stimulate or inhibit the diverse functions of important effector cells. Cytokines, sex steroids, and glucocorticoids are all important mediators of immune cell function. Sex steroid hormones themselves are able to exert their influence by acting both directly and indirectly on these T-lymphocyte subpopulations. Such direct effects are believed to be mediated classically through nuclear steroid receptors, while indirect regulation is accomplished through membrane bound receptors that interact with the cytokines and thymic hormones as well as many other substances not considered in this discussion (such as Growth Hormone, Prolactin, Thyroid Hormone, Insulin, Neurotransmitters, etc.).

While the classic theory by which sex steroids function at target cells depends on the binding of these hormones at nuclear steroid receptors, the reader should not close his or her mind to the possibility that other methods of binding may exist. The authors realize that to propose a non-cytoplasmic/nuclear binding mechanism is still mildly heretical. However, as early as 1977 Pietras and Szego[39] suggested that estrogen could bind specifically to surface receptors on endometrial cells. More recently Farnsworth[40] has suggested that at least for the androgens, membrane interactions involving sodium-potassium ATPase may mediate some regulatory events in prostate cells. If this is the case, then it provides yet another mechanism by which sex steroids could directly regulate immune effector cells that may not possess classical cytoplasmic/nuclear steroid receptors.

One further level of control between lymphocytes and the endocrine system may exist via the release of lymphokine-like factors that promote a direct stimulation of steroid production by granulosa cells. Reports by Hughes et al.[41] and Emi et al.[42] indicate that a progesterone stimulatory factor elaborated by mitogen stimulated T-cells from the spleen[41] or by allogenic peripheral blood lymphocytes[42] can stimulate granulosa cells to generate progesterone.

In addition to the above, it is now clear that down regulation of an ongoing immune response is provided by the adrenal axis. Accordingly, feedback of cytokines into this system activates the release of adrenal steroid which then function effectively to quench immune cell activation.[37,42,47] While such an effect is envisioned to be primarily important during potentially excessive immune responses (initiated by polyclonal activators, sepsis, etc.), it is not unreasonable to propose that a low level of continued negative feedback regulation may be continuously supplied through this pathway. Such down-regulation must function in conjunction with other processes (suppressor T-cells, anti-idiotypic networks, etc.) to provide exquisite control of an ongoing immune response.

Finally, there is mounting evidence that effector lymphocytes may be capable of producing such hormones as ACTH, Corticotropin, Thyrotropin, Prolactin, GH,[43] while direct innervations exist between the autonomic nervous system and effector lymphocytes.[44] Thus, on a global level bidirectional communication takes place between the immune system and other systems of the body. The outcome of these complex interactions is the maintenance of a disease-free homeostasis.

REFERENCES

1. A.S. Ahmed, W.J. Penhale and N. Talal, Sex hormones, immune response and autoimmune disease, *Am. J. Path.* 121:531 (1985).
2. I. Screpanti, A. Gulino and J.R. Pasqualini, The fetal thymus of guinea pig as an estrogen target organ, *Endocrinology* 111:1522 (1982).
3. A. Gulino, I. Screpanti, M.R. Torrisi and L. Frati, Estrogen receptors and estrogen sensitivity of fetal thymocytes are restricted to blast lymphoid cells, *Endocrinology* 117:47 (1985).
4. J.J.A.M. Weusten, M.A. Blankenstein, F.H.J. Gemlig-Meyling, H.J. Schuurman, L. Kater and J.H.H. Thijssen, Presence of oestrogen receptors in human blood mononuclear cells and thymocytes, *Acta Endocrinol.* 112:409 (1986).
5. L. Danel, G. Souweine, J.C. Monier and S. Saez, Specific estrogen binding sites in human lmphoid cells and thymic cells, *J. Steroid Biochem.* 18:559 (1983).
6. P.B. Pearce, A.K. Khalid and J.W. Funder, Androgens and the thymus, *Endocrinology* 109:1073 (1981).
7. W.J. Kovacs and N.J. Olsen, Androgen receptors in human thymocytes, *J. Immunol.* 139:490 (1987).
8. M.J. Dauphinee, S. Kipper, K. Roskos, D. Wofsky and N. Talal, Androgen treatment of NZB/W mice enhances IL-2 production, *Arthritis Rheum.* 24:(Suppl)S64 (1981).
9. E. Pardes, J.E. De Yampey, D.F. Moses and A.F. De Nicola, Regulation of glucocorticoid receptors in human mononuclear cells: Effects of glucocorticoid treatment, Cushing's disease and ketoconazole, *J. Steroid Biochem. Molec. Biol.* 39:233 (1991).
10. E. Girardin, M. Garoscio-Cholet, H. Dechaud, H. Lejeune, E. Carriert, J. Tourniarire and M. Pugeat, Glucocorticoid receptors in lymphocytes in anorexia nervosa, *Clin. Endocrinol.* 35:79 (1991).
11. H. Tanaka, H. Akama, Y. Ichikaw, M. Homma and H. Oshima, Glucocorticoid receptors in normal leukocytes: Effects of age, gender, season and plasma concentrations, *Clin. Chem.* 37:1715 (1991).
12. B.B. Xu, Z.M. Liu and Y. Zhao, A study in circadian rhythm in glucocorticoid receptor, *Special Neuroendocrine Systems, Neuroendocrinology* 53(Suppl 1):31 (1991).
13. W.C. McGimsey, J.A. Cidlowski, W.E. Stumpf and M. Sar, Immunocytochemical localization of

glucocorticoid receptor in rat brain, pituitary, liver, and thymus with two new polyclonal antipeptide antibodies, *Endocrinology* 129:3064 (1991).

14. C.J. Grossman, L.J. Sholiton and P. Nathan, Rat thymic estrogen reeptor I. Preparation, location and physicochemical properties, *J. Steroid Biochem.* 11:1233 (1979).

15. C.J. Grossman, L.J. Sholiton, G.C. Blaha and P. Nathan, Rat thymic estrogen receptor II. Physiological properties, *J. Steroid Biochem.* 11:1241 (1979).

16. C.J. Grossman, P. Nathan, B.B. Taylor and L.J. Sholiton, Rat dihydrotestosterone receptor: Preparation, location and physicochemical properties, *Steroids* 34:539 (1979).

17. P. Pearce, B.A.K. Khalid and J.W. Funder, Androgens and the thymus, *Endocrinology* 109:1073 (1981).

18. P.T. Pearce, B.A.K. Khalid and J.W. Funder, Progesterone receptors in   rat thymus, *Endocrinology* 113:1287 (1983).

19. H. Fujii-Hanamoto, C.J. Grossman, G.A. Roselle and C.L. Mendenhall, Nuclear progestin receptors in rat thymic tissue, *Thymus* 15:31 (1990).

20. C.J. Grossman, L.J. Sholiton and G.A. Roselle, Dihydrotestosterone regulation of thymocyte function in the rat - mediation by serum factors, *J. Steroid Biochem.* 19:1459 (1983).

21. C.J. Grossman, L.J. Sholiton and G.A. Roselle, Estradiol regulation of thymic lymphocyte function in the rat: mediation by serum thymic factors, *J. Steroid Biochem.* 16:683 (1982).

22. K. Sakabe, I. Kawashima, K. Seiki and H. Fujii-Hanamoto, Hormone and immune response, with special reference to steroid hormone 2. Sex steroid receptors in rat thymus, *Tokai J. Clin. Med.* 15:201 (1990).

23. I. Kawashima, K. Sakabe, K. Seiki, H. Fujii-Hanamoto, A. Akatsuka and H. Tsukamoto, Localization of sex steroid receptor cells, with special reference to thymulin (FTS)-producing cells in female rat thymus, *Thymus* 18:79 (1991).

24. C.J. Grossman, Regulation of the immune system by sex steroids, *Endocrine Rev.* 3:436 (1984).

25. C.J. Grossman, Interactions between the gonadal steroids and the immune system, *Science* 227:257 (1985).

26. C.J. Grossman and G.A. Roselle, The control of immune response by endocrine factors and the clinical significance of such regulation, *Prog. Clin. Biochem. Med.* 4:9 (1986).

27. C.J. Grossman, Personal communication.

28. B.D. Greenstein, F.T.A. Fitzpatrick, I.M. Adcock, M.D. Kendall and M.J. Wheeler, Reappearance of the thymus in old rats after orchidectomy: Inhibition of regeneration by testosterone, *J. Endocrinol.* 110:417 (1986).

29. F.T.A. Fitzpatrick and B.D. Greenstein, Effects of various steroids on the thymus, spleen, ventral prostate and seminal vesicle in old orchidectomized rats, *J. Endocrinol.* 113:51 (1987).

30. B.D. Greenstein, F.T.A. Fitzpatrick, M.D. Kendall and M.J. Wheeler, Regeneration of the thymus in old male rats treated with a stable analogue of LHRH, *J. Endocrinol.* 112:345 (1987).

31. S.A. Ahmed, M.J. Dauphinee and N. Talal, Effects of short-term administration of sex hormones on normal and autoimmune mice, *J. Immunol.* 134:204 (1985).

32. I. Screpanti, S. Morrone, D. Meco, A. Santoni, A. Gulino, R. Paolini, A. Crisanti, B.J. Mathieson and L. Frati, Steroid sensitivity of thymocyte subpopulations during intrathymic differentiation. Effects of 17 B-estradiol and dexamethasone on subsets expressing T cell antigen receptor of IL-2 receptor, *J. Immunol.* 142:3378 (1989).

33. P.-C. Ho, G.W.K. Tang and J.W.M. Lawton, Lymphocyte subsets in patients with oestrogen deficiency, *J. Reprod. Immunol.* 20:85 (1991).

34. S. Aboudkhil, J.P. Bureau, L. Garrelly and P. Vago, Effects of castration, Depo-testosterone and cyproterone acetate on lymphocyte T-subsets in mouse thymus and spleen, *Scand. J. Immunol.* 34:647 (1991).

35. N.J. Olsen, M.B. Watson, G.S. Henderson and W.J. Kovacs, Androgen deprivation induces phenotypic and functional changes in the thymus of adult male mice, *Endocrinology* 129:2471 (1991).

36. I. Screpanti, D. Meco, S. Morrone, A. Gulino, B.J. Mathieson and L. Frati, In vivo modulation of the distribution of thymocyte subsets: Effects of estrogen on the expression of different T-cell receptor VB gene families in CD4⁻, CD8⁻ thymocytes, *Cell. Immun.* 134:414 (1991).

37. C.J. Grossman, G.A. Roselle and C.L. Mendenhall, Sex steroid regulation of autoimmunity, *J. Ster. Biochem. Mol. Biol.* 40:649 (1991).

38. Y. Weinstein and Z. Berkovich, Testosterone effect on bone marrow, thymus and suppressor T cells in the (NZB x NZW)F1 mice: its relevance to autoimmunity, *J. Immunol.* 126:998 (1981).

39. R.J. Pietras and C.M. Szego, Specific binding sites for oestrogen at the outer surface of endothelial cells, *Nature* 265:69 (1977).

40. W.E. Farnsworth, The prostate plasma membrane as an androgen receptor, *Membrane Biochem.* 9:141 (1991).

41. F.M. Hughes, C.M. Pringle and W.C. Gorospe, Production of progestin-stimulating factor(s) by enriched populations of rat T and B lymphocytes, *Biol. Reprod.* 44:922 (1991).

42. N. Emi, H. Kanzaki, M. Yoshida, K. Takakura, M. Kariya, N. Okamoto, K. Imai and T. Mori, Lymphocytes stimulate progesterone production by cultured human granulosa luteal cells, *Am. J. Obste. Gynecol.* 165:1469 (1991).

43. J.E. Blalock, Production of neuroendocrine peptide hormones by the immune system, *Prog. Allergy* 43:1 (1988).

44. S.Y. Felton, D.L. Felton, D.L. Bellinger, S.L. Carlson, K.D. Ackerman, K.S. Madden, J.A. Olschowka and S. Livnat, Noradrenergic sympathetic innervation of lymphoid organs, *Prog. Allergy* 43:14 (1988).

GLUCOCORTICOID REGULATION OF MUCOSAL IMMUNITY: EFFECT OF DEXAMETHASONE ON IgA AND SECRETORY COMPONENT (SC)

Charles R. Wira and Richard Rossoll

Department of Physiology
Dartmouth Medical School
Lebanon, NH 03756

INTRODUCTION

Immunoglobulin A (IgA), the predominant immunoglobulin at mucosal surfaces of the body, is produced on a daily basis in amounts greater than all other immunoglobulins combined[1]. As the first line of defense, IgA at mucosal surfaces protects against bacterial and viral infection by antibody interactions which either destroy pathogens or block their adherence to prevent potentially infective agents from entering the body[2]. To reach mucosal sites, IgA binds to secretory component (SC), the cytoplasmic portion of the polymeric Ig receptor that is synthesized by rat hepatocytes and epithelial cells lining the mucosal surfaces of the body[3-6]. Following binding, polymeric IgA (pIgA) is transported from serum into the gastrointestinal tract via bile and from tissues into secretions of the respiratory, gastrointestinal and urogenital tracts[7-10]. Transport through hepatocytes and epithelial cells is similar in that IgA binds covalently to SC and following endocytosis, is released as pIgA-SC complex (secretory IgA)[10,11].

Previously, we have demonstrated that IgA, SC and IgG at selected mucosal surfaces are under sex hormone and cytokine control and that regulation varies from site to site[12,13]. In the female reproductive tract, for example, immunoglobulins and SC are stimulated by estradiol and inhibited by progesterone in the uterus, and inhibited by both of these hormones in cervico-vaginal secretions[14-17]. In other studies, androgens have been demonstrated to increase SC and IgA levels in lacrimal secretions[18,19] and in the male reproductive tract[20]. In other studies, SC production by human colon carcinoma cell lines were shown to be enhanced by interferon-γ and tumor necrosis factor-a[21,22]. More recently, we have shown that SC and IgA levels in the reproductive tract increase in response to interferon-γ and interleukin-6[23,24].

Our objectives in the studies to be presented were: a) to establish whether glucocorticoids have an effect on IgA and SC levels in serum and at mucosal surfaces, b) to examine the possible mechanisms though which glucocorticoids exert their effects and c) to determine whether glucocorticoids might have an immunoregulatory role in redistributing IgA from mucosal surfaces to serum.

RESULTS

In Vitro Effect of Dexamethasone and Estradiol on Secretory Component Production by Hepatocytes

To examine the role of glucocorticoids in regulating the mucosal immune system, initial studies were undertaken to determine whether SC, which is known to be synthesized hepatocytes in the rat [42], was under hormonal control. To evaluate this directly, isolated hepatocytes were prepared and incubated in the presence of hormones. To eliminate the possible effect that dexamethasone, a synthetic glucocorticoid, might influence hepatocyte attachment, flasks were incubated without hormone for 24 hr, rinsed with media, and replaced with an equal volume of media either with or without dexamethasone. As shown in Figure 1, when cells were incubated with dexamethasone (10^{-7} M), SC levels in the media increased significantly relative to control cells over the 5 day period of incubation. In other studies (not shown), we found that cycloheximide, a known inhibitor of protein synthesis, on addition to the incubation media, blocked the further SC production by dexamethasone-stimulated hepatocytes. In contrast, when hepatocytes were incubated with estradiol (10^{-6}M), SC release was not affected (Figure 1). In other studies, estradiol added at concentrations from 10^{-10} to 10^{-6}M had no effect on SC production. However, when added along with dexamethasone, estradiol antagonized the stimulatory effect of dexamethasone on hepatocyte SC production. These findings suggest that estradiol may regulate hepatocyte SC production in the presence of glucocorticoids, even though by itself it has no effect.

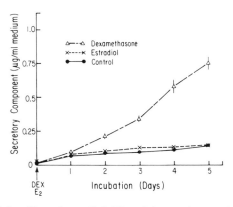

Figure 1. Time course of the effect of estradiol (X) and dexamethasone (△) on SC accumulation in media. Isolated hepatocytes were incubated at 37°C for 24 hr. After incubation, non-attached cells were removed by washing and flasks were reconstituted with incubation media containing either estradiol (10^{-6} M) or dexamethasone (10^{-7}M). Control cells (C) received incubation media without hormone. Each value represents the mean ± SE of five determinations. From[29].

To more fully characterize the effect of dexamethasone, hepatocyte SC production was calculated using several parameters. As seen in Table 1, irrespective of whether results are expressed in terms of volume of incubation media, cell number, or cell protein, dexamethasone had a pronounced stimulatory effect on SC production relative to controls. These findings demonstrate that the effect of dexamethasone on hepatocyte SC production is unique in that it does not apply to all proteins and is not due to an effect that accompanies general cell growth.

Table 1. Effect of dexamethasone on SC production by isolated hepatocytes in culture.

Secretory Component	$\mu g/ml$	$\mu g/1 \times 10^6$ cells	$\mu g/$ mg protein
Control	0.55 ± 0.049	0.59 ± 0.05	1.66 ± 0.32
Dexamethasone	1.64 ± 0.164^a	1.92 ± 0.19^b	4.30 ± 1.02^a

Isolated hepatocytes (5 dishes/well) were incubated for 24 hr prior to rinsing with PBS to remove non-adherent cells. Incubation media with or without dexamethasone (10^{-7}M) was added before incubation at 37 ° C for an additional 5 days.
[a] Significantly ($p < 0.001$) greater than control; [b] Significantly ($p < 0.02$) greater than control.

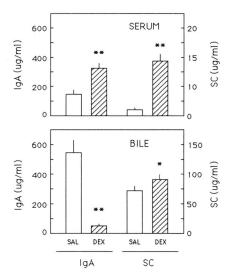

Figure 2. Effect of dexamethasone on IgA and SC levels in serum and bile. Ovariectomized rats were treated with dexamethasone (2 mg/day) or saline (0.1 ml) for 3 days. Twenty-four hr after the last injection, bile and serum were collected and analyzed for SC and IgA. Bar represents the means of 7 animals/group. Vertical lines on bars indicate the SE. **, Significantly (p < 0.001) different than saline group [28].

Effect of Dexamethasone on IgA and SC Levels in Serum, Bile, and Saliva

To examine the in vivo effect on dexamethasone in rats, levels of IgA and SC in serum and bile were analyzed following hormone treatment. As shown in Figure 2, when administered daily for 3 days, dexamethasone had a stimulatory effect on the levels of IgA and SC in serum. IgA levels increased approximately 2-fold while

SC levels increased 6- to 7-fold, relative to saline controls. To examine the effect of dexamethasone on bile IgA and SC, bile ducts were cannulated and allowed to flow for 1 hr prior to the collection of bile during the second hour. As seen in Figure 2, dexamethasone had a pronounced inhibitory effect on bile IgA levels relative to that seen in saline-treated animals. In contrast, SC levels remained unchanged in those animals that received dexamethasone. These findings indicated that, in response to dexamethasone, bile IgA was lowered at a time that coincided with an increase in serum IgA.

To more fully characterize the effect of dexamethasone on bile SC levels, SC output in terms of bile concentration was compared to the amount of SC released into bile within a given time period. When calculated in terms of concentration, SC levels in bile from dexamethasone-treated animals were not significantly different from controls (Figure 3). In contrast, when bile flow was used to calculate SC production/hr, we found that significantly more SC was produced by dexamethasone-treated animals than by saline controls. This increase was accounted for by the observation that bile output/hr is greater in dexamethasone-treated animals than in saline controls. The net effect is that SC in bile from dexamethasone-treated animals, when expressed as μ g/hr, was approximately 80% greater than that seen in the bile of saline controls. These findings indicate that dexamethasone increases both bile flow and hepatocyte SC production. Further, these observations are consistent with our earlier in vitro finding that dexamethasone stimulates hepatocyte SC production (Table 1).

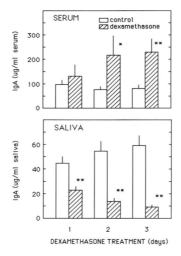

Figure 3. Time course of the effect of dexamethasone on the levels of IgA in serum and saliva. Ovariectomized rats were treated with dexamethasone (1 mg/day) or saline (0.1 ml) for each of 1, 2, or 3 days. Twenty-four hrs after the last injection, saliva and serum were collected and analyzed for IgA. Bars represent the mean $\pm$ SE of 7-8 animals/group. *, significantly (P<0.01) different from control groups. **, significantly (P<0.001) different from control groups[25].

As a part of this study, saliva was analyzed to determine whether mucosal secretions were also influenced by glucocorticoid treatment. As seen in Figure 3, when non-adrenalectomized animals were treated with daily injections of dexamethasone (1 mg/day) or normal saline for 3 days, salivary levels of IgA in dexamethasone-treated animals were significantly lower than those measured in saline controls after 1, 2 and 3 days of hormone treatment. In contrast, IgA levels in serum increased 24 hr after 2 and 3 days of hormone treatment. These findings demonstrated that saliva and bile IgA levels are affected in a similar way by glucocorticoid treatment. Further, since the decline of IgA in saliva preceded the increase measured in serum, this suggests that IgA, in response to dexamethasone, may be redirected from mucosal surfaces to serum. In other studies (not shown), we found that dexamethasone has a similar inhibitory effect on vaginal IgA levels[25]. In an earlier study, dexamethasone had a similar pronounced inhibitory effect on uterine IgA levels in ovariectomized rats that were treated with estradiol[16]. Overall, these studies indicate that glucocorticoids suppress IgA levels at mucosal surfaces at a time that coincides with increases in serum IgA levels.

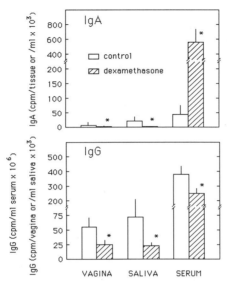

Figure 4. Effect of dexamethasone on levels of specific IgA and IgG antibodies in vaginal fluid, saliva, and serum of immunized animals after the injection (primary: day 0; Peyer's patches) and boost (secondary: day 13; intrauterine) with SRBC. Animals were ovariectomized one week before immunization and received either dexamethasone (1 mg/day) or saline daily for 3 days prior to sacrifice 24 hr after the last injection on day 26 post primary immunization. Bars represent mean ± SE of 4 animals per group. *, significantly (P<0.01) different from control groups[25].

Effect on Dexamethasone on Specific IgA and IgG Antibodies in Serum, Saliva and Vaginal Secretions

To determine whether dexamethasone had an effect on the accumulation of specific antibodies in vaginal fluid, saliva, and serum, ovariectomized rats were immunized by injecting sheep erythrocytes (SRBC) directly into Peyer's patches. Animals were then boosted by injecting SRBC into the uterine lumen 13 days after primary immunization. Rats were treated with dexamethasone or saline for 3 consecutive days before sacrifice 24 hr after on day 26 post primary immunization. This immunization approach has been used previously in our laboratory to demonstrate that antibodies in the female reproductive tract are derived, in part, from the gastrointestinal tract as well as produced locally in the uterus in response to antigenic exposure[26,27]. As seen in Figure 4, when specific antibodies are measured, significantly more IgA antibodies were present in the sera of dexamethasone-treated rats than in saline-treated animals. In contrast, IgA-antibody levels in vaginal and salivary secretions were reduced in those animals that received dexamethasone. These findings indicate that dexamethasone alters specific IgA-antibody levels in serum and secretions in a way that is comparable to its effects on total IgA levels. When IgG antibody levels were analyzed (Figure 4), we found that dexamethasone had an inhibitory effect on their presence in vaginal secretions, saliva and serum. In the latter case, dexamethasone lowered serum IgG levels by approximately 50%.

Analysis of SC and IgA in Rat Serum

To more fully understand the influence of dexamethasone on serum IgA levels, studies were undertaken to examine in detail the influence of this hormone on SC in serum. As seen in Figure 5, when animals were injected daily for 1, 2 or 3 days with dexamethasone and killed 24 hrs later, the levels of SC in serum increased significantly. In other studies designed to analyze the dose-response relationship of dexamethasone to SC and IgA levels in serum (not shown), we found that dexamethasone given in increasing doses (0.01 to 1.0 mg/day) increased serum SC and IgA levels with treatments between 0.1 mg to 1.0 mg dexamethasone[28]. To determine whether an intact adrenal gland influenced the response of SC to dexamethasone, adrenalectomized and non-adrenalectomized rats were treated with dexamethasone (1 mg/day) for 3 days. We found that dexamethasone increased SC levels of serum, relative to saline controls, with no significant differences measured between adrenalectomized and non-adrenalectomized animals receiving dexamethasone (1 mg/day).

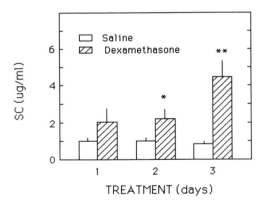

Figure 5. Time course of the effect of dexamethasone on the levels of SC in serum. Ovariectomized rats were treated with dexamethasone (2 mg/day) or saline (0.1 ml) for 1, 2 or 3 days. Serum was collected 24 h after the last injection and analyzed for SC. Bars represent the mean $\pm$ SE of seven to eight animals/group. *, Significantly ($p < 0.05$) greater than control; **, significantly ($p < 0.01$) greater than control[28].

High performance liquid chromatography (HPLC) was carried out to examine the molecular forms of IgA in serum and to determine whether SC in serum is found in association with IgA. When pooled sera was applied to a gel filtration column and individual fractions analyzed (Figure 6), chromatographic separation of serum from saline- and dexamethasone-treated animals resulted in the appearance of two peaks of IgA. The first peak eluted near the column exclusion volume and corresponded in size to polymeric IgA. The second peak eluted in a region that corresponded in size to monomeric IgA. In other studies, we found that the increase in serum IgA levels in response to dexamethasone treatment was due primarily to an increase in polymeric IgA, although monomeric IgA did increase to a limited extent. As a part of this study we also assayed individual fractions for SC. As seen in Figure 6, SC eluted as a single peak irrespective of whether serum was from saline- or dexamethasone-treated animals. Further, fractions containing SC were found exclusively in those fractions that contained polymeric IgA with no SC detected in those fractions that contained monomeric IgA. Since SC in bile and external secretions is known to range in size from M_r = 50,000 to 90,000 (for review, see[1]), these results suggest that all of the SC detected in serum was associated with polymeric IgA. In other studies (not shown), immunoblot analysis of SC in serum from saline- and dexamethasone- treated animals indicated that irrespective of hormone treatment, two closely spaced protein bands of SC were found with molecular masses of 29 and 27 kDa[28]. These results indicate that the 29/27 kDa doublet band of SC is a constitutive serum protein which increases in response to dexamethasone treatment.

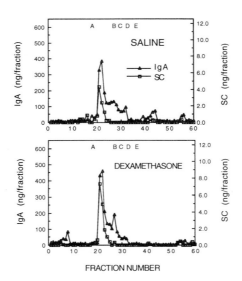

Figure 6. Analysis of IgA and SC in sera from saline (upper panel)- and dexamethasone (lower panel)-treated rats by HPLC. Pooled sera was prepared from four animals/group that were treated either with dexamethasone (1 mg/day) or saline (0.1 ml) for 3 days and killed 24 h after the last injection. Sera samples were applied to a TSK-gel column and eluted as described in Materials and Methods prior to being assayed. Letters indicate the elution points of standards with known M_r: A (thyroglobulin, 669,000), B (catalase, 240,000), C (albumin, 67,000), D (ovalbumin, 43,000), E (soybean trypsin inhibitor, 20,400)[28].

CONCLUSIONS

The present studies demonstrate that glucocorticoids regulate the secretory immune system at mucosal surfaces of the body. In early studies, we found that dexamethasone had a pronounced stimulatory effect on SC production by isolated rat hepatocytes in culture[29]. This observation led to in vivo studies on the effects of glucocorticoids on IgA levels at mucosal sites. In response to dexamethasone, IgA levels in saliva, uterine and vaginal secretions are lowered at a time which coincided with increases in polymeric IgA in serum. In other studies, we found that SC levels in blood, bile and external secretions also varied with hormone treatment. Dexamethasone increased, in a dose- and time-dependent manner, both SC levels in serum and the production by the liver of SC in bile. In contrast, levels of IgA in bile were lowered with hormone treatment. When SC in blood was analyzed, we found that it was associated exclusively with polymeric IgA and not with monomeric IgA. These findings led to the conclusion that the elevation of polymeric IgA in serum following dexamethasone treatment is most likely due to the presence of SC in blood, which on binding to polymeric IgA, decreases the IgA clearance from blood into bile.

Since mucosally stimulated anti-SRBC IgA antibodies accumulate in blood and are depleted at mucosal surfaces following dexamethasone treatment, these findings led to the conclusion that glucocorticoids stimulate the redistribution of IgA from mucosal surfaces to the blood. Further, these findings raise the possibility, as discussed elsewhere[25], that through stress and the consequent outpouring of glucocorticoids from the adrenal, IgA in secretions is shifted to systemic sites, possibly to confer protection after mucosal immune defenses have been breached. Whether glucocorticoids trigger an immune cell redistribution or a shift in antibody movement away from mucosal surfaces, remains to be established.

That glucocorticoids influence the mucosal immune system in the gastrointestinal tract has recently been demonstrated by Alverdy[30]. Following the administration of dexamethasone to rats, levels of IgA were significantly reduced throughout the gastrointestinal tract. In studies to analyze the impact of this decline, dexamethasone-treated animals were found to have increased in bacterial adherence and increased bacterial translocation to the mesenteric lymph nodes. Increases in adherence were found to be most pronounced in the cecum[30]. In other studies, dexamethasone administration was associated with alterations in intestinal permeability as measured by a decrease in transepithelial resistance[31]. Changes in permeability occurred as a consequence of bacterial-mucosal cell interactions which appeared to be secondary to the intestinal decline in IgA. Whether these events are deleterious, however, remains to be established. Since bacteria make direct contact with intestinal epithelial cells, which process and present antigen as the first step in the induction of an immune response[32], it may be that local immunity is enhanced under these conditions. Alternatively, as discussed by Alverdy[31], since bacteria translocate to the mesenteric nodes in dexamethasone-treated animals, low levels of bacteria in the mesenteric nodes may enhance immune defenses whereas persistence or dissemination of bacteria into the nodes may be harmful[30]. Our finding that glucocorticoids deplete IgA at multiple mucosal sites, suggests that bacterial adherence and dissemination may occur at sites other than the gastrointestinal tract.

The source of SC found in serum remains to be identified. With the finding that hepatocytes synthesize SC[6] and our observation that glucocorticoids given in vitro increase SC production[29], it appears likely that some SC in serum is of liver origin. Since our studies indicate that serum SC levels increase in response to dexamethasone given *in vivo*, these findings suggest that some SC is synthesized and secreted into blood and bile by hepatocytes. Dexamethasone is known to influence SC production at several sites in the body, but effects vary with the site examined. For example, glucocorticoids lowered SC levels in cervico-vaginal secretions[17], but had no effect on SC levels in uterine secretions of estradiol-stimulated animals[16]. In other studies[28], we found that saliva SC levels and liver SC production (Table 1) were increased with dexamethasone treatment. Given the selective actions of glucocorticoids, an additional source of serum SC might be the epithelial cells that line mucosal surfaces. Whether glucocorticoids influence the directional release of SC from secretions to blood, possibly along with polymeric IgA, as originally proposed by Brandtzaeg[33], remains to be examined.

ACKNOWLEDGEMENTS

The authors wish to express their appreciation to Dr. Brian Underdown (McMaster University, Hamilton, Ontario) for his generous gifts of SC and antisera used in these studies. Supported by Research Grant AI-13541 from NIH.

REFERENCES

1. J. Mestecky and J.R. McGhee, Immunoglobulin A (IgA): Molecular and cellular interactions involved in IgA biosynthesis and immune response, *Adv. Immunol.* 40:153 (1987).
2. P.C. McNabb and T.B. Tomasi, Host defense mechanisms at the mucosal surface, *Ann. Rev. Microbiol.* 138:976 (1981).
3. P. Brandtzaeg, Mucosal and glandular distribution of immunoglobulin components. Differential localization of free and bound SC in the secretory epithelial cells, *J. Immunol.* 112:1553 (1974).
4. S.S. Crago, R. Kulhavy, S.J. Prince, and J. Mestecky, Secretory component on epithelial cells is a surface receptor for polymeric immunoglobulins, *J. Exp. Med.* 147:1832 (1978).
5. D.J. Socken, K.N. Jeejeebhoy, H. Bazin, and B.J. Underdown, Identification of secretory component as an IgA receptor on rat hepatocytes, *J. Exp. Med.* 50:1538 (1979).
6. J.L. Zevenbergen, C. May, J.C. Watson, and J.P. Vaerman, Synthesis of secretory component by rat hepatocytes in culture, *Scand. J. Immunol.* 11:93 (1980).
7. D.R. Tourville, R.H. Adle, J. Bienenstock, and T.B. Tomasi, The human secretory immunoglobulin system: immunohistological localization of A, secretory "piece", and lactoferrin in normal human tissues, *J. Exp. Med.* 129:411 (1969).
8. G.D.F. Jackson, I. Lemaître-Coelho, J. Vaerman, H. Bazin, and A. Beckers, Rapid disappearance from serum of intravenously injected rat myeloma IgA and its secretion into bile, *Eur. J. Immunol.* 8:123 (1978).
9. M.M. Fisher, B. Nagy, H. Bazin, and B. Underdown, Biliary transport of IgA: Role of secretory component, Proc. Natl. Acad. Sci. USA, *Proc. Natl. Acad. Sci. USA.* 76: 2008 (1979).
10. W.R. Brown, K. Isobe, P.K. Nakane, and B. Pacini, Studies on the translocation of immunoglobulins across intestinal epithelium. IV. Evidence for binding of IgA and IgM to secretory component in intestinal epithelium, *Gastroenterology* 73:1333 (1977).
11. B.M. Mullock, R.H. Hinton, M. Dobrota, J. Peppard, and E. Orlans, Endocytic vesicles in liver carry polymeric IgA from serum to bile, *Biochim. Biophys. Acta.* 587:381 (1979).
12. C.R. Wira and C.P. Sandoe, Sex steroid hormone regulation of immunoglobulin G (IgG) and A (IgA) in rat uterine secretions, *Nature* 268:534 (1977).
13. D.A. Sullivan and C.R. Wira, Hormonal regulation of immunoglobulins in the rat uterus: Uterine response to multiple estradiol treatments, *Endocrinology* 114:650 (1984).
14. C.R. Wira and C.P. Sandoe, Hormone regulation of immunoglobulins: Influence of estradiol on IgA and IgG in the rat uterus, *Endocrinology* 106:1020 (1980).
15. D.A. Sullivan and C.R. Wira, Estradiol regulation of secretory component in the female reproductive tract, *J. Steroid Biochem.* 15:439 (1981).
16. D.A. Sullivan, B.J. Underdown, and C.R. Wira, Steroid hormone regulation of free secretory component in the rat uterus, *Immunology* 49:379 (1983).
17. C.R. Wira and D.A. Sullivan, Estradiol and progesterone regulation of IgA, IgG and secretory

component in cervico-vaginal secretions of the rat, *Biol. Reprod.* 32: 90 (1985).

18. D.A. Sullivan and M.R. Allansmith, Hormonal influence on the secretory immune system of the eye: androgen modulation of IgA levels in tears of rats, *J. Immunol.* 134:2978 (1985).

19. D.A. Sullivan and M.R. Allansmith, Hormonal influence on the secretory immune system of the eye: endocrine interactions in the control of IgA and secretory component levels in tears of rats, *Immunology* 60:337 (1987).

20. J.E. Stern, S. Gardner, D. Quirk, and C.R. Wira, Secretory immune system of the male reproductive tract: effects of dihydrotesterone and estradiol on IgA and secretory component levels, *J. Reprod. Immunol.* 22:73 (1992).

21. L.M. Sollid, D. Kvale, P. Brandtzaeg, G. Markussen, and E. Thorsby, Interferon-γ enhances expression of secretory component, the epithelial receptor for polymeric immunoglobulins, *J. Immunol.* 138:4303 (1987).

22. D. Kvale, D. Lovhaug, L.M. Sollid, and P. Brandtzaeg, Tumor necrosis factor- up-regulates expression of secretory component, the epithelial receptor for polymeric Ig, *J. Immunol.* 140:3086 (1988).

23. R.H. Prabhala and C.R. Wira, Cytokine regulation of the mucosal immune system: in vivo stimulation by interferon-gamma of secretory component and immunoglobulin A in uterine secretions and proliferation of lymphocytes from spleen, *Endocrinology* 129:2915 (1991).

24. C.R. Wira, B. O'Mara, J. Richardson, and R. Prabhala, The mucosal immune system in the female reproductive tract: Influence of sex hormones and cytokines on immune recognition and responses to antigen, *Vaccine Research* 1:151 (1992).

25. C.R. Wira, C.P. Sandoe, and M.G. Steele, Glucocorticoid regulation of the humoral immune system. I. In vivo effects of dexamethasone on IgA and IgG in serum and at mucosal surfaces, *J Immunol.* 144:142 (1990).

26. C.R. Wira and C.P. Sandoe, Specific IgA and IgG antibodies in the secretions of the female reproductive tract: effects of immunization and estradiol on expression of this response in vivo, *J. Immunol.* 138:4159 (1987).

27. C.R. Wira and C.P. Sandoe, Effect of uterine immunization and oestradiol on specific IgA and IgG antibodies in uterine, vaginal and salivary secretions, *Immunology* 68:24 (1989).

28. C.R. Wira and R.M. Rossoll, Glucocorticoid regulation of the humoral immune system. Dexamethasone stimulation of secretory component in serum, saliva, and bile, *Endocrinology* 128:835 (1991).

29. C.R. Wira and E.M. Colby, Regulation of secretory component by glucocorticoids in primary cultures of rat hepatocytes, *J. Immunol.* 134:1744 (1985).

30. J. Alverdy and E. Aoys, The effect of glucocorticoid administration on bacterial translocation, *Ann. Surgery* 214:719 (1991).

31. J. Spitz, G. Hecht, M. Taveras, E. Aoys, and J. Alverdy, The effect of dexamethasone administration on intestinal permeability: The role of bacterial adherence, *Gastroenterology* 106:35 (1994).

32. L. Mayer and R. Shlien, Evidence for function of Ia molecules on gut epithelial cells in man, *J. Exp. Med.* 166:1471 (1987).

33. P. Brandtzaeg, Human secretory immunoglobulins. V. Occurrence of secretory piece in human serum, *J. Immunol.* 106:318 (1971).

ANDROGEN REGULATION OF OCULAR MUCOSAL- AND AUTO-IMMUNITY

David A. Sullivan, Zhiyan Huang, Ross W. Lambert, L. Alexandra
Wickham, Masafumi Ono, Janethe D.O. Pena, and Jianping Gao

Department of Ophthalmology, Harvard Medical School,
and Schepens Eye Research Institute
20 Staniford Street, Boston, MA   02114

INTRODUCTION

At present, it is known that androgens exert a tremendous impact on the development, expression and function of the immune system in both health and disease. In brief, these hormones have been shown to: [a] influence the maturation, proliferation and/or activity of pluripotent stem cells, B cells, T cells and macrophages; [b] modulate the synthesis, secretion and/or action of antibodies, cytokines and growth factors; [c] alter the production of autoantibodies and the formation of immune complexes; [d] affect the immune response to, and clearance of, antigens; [e] control the generation of thymic factors and secretory component (SC), the IgA antibody receptor; and [f] modify the rejection of allografts, the extent of graft vs. host disease, the magnitude of inflammation, and the onset and severity of autoimmune disease.[e.g.1-6] The precise nature of these androgen actions is exceedingly dependent upon the specific target cell, the local microenvironment, and the particular immunological process, and may result in enhancement, inhibition, or no effect, on immune expression.[3] Nevertheless, it is quite apparent that androgens control multiple aspects of humoral, cell-mediated, mucosal and auto-immunity.

Recently, research has also demonstrated that androgens have a profound regulatory influence on the secretory immune system of the eye.[7] In addition, studies have shown that androgens cause a marked suppression of autoimmune sequelae in lacrimal glands of mouse models of Sjögren's syndrome.[8] Of interest, these hormone actions may be achieved through direct or indirect effects, and may be significantly modified by neurotransmitters and/or cytokines.[7-9] The purpose of this article is to examine the extent, and possible underlying mechanisms, of this androgen control of ocular mucosal immunity and lacrimal gland autoimmune disease.

Advances in Psychoneuroimmunology, Edited by
I. Berczi and J. Szélenyi, Plenum Press, New York, 1994

ANDROGEN ACTION ON THE SECRETORY IMMUNE SYSTEM OF THE EYE

The ocular secretory immune system is designed to protect the anterior surface of the eye against microbial infection and toxic challenge.[7,10] This immunological function is mediated primarily through secretory IgA (sIgA) antibodies, which originate from plasma cells in the lacrimal gland[7] and are known to prevent viral invasion, interfere with bacterial colonization, inhibit parasitic infestation and suppress antigen-related damage in mucosal sites.[11,12] Thus, the secretory immune system of the eye plays a critical role in the defense against inflammatory processes and infectious disease, thereby promoting ocular surface integrity and the maintenance of visual acuity.

During the past 10 years, it has become recognized that the expression and the activity of the ocular secretory immune system may be dramatically influenced by androgens. In rats, androgens induce a striking increase in the synthesis and secretion of SC by lacrimal gland acinar epithelial cells, augment the concentration of IgA in lacrimal tissue and significantly enhance the transfer and accumulation of SC and IgA, but not IgG, in tears.[13-20] These hormone effects, which may be elicited by a variety of androgen analogues,[21] are not duplicated by treatment with estrogens, progestins, glucocorticoids or mineralocorticoids,[13,15,17] and may explain the distinct, gender-related differences in the rat ocular secretory immune system.[13-15,22-25] Furthermore, the mucosal immune actions of androgens appear to be somewhat unique to the eye: androgen exposure does not alter IgA or SC content in salivary, respiratory, intestinal, uterine or bladder tissues[21] and actually diminishes secretory immunity in the mammary gland.[26] Of interest, androgens may also influence the extent of viral replication in lacrimal gland acinar cells.[27] As concerns other species, androgen treatment results in a significant increase in the lacrimal gland output of IgA in several strains of mice,[28] but this hormone's ocular immune impact in humans remains to be elucidated.

The influence of androgens on the immune system of the lacrimal gland may be significantly modified by interruption of the hypothalamic-pituitary axis, by diabetes, or by extirpation of the thyroid or adrenal glands. Thus, hypophysectomy or ablation of the anterior pituitary significantly decreases the number of IgA plasma cells in lacrimal tissue, reduces the acinar cell production of SC, elicits a marked decline in the levels of tear IgA and SC and almost completely inhibits androgen action on ocular mucosal immunity in vivo.[16,29,30] Furthermore, this endocrine disruption has a significant effect on lacrimal gland structure and function, causing both acinar cell atrophy[31] and diminished tear volume.[32] At present, the physiological mechanisms responsible for the hypothalamic-pituitary modulation of the secretory immune system of the eye have yet to be clarified,[18] but may involve numerous endocrine, neural and immune pathways[33-36] that are known to modify androgen action, lacrimal gland function and mucosal immunity.[3,37] The impact of hypophysectomy, though, does not irreparably impair acinar cell responsiveness to androgens, given that dihydrotestosterone exposure induces an increased, but limited, SC production by these cells in vitro, after their removal from hypophysectomized rats.[17] With regard to diabetes, this condition leads to a striking reduction in the density of IgA-containing cells in lacrimal tissue, a significant decrease in the basal and androgen-induced amounts of IgA and SC in tears, and a pronounced diminution in lacrimal gland size and tear output.[16] These diabetic sequelae are most likely due to the loss of insulin, which is essential for optimal SC synthesis by acinar epithelial cells,[19] and is apparently necessary for maximal androgen action on target tissues.[38] Of interest, the presence of the thyroid and adrenal glands is also important to permit the full extent of androgen-related effects on the secretory immune system of the eye.[28]

The androgen regulation of SC synthesis by lacrimal gland acinar cells may also be enhanced, or suppressed, by neurotransmitters and lymphokines. Research has shown that VIP and the b-adrenergic agent, isoproterenol, augment, whereas the cholinergic agonist, carbamyl choline, reduces, androgen-stimulated and basal SC output by rat acinar cells.[18-20] In addition, IL-1α, IL-1β and tumor necrosis factor-a (TNF-α) all heighten the androgen-related and constitutive acinar cell synthesis and secretion of SC.[18] These neuroimmune effects on SC production may be mediated through the control of intracellular adenylate cyclase and cAMP activity. In support of this hypothesis, VIP, adrenergic agonists, IL-1 and TNF-α are known to increase,[39,40] and cholinergic transmitters possibly decrease,[41] the elaboration of cellular cyclic AMP. Moreover, treatment of lacrimal gland acinar cells with cyclic AMP analogues (e.g. 8-bromoadenosine 3':5'-cyclic monophosphate), cyclic AMP inducers (e.g. cholera toxin and PGE_2) or phosphodiesterase inhibitors (e.g. 3-isobutyl-1-methylxanthine) may promote SC synthesis in the absence, and/or presence, of androgens.[18] This cAMP influence on SC generation, although pronounced in the lacrimal gland, is not necessarily duplicated in other mucosal epithelial cells.[20]

The mechanism by which androgens regulate SC synthesis by the lacrimal gland may well involve hormone association with specific, nuclear receptors in acinar epithelial cells, adherence of these androgen/receptor complexes to genomic acceptor sites and the stimulation of both androgen receptor and SC mRNA transcription and translation. Thus, saturable, high-affinity and androgen-specific receptors, which bind to DNA, have been identified in lacrimal tissue,[42,43] exist almost exclusively within epithelial cell nuclei,[44] and are up-regulated by androgen administration.[44] Furthermore, androgen exposure significantly increases SC mRNA levels in the lacrimal gland,[45] and the androgen-induced output of SC by acinar cells may be prevented by androgen receptor (e.g. cyproterone acetate), transcription (e.g. actinomycin D) or translation (e.g. cycloheximide) antagonists.[14,20]

In contrast, the processes underlying androgen action on IgA in the lacrimal gland are less clear. Testosterone appears to increase IgA production in lacrimal tissue,[16] but this effect is not paralleled by a consistent alteration in the density of IgA-containing cells,[16] and is apparently not direct, given that mature lymphocytes do not contain androgen receptors.[46] One possible mechanism, though, is that androgen action may be indirect, and mediated through the stimulation of transforming growth factor-β 1 (TGF-β 1) and IL-6 production by lacrimal gland epithelial cells: these cytokines, which are synthesized by epithelial cells[47-50] are known to augment lymphocytic IgA secretion[52,53] and their production may be regulated by sex steroids.[e.g. 47,48] In support of this hypothesis, recent research has shown that TGF-β 1 and IL-6 mRNAs exist in acinar epithelial cells of the rat lacrimal gland,[53] and preliminary studies have indicated that TGF-β 1 mRNA levels in the lacrimal gland may be enhanced by androgen treatment (Z. Huang, J. Gao and D.A. Sullivan, unpublished data). Consequently, if these cytokines are translated and secreted, they could theoretically play a major role in the paracrine control of IgA, as well as potentially modulate the androgen-induced rise in IgA output by the lacrimal gland. These functional effects are possible, given that cytokines (e.g. IL-6) are released through the basal, as well as apical, membranes of epithelial cells.[48] As an additional consideration, TGF-β 1 might also act in an autocrine manner to modulate epithelial SC synthesis.[e.g. 54]

Overall, these findings demonstrate that androgens control the lacrimal gland production of both SC and IgA in experimental animals, which may lead to augmented antibody transport to tears and improved ocular surface defense.

ANDROGEN EFFECTS ON AUTOIMMUNE DISEASE IN THE LACRIMAL GLAND

Throughout the world, countless millions of individuals suffer from tear film dysfunctions, which are collectively diagnosed as keratoconjunctivitis sicca (KCS) or, simply, dry eye.[55] One of the greatest single causes of KCS is Sjögren's syndrome.[55] This autoimmune disorder, which occurs almost exclusively in females and involves multisystem disorders, is associated with a progressive lymphocytic infiltration into the perivascular and periductal areas of main and accessory lacrimal glands, an immune-related, extensive disruption of lacrimal acinar and ductal cells and KCS.[9,56] Furthermore, Sjögren's syndrome appears to have an extremely severe influence on ocular surface integrity and visual acuity: if unmanaged with artificial tear substitutes or tear film conservation therapy, this disease may lead to corneal desiccation, ulceration and perforation, an increased incidence of infectious disease, and ultimately, significant visual impairment and blindness.[55]

The pathogenesis of Sjögren's syndrome is unknown, but may involve the interplay of numerous factors, including those of endocrine, neural, viral, environmental and genetic origin.[1,2] At present, a perception is that the ocular manifestations of Sjögren's syndrome may be clinically irreversible,[57] an eye disorder to be controlled, yet not cured.[58]

However, recent research has indicated that the endocrine system may play a major role in the onset, development, as well as potential therapy, of the immune-related, lacrimal gland defects in Sjögren's syndrome. For example, estrogens and prolactin may be involved in the etiology, acceleration and/or amplification of Sjögren's syndrome.[59-61] In contrast, androgens may provide a protective influence: systemic androgen administration to animals or humans after the onset of Sjögren's syndrome may lead to a significant suppression of autoimmune sequelae in the lacrimal gland, and/or an apparent reduction in ocular signs and symptoms.[8,62-68]

Thus, testosterone administration to female mouse models (MRL/Mp-lpr/lpr [MRL/lpr] and NZB/NZW F1 [F1]) of Sjögren's syndrome causes a precipitous, time-dependent decrease in the extent of lymphocyte infiltration in lacrimal tissue.[62-64] This hormone action, which may be duplicated by therapy with androgen analogues (i.e. anabolic or androgenic) or cyclophosphamide, but not by treatment with estradiol, danazol, cyclosporine A, dexamethasone or experimental non-androgenic steroids,[65,69] involves a significant decline in both the area and number of focal infiltrates in the lacrimal gland.[62,63] More specifically, androgen exposure leads to a pronounced diminution in the density of diverse inflammatory cell populations in lacrimal tissue, including total (Thy 1.2$^+$), helper (L3T4$^+$), suppressor/cytotoxic (Lyt 2$^+$), and immature (B220$^+$) T cells, total B cells (surface IgM$^+$) and Class II antigen-positive (Ia$^+$) lymphocytes.[64] This anti-inflammatory effect, which may be observed as early as 17 days after the initiation of therapy,[62,64] seems to be site-specific: androgens reduce lymphoid infiltration in lacrimal, as well as salivary, glands, but do not ameliorate the extent of inflammation in lymphatic (i.e. superior cervical and mesenteric) and splenic tissues.[65] In addition, the processes involved in the androgen reduction of lacrimal and salivary gland disease appear to be different.[8] In contrast, cessation of androgen treatment (e.g. after 6 weeks) to female MRL/lpr mice results in a slight rise in the extent of lacrimal gland inflammation.[69]

This immunosuppressive action of androgens on lacrimal gland autoimmune disease is also paralleled by a significant increase in the functional activity (e.g. protein and IgA secretion) of lacrimal tissue.[28] Given these findings, it is not surprising that three clinical, although uncontrolled, studies reported that systemic androgen administration alleviated dry eye signs and symptoms, and stimulated tear flow, in Sjögren's patients.[66-68] These latter results are of particular interest, in that male and female patients with systemic lupus erythematosus, who often also suffer from secondary Sjögren's syndrome, have significantly decreased serum androgen concentrations.[70-72] Indeed, the relatively high estrogen and low androgen levels in these patients may be significant predisposing factors to disease progression.[2,71]

The mechanism(s) involved in the androgen-induced suppression of lacrimal autoimmune disease in Sjögren's syndrome remains to be elucidated. In past years, a number of hypotheses have been advanced in an attempt to explain the ability of androgens to modulate systemic immunity or to ameliorate autoimmune disease. These proposed explanations focused primarily upon initial androgen interaction with the thymus, hypothalamic-pituitary axis, bone marrow and/or spleen, although other interpretations emphasized the action of androgens on lymphocytes, Ia expression, immune complex formation or clearance, genetic factors, or the central nervous system.[4,73-97] However, these postulates do not appear to account for the androgen-induced reversal of autoimmune sequelae in lacrimal tissue.[8] Rather, current evidence suggests that this hormone action is a unique, tissue-specific effect, that is initiated through androgen association with receptors in lacrimal gland epithelial cells.[8,98] In addition, it may well be that this androgen interaction then causes the altered expression and/or activity of epithelial cytokines in lacrimal tissue, resulting in the contraction of immunopathological lesions and an improvement in glandular function.[9,99]

In support of this hypothesis, the anti-inflammatory action of androgens in lacrimal tissue appears to be mediated not through lymphocytes, but rather through epithelial cells,[98] which serve as active cellular participants in the glandular inflammation in Sjögren's syndrome,[100] and which type of cell is known to secrete numerous cytokines (e.g. TGF-β).[49] In addition, as noted above, our preliminary data suggest that androgens may increase the mRNA levels of the immunosuppressive cytokine, TGF-β 1, in the lacrimal gland (Z. Huang, J. Gao and D.A. Sullivan, unpublished data). This cytokine is thought to play a protective role in Sjögren's syndrome, and increased expression of TGF-β mRNA has been correlated with reduced inflammation in exocrine glands of Sjögren's syndrome patients.[101] In contrast, the absence of TGF-β 1 leads to a pronounced lymphocytic infiltration into both lacrimal and salivary glands.[102] For comparison, androgens have been shown to induce the expression of TGF-β 1[6] or other immunosuppressive factors in other sites.[103] Moreover, hormones are known to control cytokine production in many tissues[47,48,104,105] and sex steroid modulation of cytokine synthesis is believed to account for the pronounced sexual dimorphism in the incidence of autoimmune disease.[106]

To test the above hypothesis, our ongoing studies are designed to identify the epithelial cell cytokines that may mediate, or be involved, in the androgen-induced suppression of lacrimal gland inflammation in Sjögren's syndrome. Such analysis may help lead to the development of unique, therapeutic strategies to safely and effectively treat this chronic, extremely uncomfortable and vision-threatening disease.

ACKNOWLEDGMENTS

This research review was supported by NIH grant EY05612.

REFERENCES

1. S.A. Ahmed, and N. Talal, Sex hormones and the immune system, *Bailliere's Clin. Rheum.* 4:13 (1990).

2. F. Homo-Delarche, F. Fitzpatrick , N. Christeff, E.A. Nunez, J.F. Bach, and M. Dardenne, Sex steroids, glucocorticoids, stress and autoimmunity, *J. Ster. Biochem. Mol. Biol.* 40:619 (1991).

3. D.A. Sullivan, Hormonal influence on the secretory immune system of the eye, *in*: "The Neuroendocrine-Immune Network," S. Freier, ed., CRC Press, Boca Raton, FL, pp 199-238 (1990).

4. A.H.W.M. Schuurs, and H.A.M. Verheul, Effect of gender and sex steroids on the immune response, *J. Steroid Biochem.* 35:157 (1990).

5. A.B. McCruden, and W.H. Stimson, Sex hormones and immune function, *in*: "Psychoneuroimmunology," R. Ader, D. Felten, and N. Cohen, eds., Acad. Press, San Diego, CA (1991).

6. N.J. Olsen, P. Zhou, H. Ong, and W.J. Kovacs, Testosterone induces expression of transforming growth factor-beta 1 in the murine thymus. *J. Steroid Biochem. Mol. Biol.* 45:327 (1993).

7. D.A. Sullivan, Ocular mucosal immunity, *in*: "Mucosal Immunology," P.L. Ogra, J. Mestecky, M.E. Lamm, W. Strober, J. McGhee, and J. Bienenstock, eds., Academic Press, Orlando, FL (1994).

8. D.A. Sullivan, and E.H. Sato, Potential therapeutic approach for the hormonal treatment of lacrimal gland dysfunction in Sjögren's syndrome, *Clin. Immunol. Immunopath.* 64:9 (1992).

9. D.A. Sullivan, Possible mechanisms involved in the reduced tear secretion in Sjögren's syndrome, *in*: "Proceedings of the IVth International Symposium on Sjögren's Syndrome," Kluger Press, Amsterdam, in press (1994).

10. M.G. Friedman, Antibodies in human tears during and after infection, *Surv. Ophthalmol.* 35:151 (1990).

11. N.K. Childers, M.G. Bruce, and J.R. McGhee, Molecular mechanisms of immunoglobulin A defense, *Annu. Rev. Microbiol.* 43:503 (1989).

12. T.T. MacDonald, S.J. Challacombe, P.W. Bland, C.R. Stokes, R.V. Heatley, and A. McI Mowat, eds. "Advances in Mucosal Immunology," Kluwer Academic Publishers, London (1990).

13. D.A. Sullivan, K.J. Bloch, and M.R. Allansmith, Hormonal influence on the secretory immune system of the eye: androgen regulation of secretory component levels in rat tears, *J. Immunol.* 132:1130 (1984).

14. D.A. Sullivan, K.J. Bloch, and M.R. Allansmith, Hormonal influence on the secretory immune system of the eye: androgen control of secretory component production by the rat exorbital gland, *Immunology* 52:239 (1984).

15. D.A. Sullivan, and M.R. Allansmith, Hormonal influence on the secretory immune system of the eye: androgen modulation of IgA levels in tears of rats, *J. Immunol.* 134:2978 (1985).

16. D.A. Sullivan, and L.E. Hann, Hormonal influence on the secretory immune system of the eye: endocrine impact on the lacrimal gland accumulation and secretion of IgA and IgG, *J. Steroid Biochem.* 34:253 (1989).

17. D.A. Sullivan, R.S. Kelleher, J.P. Vaerman, and L.E. Hann, Androgen regulation of secretory component synthesis by lacrimal gland acinar cells in vitro, *J. Immunol.* 145:4238 (1990).

18. R.S. Kelleher, L.E. Hann, J.A. Edwards, and D.A. Sullivan, Endocrine, neural and immune control of secretory component output by lacrimal gland acinar cells, *J. Immunol.* 146:3405 (1991).

19. L.E. Hann, R.S. Kelleher, and D.A. Sullivan, Influence of culture conditions on the androgen control of secretory component production by acinar cells from the lacrimal gland, *Invest. Ophthalmol. Vis. Sci.* 32:2610 (1991).

20. R.W. Lambert, R.S. Kelleher, L.A. Wickham, J.P. Vaerman, and D.A. Sullivan, Neuroendocrinimmune modulation of secretory component production by rat lacrimal, salivary and intestinal epithelial cells, *Invest. Ophthalmol. Vis. Sci.*, in press (1994).

21. D.A. Sullivan, L.E. Hann, and J.P. Vaerman, Selectivity, specificity and kinetics of the androgen regulation of the ocular secretory immune system, *Immunol. Invest.* 17:183 (1988).

22. D.A. Sullivan, E.B. Colby, L.E. Hann, M.R. Allansmith, and C.R. Wira, Production and utilization of a mouse monoclonal antibody to rat IgA: identification of gender-related differences in the secretory immune system, *Immunol. Invest.* 15:311 (1986).

23. D.A. Sullivan, and M.R. Allansmith, The effect of aging on the secretory immune system of the eye. *Immunology* 63:403 (1988).

24. L.E. Hann, M.R. Allansmith, and D.A. Sullivan, Impact of aging and gender on the Ig-containing cell profile of the lacrimal gland, *Acta Ophthalmologica* 66:87 (1988).

25. D.A. Sullivan, L.E. Hann, L. Yee, and M.R. Allansmith, Age- and gender-related influence on the lacrimal gland and tears, *Acta Opthalmologica* 68:189 (1990).

26. P. Weisz-Carrington, M.E. Roux, M. McWilliams, J.M. Phillips Quagliata, and M.E. Lamm, Hormonal induction of the secretory immune system in the mammary gland, *Proc. Natl. Acad. Sci. U.S.A.* 75:2928 (1978).

27. Z. Huang, R.W. Lambert, L.A. Wickham, and D.A. Sullivan, Influence of the endocrine environment on herpes virus infection in rat lacrimal gland acinar cells, *in*: "Lacrimal Gland, Tear Film and Dry Eye Syndromes: Basic Science and Clinical Relevance," D.A. Sullivan, B.B. Bromberg, M.M. Cripps, D.A. Dartt, D.L. MacKeen, A.K. Mircheff, P.C. Montgomery, K. Tsubota, and B. Walcott, eds., *Adv. Exp. Med. Biol.*, in press (1994).

28. D.A. Sullivan, J. Edwards, C. Soo, B.D. Sullivan, F.J. Rocha, and E.H. Sato, Influence of steroids and immunosuppressive compounds on tear IgA levels in a mouse model of Sjögren's syndrome, *Invest. Ophthalmol. Vis. Sci. Suppl.* 33:845 (1992).

29. D.A. Sullivan, and M.R. Allansmith, Hormonal influence on the secretory immune system of the eye: endocrine interactions in the control of IgA and secretory component levels in tears of rats. *Immunology* 60:337 (1987).

30. D.A. Sullivan, Influence of the hypothalamic-pituitary axis on the androgen regulation of the ocular secretory immune system, *J. Steroid Biochem.* 30:429 (1988).

31. M. Martinazzi, Effetti dell'ipofisectomia sulla ghiandola lacrimale extraorbitale del ratto, *Folia Endocrinol.* 150:120 (1962).

32. D.A. Sullivan, and M.R. Allansmith, Hormonal modulation of tear volume in the rat, *Exp. Eye Res.* 42:131 (1986).

33. Y. Hosoya, M. Matsushita, and Y. Sugiura, A direct hypothalamic projection to the superior salivatory nucleus neurons in the rat. A study using anterograde autoradiographic and retrograde HRP methods, *Brain Res.* 266:329 (1983).

34. J.D. Wilson, and D.W. Foster, eds. "Williams Textbook of Endocrinology," W.B. Saunders Company, Philadelphia, PA (1985).

35. I. Berczi, Neurohormonal-immune interaction, *in*: "Functional Endocrine Pathology," K. Kovacs, and S. Asa, eds., Blackwell Scientific Publications Inc., United Kingdom, p. 990 (1990).

36. I. Berczi, and E. Nagy, Effects of hypophysectomy on immune function, *in*: "Psychoneuroimmunology II," R. Ader, D. L. Felten, and N. Cohen, eds., Academic Press, San Diego, CA., p. 339 (1990).

37. A.D. Mooradian, J. E. Morley, and S. G. Korenman, Biological actions of androgens, *Endocrine Rev.* 8:1 (1987).

38. F.L. Jackson, and J.C. Hutson, Altered responses to androgen in diabetic male rats, *Diabetes* 33:819 (1984).

39. P. Mauduit, G. Herman, and B. Rossignol, Protein secretion induced by isoproterenol or pentoxifylline in lacrimal gland: Ca^{2+} effects, *Am. J. Physiol.* 246:C37 (1984).

40. D. Dartt, Signal transduction and control of lacrimal gland protein secretion: a review. *Curr. Eye Res.* 8: 619 (1989).

41. J.E. Jumblatt, G. T. North, and R. C. Hackmiller, Muscarinic cholinergic inhibition of adenylate cyclase in the rabbit iris-ciliary body and ciliary epithelium. *Invest. Ophthalmol. Vis. Sci.* 31:1103 (1990).

42. M. Ota, S. Kyakumoto, and T. Nemoto, Demonstration and characterization of cytosol androgen receptor in rat exorbital lacrimal gland, *Biochem. Internat.* 10:129 (1985).

43. J.A. Edwards, R.S. Kelleher, and D.A. Sullivan, Identification of dihydrotestosterone binding sites in the rat lacrimal gland, *Invest. Ophthalmol. Vis. Sci. Suppl.* 31:541 (1990).

44. F.J. Rocha, L.A. Wickham, J.D.O. Pena, J. Gao, M. Ono, R.W. Lambert, R.S. Kelleher, and D.A. Sullivan, Influence of gender and the endocrine environment on the distribution of androgen receptors in the lacrimal gland. *J. Steroid Biochem. Mol. Biol.* 46:737 (1993).

45. J. Gao, R.W. Lambert, and D.A. Sullivan, Endocrine control of secretory component mRNA levels in the rat lacrimal gland, *Invest. Ophthalmol. Vis. Sci. Suppl.* 34:1485 (1993).

46. L. Danel, M. Menouni, J. Cohen, J. Magaud, G. Lenoir, J. Revillard, and S. Saez, Distribution of andro-gen and estrogen receptors among lymphoid and haemopoietic cell lines, *Leukemia Res.* 9:1373 (1985).

47. S.A. Robertson, M. Brannstrom, and R.F. Seamark, Cytokines in rodent reproduction and the cytokine-endocrine interaction, *Curr. Opin. Rheumatol.* 4:585 (1992).

48. A.L. Jacobs, P.B. Sehgal , J. Julian, and D.D. Carson, Secretion and hormonal regulation of interleukin-6 production by mouse uterine stromal and polarized epithelial cells cultured in vitro, *Endocrinology* 131:1037 (1992).

49. S. Suemori, C. Ciacci, and D.K. Podolsky, Regulation of transforming growth factor expression in rat intestinal epithelial cells, *J. Clin. Invest.* 87:2216 (1991).

50. D.W. McGee, K.W. Beagley, W.K. Aicher, and J.R. McGhee, Transforming growth factor-b and IL-1b act in synergy to enhance IL-6 secretion by the intestinal epithelial cell line, IEC-, *J. Immunol.* 151:970 (1993).

51. S.S. Chen, and Q. Li, Transforming growth factor-b1 (TGF-b1) is a bifunctional immune regulator for mucosal IgA responses, *Cell. Immunol.* 128:353 (1990).

52. J.R. McGhee, J. Mestecky, C.O. Elson, and H. Kiyono, Regulation of IgA synthesis and immune response by T cells and interleukins, *J. Clin. Immunol.* 9:175 (1989).

53. M. Ono, Z. Huang, L.A. Wickham, J. Gao, and D.A. Sullivan, Analysis of androgen receptors and cytokines in lacrimal glands of a mouse model of Sjogren's syndrome. *Invest .Ophthalmol. Vis. Sci. Suppl.*, in press (1994).

54. D.W. McGee, W.A. Aicher, J.H. Eldridge, J.V. Peppard, J. Mestecky, and J.R. McGhee, Transforming growth factor-b enhances secretory component and major histocompatibility complex class I antigen expression on rat IEC-6 intestinal epithelial cells, *Cytokine* 3:543 (1991).

55. J.P. Whitcher, Clinical diagnosis of the dry eye, *Internat. Ophthalmol. Clin.* 27:7 (1987).

56. R.I. Fox, ed., "Sjogren' s Syndrome," *Rheum. Dis. Clin. N.A.* vol. 18 (3) (1992).

57. K.F. Tabbara, Sjogren's Syndrome, *in*: "The Cornea. Scientific Foundations and Clinical Practice," G. Smolin, and R.A. Thoft, eds., Little, Brown and Co, Boston, MA, pp 309-314 (1983).

58. M.C. Kincaid, The eye in Sjogren's syndrome. *in*: "Sjogren's Syndrome. Clinical and Immunological Aspects," N. Talal, H.M. Moutsopoulos, and S.S. Kassan SS, eds., Springer Verlag, Berlin, pp 25-33 (1987).

59. S.A. Ahmed, T.B. Aufdemorte, J.R. Chen, A.I. Montoya, D. Olive, and N. Talal, Estrogen induces the development of autoantibodies and promotes salivary gland lymphoid infiltrates in normal mice, *J. Autoimmunity* 2:543 (1989).

60. H. Carlsten, A. Tarkowski, R. Holmdahl, and L.A. Nilsson, Oestrogen is a potent disease accelerator in SLE-prone MRL lpr/lpr mice, *Clin. exp. Immunol.* 80:467 (1990).

61. R. McMurray, D. Keisler, K. Kanuckel, S. Izui, and S.E. Walker, Prolactin influences autoimmune disease activity in the female B/W mouse. *J. Immunol.* 147:3780 (1991).

62. H. Ariga, J. Edwards, and D.A. Sullivan, Androgen control of autoimmune expression in lacrimal glands of MRL/Mp-lpr/lpr mice, *Clin. Immunol. Immunopath.* 53:499 (1989).

63. A.C. Vendramini, C.H. Soo, and D.A. Sullivan, Testosterone-induced suppression of autoimmune disease in lacrimal tissue of a mouse model (NZB/NZW F1) of Sjogren's syndrome, *Invest. Ophthalmol. Vis. Sci.* 32:3002 (1991).

64. E.H. Sato, H. Ariga H, and D.A. Sullivan, Impact of androgen therapy in Sjogren's syndrome: Hormonal influence on lymphocyte populations and Ia expression in lacrimal glands of MRL/Mp-lpr/lpr mice, *Invest. Ophthalmol. Vis. Sci.* 33:2537 (1992).

65. E.H. Sato, and D.A. Sullivan, Comparative influence of steroid hormones and immunosuppressive agents on autoimmune expression in lacrimal glands of a female mouse model (MRL/Mp-lpr/lpr) of Sjogren' s syndrome, *Invest. Ophthalmol. Vis. Sci.*, in press (1994).

66. M. Appelmans, La Keratoconjonctivite seche de Gougerot-Sjogren, *Arch. 'Ophtalmologie* 81:577 (1948).

67. R. Bruckner, Uber einem erfolgreich mit perandren behandelten fall von Sjogren'schem symptomen komplex, *Ophthalmologica* 110:37 (1945).

68. A. Bizzarro, G. Valentini, G. Di Marinto, A. Daponte, A. De Bellis, and G. Iacono, Influence of testosterone therapy on clinincal and immunological features of autoimmune diseases associated with Klinefelter's syndrome, *J. Clin. End. Metab.* 64:32 (1987).

69. F.J. Rocha, E.H. Sato, B.D., Sullivan, and D.A. Sullivan, Comparative efficacy of androgen analogues in suppressing lacrimal gland inflammation in a mouse model (mRL/lpr) of Sjogren' s syndrome, *in*: "Lacrimal Gland, Tear Film and Dry Eye Syndromes: Basic Science and Clinical Relevance," D.A. Sullivan, B.B. Bromberg, M.M. Cripps, D.A. Dartt, D.L. MacKeen, A.K. Mircheff, P.C. Montgomery, K. Tsubota, and B. Walcott, eds., *Adv. Exp. Med. Biol.*, in press (1994).

70. C. Lavalle, E. Loyo, R. Paniagua, J.A. Bermudez, J. Herrera, A. Graef, D. Gonzalez-Barcena, and A. Fraga, Correlation study between prolactin and androgens in male patients with Systemic Lupus Erythematosus, *J. Rheum.* 14:268 (1987).

71. R.G. Lahita, The importance of estrogens in SLE, *Clin. Immunol. Immunopath.* 63:17 (1992).

72. R.G. Lahita, H.L. Bradlow, E. Ginzler, S. Pang, and M. New, Low plasma androgens in women with systemic lupus erythematosus, *Arth. Rheum.* 30:241 (1987).

73. J. Comsa, H. Leonhardt, and H. Wekerle H, Hormonal coordination of the immune response, *Rev. Physiol. Biochem. Pharmacol.* 92:115 (1982).

74. D. Eidenger, and T.J. Garrett, Studies of the regulatory effects of the sex hormones on antibody formation and stem cell differentiation, *J. Exp. Med.* 136:1098 (1972).

75. Y. Lebranchu, P. Bardos, and M.A. Bach, Thymic function in NZB mice. IV. Role of thymus secretion and sex factors in the expression of suppression, *Clin. Immunol. Immunopathol.* 23:563 (1982).

76. J.R. Roubinian, N. Talal, J.S. Greenspan, J.R. Goodman, and P.K. Siiteri, Effect of castration and sex hormone treatment on survival, anti-nucleic acid antibodies, and glomerulonephritis in NZB/NZW F1 mice, *J. Exp. Med.* 147:1568 (1978).

77. N. Talal, and S.A. Ahmed, Sex hormones and autoimmune disease, *Int. J. Immunotherapy* 3:65 (1987).

78. W. Savino, E. Bartoccioni, F. Homo-Delarche, M. Gagnerault, T. Itoh, and M. Dardenne, Thymic hormone containing cells-IX. Steroids in vitro modulate thymulin secretion by human and murine thymic epithelial cells, *J. Steroid. Biochem.* 30:479 (1988).

79. E.J. Goldsteyn, and M.J.L. Fritzler, The role of the thymus-hypothalamus-pituitary-gonadal axis in normal immune processes and autoimmunity, *J. Rheumatol.* 14:982 (1987).

80. C.J. Grossman, Are there underlying immune-neuroendocrine interactions responsible for immunological sexual dimorphism?, *Prog. NeuroEndocrinImmunology* 3:75 (1990).

81. N.J. Olsen, M.B. Watson, and W.J. Kovacs, Studies of immunological function in mice with defective androgen action. Distinction between alterations in immune function due to hormonal insensitivity and alteration due to other genetic factors, *Immunology* 73:52 (1991).

82. Y. Weinstein, and Z. Berkovich, Testosterone effect on bone marrow, thymus and suppressor T cells in the (NZB/NZW) F1 mice: its relevance to autoimmunity, *J. Immunol.* 126:998 (1981).

83. K.A. Melez, W.A. Boegel, and A.D. Steinberg, Therapeutic studies in New Zealand mice. VII. Successful androgen treatment of NZB/NZW F1 females of different ages, *Arthritis Rheum.* 23:41 (1980).

84. N. Talal, H. Dang, S.A. Ahmed, E. Kraig, and M. Fischbach, Interleukin 2, T cell receptor and sex hormone studies in autoimmune mice, *J. Rheum.* (suppl. 13) 14:21 (1987).

85. S.A. Ahmed, W.J. Penhale, and N. Talal, Sex hormones, immune responses and autoimmune diseases, *Am. J. Pathol.* 121:531 (1985).

86. J.P. Michalski, C.C. McCombs, J.R. Roubinian, and N. Talal, Effect of androgen therapy on survival and suppressor cell activity in aged NZB/NZW F1 mice, *Clin. exp. Immunol.* 52,229 (1983).

87. G. Holdstock, B.F. Chastenay, and E.L. Krawitt, Effects of testosterone, oestradiol and progesterone on immune regulation, *Clin. exp. Immunol.* 47:449 (1982).

88. J.B. Allen, D. Blatter, G.B. Calandra, and R.L. Wilder, Sex hormonal effects on the severity of streptococcal cell wall-induced polyarthritis in the rat, *Arthritis Rheum.* 26:560 (1983).

89. S.A. Ahmed, M.J. Dauphinee, and N. Talal, Effects of short-term administration of sex hormones on normal and autoimmune mice, *J. Immunol.* 134:204 (1985).

90. N. Talal, S.A. Ahmed, and M. Dauphinee, Hormonal approaches to immunotherapy of autoimmune diseases, *N.Y. Acad. Sci.* 475:320 (1986).

91. S.A. Ahmed, P.R. Young, and W.J. Penhale, Beneficial effect of testosterone in the treatment of chronic autoimmune thyroiditis in rats, *J. Immunol.* 136:143 (1986).

92. A.D. Steinberg, J.B. Roths, E.D. Murphy, R.T. Steinberg RT, and E.S. Raveche, Effects of thymectomy or androgen administration upon the autoimmune disease of MRL/Mp-lpr/lpr mice, *J. Immunol.* 125:871 (1980).

93. H. Carlsten, R. Holmdahl, A. Tarkowski, and L.A. Nilsson, Oestradiol- and testosterone-mediated effects on the immune system in normal and autoimmune mice are genetically linked and inherited as dominant traits, *Immunology* 68:209 (1989).

94. J.I. Morton, D.A. Weyant, B.V. Siegel, and B.L. Golding, Androgen sensitivity and autoimmune disease. I. Influence of sex and testosterone on the humoral immune response of autoimmune and non-autoimmune mouse strains to sheep erythrocytes, *Immunology* 44:661 (1981).

95. J.E. Brick, D.A. Wilson, and S.E. Walker, Hormonal modulation of responses to thymus-independent and thymus-dependent antigens in autoimmune NZB/W mice, *J. Immunol* 134:3693 (1985).

96. J.E. Brick, S.E. Walker, and K.S. Wise, Hormone control of autoantibodies to calf thymus nuclear extract (CTE) and DNA in MRL-lpr and MRL-+/+ mice, *Clin. Immunol. Immunopathol.* 46:68 (1988).

97. Z.M. Sthoeger, N. Chiorazzi, and R.G. Lahita, Regulation of the immune response by sex hormones. I. In vivo effects of estradiol and testosterone on pokeweed-mitogen-induced human B-cell differentiation, *J. Immunol.* 141:91 (1988).

98. M. Ono, F.J. Rocha, E.H. Sato, and D.A. Sullivan, Distribution and endocrine regulation of androgen receptors in lacrimal glands of the MRL/lpr mouse model of Sjögren's syndrome, *in*: "Proceedings of the IVth International Symposium on Sjögren's Syndrome," Kluger Press, Amsterdam, in press (1994).

99. D.A. Sullivan, H. Ariga, A.C. Vendramini, F.J. Rocha, and E.H. Sato EH, Androgen-induced suppression of autoimmune disease in lacrimal glands of mouse models of Sjögren's syndrome, *in*:

"Lacrimal Gland, Tear Film and Dry Eye Syndromes: Basic Science and Clinical Relevance," D.A. Sullivan, B.B. Bromberg, M.M. Cripps, D.A. Dartt, D.L. MacKeen, A.K. Mircheff, P.C. Montgomery, K. Tsubota, and B. Walcott, eds., *Adv. Exp. Med. Biol.*, in press (1994).

100. R.I. Fox, and I. Saito, Sjogren's syndrome: immunologic and neuroendocrine mechanisms, *in*: "Lacrimal Gland, Tear Film and Dry Eye Syndromes: Basic Science and Clinical Relevance," D.A. Sullivan, B.B. Bromberg, M.M. Cripps, D.A. Dartt, D.L. MacKeen, A.K. Mircheff, P.C. Montgomery, K. Tsubota, and B. Walcott, eds., *Adv. Exp. Med. Biol.*, in press (1994).

101. H. Dang, K. Lazaridis, T. Aufdemorte, H. McGuff, and N. Talal, PCR analysis of cytokines produced in salivary glands of Sjogren's syndrome, *in*: "Proceedings of the IVth International Symposium on Sjogren's Syndrome," Kluger Press, Amsterdam, in press (1994).

102. M.M. Shull, I. Ormsby, A.B. Kier, S. Pawlowski, R.J. Diebold, M. Yin, R. Allen, C. Sidman, G. Proetzel, D. Calvin, N. Annunziata, and T. Doetschman, Targeted disruption of the mouse transforming growth factor-b1 gene results in multifocal inflammatory disease, *Nature* 359:693 (1992).

103. E. Nagy, I. Berczi, and E. Sabbadini, Endocrine control of the immunosuppressive activity of the submandibular gland, *Brain Beh. Imm.* 6:418 (1992).

104. J. Krueger, A. Ray, I. Tamm, and P.B. Sehgal, Expression and function of interleukin-6 in epithelial cells, *J. Cell. Biochem.* 45:327 (1991).

105. M. De, T.R. Sanford, and G.W. Wood, Interleukin-1, interleukin-6, and tumor necrosis factor a are produced in the mouse uterus during the estrous cycle and are induced by estrogen and progesterone, *Dev. Biol.* 151:297 (1992).

106. N. Sarvetnick, and H.S. Fox, Interferon-gamma and the sexual dimorphism of autoimmunity, *Mol. Biol. Med.* 7:323 (1990).

IMMUNOREGULATION BY SEX HORMONE ANALOGUES

A.H.W.M. Schuurs, G.H.J. Deckers, H.A.M. Verheul

Organon Scientific Development Group
P.O. Box 20
5340 BH Oss, The Netherlands

SUMMARY

Evidence is presented suggesting that endocrine and immunomodulating activities of estrogens, progestogens and androgens are often not correlated. The consequences of such a possible lack of correlation is discussed. The possibility to dissociate endocrine and immunomodulating activities may lead to novel compounds for treating autoimmune diseases.

INTRODUCTION

There is ample evidence that gender, gonadectomy, hypogonadism and sex hormones affect the performance of the immune system. This is clearest for androgens which can diminish both cell-mediated and humoral hetero- and auto-immune responses. For estrogens the situation is somewhat more complicated for various reasons: (i) in some studies, non-physiological, even toxic doses have been used so that it is difficult to draw conclusions; (ii) it is uncertain whether non-steroidal estrogens, such as diethylstilbestrol (DES) have comparable effects on the immune system; (iii) chronic and cyclic administration of estrogen has different effects on the immune system.[1] Few studies have been performed with progestogens. These steroids seem to have effects which are strictly based on locally high concentrations; one may wonder to what extent binding to the glucocorticoid receptor could explain its action on the immune system.

Several recent reviews give an adequate account of the present knowledge in this area.[2,3,4] In this paper we intend to concentrate on the question as to what extent hormonal potencies of sex steroids are correlated with their effects on the immune system. For this purpose we shall review only those studies in which the effects of two or more estrogens, progestogens or androgens on the immune system have been compared in one experimental set-up.

Advances in Psychoneuroimmunology, Edited by
I. Berczi and J. Szélenyi, Plenum Press, New York, 1994

ESTROGENS

Nicol et al.[5] studied the effects of various steroidal and non-steroidal estrogens on the clearance of carbon particles in male mice as a measure of phagocytosis. Their results with the steroidal estrogens are summarized in Table 1; the lack of correlation between estrogenic activities and effects on phagocytosis is evident. The authors state: "RES (reticuloendothelial system) stimulation does not rise in step with estrogenicity, and estrogenic compounds have two separate biological activities - one which stimulates RES to raise body defence, and one which acts on the reproductive organs; these two activities are apparently unrelated, although shared by the same molecule." A limitation of this study, however, is the use of very high doses: 0.5 mg of estrogen per mouse (of 20-26 mg) once daily for 6 days so that it is questionable to what extent these data can be extrapolated to physiological estrogen levels.

Table 1. The effect of various estrogens on phagocytosis in mice.[5]

Treatment	$10^3/K$	Inverse estrogenic activity	
		Nicol	own data[a]
E1	15	2.5	10
E2 benzoate	12	0.1	1
E3	21	10	167
EE	13	0.1	1
equilin	13	2.5	50
equilenin	20	20	n.t.

K: phagocytic index
[a] Allen-Doisy test

Clearance of antibody-coated erythrocytes by guinea pigs and the effect of three estrogens in physiological concentrations was studied by Schreiber et al.[6] The authors tested 17β-estradiol (E2), estriol (E3) and 16β-estradiol (1,3,5(10)-estratrien-3,16β-diol) and found increases of clearance which are not at all related to the estrogenic activities of these compounds (Table 2). The authors conclude that "similar results are attained with structural analogs of steroid hormones which have potentially minimal end organ effects."

Table 2. Percentage increase of clearance in guinea pigs of radiolabelled antibody-sensitised erythrocytes by some estrogens.[6]

Treatment (1 mg daily, 5 days)	% Increase of clearance	estrogenic activity
E3 (3, 16α, 17β)	26	0.006[a]
E2 (3, 17β)	141	1[a]
"E2" (3, 16β)	107	1/100 x E3[b]

[a] Allen-Doisy test data from our own laboratory
[b] Data from Schreiber et al.[6]

Table 3. Effects of androgen-like steroids on the development of the bursa of Fabricius.

A.

Steroid	Bursa weight (mg/100 g body wt)[a]	Potency order[b]		
		SV	VP	MLA
17α-ethyl-19-nor-T[c]	d	3		1
DHT	1	1	2	4
Androsterone	11	5	1	5
Androstane-3,17-dione	36	6	3	6
19-nor-T	42	4	5	2
T	65	2	4	3
None	104			
Vehicle	117			

[a] Aspinall et al.,[11] table 2, column 0.2 mg.

[b] Effect of the test compounds on the seminal vesicles (SV), ventral prostate (VP) and levator ani muscle (MLA) of castrated immature male rats after daily s.c. administration for one week (Hershberger test; cf. de Visser et al.[12]); the ranking order of potency is given for each target organ.

[c] T = testosterone

[d] Stated to be "considerably more potent than DHT."

B.

Steroid	Potency order Bursa development inhibition[a]	Potency order[b]		
		SV	VP	MLA
Mibolerone	1			
17-ethyl-19-nor-T	2	2		1
DHT	3	1	2	3
Androsterone	4	4	1	4
Androstane-3,17-dione	5	5	3	5
19-nor-T	6	3	4	2
T-cyclopentylpropionate	7			

[a] Bhanushali et al.[13]

[b] See under A.

Pfeifer and Patterson[7] compared the effects of estrone (E1), E2, 2-hydroxyestrone (2-OH E1) and 2-methoxyestrone (2-OCH3 E1) on macrophage induced cytostasis of cultured tumour cells. Preincubation of macrophages with 2-OH E1 in concentrations of 1-100 μ mol/1 caused significant inhibition of cytostasis, the other estrogens being inactive or less active. If macrophages were incubated with supernatant from spleen lymphocytes containing macrophage-activating factor, prior to incubation with estrogen, all four estrogens stimulated cytostasis, E2 being the least active in this respect. These effects are unrelated to the relative binding affinities of these steroids to the estrogen receptor (ER); these binding affinities are: E2: 100%, E1: 11%, 2-OH E1: 2%, 2-OCH3 E1: 0.01%.[8] Pfeifer and Patterson[9,10] also found 2-OH E1 to be active in suppressing spleen lymphocyte proliferation and lymphocyte-mediated growth inhibition of tumour cells; the other estrogens were less active. Estrogens were found to affect PHA-induced lymphocyte agglutination within minutes, suggesting a modulation of cell function through a cell surface rather than a nuclear receptor.

ANDROGENS

The important role of the bursa of Fabricius for the humoral immunity in birds was discovered by Glick et al.[14] who found that surgically bursectomised chickens failed to produce antibodies upon antigenic stimulation. Warner and Burnet[15] and Aspinall et al.[11] found that androgens, notably testosterone and 5α-dihydrotestosterone (DHT), inhibit the development of the bursa of Fabricius. Remarkably, Meyer et al.[16] had already found an inhibition of bursal development by 19-nortestosterone (nandrolone), a less virilising steroid but with a more pronounced anabolic activity.

Aspinall et al.[11] compared the effect of various steroids administered in ovo on the development of the bursa. Table 3A presents some of their data combined with the order of androgenic and anabolic potencies as found in immature castrated male rats in our laboratories (Deckers, unpublished data). In particular the effects of seminal vesicles and levator ani muscle do not seem to correlate with effects on the bursa. A similar comparison was made for data presented by Bhanushali et al.[13]; Table 3B shows the lack of correlation.

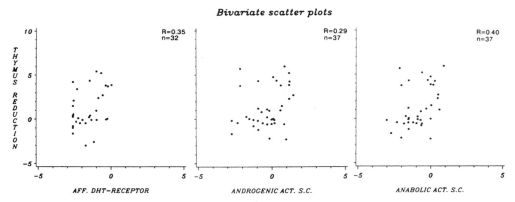

Figure 1. Lack of correlation between thymus weight and various androgenic parameters: *in vitro*: relative binding affinity of steroid to the androgen receptor, *in vivo*: the weights of the ventral prostate and of the levator ani muscle as androgenic and anabolic parameters, respectively, after subcutaneous administration of steroid. n = number of steroids tested.

In our laboratories, we confirmed this lack of correlation by comparing the activities of more than 50 androgens and steroids structurally related to androgens in the Hershberger test (steroids administered orally and subcutaneously) and their binding to the androgen receptor with their effect on the developing bursa; correlation coefficients (R^2) varied between 0.06 and 0.30.[17,18] We also studied the effects of about 35 steroids on the thymus in immature male and female rats. These steroids were subcutaneously administered once every day for 7 days; on day 8 the animals were autopsied and the weights of thymus and various other organs, notably the adrenals, were determined. Figure 1 shows the bivariate scatter plots for various androgenic parameters versus the reduction of thymus weights. None of the compounds tested in this series had a substantial effect on the adrenals indicating the absence of glucocorticoid activities.

The effects of testosterone, 19-nortestosterone (nandrolone) and their decanoate esters on autoimmune disease were compared in two studies with NZB/NZW F1 mice.[19,20] Nandrolone (decanoate) was more efficient in suppressing and delaying disease symptoms than testosterone (decanoate). The former has about half the effect of the latter on seminal vesicle and ventral prostate weights, indicating no correlation between immunomodulating and virilising properties.

PROGESTOGENS

Hulka et al.[21] found different effects of progestogens on cell-mediated and humoral responses. As is evident from Table 4, the two 17-substituted nortestosterone compounds delayed rejection while the three pregnanes were inactive; the humoral response was inhibited by progesterone, medroxyprogesterone acetate and norethynodrel. Neither effect correlates with the progestational potencies as measured in the Clauberg assay.

Table 4. The effect of various progestogens on allograft rejection and antibody response in rabbits.[21]

Treatment[a]	Allograft rejection (average, days)	Ab response against BSA (mg/ml serum)	Relative potencies[b]
None	9	10.4	
P	9.2	3.2	1
17α -OH-P capr.	9.8	9.3	1
Medroxy-P ac.	9.2	0.9	32
Norethynodrel	12.8	5.8	0.25
Norethindrone	13	3.4	4
Cortisone[c]	>21	0.6	

[a] P = progesterone; capr. = caproate; ac. = acetate
[b] As determined in the Clauberg test with the steroids subcutaneously administered; potency assigned to P is unity. The differences in dosage schemes have not been taken into account; if this had been done, the discrepancy between effects on the immune and endocrine systems would have been even more extreme.
[c] Positive control

Table 5. Effect of three progestogens on PHA-induced lymphocyte proliferation.[22]

Treatment	ID50 (nmol/l)	Relative binding to PR (%)[a]
P[b]	20	20
17α -OH-P	>300	<1
20α -OH-P	30	0.24

[a] PR = progesterone receptor; Org 2058: 100%
[b] P = progesterone

A small study with progestogens in the lymphocyte proliferation test by Mori et al.[22] showed a lack of correlation between proliferation inhibition and binding to the progesterone receptor (Table 5). A limitation of this study, however, is the high concentrations needed to obtain an effect on the lymphocyte proliferation, concentrations which are unlikely to be reached in vivo.

Like for androgens (see above) we tested the effects of steroids with progestational activity on the development of the bursa of Fabricius. For 50 compounds the binding to the progesterone receptor was determined; 77 and 97 steroids were assayed in the Clauberg assay, administered subcutaneously and orally respectively; correlation coefficients (R^2) of around 0.10 were found.[17,18]

Two studies with synthetic progestogens in the autoimmune disease model NZB/NZW F1 (B/W) mice are summarized in Table 6 from which it is evident that disease-modifying effects and progestational activities are not related.

Table 6. Effect of progestogens on the course of autoimmune phenomena in NZB/W mice.

A.[23,24]

	Lynestrenol	Ethylestrenol	Desogestrel
Mortality	↓	=	=
Proteinuria	↓	=	=
Sialoadenitis	↓	↓	=
Anti-DNA	↓	↓	=
Clauberg, rel. potency (oral)	1	2	16

B.[25]

	6α -CH$_3$-OHP	Norethindrone	Norgestrel
Mortality	=	=	=
Glom. nephritis	=	=	=
Anti-DNA	=	↓	↓
Clauberg, rel. potency (oral)	2	1	8

IS THERE A GENERAL LACK OF CORRELATION BETWEEN IMMUNOMODULATING AND ENDOCRINE PROPERTIES OF SEX STEROIDS?

The data summarised above suggest that immunomodulating and endocrine properties of the three types of sex steroids are not related.

The evidence is weak for estrogens where only a few scattered data related to phagocytosis suggest this lack of correlation. Obviously, more data are required to allow firmer conclusions. Three conditions should be met in such studies: non-physiological and toxic doses as have been used in some earlier studies should be avoided, one should take into account possible differences between cyclic and chronic administration (cf. Erbach and Bahr[1]), one should distinguish between steroidal and non-steroidal estrogens since it cannot be excluded that they behave differently.

The situation with progestogens is influenced by the fact that "substances described as progestogens differ greatly in their additional properties".[26] This statement relates to various endocrine properties: (anti-)estrogenic, (anti-)androgenic and glucocorticoid. The presence or absence of these activities is determined to a large extent by the chemical groups or progestogens: e.g. 17α-hydroxypregnanes, 19-nor-testosterone derivatives). For immunomodulating properties of progestogens this is probably equally true.

For androgen-derived steroids the discrepancy between endocrine and immunomodulating activities has been extensively shown. However, here too, more data would be valuable: testosterone and DHT have been shown to affect various other aspects of the immune response and it is unknown whether these effects are also correlated with their virilising potency. Some examples: inhibition of pokeweed mitogen-induced B cell differentiation,[27] inhibition of in vivo humoral response to sheep erythrocytes,[28] increased susceptibility to infections by *Mycobacterium marinum*[29] or *Plasmodium chabaudi*.[30]

CONSEQUENCES OF A LACK OF CORRELATION BETWEEN IMMUNOMODULATING AND ENDOCRINE ACTIVITIES

Obviously, endocrine properties have little predictive value for immunomodulating potential of sex steroid-derived hormonomimetics. A theoretical aspect of the dissociation of the two types of activity is the question of the role of classical steroid receptors found in immune organs or cells, such as thymus, spleen and lymphocytes (cf. Homo-Delarche[3]). A related question is how sex steroids modulate the immune system if not through these classical receptors. No answers to these questions are available.

An interesting practical consequence of the lack of correlation is the possibility to devise compounds with little endocrine effects and raised immunomodulating activity, in other words, with an increased immunological/endocrinological (imm/end) ratio; this will be discussed in the next section.

EFFECTS OF STEROIDS WITH RAISED IMM/END RATIOS IN ANIMAL MODELS FOR AUTOIMMUNE DISEASE

Verheul et al.[23,24] studied the effect in female and castrated male NZB/W mice of tibolone (Org OD14), a compound with weak estrogenic, weak progestational and weak androgenic activity.[12] In a dose of 0.1 mg/day/ mouse it gave significant improvement of the lupus-related parameters proteinuria, anti-DNA levels and survival, and of the Sjögren's syndrome-like infiltration in the submandibular glands.

Org OD14 was also tested in the Obese Strain chicken model for autoimmune thyroiditis.[31] Administered in ovo by means of dipping fertilised eggs in solutions containing 40 or 120 μ g/ml it completely inhibited the formation of autoantibodies against thyroglobulin and partially inhibited lymphocyte infiltration in the thyroid. Another compound with low endocrine activities, 11α-hydroxynandrolone (Verheul et al.[17], addendum 1: compound #18), caused inhibition of autoantibody formation and thyroid infiltration in concentrations of 1-10 mg/ml, it is thus at least as active as nandrolone which, for instance, binds > 100 times more strongly to androgen receptor than 11α-hydroxynandrolone. Various steroids with low endocrine activities including Org OD14 were also administered after hatching; Org OD14 in doses of 0.4-10 mg/kg body weight/day again inhibited autoantibody formation and infiltration in the thyroid.

We recently identified another steroid devoid of androgenic, glucocorticoid and progestational activities and with very little residual estrogenic activity, Org 4094, which reduced sialoadenitis and lupus symptoms in NZB/W mice, and reduced sialoadenitis and diabetes incidence in NOD mice.[32,33]

The data in this section suggest that the possibiity to dissociate endocrine and immunomodulating activities may lead to novel therapeutic modalities for treating human autoimmune diseases.

REFERENCES

1. G.T. Erbach and J.M. Bahr, Effect of chronic or cyclic exposure to estradiol on the humoral immune response and the thymus, *Immunopharmacology* 16:45 (1988).
2. A.H.W.M. Schuurs and H.A.M. Verheul, Effect of gender and sex steroids on the immune response, *J. Steroid Biochem.* 35:157 (1990).
3. F. Homo-Delarche, F. Fitzpatrick, N. Christeff, E.A. Nunez, J.F. Bach, and M. Dardenne, Sex steroids, glucocorticoids, stress and autoimmunity, *J. Steroid Biochem. Molec. Biol.* 40:619 (1991).
4. C.J. Grossman, G.A. Roselle, and C.L. Mendenhall, Sex steroid regulation of autoimmunity, *J. Steroid Biochem. Molec. Biol.* 40:649 (1991).
5. T. Nicol, D.L.J. Bilbey, L.M. Charles, J.L. Cordingley, and B. Vernon-Roberts, Oestrogen: the natural stimulant of body defence, *J. Endocrinol.* 30:277 (1964).
6. A.D. Schreiber, F.M. Nettl, M.C. Sanders, M. King, P. Szabolcs, D. Friedman, and F. Gomez, Effect of endogenous and synthetic sex steroids on the clearance of antibody-coated cells, *J. Immunol.* 141:2959 (1988).
7. R.W. Pfeifer and R.M. Patterson, Modulation of lymphokine-induced macrophage activation by estrogen metabolites, *J. Immunopharmacol.* 7:247 (1985).
8. C. Martucci and J. Fishman, Uterine estrogen receptor binding of catecholestrogens and estetrol (1,3,5(10)-estratriene-3,15a,16a,17b-tetrol), *Steroids* 27:325 (1976).
9. R.W. Pfeifer and R.M. Patterson, Modulation of nonspecific cell-mediated growth inhibtion by estrogen metabolites, *Immunopharmacology* 10:127 (1985).
10. R.W. Pfeifer and R.M. Patterson, Modulation of lectin-stimulated lymphocyte agglutination and mitogenesis by estrogen metabolites: effects of early events of lymphocyte activation, *Arch. Toxicol.* 58:157 (1986).
11. R.L. Aspinall, R.K. Meyer and M.A. Rao, Effect of various steroids on the development of the bursa Fabricii in chick embryos, *Endocrinology* 68:944 (1961).
12. J. de Visser, A. Coert, H. Feenstra, and J. van der Vies, Endocrinological studies with (7a,17a)-17-hydroxy-7-methyl-19-norpregn-5(10)-3n-20-yn-3-one (Org OD 14), *Arzneim. Forsch./Drug Res.* 9:1010 (1984).
13. J.K. Bhanushali, K.K. Murthy and W.L. Ragland, Effect of androgens on the ontogeny of humoral immunity in chickens, *in:* "Chemical Regulation of Immunity in Veterinary Medicine," Alan R. Liss, Inc., New York (1984).
14. B. Glick, T.S. Chang, and R.G. Jaap, The bursa of Fabricius and antibody production in the

domestic fowl, *Poult. Sci.* 35:224 (1956).

15. N.L. Warner and F.M. Burnet, The influence of testosterone on the development of the bursa of Fabricius in chick embryo, *Aust. J. Biol. Sci.* 14:580 (1961).

16. R.K. Meyer, M.A. Rao, and R.L. Aspinall, Inhibition of the development of the bursa of Fabricius in the embryos of the common fowl by 19-nortestosterone, *Endocrinology* 64:890 (1959).

17. H.A.M. Verheul, E.V. Tittes, J. Kelder and A.H.W.M. Schuurs, Effect of steroids with different endocrine profiles on development, morphology and function of the bursa of Fabricius in chickens, *J. Steroid Biochem.* 25:665 (1986).

18. A.H.W.M. Schuurs, G.H.J. Deckers, L.P.C. Schot, and H.A.M. Verheul, Lack of correlation between endocrinological and immunological effects of androgen-derived steroids, *in*: "Hormones and Immunity," I. Berczi, K. Kovacs, eds., MTP Press, Lancaster, UK (1987).

19. H.A.M. Verheul, W.H. Stimson, F.C. den Hollander and A.H.W.M. Schuurs, Effects of nandrolone, testosterone and their decanoate esters on murine lupus, *Clin. Exp. Immunol.* 44:11 (1981).

20. H.A.M. Verheul, G.H.J. Deckers and A.H.W.M. Schuurs, Effects of nandrolone or testosterone decanoate on murine lupus: further evidence for a dissociation of immune and endocrine effects, *Immunopharmacology* 11:93 (1986).

21. J.F. Hulka, K. Mohr, and M.W. Lieberman, Effect of synthetic progestational agents on allograft rejection and circulating antibody production, *Endocrinology* 77:897 (1965).

22. T. Mori, H. Kobayashi, H. Nishimoto, A. Suzumi, T. Nishimura, and T. Mori, Inhibitory effect of progesterone and 20a-hydroxypregn-4-en-3-one on the phytohemagglutinin-induced transformation of human lymphocytes, *Am. J. Obstet. Gynecol.* 127:151 (1977).

23. H.A.M. Verheul, L.P.C. Schot, G.H.J. Deckers and A.H.W.M. Schuurs, Effects of tibolone, lynestrenol, ethylestrenol and desogestrel on immune disorders in NZB/W mice, *Clin. Immunol. Immunopathol.* 38:198 (1986).

24. H.A.M. Verheul, L.P.C. Schot and A.H.W.M. Schuurs, Effects of nandrolone decanoate, tibolone, lynestrenol and ethylestrenol on Sjögren's syndrome-like disorders in NZB/W mice, *Clin. Exp. Immunol.* 64:243 (1986).

25. L.W. Keisler, A.B. Kier and S.E. Walker, Effects of prolonged administration of the 19-nor-testosterone derivatives norethindrone and norgestrel to female NZB/W mice: comparison with medroxyprogesterone and ethinyl estradiol, *Autoimmunity* 9:21 (1991).

26. F. Neumann, The physiological action of progesterone and the pharmacological effects of progestogens - a short review, *Postgrad. Med. J.* 54(suppl 2):11 (1978).

27. Z.M. Sthoeger, N. Chiorazzi and R.G. Lahita, Regulation of the immune response by sex hormones. I. In vitro effects of estradiol and testosterone on pokeweed mitogen-induced human B cell differentiation, *J. Immunol.* 141:91 (1988).

28. D. Catanzano-Troutaud, D. Ardail and P.A. Deschaux, Testosterone inhibits the immunostimulant effect of thymosin fraction 5 on secondary immune response in mice, *Int. J. Immunopharmacol.* 14:263 (1992).

29. Y. Yamamoto, H. Saito, T. Setogawa and H. Tomioka, Sex differences in host resistance to *Mycobacterium marinum* infection in mice, *Infect. Immun.* 59:4089 (1991).

30. W.P.M. Benten, U. Bettenhaeuser, F. Wunderlich, E. van Vliet and H. Mossmann, Testosterone-induced abrogation of self-healing of *Plasmodium chabaudi* malaria in B10 mice: mediation by spleen cells, *Infect. Immun.* 59:4486 (1991).

31. A.H.W.M. Schuurs, H. Dietrich, J. Gruber and G. Wick, Effects of sex steroid analogs on spontaneous autoimmune thyroiditis in Obese Strain chickens, *Int. Arch. Allergy Immunol.* 97:337 (1992).

32. W.M. Bagchus, T. Coenen, A. van Doornmalen, J. Dulos, A.H.W.M. Schuurs and H.A.M. Verheul, Immunotherapy with an androgen-derived steroid (Org 4094) on autoimmune diseases in NZB/W and NOD mice, 8th Int. Cong. Immunol., Budapest, Hungary, Abstract W-87 #1 (1992).

33. H.A.M. Verheul, M. Verveld, G.H.J. Deckers, W.M. Bagchus and A.H.W.M. Schuurs, Dissociation of endocrine and autoimmunosuppresssive properties of androgen-derived steroids in the NZB/W and NOD mice, Am. College Rheumatol., 56th Annual Scientific Meeting, Atlanta, Georgia, Abstract PO331 (1992).

TAMOXIFEN AS AN IMMUNOMODULATING AGENT

E. Baral [1], E. Nagy[2], & I. Berczi [2]

[1]Manitoba Cancer Treatment & Research Foundation and [2]Department of Immunology, Faculty of Medicine, University of Manitoba Winnipeg, Manitoba, Canada R3E 0W3

INTRODUCTION

The nonsteroidal antiestrogenic agent, tamoxifen (TX), has gained wide therapeutic application for the treatment of breast carcinomas and of some other sex hormone dependent tumors. [1] Numerous investigations indicate that TX is capable of combining with the estrogen receptor of tumor cells which results in partial activation of the receptor, but without the growth promoting effect of estradiol (E2), which is the proper ligand for the receptor. For this reason TX and related drugs (ethamoxytriphethol or MER25, and clomiphene) are referred to under the collective term, nonsteroidal antiestrogens. [2] However, it was also demonstrated that nonsteroidal antiestrogens are able to inhibit the stimulatory effect of growth factors, such as epidermal growth factor or insulin, on human breast carcinoma cells in the complete absence of estrogens. Moreover, a cytotoxic effect was also observed when the antiestrogens were used at 4 μM concentration or higher. The presence of the estrogen receptor was still necessary for these effects. Therefore, Rochefort [3] suggested that these agents be named estrogen receptor targeted drugs, rather than antiestrogens. However, it is a well established clinical fact that TX is capable of inducing regression of tumors that lack the classical receptor for E2. [4-6] False negative receptor assay results were proposed as one possibility for this observation. Another possibility is a favorable influence of TX on immune host defence mechanisms. The available evidence supporting this latter possibility is discussed below.

THE EFFECT OF TAMOXIFEN ON THE IMMUNE SYSTEM

We have been interested in the possibility of immune modulation by tamoxifen for several years. Our initial observation performed with human peripheral blood lymphocytes (PBL) showed that after incubation for 24 hours with various therapeutic concentrations of TX *in vitro*, the expression of receptors for the C'3 component of complement on human B lymphocytes was diminished. In contrast, the expression of membrane receptors for sheep red blood cells, which is a T lymphocyte marker, was not changed under the same experimental conditions. [7] When human PBL was stimulated by pokeweed mitogen (PWM) for immunoglobulin secretion, the pretreatment of T cells with therapeutically achievable concentrations of TX augmented their capacity to promote IgG, but not IgM, secretion by untreated autologous B lymphocytes. On the other hand, cultures containing TX pretreated B cells and untreated T cells exhibited reduced secretion of both IgG and IgM. [8]

Paavonen and Anderson [9] also studied the effects of two antiestrogens, TX and FC-1157a, on the response of human lymphocytes to PWM. Both drugs increased significantly the number of immunoglobulin secreting cells when added at 1-100 nM concentration. This concentration of TX did not affect the proliferative response itself. Experiments with fractionated lymphocytes revealed that the antiestrogens exerted their enhancing effect by inhibiting the suppressive function of $CD8^+$ cells.

Webster et al. [10] studied 72 patients with advanced breast cancer during a randomized chemotherapy trial and hormonal manipulation. A gradual fall in IgM levels was observed as the study progressed. Nevertheless, these investigators concluded that hormonal therapy (ovariectomy and treatment with TX or androgens) had no detectable effect on the immune response. Joensuu et al. [11] studied 10 patients prior to and 3, 6 and 12 months after starting TX treatment (20 mg twice daily). Complement receptor bearing cells decreased significantly at 6 months and those with receptors for SRBC at 12 months after starting TX therapy.

We observed in rats that treatment with high doses (5 mg/kg) of TX daily during immunization with SRBC significantly inhibited the antibody response, which could be reversed by additional treatment with either prolactin or growth hormone. [12]

The next series of experiments showed that TX, at therapeutic concentrations, suppresssed significantly the proliferative response of rat spleen cells to PHA, concanavalin A (Con A), PWM and lipid A (LA), which is the active moiety of bacterial endotoxin. [13] The mixed lymphocyte reaction was similarly suppressed. [14]

The effect of agents acting on protein kinase C (PKC), calmodulin (CM) and calcium (Ca^{++}) was investigated on the mitogenic response of rat spleen cells in conjunction with TX. The PKC activator, phorbol 12-myristate 13-acetate (PMA) significantly inhibited the mitogenic response to PWM and Con A, had no effect on the response to PHA and enhanced markedly the response to LA. Chlorpromazine (CP), which is a PKC inhibitor, had no significant influence on mitogenesis. The calmodulin inhibitor, R24571, tended to enhance the PHA response and had no effect on the responses to other mitogens. The calcium ionophore, A23187, was mitogenic on its own, enhanced significantly the LA response and tended to inhibit the responses to PWM, Con A and PHA. TX inhibited significantly the response to PWM, Con A, PHA and LA, but had no influence on the mitogenic response to A23187. When TX was applied jointly with the mitogens and the PKC, CM or Ca^{++} modulating agents, the overall response was lowered by TX, but the pattern was identical to those obtained without TX. The intracellular Ca^{++} chelating agent, TMB-8, the calcium channel blocking agents, verapamil and nifedipine, and the extracellular Ca^{++} chelating agent EGTA were strong inhibitors of mitogenesis and synergized with TX. Excess Ca^{++} placed into the culture medium antagonized the inhibitory effect of TX on T lymphocyte mitogenesis, but had no influence on the inhibitory effect of this drug on B cell proliferation. Additional experiments revealed that treatment of spleen cells with TX led to an elevation of cytoplasmic free Ca^{++}, as measured by the fluorescent dye, fura-2. When Con-A stimulated cells were treated with TX, the mitogen-induced increase of cytoplasmic Ca^{++} was further elevated by the drug. Neither E2, nor verapamil caused similar alterations of Ca^{++} in normal or Con A treated rat spleen cells. Pretreatment of spleen cells with TX, E2 or VE for 4 hours prior to Con-A stimulation had no influence on the extent of mitogen-induced Ca^{++} elevation. These results do not support the idea that TX inhibits mitogenesis by acting on PKC or CM. It appears that a Ca^{++} dependent negative signal is delivered by TX, which is capable of inhibiting lymphocyte mitogenesis. The results also indicate that various signalling pathways are involved in the triggering of lymphoproliferation by the mitogens studied (unpublished).

Webster et al.[10] found no alterations in the PHA response of ovariectomized and TX treated patients with advanced breast cancer. Joensuu et al.[11] did not observe consistent changes in the response of peripheral blood lymphocytes to PHA and Con A of 10 patients studied prior to and 3, 6 and 12 months after starting TX treatment. However, it is well known that the response of peripheral blood lymphocytes to mitogens is subject to considerable variations over time even in the case of healthy donors, which makes it virtually impossible to observe drug induced changes that may result in the decrease, but not the complete elimination, of the response. Other reasons for this discrepancy between the *in vitro* and *in vivo* observations are also possible.

We observed that the development of contact sensitivity reactions in rats can be inhibited by treatment with TX in a dose dependent manner. Additional treatment with either prolactin or growth hormone reversed this immunosuppressive effect of TX.[12] These observations are compatible with the antiproliferative effect of TX in the mixed lymphocyte reaction *in vitro*.[14] Rotstein et al.[15] studied 23 breast cancer patients which were treated with TX for 1.5-2 years. A significantly lower natural killer (NK) activity was found against K562 target cells. In contrast, Berry et al.[16] found in 17 postmenopausal breast carcinoma patients treated with TX for 1 month that NK activity was significantly increased.

That TX is capable of enhancing cell mediated cytotoxicity *in vitro* was first suggested by Mandeville et al.[17] who performed NK mediated cytotoxicity assays with human peripheral blood lymphocytes using the K562 erythroleukemic cell line as target. When TX was present during the assay, significant enhancement of cytotoxicity was observed by TX from 8 different donors, 10^{-7} and 10^{-8} M concentrations being effective. TX had no cytotoxic effect on the K562 target cells. An additive effect in the augmentation of cytotoxicity was observed when the cells were exposed simultaneously to human leukocyte interferon and TX.

We isolated peripheral blood lymphocytes from 20 patients, 18 with various types of carcinomas and 2 with malignant mesenchymal tumors. Autologous tumor cells and the K562 cell line were used as targets in the ^{51}Cr release assay. Aliquots of the lymphocytes were preincubated for 15 h with various concentrations of TX (50, 100, 200, 400 ng/ml), washed and examined for target cell destruction. Untreated peripheral blood lymphocytes of 4 patients lysed autologous tumor cells. After TX treatment, 8 patients' lymphocytes lysed autologous targets. TX induced autotumor lysis in lymphocytes of 6 patients, enhanced cytotoxicity in 1, and inhibited existing cytotoxicity in 2. The optimal concentration for enhancing cytotoxicity was 200 ng/ml in 5 cases, 400 ng/ml in 2 cases, and 50 ng/ml in 1 case. High density small lymphocytes which showed specific cytotoxicity for autologous tumor cells, but did not kill K562 targets, were also studied. TX induced autotumor lysis in one of 5 cases.[5]

We also studied the effect of E2 and TX on NK cell activity of rat spleen cells using the YAC-1 murine lymphoma as target. TX enhanced and E2 inhibited cell killing if placed into cultures for the duration (5 h) of the cytotoxic reaction. In this system E2 interfered with the enhancing effect of TX if applied jointly. Pretreatment of the target cells with TX led to highly significant enhancement of cytotoxicity. E2 also enhanced target cell killing but less effectively. When target cells were treated

Figure 1. Amplification of cell mediated cytolysis by estradiol and tamoxifen.

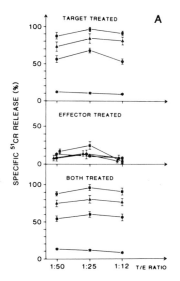

Figure 1A. Natural killer activity of rat spleen cells on the Yac-1 target. TX and E2 were used at 100 nM concentrations each and the duration of pretreatment was 4 h of target, effector cells or both, as indicated in the figure. • Control • TX ▲ E2 ■ TX+E2 ⊤ S.E.

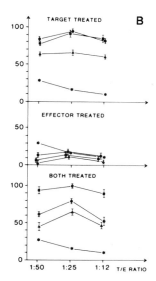

Figure 1B. Cytotoxic reaction with NK cells exposed to human recombinant IL-2 for 48 h. TX and E2 were used at 1 μM concentrations. The duration of treatment was 4 h prior to the cytotoxic reaction.

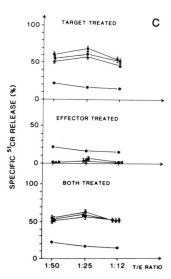

Figure 1C. Lymphokine activated killer activity of rat spleen cells treated with human recombinant IL-2 for 6 days. P815 were used as targets. TX and E2 were used at 100 nM concentrations for pretreatment for 4 h.

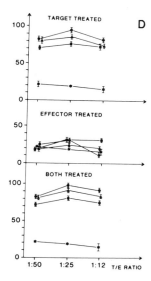

Figure 1D. Target cell killing by cytotoxic T lymphocytes induced in mixed lymphocyte reactions for 5 days. TX and E2 were used in 100 nM concentrations for 4 h treatments prior to cytotoxic reactions. The concentrations of TX and E2 used did not alter the background and maxiumum release of ^{51}Cr in any of the above systems.

with TX+E2, the level of cytotoxicity was comparable to the one obtained with TX alone (Fig. 1A). These effects on target cells were detectable at pharmacological (1 μM) and physiological (1 nM) E2 and equimolar TX concentrations. Pretreatment of effector spleen cells with TX enhanced their cytotoxic potential in some combinations but was inhibitory in others. Pharmacological levels of estradiol inhibited effector cells when applied alone, or in combination with TX. Highly significant enhancements of target cell destruction occurred if both target and effector cells were pretreated with TX (Fig. 1A). Estradiol treatment of both cell types resulted in slight enhancement or no effect on cytotoxicity. Similar results were obtained if the spleen cells were pretreated with IL-2 for 48 h. In this case the treatment of both the targets and effectors with TX and E2 for 4 h led to superior cytotoxicity compared to what was achieved by TX treatment alone (Fig. 1B). Both E2 and TX changed the kinetics of ^{3}H-thymidine incorporation by YAC-1 cells, but the cells were capable of growing in the presence of drug concentrations (1 μM) used in the cytotoxicity experiments. YAC-1 cells have no cytosolic E2 receptors, and are weakly positive for cytosolic progesterone receptors.

The effect of TX and E2 was also examined on IL-2 activated killer (LAK) cell mediated cytotoxicity using spleen cells of Fisher 344 rats as the source of effectors and the P815 murine mastocytoma cells as targets. Treatment of the target cells with either TX or E2 for 4 or 18 h rendered them highly sensitive to LAK cell mediated lysis. TX was more effective in this regard than was E2. When joint treatment was applied, cytotoxicity remained at the level of TX treatment (Fig. 1C). The cytotoxic potential of IL-2 primed LAK cells was also modified by TX and E2 but to a lesser extent. Enhancement and inhibition of cytotoxic potential could be observed in certain experimental designs. When drug treated target and effector cells were combined, high cytotoxicity characteristic of sensitized target cells was observed in most experiments. Evidence of synergism and of inhibition of cytotoxicity was also seen in some of these experiments. Target cells could be sensitized for LAK cell mediated destruction by physiological concentration (1 nM) of E2 and equimolar concentrations of TX. Neither E2 nor TX exerted a direct cytotoxic effect on P815 cells. Cytosol preparations of P815 cells had no receptors for E2 or progesterone.

The effect of E2 and TX was studied on the lytic activity of cytotoxic T lymphocytes (CTL) generated in mixed cultures of rat spleen cells, using female Fisher 344 cells as responders and female Wistar rat cells as stimulators. Con A stimulated Wistar lymphoblasts were used as target cells. CTL harvested on day 5 exerted 16-25% cytotoxicity when used at 1:12 - 1:50 target:effector cell ratios. Day 6 CTL had no cytotoxic activity. Treatment of target cells with either TX or E2, or both, at 1 μM concentrations for 4 h prior to cytotoxicity testing raised the target cell killing to 100%. Highly significant enhancements of cytotoxicity were observed also when the drugs were used at 100-, 10-, or 1 nM concentrations. Treatment of effector cells under similar conditions led to inhibition of cytotoxicity at 1 μM concentration, some enhancement at 100 nM and no effect at 10 and 1 nM. When treated target and effector cells were combined, the amplification of target cell lysis was similar in magnitude to tests with treated targets only (Fig. 1D). These results illustrate that CTL mediated cytotoxicity is amplified by physiological concentrations (1 nM) of E2 and equimolar concentrations of TX. Since even the highest concentration (1 μM) of TX tested can be obtained during cancer therapy, [1] it is possible that tumor cell sensitization to cell mediated killing plays a role in the beneficial effects of this drug.

These results indicate that immune cytolysis mediated by NK, LAK, or CTL effector cells is amplified by TX and, to a lesser extent, E2. This amplification is due to the sensitization of the target cells to immune cytolysis, whereas the treatment of effector cells by either of these agents is usually, but not always, inhibitory. When drug treated target and effector cells are combined for cytotoxic reactions, the rule is enhanced cytotoxicity for TX without exception, and usually for E2. This is also true when the cells are treated jointly with both TX and E2. None of the target cells (Yac-1, P815, Nb2) studied so far expressed conventional estrogen receptors, yet physiological concentrations of E2 were capable of sensitizing these cells for immune cytolysis. Comparable concentrations of TX were also effective. Clearly, the concentrations used are nontoxic. According to our latest observations, the sensitizing effect is dependent on active nucleic acid and protein synthesis as pretreatment of the target cells with actinomycin D or cycloheximide, intereferes with subsequent sensitization.

Screpanti et al.[18] also observed that estradiol enhances the natural killer cell susceptibility of human breast cancer cells. Treatment of the MCF-7 cell line with 50 nM estradiol for 24 h increased significantly the susceptibility of these cells to lysis by IL-2 activated or non-activated human peripheral blood lymphocytes. This increase in susceptibility reached the maximum after 3 days and was maintained throughout 10 days of treatment. A monoclonal antibody specific for CD16 abrogated completely the cytotoxic activity of effector cells against MCF-7 and K562 indicating that NK-type cells were involved. Estradiol had no effect on the susceptibility of the estrogen receptor negative breast cancer cell line BT-20 to NK cell mediated lysis.

Our results suggest that treatment of tumor bearing hosts with TX could enhance host defence by amplifying the cell mediated killing of tumor cells. Indeed, it has been observed by Kim and coworkers[19] that TX potentiates *in vivo* the antitumor activity of IL-2 in C57BL/6 mice. A weakly immunogenic fibrosarcoma, MCA-106, was injected i.v. and some of the animals received 50,000 units of IL-2 twice a day i.p. from days 3-12. Some IL-2 treated and non-treated mice also received 2 μg/ml of TX in the drinking water which was supplied *ad libitum* leading to an average daily intake of TX at 10 μg/mouse/day. At termination of the experiment on day 18 the number of pulmonary tumor nodules were reduced by 66% in the IL-2 treated group, by 32% in the TX treated group, and by 95% in the IL-2 plus TX treated animals.

In addition, there is evidence that TX antagonized the effect of E2 in the guinea pig thymus.[20] E2, estriol, and a structural estrogen analogue with minimal estrogenic activity, 1,3,5(10)-estratrien-3,16β-diol, enhanced the clearance of IgG coated erythrocytes from the circulation of guinea pigs, which could be partially inhibited by TX. On the other hand, estradiol in physiological concentrations increased the number of monocyte colonies developed from peripheral blood leukocytes of male and female donors and TX did not inhibit this effect.[21] We observed that TX upregulated the production of TNF α and downregulated IL-1 production in lipid A stimulated rat peritoneal macrophages (unpublished).

The incubation of peripheral blood monocytes from patients with breast cancer under agarose for 6 days at 37°C in 5% CO_2 results in giant cell formation. This reaction is significantly inhibited if TX is placed into the culture medium. Furthermore, the incubation of monocytes from mastectomy patients treated with TX for 3 months, showed a significant reduction in the number of giant cells compared to the samples tested before the commencement of treatment.[22] TX blocked completely the tumor promoter induced H_2O_2 production by human neutrophils in a dose dependent fashion.[23]

A case of autoimmune progesterone dermatitis did not respond to estrogen, but marked improvement occurred after TX treatment.[24] The occurrence of purpuric vasculitis has been described during TX therapy. Withdrawal of TX resulted in complete clearing of the lesions. Histological examination suggested that the vasculitis was immune complex mediated.

DISCUSSION

It is clear from the results presented in Figure 1 that the kinetics of the cytotoxic reaction are altered by TX and/or E2 treatment in that the amplifying effect was frequently strongest at 1:25 target effector ratios, not at 1:50. On the other hand, in control experiments, there was a linear decrease of isotope release in direct correlation with the decreasing number of effector cells. A possible explanation for this alteration of kinetics by the drugs used is that sensitization is an active process which requires nucleic acid and protein synthesis. It is likely that at high effector:target cell ratios a significant proportion of the target cells are damaged excessively, and for this reason the sensitizing effect is either undetectable or, perhaps, cannot take place as the metabolic pathways required for its manifestation are impaired. Our results obtained with actinomycin-D and cycloheximide suggest that sensitization by TX and E2 is indeed mediated by active signalling, which may be similar to programming for apoptosis. Indeed, the observation that TX is capable of inducing Ca^{++} influx in normal rat spleen cells is compatible with the hypothesis that active signalling might be involved. However, the Ca^{++} channel blocking agent, verapamil, did not induce Ca^{++} influx in normal spleen cells, yet was capable of inhibiting mitosis as well as sensitizing for cytotoxicity. This suggests that there might be more than one mechanism for target cell sensitization, one being through metabolic pathways, whereas the other might simply be mediated by the functional modulation of ion channels. Since evidence is increasing that steroid hormones and TX do influence ionic channels,[25-27] it is possible that both of the above pathways are involved in the sensitizing process by TX.

Killer cells, regardless of their type, secrete proteins (e.g. cytolysins, perforins) that are capable of imbedding into the target cell membrane through their hydrophobic portions, and after calcium dependent polymerization, create micropores that are highly permeable for ions and small molecules. Such channels have the capacity to upset the osmotic environment in the cytoplasm and cause lysis.[28,29] Recent evidence also indicates that enzymes capable of degrading DNA (fragmentin) are also produced by NK cells.[30] Cytotoxic T lymphocytes are capable of inducing programmed cell death, which requires the active participation of the target cell.[31] That nucleated cells may be capable of defending themselves against cytotoxic insult, by altering the function of ionic channels, is suggested by the observation that agents capable of blocking such channels enhance cytolysis.[32] Thus, it is conceivable that TX and E2 somehow interfere with the membrane changes necessary for defence against cytotoxicity.

Cytotoxic T lymphocytes are known to express classical estrogen receptors.[33,34] Therefore, it is possible that, whereas the target cells are sensitized via a putative membrane bound receptor, the effector cells are influenced by these agents via the classical estrogen receptor pathway. This hypothesis is supported by the fact that TX and E2 are not antagonistic with regards to target cell sensitization. Antagonism has been observed, however, with regards to their effect on NK cells and also on a variety of macrophage functions as pointed out earlier.

The use of cytokines (interleukins, interferon) and of various killer cells in cancer therapy is gradually gaining ground. It is conceivable that the beneficial effect of these immunostimulatory agents and of killer cells could be further amplified by agents, such as TX, which are capable of sensitizing the tumor cells to immune cytolysis. Some preliminary observations made *in vivo* lend support to this hypothesis. If feasible, this approach could advance significantly the success rate of immunological treatment of cancer. Moreover, the pharmacological enhancement of target cell lysis may also gain practical applications in the treatment of other diseases in which cell mediated immunity is fundamental to host defence.

ACKNOWLEDGEMENTS

The work discussed in this paper was supported in part by the Manitoba Health Research Council, The Zellers Foundation and the Manitoba Cancer Research and Treatment Foundation.

REFERENCES

1. J.S. Patterson and L.A. Battersby, Tamoxifen: an overview of recent studies in the field of oncology, *Cancer Treat. Rep.* 64:775 (1980).
2. L.J. Lerner and V.C. Jordan, Development of antiestrogens and their use in breast cancer: eighth Cain memorial award lecture, *Cancer Res.* 50:4177 (1990).
3. H. Rochefort, Nonsteroidal antiestrogens are estrogen-receptor-targeted growh inhibitors that can act in the absence of estrogens, *Hormone Res.* 28:196 (1987).
4. M. Baum, Novaldex Adjuvant Trial Organisation: Controlled trial of tamoxifen as single adjuvant agent in management of early breast cancer; analysis at six years, *Lancet* 1:836 (1985).
5. E. Baral and F. Vanky, Effect of tamoxifen on the cell-mediated autotumor lysis, *J. Clin. Lab. Immunol.* 22:97 (1987).
6. C.L. Vogel, D.R. East, W. Voigt and S. Thomsen, Response to tamoxifen in estrogen receptor-poor metastatic breast cancer, *Cancer* 1987; 60:1184 (1987).
7. E. Baral, H. Blomgren, S. Rotstein and L. Virving, Antiestrogen effects on human blood lymphocyte subpopulations in vitro, *J. Clin. Lab. Immunol.* 1985; 17:33 (1985).
8. E. Baral, H. Blomgren, J. Wasserman, S. Rotstein and L.V. von Stedingk, Effect of tamoxifen on pokeweed mitogen stimulated immunoglobulin secretion in vitro, *J. Clin. Lab. Immunol.* 21:137 (1986).
9. T. Paavonen and L.C. Andersson, The estrogen antagonists, Tamoxifen and FC-1157a, display oestrogen like effects on human lymphocyte functions in vitro, *Clin. Exp. Immunol.* 61:467 (1985).
10. D.J.T. Webster, G. Richardson, M. Baum, T. Priestman and L.E. Hughes, Effect of treatment on the immunological status of women with advanced breast cancer, *Brit. J. Cancer* 39:676 (1979).
11. H. Joensuu, A. Toivanen and E. Nordman, Effect of tamoxifen on immune functions, *Cancer Treat. Rep.* 70:381 (1986).
12. E. Nagy and I. Berczi, Immunomodulation by tamoxifen and pergolide, *Immunopharmacology* 12:145 (1986).
13. E. Baral, S. Kwok and I. Berczi, Suppression of lymphocyte mitogenesis by tamoxifen, *Immunopharmacology* 18:57 (1989).
14. E. Baral, S. Kwok and I. Berczi, The influence of estradiol and tamoxifen on the mixed lymphocyte reaction in rats, *Endocrinology* 146:2776 (1991).
15. S. Rotstein, H. Blomgren, B. Petrini, J. Wasserman and L.V. Von Stedingk, Influence of adjuvant tamoxifen on blood lymphocytes, *Breast Cancer Res. Treat.* 12:75 (1988).
16. J. Berry, B.J. Green and D.S. Matheson, Modulation of natural killer cell activity by tamoxifen in stage I post-menopausal breast cancer, *Eur. J. Cancer Clin. Oncol.* 23:517 (1987).
17. R. Mandeville, S.S. Ghalli and J.-P. Chausseau, In vitro stimulation of human NK activity by an estrogen antagonist (tamoxifen), *Eur. J. Cancer Clin. Oncol.* 20:983 (1984).
18. I. Screpanti, M.P. Felli, E. Toniato, D. Meco, S. Martinotti, L. Frati, A. Santoni and A. Gulino, Enhancement of natural killer-cell susceptibility of human breast-cancer cells by estradiol and v-Ha-ras oncogene, *Int. J. Cancer* 47:445 (1991).
19. B. Kim, P. Warnaka and C. Konrad, Tamoxifen potentiates in vivo antitumor activity of interleukin-2, *Surgery* 108:139 (1990).
20. L.J. Brandes, L.M. MacDonald and R.P. Bogdanovic, Evidence that the antiestrogen binding site is a histamine-like receptor, *Biochem. Biophys. Res. Commun.* 126:905 (1985).
21. H. Maoz, N. Kaiser, M. Halimi, V. Barak, A. Haimovitz, D. Weinstein, A. Simon, S. Yagel, S. Biran and A.J. Treves, The effect of estradiol on human myelomonocytic cells. 1. Enhancement of colony formation, *J. Reprod. Immunol.* 7:325 (1985).
22. A.M. Al-Sumidaie, The effect of tamoxifen and medroxyprogesterone on giant cell formation by monocytes from patients with breast cancer, *J. Cancer Res. Clin. Oncol.* 114:399 (1988).
23. W. Troll and J.S. Lim, Tamoxifen suppresses tumor promoter-induced hydrogen peroxide in human neutrophils, *Proc. Am. Assn. Cancer Res.* 32:149 (1991). (Abstract #891)
24. C.J.M. Stephens, F.T. Wojnarowska and J.D. Wilkinson, Autoimmune progesterone dermatitis responding to Tamoxifen, *Brit. J. Dermatol.* 121:135 (1989).

25. I. Nemere and A.W. Norman AW, Steroid hormone actions at the plasma membrane: induced calcium uptake and exocytotic events, *Mol. Cell. Endocrinol.* 80:C165 (1991).

26. P. Sartor, P. Vacher, P. Mollard and B. Dufy, Tamoxifen reduces calcium currents in a clonal pituitary cell line, *Endocrinology* 123:534 (1988).

27. D.A. Greenberg, C.L. Carpenter and R.O. Messing, Calcium channel antagonist properties of the antineoplastic antiestrogen tamoxifen in the PC12 neurosecretory cell line, *Cancer Res.* 47:70 (1987).

28. P.A. Henkart, Mechanism of lymphocyte mediated cytotoxicity, *Ann. Rev. Immunol.* 3:31 (1985).

29. J.D.E. Young, Killing of target cells by lymphocytes: a mechanistic view, *Physiol. Rev.* 69:250 (1989).

30. L. Shi, R.P. Kraut, R. Aebersold and A.H. Greenberg, A natural killer cell granule protein that induces DNA fragmentation and apoptosis, *J. Exp. Med.* 175:553 (1992).

31. A. Zychlinsky, L.M. Zheng, C.-C. Liu and J.D. Young, Cytotoxic lymphocytes induce both apoptosis and necrosis in target cells, *J. Immunol.* 146:393 (1991).

32. R.P. Kraut, D. Bose, E.J. Cragol Jr. and A.H. Greenberg, The influence of calcium, sodium and the Na^+/Ca^{2+} antiport on susceptibility to cytolysin/perforin-mediated cytolysis, *J. Immunol.* 144:3498 (1990).

33. J.H.M. Cohen, L. Danel, G. Cordier, S. Saez and J.P. Revillard, Sex steroid receptors and restriction of estrogen receptors to OKT8 positive cells, *J. Immunol.* 131:2767 (1983).

34. W.H. Stimson, Oestrogen and human T lymphocytes: presence of specific receptors in the T-suppressor/cytotoxic subset, *Scand. J. Immunol.* 28:345 (1988).

ENDOCRINE-IMMUNE INTERACTIONS IN PITUITARY PATHOLOGY

Sylvia L. Asa and Kalman Kovacs

Department of Pathology
Mount Sinai Hospital and St. Michael's Hospital
University of Toronto
Toronto, Ontario, Canada

The first observations of endocrine-immune interactions were made by pathologists studying mechanisms of disease at the beginning of the twentieth century. They noted the association of thymic involution with Cushing's syndrome and thymic hyperplasia associated with acromegaly, Graves' and Addison's diseases.

It has now become evident that a number of pituitary hormones affect immune function, for example, growth hormone (GH) and prolactin (PRL) enhance immunocompetence and adrenocorticotropin (ACTH) suppresses the immune response. It is also clear that immune factors play a role in the regulation of pituitary function; a number of cytokines have been implicated as stimuli or inhibitors of adenohypophysial hormone release.[1] This complex interaction between the immune and endocrine systems clearly has relevance to human disease: it can have pathogenetic and/or therapeutic relevance in allergic and infectious conditions, in autoimmune disorders and even in neoplasia. In this chapter, we will briefly review the information available to date on the advances in this field concerning pituitary pathology.

PITUITARY HORMONES AND IMMUNE FUNCTION

Pituitary GH and PRL are known to enhance immunocompetence.[2,3] These hormones appear to act in concert and lack of one or the other, such as in pituitary dwarfs with isolated GH deficiency, the immune system does not appear to compromise. However, hypophysectomized animals who lack both hormones are symptomatically immunodeficient.[4-10] Recently, a rare disorder in which both GH and PRL are absent has been described in patients with deficiency of the pituitary transcription factor Pit-1 that regulates transcription of both hormones;[11-14] it remains to be determined if these patients exhibit evidence of immunodeficiency.[13]

GH affects lymphocyte function both directly and indirectly by stimulating the production of insulin-like growth factors (IGFs). Activated lymphocytes have IGF

Advances in Psychoneuroimmunology, Edited by
I. Berczi and J. Szélenyi, Plenum Press, New York, 1994

receptors and some lymphocyte cell lines express GH receptors;[15-18] under certain experimental conditions, a mitogenic effect of GH on lymphocytes can be demonstrated.[15] There is even evidence that mononuclear leukocytes produce GH which can act in a paracrine fashion.[19,20] The trophic role of the hormone is evidenced in hypophysectomized animals; they exhibit thymic and lymphoid atrophy and a reduced immune response to stimulation, which are reversed by GH administration.[2,16]

The role of PRL in immunoregulation was proven by the restoration of immunocompetence in hypophysectomized animals after PRL administration;[5-9,21] bromocriptine, a dopamine agonist that inhibits PRL synthesis and release, ameliorates autoimmune responses in experimentally induced inflammation.[22] PRL binds to specific receptors on lymphocytes and stimulates cytokine release as well as proliferation in vitro;[3] the latter has provided a sensitive and specific bioassay of prolactin using the Nb2 rat lymphoma cell line.[15,23]

ACTH and its related derivatives of proopiomelanocortin (POMC), including ß-endorphin and alpha melanocyte-stimulating hormone (α-MSH), are also immunomodulatory.[2] The primary effects of ACTH, both direct and by stimulation of glucocorticoid production, are immunosuppressive; ACTH in high concentrations inhibits the splenic antibody response and production of interferon (IFN).[24] However, it also enhances B cell growth and differentiation and increases immunoglobulin synthesis.[24] α-MSH is a potent antagonist of interleukin-1 (IL-1)-stimulated pyrogenicity and thymocyte proliferation, but the effects are apparently not mediated by classic α-MSH receptors.[2] Human leukocytes and lymphocytes have opiate receptors and cultured lymphocytes express receptors that bind ß-endorphin but are not typical opiate receptors;[24] ß-endorphin stimulates migration of neutrophils and mononuclear cell chemotaxis but inhibits generation of lymphocyte chemotactic factors from stimulated mononuclear cells.[24] Opioid peptides also increase the production of interferons by mononuclear cells.[24] It must be noted, however, that the immunoregulatory roles of POMC-derived peptides may be mediated either directly through the hypothalamic-pituitary-adrenal (HPA) axis or in a paracrine/autocrine fashion, since POMC is reported to be synthesized by leukocytes themselves[24] and CRH is produced in peripheral inflammatory sites.[25] The systemic importance of this ACTH remains dubious,[26] however a single case report of Cushing's syndrome attributable to ectopic ACTH production by granulomatous inflammation[27] has raised the possibility that this may be significant.

Pathological Applications

Since the pituitary hormones GH and PRL enhance immunocompetence, it follows that patients with hormone deficiency or excess may exhibit dysregulation of their immune systems. As indicated above, deficiency of GH or PRL alone would not be expected to manifest in immundeficiency and pituitary dwarfs are not detectably immunocompromised. Lack of both hormones is extremely rare and has only recently been recognized; further studies will clarify the immune status of patients with this condition.

The immunologic effects of GH or PRL excess also remain uncertain. Patients with pituitary adenomas producing GH or PRL have normal immunoglobulins, leukocyte subsets, natural killer cell activity, serum IL-1 and IL-3, IL-2 receptors and IFN-gamma; IL-6 and TNFα have been reported to be slightly elevated in a subgroup of acromegalics but no clinical alterations were detected in these patients.[28] It may be that the enhancing effects of these hormones might lead to an increased tendency to autoimmune phenomena,[3] but this remains speculative.

Detailed studies of the immune systems of these patients and of animal models, such as transgenic mice who overexpress GH,[29] remain to be performed, but certainly will elucidate the true significance of these hormones as modulators of immunity.

The effects of ACTH excess in patients with Cushing's disease are well recognized; these patients exhibit full blown immunosuppression that is mainly the result of the marked glucocorticoid excess. Deficiency of ACTH theoretically would lead to a reduced immunological surveillance. This has proven true of a strain of Lewis rats with a genetic deficiency of corticotropin-releasing hormone (CRH) that results in reduced ACTH and adrenocortical function; these animals are overly susceptible to the development of experimentally-induced arthritis.[30] A model of glucocorticoid insufficiency that has been studied is the obese strain (OS) of chickens; these animals have elevated glucocorticoid binding globulins that interfere with the regulation of the HPA axis and they exhibit cyclic hematopoiesis as well as autoimmune thyroiditis.[31] However, there is as yet no model of isolated ACTH deficiency.

The clinical importance of the HPA axis in the development of autoimmunity has been suggested by the finding that patients with rheumatoid arthritis have a less pronounced diurnal variation of cortisol and less marked ACTH and adrenal responses to the stress of surgery compared to those with osteoarthritis.[32]

The effects of POMC-derived peptides are also seen in other pathological situations. These substances modulate fever induced by IL-1, IL-6 and TNF; plasma α-MSH is higher in AIDS patients than in normal controls and the level of elevation of α-MSH correlates with survival. It has therefore been suggested that α-MSH reduces cytokines which stimulate HIV expression in certain cells, modifying the progression of the disease in these patients.[33]

In situations of stress, CRH is thought to increase the release of ACTH, peripheral blood IL-1, IL-2, thereby mediating both immune and endocrine responses. This inhibition of neuroendocrine-immune function is now thought to be the mechanism by which the psychological state of an individual can alter immune surveillance, giving rise to immunosuppression and even possibly altering the development and course of cancer.[34]

IMMUNE MEDIATORS AND PITUITARY FUNCTION

The immune system is able to communicate information regarding the immune status to the brain and endocrine system. In the pituitary, this is largely due to the effects of immune mediators on hormonal regulation.[35] For example, IL-1α and β have been reported to stimulate GH, ACTH, TSH and LH release and to inhibit PRL.[36] IL-2 may stimulate ACTH. IL-6 has been shown to stimulate GH, PRL, LH and ACTH. TNF also stimulates ACTH and LH release; its reported effects on PRL are controversial.[37] IFN-gamma has been shown to stimulate ACTH release by a pituitary adenoma but to inhibit ACTH, GH and PRL release in the nontumorous gland.[37,38] IFNα-2 may stimulate[39] or suppress[40] ACTH and cortisol and suppresses TSH *in vivo*.[40] In a single report, thymosin-5 and G-CSF also stimulated ACTH release by two human pituitary corticotroph adenomas.[41] These data suggest that there is an intricate mechanism to ensure neuroendocrine modulation of the immune response. In addition, cytokines have been identified as products of the hypothalamus and pituitary,[34,42] where they likely have regulatory functions that may be involved in pathological processes.

Interleukin-1

IL-1 is able to directly stimulate GH release and inhibit PRL secretion *in vitro*;[36,43] *in vivo*, stimulation of GH release is provoked by low doses of IL-1 and inhibition is found with high doses.[44] IL-1 reduces pulsatile GH release *in vivo*. These data are complicated by the discrepant data concerning the effects of IL-1 on the hypothalamus; IL-1 can stimulate[44] or inhibit[45] hypothalamic GRH and stimulates hypothalamic somatostatin synthesis but possibly not its release.[46]

IL-1 has been shown to increase pituitary ACTH *in vivo*;[47,48] the effect is thought to be both direct[36,43,49] and via CRH release from the hypothalamus.[48,50-53] IL-1 receptors are localized to cells of the pituitary gland and are present on the rat-derived AtT-20 corticotroph cell line[54,55] and a direct effect of IL-1α and ß on ACTH secretion was observed in cultured rat pituitary cells,[43] AtT-20 cells[49] as well as in human pituitary adenomas;[41] IL-1 receptors on AtT-20 cells are upregulated by CRH,[56] indicating the complex interactions between CRH and IL-1 at the pituitary level. It has been suggested that IL-1 stimulation may play a role in the development of Cushing's disease. The stimulation of hormone release was not associated with a detectable alteration in the proliferation of AtT-20 cells.[49]

IL-1 inhibits LH release in ovariectomized monkeys but not after ovarian steroid replacement.[57] It inhibits the hypothalamic-pituitary-gonadal axis centrally with reduced c-*fos* expression in GnRH neurons[58] and by inhibiting release of GnRH, but the data have been obtained *in vivo* where the role of CRH, endogenous opioids, estrogen and progesterone is unclear. *In vitro*, it has been shown to stimulate release of LH and TSH.[36,43]

Interleukin-2

IL-2 inhibits GH release *in vitro*;[59] this may be attributed to stimulation of somatostatin release or alternatively to its inhibition of stimulated GRH release (60); IL-2 does not alter basal GRH levels.[60] It also has been shown to reduce stimulated somatostatin release. It stimulates basal PRL and TSH release but suppresses basal LH and FSH release.[59] IL-2 receptors have been detected on normal somatotrophs and lactotrophs[61] and on the GH3 cell line and IL-2 stimulates growth of that cell line.[61]

IL-2 stimulation of ACTH release is controversial[48,59] and no alteration of ACTH synthesis or release has been detected in IL-2-exposed AtT-20 cells. However, IL-2 receptors have been localized on normal rat corticotrophs.

Interleukin-6

IL-6 stimulates ACTH and the HPA axis;[48] one paper reporting stimulation of ACTH release by AtT-20 cells suggests that the effect is direct,[49] but most agree that the main action is at the hypothalamic level and is mediated via CRH.[62] There appears to be a carefully regulated feedback loop: glucocorticoids reduce IL-6 mRNA levels and secretion and ACTH, α-MSH suppress IL-6-induced CRH release.[63]

IL-6 stimulates GH and PRL release *in vivo* and directly *in vitro*.[64] In addition, IL-6 potentiates GRH- and TRH-stimulation respectively.[64] The direct nature of the effect on PRL is substantiated by the fact that it can be attenuated by dopamine in perifused isolated pituitary cells.[64] A dose responsive effect on cell proliferation has also been demonstrated with GH3 cells,[61] However, perifusion experiments have disputed this finding. IL-6 also stimulates LH release *in vitro* but little further is understood about the regulation of this action.

Tumor Necrosis Factor-α (TNF-α)

TNF-α has a direct inhibitory effect on basal and GRH-stimulated GH release as well as on basal and TRH-stimulated PRL release;[37] a direct effect of TNF-α on ACTH has been demonstrated but remains controversial[48,50,65] and it appears that it may rather be mediated centrally. Addition of recombinant TNF-α to cultures of pituitary cells results in stimulation of gonadotropin release.[65] Interestingly, TNF-α suppresses basal TSH release but enhances TRH-stimulated TSH release; it has been proposed that it plays a role in mediating the altered TSH secretion seen in nonthyroidal illness known as the "sick euthyroid syndrome".

PRODUCTION OF INTERLEUKINS BY PITUITARY CELLS

As indicated above, pituitary hormones affect immune function and may even be produced by immunocompetent cells. Likewise, immune regulators can alter hormone synthesis and release and cytokines may also be synthesized in the hypothalamus and pituitary[42] where they may have important paracrine and/or autocrine effects.

IL-6 is the cytokine that has been implicated as the most abundant and perhaps the most important immune regulator of pituitary function. IL-6 is synthesized by the rat pituitary;[66] its mRNA is localized in adenohypophysial cells and in pars nervosa.[67] It is released from rat adenohypophysis *in vitro*[67] and its release is stimulated by activators of adenylate cyclase, such as forskolin, phorbol esters, lipopolysaccharide, and by IL-1β, thought to act via the arachadonic acid pathway, and is inhibited by dexamethasone.[67,68]

The exact cell types responsible for the synthesis and release of IL-6 are still unclear. Folliculostellate cells have been implicated by the classic studies of Denef et al.[69] This cytokine is produced by aggregate cultures of pituitary cells enriched in folliculostellate cells but not in cultures devoid of that cell type;[69] the amount correlates with presence and number of folliculostellate cells.[69] A putative pituitary folliculostellate cell line, TtT/GF, secretes IL-6 in a regulated fashion.[70] However, the other adenohypophysial cell types have also been implicated. The source of IL-6 mRNA in the posterior pituitary is uncertain, but IL-6 is known to be produced by the hypothalamus.[34,71]

Pathological Applications

Interleukins and Pituitary Adenomas. IL-6 is produced by pituitary adenomas;[67,72,73] while some studies have suggested that IL-6 and its receptor are selectively expressed in ACTH and in GH-producing adenomas,[73] others have found that IL-6 expression is not restricted to any specific tumor type and its mRNA has been localized in pituitary adenomas producing ACTH, GH, PRL and in clinically nonfunctioning adenomas, many of which produce gonadotropins.[72,74]

Recent studies have also suggested that IL-2 is produced by pituitary adenomas. IL-2 mRNA has been detected in cultured human corticotroph adenomas and increases after stimulation with phorbol esters but not with CRH.[75]

These data suggest that these cytokines may play a paracrine or autocrine role in pathogenesis of pituitary tumors. However, it remains to be seen whether human pituitary adenomas respond to cytokines in the same fashion as nontumorous rat pituitary cells do. Our preliminary data suggest that IL-1ß and IL-6 exert no direct effect on hormone release *in vitro* by human pituitary adenomas of any type.[76]

Pituitary Autoimmunity. Interleukins may also play a role in the genesis of pituitary autoimmunity. Lymphocytic hypophysitis is an autoimmune disorder with a female predominance and a striking association with pregnancy.[77,78] Patients with this disorder may have anti-pituitary antibodies that account for their disease;[79] the specificity of the antibodies may alter the clinical manifestations.[80,81] The most common form of the disease is the pregnancy-associated lesion that is thought to represent an autoimmune attack on hyperplastic lactotrophs that expose altered antigens attributable to their hyperplastic nature; this lesion presents as a mass, associated with the hyperprolactinemia of pregnancy, mimicking pituitary adenoma.[77] The hyperprolactinemia itself may enhance the inflammatory response,[3] as suggested above, to exacerbate this disorder and account for its frequent onset during gestation. Selective pituitary stimulating antibodies have been suggested as a rare cause of Cushing's disease.[82] Pituitary hormone deficiency may be attributable to this disorder; patients with dwarfism have been reported to have antibodies against somatotrophs,[80] rare patients with ACTH deficiency have had anti-corticotroph antibodies,[83] and isolated gonadotropin deficiency has been associated with endocrine autoimmunity.[84]

CONCLUSIONS

It is becoming increasingly clear that there is a very important bidirectional communication between the endocrine and the immune systems and that the pituitary plays an integral role in the transmission of information both ways. Neuroendocrine-immunology provides the means to regulate physiological activities in both arenas, clearly to enhance biological adaptation. Further knowledge of these integrative mechanisms will allow us to better understand the pathophysiology that may result from abnormalities in endocrine or immune homeostasis.

REFERENCES

1. E.J. Goetzl and S.P. Sreedharan, Mediators of communication and adaptation in the neuroendocrine and immune systems, *FASEB J.* 6:2646 (1992).
2. I. Berczi, Neurohormonal immunoregulation, *Endocr. Pathol.* 1:197 (1990).
3. I. Berczi, The role of prolactin in the pathogenesis of autoimmune disease, *Endocr. Pathol.* 4:178 (1993).
4. E. Nagy and I. Berczi, Immunodeficiency in hypophysectomized rats, *Acta Endocrinol. (Copenh)* 89:530 (1978).
5. I. Berczi, E. Nagy, K. Kovacs and E. Horvath, Regulation of humoral immunity in rats by pituitary hormones, *Acta Endocrinol. (Copenh)* 98:506 (1981).
6. I. Berczi and E. Nagy, A possible role of prolactin in adjuvant arthritis, *Arthritis Rheum.* 25:591 (1982).
7. I. Berczi, E. Nagy, S.L. Asa and K. Kovacs, Pituitary hormones and contact sensitivity in rats, *Allergy* 38:325 (1983).
8. I. Berczi, E. Nagy, S.L. Asa and K. Kovacs, The influence of pituitary hormones on adjuvant arthritis, *Arthritis Rheum.* 27:682 (1984).
9. E. Nagy, I. Berczi and H.G. Friesen, Regulation of immunity in rats by lactogenic and growth hormones, *Acta Endocrinol. (Copenh)* 102:351 (1983).
10. I. Berczi and E. Nagy, Effects of hypophysectomy on immune function, *in:*

"Psychoneuroimmunology," 2nd Ed., R. Ader, D.L. Felten and N. Cohen, eds., Academic Press, Inc., (1991).

11. R.W. Pfaffle, G.E. DiMattia, J.S. Parks, M.R. Brown, J.M. Wit, M. Jansen, H. Van der Nat, J.L. Van den Brande, M.G. Rosenfeld and H.A. Ingraham, Mutation of the POU-specific domain of Pit-1 and hypopituitarism without pituitary hypoplasia, *Science* 257:1118 (1992).

12. K. Tatsumi, K. Miyai, T. Notomi, K. Kaibe, N. Amino, Y. Mizuno and H. Kohno, Cretinism with combined hormone deficiency caused by a mutation in the Pit-1 gene, *Nature Gen.* 1:56 (1992).

13. S.L. Asa, K. Kovacs, A. Halasz, A.M. Toszegi and P. Szùcs, Absence of somatotroph, lactotrophs, and thyrotrophs in the pituitary of two dwarfs with hypothyroidism: deficiency of pituitary transcription factor-1? *Endocr. Pathol.* 3:93 (1992).

14. S. Radovick, M. Nations, Y. Du, L.A. Berg, B.D. Weintraub and F.E. Wondisford, A mutation in the POU-homeodomain of Pit-1 responsible for combined pituitary hormone deficiency, *Science* 257:1115 (1992).

15. T. Tanaka, R.P.C. Shiu, P.W. Gout, C.T. Beer, R.L. Noble and H.G. Friesen, A new sensitive and specific bioassay for lactogenic hormones: measurement of prolactin and growth hormone in human serum, *J. Clin. Endocrinol. Metab.* 51:1058 (1980).

16. I. Berczi and E. Nagy, The effect of prolactin and growth hormone on hemolymphopoietic tissue and immune function, *in:* "Hormones and Immunity," I. Berczi and K. Kovacs, eds., MTP Press, Lancaster, UK (1987).

17. K.W. Kelley, Growth hormone, lymphocytes and macrophages, *Biochem. Pharmacol.* 38:705 (1989).

18. K.W. Kelley, Growth hormone in immunobiology, *in:* "Psychoneuroimmunology," 2nd Ed., R. Ader, D.L. Felten and N. Cohen, eds., Academic Press, New York (1991).

19. D.A. Weigent, J.B. Baxter, W.E. Wear, L.R. Smith, K.L. Bost and J.E. Blalock, Production of immunoreactive growth hormone by mononuclear leukocytes, *FASEB J.* 2:2812 (1988).

20. S. Varma, P. Sabharwal, J.F. Sheridan and W.B. Malarkey, Growth hormone secretion by human peripheral blood mononuclear cells detected by an enzyme-linked immunoplaque assay, *J. Clin. Endocrinol. Metab.* 76:49 (1993).

21. E. Nagy and I. Berczi, Prolactin and contact sensitivity, *Allergy* 36:429 (1981).

22. E. Nagy, I. Berczi, G.E. Wren, S.L. Asa and K. Kovacs, Immunomodulation by bromocriptine, *Immunopharmacology* 6:231 (1983).

23. R.C. Rowe, E.A. Cowden, C. Faiman and H.G. Friesen, Correlation of Nb2 bioassay and radioimmunoassay values for human serum prolactin, *J. Clin. Endocrinol. Metab.* 57:942 (1983).

24. A. Bateman, A. Singh, T. Kral and S. Solomon, The immune-hypothalamic-pituitary-adrenal axis, *Endocr. Rev.* 10:92 (1989).

25. K. Karalis, H. Sano, J. Redwine, S. Listwak, R.L. Wilder and G.P. Chrousos, Autocrine or paracrine inflammatory actions of corticotropin-releasing hormone in vivo, *Science* 254:421 (1991).

26. N.J. Olsen, W.E. Nicholson, C.R. DeBold and D.N. Orth, Lymphocyte-derived adrenocorticotropin is insufficient to stimulate adrenal steroidogenesis in hypophysectomized rats, *Endocrinology* 130:2113 (1992).

27. A.G. DuPont, G. Somers, A.C. Van Steirteghem, F. Warson and L. Vanhaelst, Ectopic adrenocorticotropin production: disappearance after removal of inflammatory tissue, *J. Clin. Endocrinol. Metab.* 58:654 (1984).

28. H. Kotzman, M. Clodi, T. Svoboda, R. Deyssig and A. Luger, Cytokine production and natural killer cell activity in patients with prolactinomas and acromegaly, *Endocrinology* (suppl.):68, (1992) (Abstract No. 68).

29. R.E. Palmiter, R.L. Brinster, R.E. Hammer, M.E. Trumbauer, N.G. Rosenfeld, N.C. Birhberg and R.M. Evans, Dramatic growth of mice that develop from eggs microinjected with metallothionein-growth hormone fusion genes, *Nature* 300:613 (1982).

30. E.M. Sternberg, W.S. Young III and R. Bernardini, A central nervous system defect in biosynthesis of corticotropin-releasing hormone is associated with susceptibility to streptococcal cell wall-induced arthritis in Lewis rats, *Proc. Natl. Acad. Sci. USA* 86:4771 (1989).

31. R. Fassler, K. Schauenstein, G. Kremer, S. Schwarz and G. Wick, Elevation of corticosteroid binding globulin in obese strain (OS) chickens: possible implications for the distributed immunoregulation and the development of spontaneous autoimmune thyroiditis, *J. Immunol.* 136:3657 (1986).

32. I.C. Chikanza, P. Petrou, G. Kingsley, G.P. Chrousos and G.S. Panayi, Defective hypothalamic response to immune and inflammatory stimuli in patients with rheumatoid arthritis, *Arthritis Rheum.* 35:1281 (1992).

33. A. Catania, L. Airaghi, M.G. Manfredi, M.C. Vivirito and I.M. Lipton, Relations among plasma concentrations of alpha-MSH, ACTH and cytokines in AIDS patients, *Endocrinology* (Suppl.):199 (1992) (Abstract No. 59).

34. S. Reichlin, Neuroendocrine-immune interactions, *N. Engl. J. Med.* 329:1246 (1993).

35. B.L. Spangelo and R.M. MacLeod, The role of immunopeptides in the regulation of anterior pituitary hormone release, *Trends Endocrinol. Metab.* 2:408 (1990).

36. E.W. Bernton, J.E. Beach, J.W. Holaday, R.C. Smallridge and N.G. Fein, Release of multiple hormones by a direct action of interleukin-1 on pituitary cells, *Science* 238:519 (1987).

37. P.E. Walton and M.J. Cronin, Tumor necrosis factor-α and interferon-gamma reduce prolactin release in vitro, *Am. J. Physiol.* 259:E672 (1990).

38. H. Vankelecom, M. Andries, A. Billiau and C. Denef, Evidence that folliculo-stellate cells mediate the inhibotyory effect of interferon- on hormone secretion in rat anterior pituitary cell cultures, *Endocrinology* 130:3537 (1992).

39. H. Müller, E. Hammes, C. Hiemke and G. Hess, Interferon-alpha-2-induced stimulation of ACTH and cortisol secretion in man, *Neuroendocrinology* 54:499 (1991).

40. C.J. Wiedermann, W. Vogel, H. Tilg, W.J. Wiedermann, M. Herold, U. Zilian, T. Wohlfarter, M. Gruber and H. Braunsteiner, Suppression of thyroid function by interferon-alpha$_2$ in man, *Arch. Pharmacol.* 343:665 (1991).

41. W.B. Malarkey and B.J. Zvara, Interleukin-1b and other cytokines stimulate adrenocorticotropin release from cultured pituitary cells of patients with Cushing's disease, *J. Clin. Endocrinol. Metab.* 69:196 (1989).

42. S.L. Asa, K. Kovacs and S. Melmed, The hypothalamic-pituitary axis, *in:* "The Pituitary," S. Melmed, ed., Blackwell Scientific Publications Inc., Boston (1994) (in press).

43. H.N. Kamerow, A. Perchick and D.E. Burstein, Immunocytochemical detection of cyclin, a proliferation-associated protein, in cytologic preparations, *Acta Cytol.* 35:491 (1991).

44. L.C. Payne, F. Obal Jr., M.R. Opp and J.M. Krueger, Stimulation and inhibition of growth hormone secretion by interleukin-1b: The involvement of growth hormone-releasing hormone, *Neuroendocrinology* 56:118 (1992).

45. J.N. Peisen, J.W. Bordeaux, K.J. McDonnell, D.E. Tracey and M.D. Lumpkin, Effects of interleukin-1 (IL-1) and IL-1 receptor antagonist on growth hormone releasing hormone and somatostatin secretion from rat hypothalamus, *Endocrinology* (suppl.):394 (1992) (Abstract No. 137).

46. D.E. Scarborough, S.L. Lee, C.A. Dinarello and S. Reichlin, Interleukin-1b stimulates somatostatin biosynthesis in primary cultures of fetal rat brain, *Endocrinology* 124:549 (1989).

47. Y. Naito, J. Fukata, T. Tominaga, Y Masui, Y. Hirai, N. Murakami, S. Tamai, K. Mori and H. Imura, Adrenocorticotropic hormone-releasing activities of interleukins in a homologous in vivo system, *Biochem. Biophys. Res. Commun.* 164:1262 (1989).

48. F. Sweep, C. Rijinkels and A. Hermus, Activation of the hypothalamus-pituitary-adrenal axis by cytokines, *Acta Endocrinol.* (Copenh) 125:84 (1991).

49. J. Fukata, T. Usui, Y. Naitoh, Y. Nakai and H. Imura, Effects of recombinant human interleukin-1α, -1β, 2 and 6 on ACTH synthesis and release in the mouse pituitary tumour cell line AtT-20, *J. Endocrinol.* 122:33 (1989).

50. B.M. Sharp, S.G. Matta, P.K. Peterson, R. Newton, C. Chao and K. Mcallen, Tumor necrosis factor-a is a potent ACTH secretagogue: Comparison to interleukin-1b, *Endocrinology* 124:3131 (1989).

51. R. Sapolsky, C. Rivier, G. Yamamoto, P. Plotsky and W. Vale, Interleukin-1 stimulates the secretioin of hypothalamic corticotropin-releasing factor, *Science* 238:522 (1981).

52. S. Tsagarakis, G. Gillies, L.H. Rees, M. Besser and A. Grossman, Interleukin-1 directly stimulates the release of corticotrophin releasing factor from rat hypothalamus, *Neuroendocrinology* 49:98 (1989).

53. F. Berkenbosch, J. Van Oers, A. Del Rey, F. Tilders and H. Besedovsky, Corticotropin-releasing factor-producing neurons in the rat activated by interleukin-1, *Science* 238:524 (1987).

54. E.T. Cunningham, Jr. and E.B. DeSouza, Interleukin 1 receptors in the brain and endocrine tissues, *Immunol. Today* 14:1171 (1993).

55. H. Kobayashi, J. Fukata, T. Tominaga, N. Murukami, M. Fukushima, O. Ebisui, H. Segawa, Y. Nakai and H. Imura, Regulation of interleukin-1 receptors on AtT-20 mouse pituitary tumour cells, *FEBS Lett.* 298:100 (1991).

56. E.L. Webster, D.E. Tracey and E.B. De Souza, Upregulation of interleukin-1 receptors in mouse AtT-20 pituitary tumor cells following treatment with corticotropin-releasing factor, *Endocrinology* 129:2796 (1991).

57. L. Xia, E. Shalts, Y.J. Feng and M. Ferin, Modulation by the ovarian steroids of the effects of interleukin-1 alpha on luteinizing hormone in the ovariectomized rhesus monkey, *Endocrinology* (suppl.):354 (1992) (Abstract No. 210).

58. S. Rivest and C. Rivier, Inhibitory influence of interleukin-1b on c-*fos* expression located in the LHRH neurons and hypothalamic LHRH release during proestrus in rats, *Endocrinology* (suppl.):258 (1992) (Abstract No. 825).

59. S. Karanth and S.M. McCann, Anterior pituitary hormone control by interleukin 2, *Proc. Natl. Acad. Sci. USA* 88:2961 (1991).

60. S. Karanth, M.C. Aguila and S.M. McCann, The influence of interleukin-2 on the release of somatostatin and growth hormone-releasing factor by the mediobasal hypothalamus, *Endocrinology* (suppl.):397 (1992) (Abstract No. 1383).

61. E. Arzt, R. Buric, G. Stelzer, J. Stalla, J. Sauer, U. Renner and G.K. Stalla, Interleukin involvement in anterior pituitary cell growth regulation: Effects of IL-2 and IL-6, *Endocrinology* 132:459 (1993).

62. G. Wick, Y. Hu, S. Schwarz and G. Kroemer, Immunoendocrine communication via the hypothalamo-pituitary-adrenal axis in autoimmune diseases, *Endocr. Rev.* 14:539 (1993).

63. K. Lyson and J.M. Lipton, Alpha-melanocyte-stimulating hormone and adrenocorticotropic hormone inhibit interleukin-6-induced corticotropin-releasing factor release *in vitro*, *Endocrinology* (suppl.):380 (1992) (Abstract No. 1315).

64. B.L. Spangelo, A.M. Judd, P.C. Isakson and R.M. MacLeod, Interleukin-6 stimulates anterior pituitary hormone release *in vitro*, *Endocrinology* 125:575 (1989).

65. M. Yamaguchi, M. Sakata, N. Matsuzaki, K. Koike, A. Miyake and O. Tanizawa, Induction by tumor necrosis factor-alpha of rapid release of immmunoreactive and bioactive luteinizing hormone from rat pituitary cells in vitro, *Neuroendocrinology* 52:468 (1990).

66. P. Carmeliet, H. Vankelecom, J. van Damme, A. Billiau and C. Denef, Release of interleukin-6 from anterior pituitary cell aggregates: Developmental pattern and modulation by glucocorticoids and forskolin, *Neuroendocrinology* 53:294 (1991).

67. B. Velkeniers, R. Hooghe, Y. Hongqu, et al, Interleukin-6 in the pituitary gland, *J. Endocrinol. Invest.* 14 (suppl.4):189 (1991) (Abstract).

68. B.L. Spangelo and R.M. Wright RM, Arachadonic acid stimulates interleukin-6 release from rat anterior pituitary cells in vitro, *Endocrinology* (suppl.):179 (1992) (Abstract No. 512).

69. H. Vankelecom, P. Carmeliet, J. van Damme, A. Billiau and C. Denef, Production of interleukin-6 by folliculo-stellate cells of the anterior pituitary gland in a histiotypic cell aggregate culture system, *Neuroendocrinology* 49:102 (1989).

70. H. Matsumoto, C. Koyama, T. Sawada, K. Koike, K. Hirota, A. Miyake, A. Arimura and I. Inoue, Pituitary folliculo-stellate-like cell line (TtT/GF) responds to novel hypophysiotropic peptide (pituitary adenylate cyclase-activating peptide), showing increased adenosine3',5'-monophosphate and interleukin-6 secretion and cell proliferation, *Endocrinology* 133:2150 (1993).

71. S.L. Asa, K. Kovacs snf S. Melmed, The hypothalamic-pituitary axis, *in:* "The pituitary," S. Melmed, ed., Blackwell Scientific Publications, Boston (1994) (in press).

72. T.H. Jones, S. Justice, A. Price and K. Chapman, Interleukin-6 secreting human pituitary adenomas *in vitro, J. Clin. Endocrinol. Metab.* 73:207 (1991).

73. B. Velkeniers, P. Vergani, J. Trouillas, J. D'Haens, R.J. Hooghe and E.L. Hooghe-Peters, Expression of IL-6 mRNA in normal rat and human pituitaries and in human pituitary adenomas, *J. Histochem. Cytochem.* 42:67 (1994).

74. S. Tsagarakis, G. Kontogeorgos, P. Giannou, N. Thalassinos, J. Wooley, G.M. Besser and A. Grossman, Interleukin-6 a growth promoting cytokine, is present in human pituitary adenomas: an immunocytochemical study, *Clin. Endocrinol.* (Oxf) 37:163 (1992).

75. E. Arzt, U. Renner, K. Lechner, G. Stelzer, O.A. Muller and G.K. Stalla, Expression of interleukin-2 mRNA in human corticotropic adenoma cells *in vitro, J. Endocrinol. Invest.* 14 (suppl.4):188 (1991) (Abstract).

76. L. Ramyar, S.L. Asa, W. Singer and K. Kovacs, Cytokines do not alter hormone release by human pituitary adenomas in vitro, *Endocrinology* (suppl.):(1993) (Abstract No. 1569).

77. S.L. Asa, J.M. Bilbao, K. Kovacs, R.G. Josse and K. Kreines, Lymphocytic hypophysitis of pregnancy resulting in hypopituitarism: a distinct clinicopathologic entity, *Ann. Intern. Med.* 95:166 (1981).

251

78. F. Cosman, K.D. Post, D.A. Holub and S.L. Wardlaw, Lymphocytic hypophysitis. Report of 3 new cases and review of the literature, *Medicine* 68:240 (1989).

79. P. Crock, M. Salvi, A. Miller, J. Wall and H. Guyda, Detection of anti-pituitary autoantibodies by immunoblotting, *J. Immunol. Methods* 162:31 (1993).

80. A. Pouplard, Pituitary autoimmunity, *Horm. Res.* 16:289 (1982).

81. G.F. Bottazzo and D. Doniach, Pituitary autoimmunity: A review, *J. Royal Soc. Med.* 71:433 (1978).

82. W.A. Scherbaum, M. Gluck, U. Schrell, R. Fahlbusch and E.F. Pfeiffer, Autoantibodies to pituitary corticotropin-producing cells: Possible marker for unfavourable outcome after pituitary microsurgery for Cushing's disease, *Lancet* I:1394 (1987).

83. M.D. Jensen, B.S. Handwerger, B.W. Scheithauer, P.C. Carpenter, R. Mirakian and P.M. Banks, Lymphocytic hypophysitis with isolated corticotropin deficiency, *Ann. Intern. Med.* 105:200 (1986).

84. A.L. Barkan, R.P. Kelch and J.C. Marshall, Isolated gonadotrope failure in the polyglandular autoimmune syndrome, *N. Engl. J. Med.* 312:1535 (1985).

NEUROENDOCRINE-IMMUNE INTERACTIONS IN THE GASTROINTESTINAL TRACT

Hiroshi Nagura, Mitsuo Kimura, Mituru Kubota and Noriko Kimura

Department of Pathology
Tohoku University School of Medicine
Sendai 980 Japan

INTRODUCTION

Gastrointestinal function is modulated by a complex series of neurohormonal interrelations. There is a great deal of evidence for the neurohormonal influence on various aspects of the immuno-inflammatory response of the gastrointestinal mucosa. [1-4] Psychological stress, for example, can influence the symptoms of inflammation in such diseases as inflammatory bowel disease(IBD). [5] The intestine contains highly specialized subdivisions of both the neuroendocrine and immune systems, and the mucosal surface is especially well designed to allow for interactions between nerves and the cells of the immune system.

Interaction between the neuroendocrine and immune systems has been recognized for many years, and it has recently become apparent that several neuropeptides originally described as neurotransmitters, or gut hormones, play important regulatory roles in the intestinal mucosa. [1-4] The intestinal mucosa contains significant amounts of the neuropeptide neurotransmitters, and three of these peptides, vasoactive intestinal peptide(VIP), substance P(SP), and somatostatin(SOM), have been shown to regulate the function of immune effector cells in the intestinal mucosa including gut-associated lymphoid tissue(GALT). [2,3]

Abnormalities in neuroendocrine-immune interactions may be important in disease pathogenesis of immuno-inflammatory disorders in the intestine, including IBD. [6-9] The objective of this brief review is to present the anatomic basis for nerve-immune interactions and for disease pathogenesis of IBD which stems from the abnormalities of these interactions.

THE NEUROENDOCRINE SYSTEM IN THE INTESTINE

The intestinal tract includes major endocrine organs and profuse connection with the nervous system, that is, this intestinal neuroendocrine system consists of an extensive network of the enteric nervous system and specialized epithelial endocrine cells, which constitute a "diffuse neuroendocrine system."[2-4,10] This synthesizes and secretes neuropeptides or neurotransmitters. The enteric nervous system is integrated with the autonomic nervous system, which is made up of sympathetic and parasympathetic systems and sensory afferent nerves.[10] These nerve cells lie within the myenteric and submucosal plexuses, and their processes are distributed throughout the intestinal wall.[10] The intestinal tract also contains abundant immune effector cells, including GALT,[4] and enteric nerve processes extend to, contact or lie adjacent to the immune effector cells, smooth muscle cells and vascular networks as well as the epithelial endocrine and nonendocrine cells (Fig.1).[1-4,9]

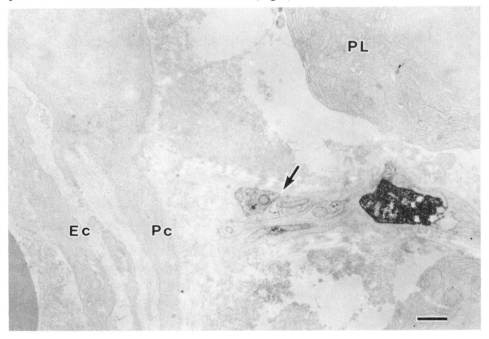

Figure 1. Immunohistochemical localization of vasoactive intestinal peptide (VIP) in the colonic lamina propria. A VIP-nerve (→) lie adjacent to plasma cell (PL) and pericytes (Pc) of a blood vessel. Ec: endothelial cells. (- : 11 m)

Innervation of Intestinal Lamina Propria

There is a dense distribution of neuropeptide-containing nerves throughout the intestinal lamina propria. The three most abundantly used neuropeptides in the enteric nerve are VIP, SP, and SOM, which have been shown to be potent immunoregulatory molecules. VIP-containing nerve fibers (VIP-nerves) are the most prevalent nerves in the lamina propria, and their distribution is divided into three patterns,[9] i.e.,(1) dense networks around crypts, (2) along capillaries, and (3) networks without relation to crypts and capillaries, but with occasional relation to immune effector cells. Immunoelectron microscopy demonstrates intimate nerve membrane-lymphocyte contacts. SP-nerves show a similar, but less dense pattern in the lamina propria. SOM-nerves are much less frequent around crypts, and are almost absent within the villi. SP-nerves are also present in axons adjacent to mast cells, which are abundant the intestinal mucosa. According to Ottaway [3] the lenght of nerve fibers in human intestine was estimated to be about $2m/mm^3$ of tissue, and calculations suggest that about half of the mucosal tissue consists of 9 μm nerve fibers. This indicates that a large proportion of lymphoid cells are within one cell distance of nerve fibers in the lamina propria and can reach proximity to nerves by chance alone. The mucosal localization of these neuropeptides suggests that their release locally can directly affect the function of epithelial cells, endothelial cells and immunocytes.

Innervation of Peyer's Patches

The most characteristic feature of the mucosal immune system is its preferential utilization of the IgA immunoglobulin class for its B-cell responses to antigenic challenge throughout the mucosa, where the lymphoid apparatus called GALT is located. [4,11] GALT consists of organized lymphoid tissue, a diffuse collection of lymphocytes and IgA plasma cells within the lamina propria, and lymphocytes within villous epithelial cells. The organized lymphoid tissues, called Peyer's patches are sites of the induction of mucosal immune responses and also of the generation of specific hyporesponsiveness to mucosal antigens leading to oral tolerance.

VIP-nerves are abundant in areas that surround the specialzed high endothelial postcapillary venules(PCV) through which T and B-cells extravasate from the systemic circulation into the intestinal mucosa. SP-nerves also come into close proximity with PCV. [12] The high endothelial cells of PCV express a variety of cell adhesion molecules, such as intercellular adhesion molecule-1 (ICAM-1), CD54 and E-selectin, which mediate the specific binding of lymphocytes to endothelial cells and their subsequent migration to the intestinal mucosa. VIP seems to decrease lymphocyte traffic, whereas SP induces the expression of E-selectin by high endothelial cells of PCV, and stimulates adhesion and migration of lymphocytes and leucocytes. Intestinal follicular areas and their germinal centers comprised of B-cell-dependent areas, are scarcely innervated.

NEUROENDOCRINE REGULATION OF MUCOSAL IMMUNE RESPONSES

Cells of the mucosal immune system must be continuously exposed to locally and systemically derived neuropeptide signals in the microenvironment of the mucosa. They express receptors for these neurophysiologic signals. [2,3] The demonstration of neuropeptide receptors on immunocytes supports the notion that functional interation between immune effector cells and neuropeptide-containing nerves exist. [2,3,14,15] Neuropeptides have been shown to influence a variety of immune reactions, including lymphocyte proliferation and activation, antibody synthenis, natural killer function, macrophage activation, and histamine release by mast cells. [2,3,14-17]

Neuropeptide Receptors on Immunocytes in the Intestine

Specific binding of the neuropeptides, VIP, SP and SOM with immunocytes in the intestinal mucosa has added a further evidence to the notion that functional interaction between mucosal immune responses and nerves can exist. These neuropeptides can be released by nerve endings in close proximity to lymphocytes, plasma cells, mast cells, granulocytes, cells of a macrophage lineage in the lamina propria and lymphocytes in the intraepithelial space. Specific cell surface receptors for these neuropeptides have been described on these immune effector cells, but the binding ability varies within the lymphocyte subset. [14,15] There are also organ/age-related differences in binding capabilities. Thus, the demonstration of neuropeptide receptors on various immune effector cells in the intestinal mucosa supports the idea of neurohormonal regulation of mucasal immune responses to a variety of antigens in the intraluminal environment of the intestine.

Lymphocytes of both the B- and T cell lineage express neuropeptide receptors. [2,3,18] Specific cell surface receptors for SP have been shown predominantly on T-helper CD4+ cells, isolated from Peyer's patches in significantly greater numbers when compared with spleen cells. In addition, SOM binds onto a greater percentage of cells from Peyer's patches than to those from spleen, in which the T-suppressor/cytotoxic phenotye is predominant. VIP- and SP-binding receptors were localized to the intrafollicular T cell regions and germinal centers of Peyer's patches (Fig.2). Immunoelectron microscopically dendritic cells and cells of the macrophage lineage were shown to express the VIP receptors, which were absent in similar cells of neonates before the mucosal immune system matures. [19] The germinal centers of Peyer's patches provide a critical microenvironment for antigen and T-helper cell dependent B cell class switching and differentiation. This raises the possibility that interaction with antigen may also induce the expression of neuropeptide receptors on the surface of follicular dendritic cells, which participate in the regulation of the immune response. Organ-, cell-, and phenotype-specific differences of neuropeptide receptor expression may allow us to explain their specific effects seen in the intestinal immune response.

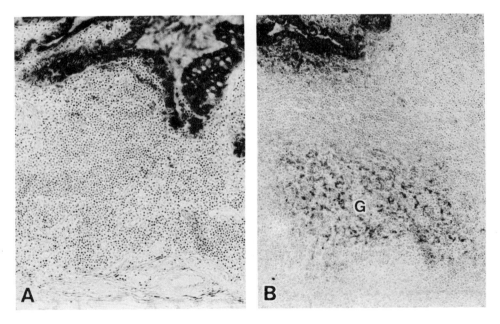

Figure 2. Immunohistochemical localization of vasoactive intestinal peptide (VIP) receptors in the human neonatal (A) and adult (B) ileal mucosa. Immunoreactivity for VIP receptor is present on follicular dendritic cells in the germinal center of Peyer's patches in the adult (G). No immunoreactivity is found in the neonate before the mucosal immune system matures.

Table 1. Neuropeptide receptors and neuropeptide effects on immune effector cells.

neuropeptide	receptor expression	lymphocyte proliferation	antibody production
VIP	PBMNC	T cell ↓	spleen IgA ↑
	T cells	(CD4$^+$>CD8$^+$) ↓	IgG →
	(CD4$^+$>CD8$^+$, MLN)	B cells ↓	PP IgA ↓
	B cells		IgM ↑
	FDC	spleen ↓	
	macrophage	MLN ↓	
SP	PBMNC	T cell ↑	spleen IgA ↑
	T cells		IgM ↓
	(CD4$^+$>CD8$^+$)	Spleen ↑	PP IgA ↑
	B cells	MLN ↑	(PP >spleen)
	(PP > spleen)	PP ↑	IgM ↓
SOM	PBMNC	PBMNC ↓	spleen IgA ↓(↑)*
	T cell		IgM ↓
	(CD4$^+$<CD8$^+$)	spleen ↓ (↑)*	PP IgA ↓(↑)*
	(PP >spleen)	MLN ↓ (↑)*	

PBMNC: peripheral blood mononuclear cells, PP: Peyer's Patch, MLN: mesenteric lymph node, FDC: follicular dendritic cells
*Immunoglobulin production after the administration of SOM by cells retrieved from the tissues listed.
↓: inhibition, ↑: stimulation, →: no effect

Neuropeptide Effects on Immune Effector Cells

There is a great deal of evidence for an extensive network of local regulatory effects of immune effector cells and neuropeptides such as VIP, SP, and SOM in the intestine. These neuropeptides can modulate IgA secretion and IgA lymphoblast migration.[12] Antibody synthesis by concanavalin A-stimulated murine B lymphocytes is enhanced by SP, and the effect is more marked on cells derived from Peyer's patches than on those from spleen.[18] In addition, the effect was isotype-specific, i.e. IgA synthesis was increased more than that of IgM, whereas there is little influence on IgG synthesis.[18] VIP affects IgA synthesis more than that of IgM or IgG. Immunoglobulin synthesis by Peyer's patch B cells is inhibited and by splenocytes is stimulated by VIP.[18] VIP inhibits the proliferative response of T cells from various tissues, and suppresses the activation of CD4+ T cells. SP can affect B cells directly or indirectly through the regulation of cytokine production by T cells, such as IL-4, IL-5, or IL-6.[2] The neuropeptide effect on the mucosal immune system may also occur through effects on other nonimmune cells, such as vascular endothelial cells and villous epithelial cells, as well as on antigen-presenting cells, such as macrophages and dendritic cells. Table 1 summarizes the recently identified receptors for VIP, SP and SOM on immune effector cells and the possible role of these neuropeptides in immune function. Details of the mechanism for neuropeptide-modulation of immune effector cell functions have been reviewed by Bienenstock, Ottaway and their colleagues. [2,3]

Neuropeptide Effects on Lymphoid Cell Migration

The intestinal mucosa is a major site of lymphoid cell extravasation through the specialized post-capillary high endothelial venules of Peyer's patches [20] and through the dilated venules with hypertrophied endothelial cells in the lamina propria of inflamed mucosa. [4,13] The effect of neuropeptides on such lymphocyte migration is well substantiated. Neuropeptide-containing nerves, particularly VIP and SP nerves, are abundant in close proximity to PCV, and proliferate in areas surrounding venules with hypertrophic tall endothelial cells in the inflamed mucosa. Lymphocyte migration is governed by adhesive surface molecules on lymphocytes and endothelial cells, and by the interaction of VIP receptor-bearing lymphocytes with VIP in the vicinity of PCV. VIP decreases lymphocyte traffic into tissues, [21] whereas SP stimulates migration and lymphocyte adhesion to endothellial cells. E-selectin is expressed exclusively by high endothelial venules, including PCV. Inflammatory lesions in the intestinal mucosa and granulation tissue in the ulcer base are rich in venules with hypertrophic tall endothelial cells where neutrophils and lymphocytes extravasate at a high frequency. [13] Expression of E-selectin by these endothelial cells is enhanced by SP. [22]

NEUROIMMUNE REGULATION AND INTESTINAL INFLAMMATION

Ulcerative colitis (UC) and Crohn's disease, known as idiopathic chronic inflammatory diseases, are characterized by prolonged clinical course, and inflammation and ulceration of the mucosa. [4,13] Spontaneous exacerbations and remissions are also characteristic for these diseases. Regardless of the initiating factors, disorders of immunoregulation must be involed in the pathophysiology of IBD. [4,7-9,13] Most investigations have focused on the immunopathogenesis of IBD. In addition there are indications that alteration of the enteric nervous system could paticipate in the pathophysiology of IBD. [7-9] Degeneration and subsequent-proliferation of nerves in the mucosa in IBD have been demonstrated by the immunocytochemical detection of VIP and SP.

Abnormal Distribution of Nerves in the Inflamed Intestinal Mucosa

In the normal intestinal mucosa, a significant number of VIP-, and SP-nerves were intimately associated with glands, blood vessels, and immune effector cells. The distribution of these peptidergic nerves in the lamina propria follow three patterns, as previously described. In severe inflammatory lesions of the active UC mucosa, VIP-containg and SP-nerves decrease. These nerves are almost absent around crypt abscess. [9,13] In hypervascular areas of the active UC mucosa, increasing VIP- and SP-innervations can be observed along the proliferating dilated venule-like blood vessels. This shows that enteric nerves degenerate in the severely inflamed mucosa, and subsequently regenerate and proliferate simultaneously with the regeneration and ploliferation of structural components of the colonic mucosa, such as blood vessels and colonic glands (Fig.3). In the UC mucosa, venules with swollen endothelial cells are increased, and express several adhesion molecules such as E-selectin and ICAM-1.

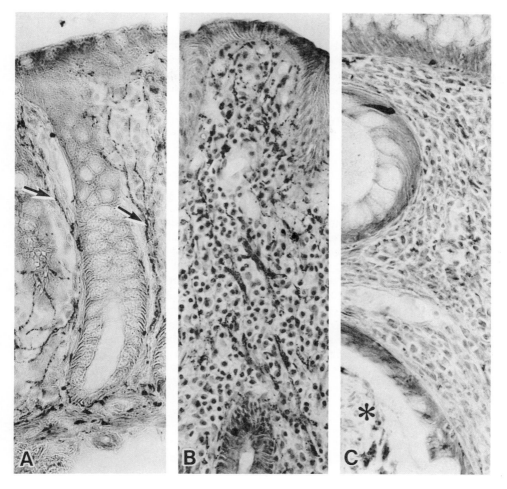

Figure 3. Immunohistochemical distribution of vasoactive intestinal peptide (VIP)-containing nerves in the colonic mucosa from the control (A) and ulcerative colitis patient (B,C).

A: VIP-nerves are present in close proximity to colonic glands, capillaries and mononuclear cells in the lamina propria (→).

B: In hypervascular lesions of the inflamed mucosa, dense and irregular networks of VIP-nerves are present associated with proliferative venule-like blood vessels.

C: VIP-nerves around crypt abscess(*) are almost absent in the severe inflammatory lesion.

It is interesting that the expression of these adhesion molecules by endothelial cells in inflammatory lesions is modulated by neuropeptides. In the resolving and quiescent UC mucosas, these neuropeptide-containing nerves still decrease. In the uninvolved mucosa of UC patients, however, they do not change their distribution and concentration. These immunohistochemical findings suggest that the change of VIP-and SP-nerve distribution and concentration in the UC mucosa may be the consequence of mucosal inflammation and damage. Alterations in innervation are correlated to the severity of inflammation and with structural changes in the distribution of colonic glands and blood vessels. VIP and SP are immuno-inflammatory regulators as described above. We hypothesize that the abnormalities the UC mucosa.[4,13] This may cause the mucosal damage and chronicity of the disease. However, reports on neuropeptide localization in the IBD mucosa are still conflicting, though it is generally accepted that VIP and SP are immuno-inflammatory modulators and that the onset or exacerbation of IBD symptoms is affected by emotional stress.[5] Many problems remain to be solved in the relationships between immunoregulatory neuropeptides and the mucosal lesions of inflammatory bowel disease. The characterization of neuropeptides and their receptors on tanget cells will permit an analysis of their role in the regulation of lymphocyte traffic, regional immunological responses and inflammation, and hypersensitivity reactions restricted to distinct tissue compartments.[23]

CONCLUSIONS

In the intestine a dense network of neuropeptide-containing nerves and abundant well organized lymphoid tissue and immune effector cells are present. The nerve fibers are intimately associated with glands, blood vessels and immune effector cells. Neuropeptides, such as VIP, SP, and SOM, are released by nerve endings in close proximity to immune effector cells, and participate in the regulation of immuno-inflammatory functions of the intestinal mucosa. Specific cell surface receptors for these neuropeptides are present on these immunocytes. These neuropeptide-containing enteric nerves often show a close relationship with blood vessels, particularly with high endothelial venules, where several cell adhesion molecules are expressed, and lymphocyte and leucocyte extravasation takes place. Abnormalities of peptidergic innervation may lead to the disorder of immuno-regulation and cause mucosal damage and prolonged inflammation.

ACKNOWLEDGEMENTS

This work was supported in part by grants from the Ministry of Welfare of Japan. The authors wish to thank Mrs. M. Komori and A. Tsukano for secretarial assistance.

REFERENCES

1. M.S. O'Dorisio, Neuropeptides and gastrointestinal immunity, *Am. J. Med.* 81:74 (1986).
2. K. Croitoru, P.B. Ernst, A.M. Stanisz, R.H. Stead and J. Bienenstock, Neuroendocrine regulation of the mucosal immune System, *in:* "Immunology and Immunopathology of the Liver and Gastrointestinal Tract," S.R. Targan, and F. Shanahan, eds., Igaku Shoin, New York, (1990).
3. C.A. Ottaway, Neuroimmunomodulation in the intestinal mucosa, *Gastrointest. Clin. North. Am.* 20, 511 (1991).
4. H. Nagura, Mucosal defense mechanism in health and disease: Role of the mucosal immune system, *Acta Pathol. Jpn.* 42:387 (1992).
5. D.A.Drossman, Psychosocial aspects of ulcerative colitis and Crohn's disease, *in:* "Inflammatory Bowel Disease," R.G. Shorter, ed., Lea Febiger, Philadelphia, (1988).
6. G-L, Ferri, T.E. Adrian, M.A. Ghatel, D.J. O'Shaughnessy, L. Plobert, Y.C. Lee, A.M.J. Buchan, J.M. Polak and S.R. Bloom, Tissue localization and relative distribution of regulatory peptides in separated layer from the human bowel, *Gastroenterolology* 84:717 (1987)
7. T.R. Koch, J.A. Carney, and V.L.W. Go, Distribution and quantitation of gut neuropeptides is normal intestine and inflammatory bowel disease, *Dig. Dis. Sci.* 22:369 (1987).
8. Y. Kubota, R.E. Petras, C.A. Ottaway, R. Tubbs, R. Farmer and C. Fiocchi, Colonic vasoactive intestinal peptide nerves in inflammatory bowel disease, *Gastroenterology* 102:1242 (1992).
9. M. Kimura, N. Hiwatashi, T. Masuda, H. Nagura and T. Toyota, Immunohistochemical distribution

of VIP-containing nerve fibers in colonic mucosa from patients with ulcerative colitis, *Biomed. Res.* 13 (Suppl.): 81 (1992).

10. G. Gabella, Innervation of the gastrointestinal tract. *Int. Rev. Cytol.* 59:129 (1979).

11. P. Brandtzaeg, T.S. Halstensen, K. Kett, K. Krajci, D. Kvale, T.O. Rognum, H. Scott and L.M. Sollid, Immunobiology and immunopathology of human gut mucosa. Humoral immunity and intraepithelial lymphocytes, *Gastroenterology* 97:1562 (1989).

12. C.A. Ottaway, Migration of lymphocytes within the mucosal immune system, *in*: "Immunology and Immunepathology of the Liver and Gastrointestinal Tract, " S.R. Targan, and F. Shanahan, eds., Igaku Shoin, New York (1987).

13. S. Nakamura, H. Ohtani, Y. Watanabe, K. Fukushima, T. Matsumoto, A. Kitano, K. Kobayashi and H. Nagura, In situ expression of the cell adhesion molecules in inflammatory bowel disease. Evidence of immunologic activation of vascular endothelial cells, *Lab. Invest.* 69:77 (1993).

14. M.S. O'Dorosio, T, O'Dorisio, C.L. Wood, J. Bresnahan, M. Beattie and L. Campolito, Characterization of vasoactive intestinal peptide receptors in nervous and immune systems, *Ann. N.Y.Acad. Sci.* 527:257 (1988).

15. A.M. Stanisz, R. Scicchitano, P. Dazin, J. Bienenstock and D.G. Payan, Distribution of substace P receptors on murine spleen and Peyer's Patch T and B cells, *J. Immunol.* 139:749 (1987).

16. M. Rola-Pleszczynski, D. Bolduc, and S. Pierre, The effects of vasoactive intestinal peptide on human natural killer cell function, *J. Immunol.* 135:2569 (1985).

17. F. Shanahan, J. Denburg, J. Fox, J. Bienenstock and D. Befus, Mast cell heterogeneity: Effects of neuroenteric peptides on histamine release, *J.Immunol.* 135:1331 (1985).

18. A.M. Staniz, D. Befus, and J. Bienenstock, Differential effects of vasoactive intestinal peptide, substance P, and somatostatin on immunoglobulin synthesis and proliferation by lymphocytes from Peyer's patch, mesenteric lymphonode and spleen, *J.Immunol.* 136:152 (1986).

19. M. Kubota, H. Ohtani, N. Kimura, S. Matsumoto and H. Nagura, Immunohistochemical analysis of VIP-receptor expression on immunocytes in the development of intestinal mucosa, *Proc. Jpn. Soc. Immunol.* 23:675 (1993).

20. M. Kotani, Y. Nawa, H. Fujii, T. Fukumoto, M. Miyamoto and A. Yamashita, Postcapillary venules as the pathway of migrating B lymphocytes, *Cell Tissue Res.* 152:299 (1974).

21. T.C. Moore, C.H. Spruck, and I Said, Depression of lymphocyte traffic in sheep by vasoactive intestinal peptide (VIP), *Immunology* 64:475 (1988).

22. W.L. Matis, R.M. Lavker, and G.F. Murphy, Substance P induces the expression of an endothelial-leukocyte adhesion molecule by microvascular endothelium, *J. Invest. Dermatol.* 94:492 (1993).

23. D.G. Payan, J.P. McGillis, F.K. Renold, M. Mitsuhashi and D.J. Goetzl, Immunomodulating properties of neuropeptides, *in*: "Hormones and Immunity," I. Berczi, and K. Kovacs, eds., MTP Press, Lancaster (1987).

SUBSTANCE P, IMMUNITY AND INFLAMMATION OF THE INTESTINAL MUCOSA

Andrzej M. Stanisz

Department of Pathology
and Intestinal Diseases Research Programme
McMaster University
Hamilton, Ontario L8N 3Z5
Canada

BACKGROUND

The concept of "neurogenic inflammation" is now receiving more attention in the literature. It has been discussed extensively in the context of polyarthritis and inflammation of the eye, skin and respiratory tract; however, the putative role(s) of neuropeptides in inflammatory reactions in the gastrointestinal tract has received comparatively little attention. Neurogenic inflammation usually describes reactions that include vasodilatation, plasma extravasation, and smooth muscle contraction due to neuronal activation and release of mediators from unmyelinated afferent (sensory) nerve endings. In addition, activation of axon collaterals associated with these afferent nerve endings results in further release of inflammatory mediators. There is a considerable body of evidence suggesting that neuropeptides play an important role in mediating this response. Enteric nerve plexuses ramify extensively in the lamina propria, with some fibres projecting into the sub-mucosa.[1,2] Several neuropeptides [e.g. Substance P (SP), Vasoactive Intestinal Peptide (VIP), Somatostatin (SOM) and Calcitonin Gene-Related Peptide (CGRP)] have been localized within the gastrointestinal tract.[3,4] Accompanying these events there is cellular infiltration and activation of mast cells, monocytes, neutrophils, eosinophils and lymphocytes; all of which have been implicated as playing an important role in inflammatory processes. Direct membrane/membrane contact between mast cells, eosinophils or plasma cells and nerves often occurs, suggesting that a supply of neurotransmitters to inflammatory cells is likely to be readily available.[5] It has been suggested that *all* of the inflammatory cells in the intestinal mucosa of rats may be structurally innervated.[6] These data strongly support the concept of bi-directional cellular signalling between the immune and neuroendocrine systems.

Of all known neuropeptides present in the gut, SP has been particularly implicated in the pathogenesis of various inflammatory conditions. Substance P is a prime candidate in neurogenic inflammation, as it has been shown to affect many aspects of inflammatory responses including macrophage and neutrophil activation and modulation of a range of immune events.[7,8] Additional evidence suggesting that SP may mediate neurogenic inflammation arises from its ability to increase vascular permeability, potentiate histamine activity, and cause bronchoconstriction and bronchosecretion.[9,10] During chronic inflammation of the intestine, SP levels and expression of SP-receptors are significantly increased.[11,12] The implication of such broad activity of SP is that it is likely to play a major role in the initiation and/or potentiation and direction of the inflammatory process in general.

Substance P

Substance P is an 11-amino acid peptide (for review see ref. 13) which belongs to the tachykinin family, defined by their common pharmacological properties and conserved carboxyl-terminal sequences, -Phe-X-Gly-Leu-Met-NH_2, where X is a branched aliphatic or aromatic residue. The principal biological activities of SP reside in its carboxyl sequence. The mammalian nervous system produces three tachykinins: substance P, substance K (SK) and neuromedin K (NMK). However, SP production from non-neuronal cells such as eosinophils,[14,15] macrophages[16] and endocrine cells in the intestine has also been demonstrated. Substance P and SK are derived from the same gene. Specific surface receptors for these tachykinins have been recently described in neuronal tissues, and designated as NK1, NK2 and NK3 for SP, SK and NMK, respectively. In peripheral tissues, including cells of the immune system, both NK1 and NK2 receptors have been found to be present.[17,18] There is significant homology between the receptors and various tachykinins are able to bind and activate different receptors. For example, SK and NMK at high concentrations (10^{-7} M) can stimulate NK1 receptor activity. All 3 receptor types are coupled to G-proteins and evoke inositol phospholipid hydrolysis upon activation,[19] but also in some cases can evoke an increase in cyclic AMP. In investigating the physiological role of tachykinins, Spantide and other SP antagonists were initially used, but their value has been undermined by their non-specific effects on a variety of cells, cross-reaction with other receptors and partial agonist activity. Similarly, the neurotoxin capsaicin, which is selective for unmyelinated sensory neurons, releases and consequently depletes both SP and SK, as well as other neuropeptides. However, specific antagonists of NK1 and NK2 but not NK3 receptors have recently been described and their introduction should make physiological studies possible and more illustrative of precise tachykinin function in a number of biological pathways.[20,21] The effects of one such antagonist on gastrointestinal immune function during inflammation will be discussed later.

It is also important to remember that SP could apparently activate cells directly via a non-receptor mediated mechanism. A classical example of this phenomenon is the activation of mast cells by SP in the absence of SP-receptors on these cells.[22] It was suggested that SP inserts its carboxyl terminus directly into the membrane.[23] Possibly the same mechanism could lead to activation of neutrophils[24] and some lymphocytes.[25]

SUBSTANCE P, INFLAMMATION AND THE IMMUNE SYSTEM

Substance P and other neuropeptides have been implicated in the pathogenesis of various chronic inflammatory conditions. SP induces the margination and adherence of monocytes to venule endothelium, induces lysosomal enzyme release and stimulates macrophage phagocytosis.[26] Substance P also evokes a number of responses from monocytes/macrophages including: the generation of thromboxane A2, oxygen free radicals and hydrogen peroxide; the down regulation of membrane-associated 5'-nucleotidase; and the stimulation of synthesis and release of arachidonic acid metabolites.[27] Production and release of PGE_2 and IL-1 are also increased by SP (7). Substance P has also been shown to promote mast cell degranulation[28] and function as a neutrophil activator and chemoattractant.[29] Stimulatory effects of SP, mediated possibly by IL-3 and GM-CSF, on human bone marrow cell hematopoiesis has also been documented.[30] We have shown that SP can significantly increase, among other factors, IL-6 and IL-8 synthesis by human fibroblasts.[31] More recently it has been shown that effects of SP on lymphocyte function are mediated via stimulation of IL-2 production by SP.[32]

Lymphoid cell migration is also affected by SP. Moore et al. have shown that SP can alter the flow of various populations of lymphocytes through sheep popliteal nodes.[33] These results have important implications in disease processes since the manipulation of the recruitment of various subsets of lymphocytes by neuropeptides could quantitatively and qualitatively modulate the inflammatory process and regulate cell-mediated immunity. From these results, it was postulated that SP might promote antigen processing, as well as mediate the inflammatory response. Indeed, a direct effect of SP on plasma cells has been postulated by Helme and colleagues who showed that the plaque forming response to sheep red blood cell antigens in draining lymph nodes was virtually abolished in animals treated with capsaicin neonatally.[34] In these experiments, the ability to form antigen-specific plaques was restored following the local simultaneous injection of SP and antigen. Substance P is also a late-acting B lymphocyte differentiation cofactor and promotes IgA synthesis.[35]

The physiological effects of SP on lymphocytes are mediated via specific (NK1 and/or NK2) receptors. Payan and colleagues have previously shown that there are specific surface membrane receptors for SP present on human lymphocytes.[17] We have significantly extended this work to the murine system where we described SP receptors on both T and B cells.[36]

SUBSTANCE P IN INFLAMMATORY DISORDERS

It has been documented that the pathophysiology of asthma, urticaria, chronic arachnoiditis, inflammatory bowel disease, rheumatoid arthritis and other chronic syndromes may be associated with or aggravated by SP. For example, levels of SP in the synovial fluid and serum of rheumatoid arthritis patients are significantly higher than in controls.[37] The presence of SP-immunoreactive nerves in the synovium has been confirmed[38] in various animal models of arthritis. In rats, once the nerves containing SP are abolished by capsaicin treatment, the arthritic limb is spared an inflammatory bout; denervation appears to relieve the joint swelling and pain as well. To the contrary, worsening signs of arthritis in rats are seen after intra-articular injection of SP. This can be prevented by co-treatment with an anti-SP antibody.[39]

SUBSTANCE P AND MUCOSAL IMMUNE FUNCTION

In contrast to other tissues, relatively little is known about the mechanism of neurogenic inflammation within the gastrointestinal tract. Idiopathic inflammatory bowel disease (IBD) is a generic term that refers to chronic inflammatory diseases of the intestine that are of unknown etiology, principally ulcerative colitis and Crohn's disease. Ulcerative colitis is an inflammatory, ulcerating process of the colon; Crohn's disease, an inflammation of the intestine characterized by granulomatous inflammatory lesions throughout the entire gut wall, may involve any part of the intestine, but primarily attacks the distal small intestine and colon. Mucosal and sub-mucosal concentrations of VIP are significantly decreased in Crohn's disease and ulcerative colitis, as quantitated by radioimmunoassay. In contrast, SP levels are significantly increased in left-sided ulcerative colitis.[11,12] In a recent report[40] it was suggested that SP levels are decreased in gut during acute inflammation. However, the same authors concluded later that the apparent decrease in SP levels is not associated with a decrease in mRNA level (ie. production) but due to increased release of SP from extrinsic nerve fibres during the inflammation.[41] Additionally, changes in neuropeptide receptor status were examined and SP, but not SK or NMK, binding sites were significantly increased. Though this study describes an important phenomenon during a pathological condition, no functional or physiological implications were assessed.

In our earlier work, we described the presence of SP receptors on both T and B murine mucosal lymphocytes. We also characterized the kinetics of SP binding sites on these cells.[36] We reported that not only cell proliferation, but also immunoglobulin synthesis, is significantly altered by SP for both B and T cells. These effects are isotype- and organ-specific; that is, IgA synthesis is affected most and lymphocytes from mucosal sites respond to a greater extent than those isolated from peripheral lymphoid organs.[42] We were able to fully confirm these *in vitro* observations *in vivo*, using a constant delivery system of SP (via miniosmotic pumps).[43] In addition, we have shown both *in vitro* and *in vivo* that SP enhanced an antigen specific (anti-SRBC) immunoglobulin response.[44] Natural killer cell activity of lymphocytes from the compartmentalized population of intestinal epithelial lymphocytes (IEL) but not those isolated from spleens was significantly enhanced by SP both *in vivo* and *in vitro*.[45]

We have recently described increased levels of SP in myenteric plexuses of *Trichinella spiralis*-infected rats.[46] Mice infected with *T. spiralis* demonstrate significantly increased levels of serum and whole-gut tissue homogenate SP, correlating well with the progress of inflammation.[47] Lymphocytes isolated from animals at the peak of inflammation fail to respond to *in vitro* stimulation with exogenous SP when proliferation and immunoglobulin synthesis was examined. This effect was seen over a variety of SP concentrations.

In more recent studies constant delivery of anti-SP antibodies via mini-osmotic pumps, following parasitic infection with *T. spiralis* resulted in normal tissue levels of SP, a normal lymphocyte response to exogenous SP, and reduced inflammation in comparison with infected animals not treated with anti-SP antibody.[47] Further, we used a specific SP antagonist, CP96,345 (Pfizer), acting via NK1 receptor, in order to inhibit proinflammatory effects of SP. This was administered subcutaneously at the rate of 10^{-7}M/day over a 14 day period using the same micro-osmotic pump delivery system.[48] In those animals dramatic accumulation of SP was witnessed, however a significant reduction in the degree of inflammation was seen in terms of restoration of normal gut morphology, lymphocyte function, and reduction in the amount of myeloperoxidase (a good marker of inflammation). Therefore, blocking the NK1 receptor was sufficient to inhibit the progress of inflammation. In addition, when SMS202,995 (Sandoz), an analogue of somatostatin which usually has opposite physiological effects to SP, was used under the same experimental conditions a dramatic inhibition in SP levels, associated with inhibition of all measured parameters of the inflammation, was documented. These results together suggest that by inhibiting SP synthesis/release (SMS202,995) or by blocking the SP receptor (CP96,345) a significant clinical improvement in inflammatory disease could be achieved.

CONCLUSIONS

Substance P is released locally from nerves at the site of inflammation.[1] It can then activate synthesis and/or release of various pro-inflammatory mediators from different cells such as MMCP II from mast cells,[2] MPO from neutrophils,[3] Il-1 from macrophages.[4] In addition, in both T and B lymphocytes proliferative response is affected by SP. Substance P is at the top of the long list of neuropeptides that may be involved in inflammation. Collectively, the above studies suggest SP involvement in the inflammatory reaction. However, no indication of cause (evoking inflammation) or effect (involvement in gut recovery) is given.

REFERENCES

1. J. Christensen and D. Wingate, eds., "Guide to Gastrointestinal Motility," Wright, Bristol UK (1983).
2. J.B. Furness and M. Costa, Types of nerves in the enteric nervous system, *Neuroscience* 5:1 (1980).
3. E. Ekblad, C. Winther, R. Ekman, R. Hakanson and F. Sundler, Projections of peptide-containing neurons in rat small intestine, *Neuroscience* 20:169 (1987).
4. J.M. Polak and S.R. Bloom, Distribution and tissue localization of VIP in the central nervous system and seven peripheral organs, *in:* "Vasoactive Intestinal Peptide," S.I. Said, ed., Raven Press, New York (1982).
5. R.H. Stead, M.F. Dixon, N.H. Bramwell, R.H. Riddell and J. Bienenstock, Mast cells are closely apposed to nerves in the human gastrointestinal mucosa, *Gastroenterology* 97:575 (1989).
6. R.H. Stead, M. Tomioka, G. Quinonez, G.T. Simon, S.Y. Felten and J. Bienenstock, Intestinal mucosal mast cells in normal and nematode-infected rat intestines are in intimate contact with peptidergic nerves. *Proc. Natl. Acad. Sci. USA* 84:2975 (1987).
7. R.H. Stead, J. Bienenstock and A.M. Stanisz, Neuropeptide regulation of mucosal immunity, *Immunol. Rev.* 100:333 (1987).
8. A.M. Stanisz, J. Bienenstock and A. Agro, Neuromodulation of mucosal immunity. *Regional Immunol.* 2:414 (1989).
9. F. Lembeck and P. Holzer, Substance P as a neurogenic mediator of antidromic vasodilatation and neurogenic plasma extravasation. *Naunyn-Schmiedeberg's Arch. Pharmacol.* 310:175 (1979).
10. J.P. Barnes, F.K. Chung and C.P. Page, Inflammatory mediators in asthma, *Pharmacol. Rev.* 40:49 (1988).
11. P.W. Mantyh, M.D. Cotton, C.G. Boehme, M.L. Welton, E.P. Passaro, J.E. Maggio and S.R. Vigna, Receptors for sensory neuropeptides in human inflammatory disease: implications for the effector role of sensory neurons, *Peptides* 10:627 (1989).
12. C.R. Mantyh, T.S. Gates, R.P. Zimmerman, M.L. Welton, E.P. Passard, S.R. Vigna, J.E. Maggio, L. Kruger and P.W. Mantyh, Receptor binding sites for substance P, but not substance K or neuromedin K, are expressed in high concentrations by arterioles, venules, and lymph nodules in surgical speciments obtained from patients with ulcerative colitis and Crohn's disease, *Proc. Natl. Acad. Sci. USA* 85:3235 (1988).
13. B. Pernow, Substance P, *Pharmacol. Rev.* 35:85 (1983).
14. J.V. Weinstock, A. Blum, J. Walder and R. Walder, Eosinophils from granulomas in murine Schistosomiasis Mansoni produce substance P, *J. Immunol.* 141:961 (1988).
15. G.A. Neil, A. Blum and J.V. Weinstock, Substance P but not vasoactive intestinal peptide modulates immunoglobulin secretion in murine schistosomiasis, *Cell Immunol.* 135:394 (1991).
16. D.W. Pascual and K.L. Bost, Substance P production by macrophage cell lines: A possible autocrine function for this neuropeptide, *Immunology* 71:52 (1990).
17. D.G. Payan, D.R. Brewster and E.J. Goetzl, Stereo-specific receptors for substance P on cultured IM-9 lymphoblasts, *J. Immunol.* 133:3260 (1984).
18. A. Eglezos, P.V. Andrews, R.L. Boyd and R.D. Helme, Tachykinin-mediated modulation of the primary antibody response in rats: evidence for mediation by an NK-2 receptor, *J. Neuroimmunol.* 32:11 (1991).
19. P. Parnet, M. Mitsuhashi, C.W. Turck, B. Kerdelhue and D.G. Payan, Tachykinin receptor cross-talk. Immunological cross-reactivity between the external domains of the substance K and substance P receptors, *B.B.I.* 5:73 (1991).
20. R.M. Snider, J.W. Constantine, J.A. Lowe III, K.P. Longo, W.S. Lebel, H.A. Woody, S.E. Drozda,

20. R.M. Snider, J.W. Constantine, J.A. Lowe III, K.P. Longo, W.S. Lebel, H.A. Woody, S.E. Drozda, M.C. Desai, F.J. Vinick, R.W. Spencer and H.J. Hess, A potent nonpeptide antagonist of substance P (NK-1) receptor, *Science* 251:435 (1991).

21. C.A. Maggi, S. Giuliani, L. Ballati, A. Lecci, S. Manzini, R. Patacchini, A.R. Renzetti, P. Rovero, L. Quartara and A. Giachetti, In vivo evidence for tachykininergic transmission using a new NK-2 receptor-selective antagonist MEN10,376, *J. Pharmacol. Exp. Ther.* 257:1172 (1991).

22. M. Mousli, J.L. Bueb, C. Bronner, B. Rouot and Y. Landry, G protein activation: a receptor independent mode of action for cationic amphiphilic neuropeptides and venom peptides, *TiPS.* 11:358 (1990).

23. H. Duplaa, O. Convert, A.M. Sautereau, J.F. Tocanne and G. Chassaing, Binding of substance P to monolayers and vesicle made of phosphatidylcholine and/or phosphatidylserine, *Biochim. Biophys. Acta.* 1107:12 (1992).

24. A. Wozniak, R. Scicchitano, W.H. Betts, G. McLennan, The effect of substance P on neutrophil function in normal and asthmatic subjects, *Ann. NY Acad. Sci.* 650:154 (1992).

25. A. Kavelaars, F. Jeurissen, J. von Frijtag, D. Kunzel, J.H. van Roijen, G.T. Rijkers and C.J. Heijnen, Substance P induces a rise in intracellular calcium concentration in human T lymphocytes in vitro: evidence of a receptor-independent mechanism, *J. Neuroimmunol.* 42:61 (1993).

26. Z. Bar-Shavit, R. Goldman, Y. Stabinsky, P. Gottlieb, M. Fridkin, V.I. Teichborg and S. Blumberg, Evidence of phagocytosis - a newly found activity of substance P residing in its N-terminal tetrapeptide sequence, *Biochem. Biophys. Res. Com.* 94:1445 (1980).

27. H.P. Hartung, K. Wolters and K.V. Toyka, Substance P: binding properties and studies on cellular responses in guinea-pig macrophages, *J.Immunol.* 136:3856 (1986).

28. W. Piotrowski and J.C. Foreman, On the action of substance P, somatostatin and vasoactive intestinal polypeptide on rat peritoneal mast cells and human skin, *Arch. Pharmacol.* 331:364 (1985).

29. I. Hafstrom, H.Gyllenhammar, J. Palmblad and B. Ringertz, Substance P activates and modulates neutrophil oxidative metabolism and aggregation, J. Rheumatol. 16:1033 (1989).

30. J. Gauldie, A. Xaubet, T. Otoshi, M. Jordana and A. Stanisz, Neuropeptide mediated relese of Il-6 and other cytokines from human fibroblasts. 7th Int. Cong. Immunol. West Berlin. (1989).

31. P. Rameshwar, D. Ganea and P. Gascon, In vitro stimulatory effects of substance P on hematopoiesis, *Blood.* 81:391 (1993).

32. P. Rameshwar, P. Gascon and D. Ganea, Immunoregulatory effects of neuropeptides. Stimulation of interleukin-2 production by substance P, *J. Neuroimmunol.* 37:65 (1992).

33. T.C. Moore, Modification of lymphocyte traffic by vasoactive neurotransmitter substances, *Immunology* 52:511 (1984).

34. R.D. Helme, A. Eglezos, G.W. Dandie, P.V. Andrews and R.L. Boyd, The effect of substance P on the regional lymph node antibody response to antigenic stimulation in capsaicin-pretreated rats, *J.Immunol.* 139:3470 (1987).

35. K.L. Bost, D.W. Pascual, Substance P: a late-acting B lymphocyte differentiation cofactor, *Am. J. Physiol.* 262:C537 (1992).

36. A.M. Stanisz, R. Scicchitano, P. Dazin, J. Bienenstock and D.G. Payan, Distribution of substance P receptors on murine spleen and Peyer's patch T and B cells, *J. Immunol.* 139:749 (1987).

37. K.W. Marshall, B.Chiu and R.D. Inman, Substance P and arthritis: analysis of plasma and synovial fluid levels, *Arthritis Rheumat.* 33:87 (1990).

38. M. Gronblad, Y.T. Konttinen, O. Korkala, P. Liesi, M. Hukkanen and J.M. Polak. Neuropeptides in synovium of patients with rheumatoid arthritis and osteoarthritis, *J. Rheumatol.* 15:1807 (1988).

39. J.D. Levine, D.H. Collier, A.I. Basbaum, M. A. Moskowitz and C.A. Helms, Hypothesis: the nervous system may contribute to the pathophysiology of rheumatoid arthritis, *J. Rheumatol.* 12:406 (1985).

40. V.E. Eysselein, M. Reinshagen, F. Cominelli, C. Sternini, W. Davis, A. Patel, C. C. Nast, D. Bernstein, K. Anderson, H. Khan and W.J. Snape Jr., Calcitonin-related gene peptide and substance P decrease in rabbit colon during colitis, *Gastroenterology* 101:1211 (1991).

41. M. Reinshagen, A. Patel, M. Sottili, W. Davis and V.E. Eysselein, Regulation of substance P gene expression in experimental colitis, *Gastroenterology* 102:A505 (1992).

42. A.M. Stanisz, A.D. Befus and J. Bienenstock, Differential effect of vasoactive intestinal peptide, substance P and somatostatin on immunoglobulin synthesis and cell proliferation by lymphocytes from spleen, Peyer's patches and mesenteric lymph nodes, *J. Immunol.*136:152 (1986).

43. R. Scicchitano, A.M. Stanisz and J. Bienenstock, In vivo immunomodulation by the neuropeptide substance P, *Immunology* 63:733 (1988).
44. I. Padol, A. Agro and A.M. Stanisz, Stimulatory effect of the neuropeptide substance P on antigen-specific immune response in mice, *P.N.E.I.* 3:277 (1990).
45. K. Croitoru, P.B. Ernst, J. Bienenstock, I. Padol and A.M. Stanisz, Selective modulation of the natural killer activity of murine intestinal intraepithelial leukocytes by the neuropeptide substance P, *Immunology* 71:196 (1990).
46. M.G. Swain, A. Agro, P. Blennerhassett, A.M. Stanisz and S.M. Collins, Increased levels of substance P in the myenteric plexus of Trichinella-infected rats, *Gastroenterology* 102:1913 (1992).
47. A. Agro and A.M. Stanisz, Depletion of substance P levels reduces intestinal inflammation and restores lymphocyte reactivity to substance P in Trichinella spiralis infected mice, *Reg. Immunol.* 5:120 (1993).
48. A.M. Stanisz, Neuroimmunomodulation in the gastrointestinal tract, *Ann. N.Y. Acad. Sci. USA* (in press).

EXPRESSION OF CHOLINERGIC STRUCTURES ON LYMPHOCYTES

Judith Szelenyi and Nguyen Thi Hu Ha

National Institute of Haematology
Blood Transfusion and Immunology
Budapest, Hungary

INTRODUCTION

A number of receptors, surface markers and enzymes are known to be expressed commonly on cells of nervous-, neuroendocrine- and immune-systems. The presence of cholinergic structures, i.e. muscarinic (mAChR) and nicotinic (nAChR) receptors as well as acetylcholinesterase (AChE) on peripheral blood lymphocytes (PBLs) was reported by several authors.[1-4] As might be expected, lymphocytes respond to parasympathetic signals. mAChR agonists increase PBL motility and chemotaxis,[5,6] rosette formation[7] and proliferation[2] while unilateral parasympathectomy results in decreased plaque-forming cell response in submandibular lymph nodes.[8] *In vitro* inhibition of AChE results in decreased T-cell rosette formation[9] while the *in vivo* administration of physostigmine increases the number of plaque-forming cells.[10]

As it is well known in immunology, the expression of certain surface markers on lymphocytes is influenced by several factors. First of all, the distribution of a given structure might be unequal on different lymphocyte populations/subsets. Strom et al.[11] reported that specific muscarinic-type receptors are associated exclusively with T-lymphocytes.

The appearance and ligand binding characteristics of surface receptors are modulated by the environment, i.e. their expression and activity can be changed by certain drugs, ions or internal factors. The effect of cholinergic agonists and AChE inhibitors were studied *in vivo* and *in vitro*.[7,9,10] Lymphocyte activation results in a number of changes in the membrane, in the expression of certain receptors and/or enzymes according to the state of differentiation. Expression of mAChR on maturing rat thymocytes was described by Maslinski et al.[12] Previously we compared human peripheral blood lymphocytes and thymocytes, and found that thymocytes had lower AChE activity, lacked mAChR, whereas nAChR values were normal.[13]

Little is known about the age-related physiological changes of extraneuronal cholinergic structures. There are data published on the decreased functional activity of T-cells and on the decreased ratio of T/B lymphocytes in the elderly.[14,15] Since

Advances in Psychoneuroimmunology, Edited by
I. Berczi and J. Szélenyi, Plenum Press, New York, 1994

there are well known changes in cholinergic transmission in certain brain regions of normal aged people, it seemed to be worthwhile to follow the expression of the elements of the cholinergic system on peripheral blood lymphocytes (PBL) during aging. We have assumed that not only the defects of the immune system, but some other disorders - most notably the neurological ones, where the cholinergic transmission is damaged - influence the presence and activity of cholinergic structures on PBLs.

In the present work, some new data are given on the possible role of cholinergic structures in activation and differentiation of lymphocytes in healthy aged and in some neurologically affected persons.

RESULTS AND DISCUSSION

The expression and role of cholinergic structures on peripheral blood lymphocytes in health and disease has not been studied adequately. We demonstrated an uneven distribution of AChE in lymphocytes, i.e. a significantly higher activity was found on T-cells than on their B-cell counterparts.[4] Since the T/B cell ratio is decreasing during physiological aging, it was interesting to study whether this was reflected in a decreased AChE activity as well.

As it is shown in Table 1 and Figure 1, the AChE activity remained unchanged in various age groups while there was a significant age-dependent decrease in both mAChR and nAChR.

Table 1. Cholinergic structures of PBLs from young and aged subjects.

Parameter studied	n	Age group (years)	Mean age[a] (years)	Activity/Concentration (Mean values $\pm$ SE)	P
AChE	36	<50	35.5 ± 1.3	21.3 ± 0.7[a]	>0.5
	29	>60	70.5 ± 1.5	20.5 ± 1.0[a]	
mAChR	50	<50	33.7 ± 1.1	0.105 ± 0.003[b]	<0.001
	32	>60	69.6 ± 1.4	0.078 ± 0.004[b]	
nAChR	10	<50	35.2 ± 2.8	0.54 ± 0.04[b]	<0.01
	12	>60	70.3 ± 1.9	0.34 ± 0.04[b]	

[a] pM/min/10^6 cells; [b] pM/10^6 cells

Alteration of the cholinergic system in certain brain areas is one of the well-known characteristics of aging and also found in senile dementia of Alzheimer type (SDAT). Therefore, the expression of the cholinergic structure on PBLs was compared in young, healthy aged, and SDAT groups (Figure 2).

This study revealed that there was no significant change in AChE activity in the healthy aged group; however, significantly lower enzyme activity was found on PBLs from the SDAT patients. The number of nAChR represented by the Bmax values, was decreased in the elderly and the SDAT groups while KD values remained normal. In spite of these facts, mAChR binding in the SDAT group was characterized by a decreased affinity (i.e. higher KD), while the affinity of mAChR in the aged group remained unchanged accompanied by a decreased Bmax value (Figure 2).

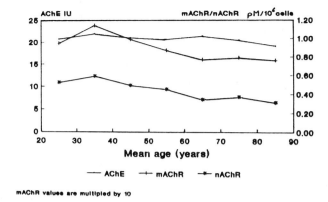

Figure 1. Expression of the cholinergic structures in various age-groups. AChE, acetylcholinesterase given in international units (IU). mAChR = muscarinic acetylcholine receptor, nAChR = nicotinic acetylcholine receptor. Both values are given in picomoles (pM) per 10^6 cells. The values for mAChR have been multiplied by a factor of 10.

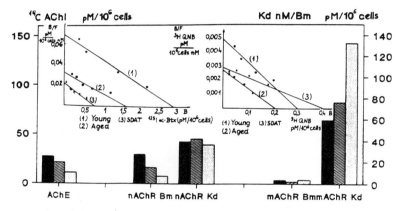

Figure 2. Expression of the cholinergic structures on peripheral blood lymphocytes of young and aged individuals and of patients with Alzheimer's disease.
^{14}CAChl = radioactive acetylcholine, pM = picomole, Kd = dissociation constant, Bm = maximum binding, AChE = acetylcholine esterase, nAChR = nicotinic acetylcholine receptor, mAChR = muscarinic acetylcholine receptor, SDAT = senile dementia of the Alzheimer type, ^{125}I- Btx = ^{125}Iα -bungarotoxin; ^{3}H-QNB = ^{3}H-Quinuclidinyl benzoate.

The decreased mAChR binding in SDAT is in accordance with the observations of other authors,[16,17] who suggested that cholinergic muscarinic binding by PBLs might serve as a useful peripheral marker reflecting neurological changes due to drugs, aging or disease. The data presented here on the decreased number of mAChR binding sites of healthy aged is in contradiction with the increased mAChR binding capacity demonstrated by Rabey et al.[16] Technical differences could account for this discrepancy. An increased mAChR binding in the aged would be difficult to explain, considering the decreased number and functional activity of T-cells in this group.

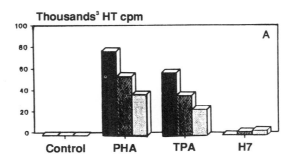

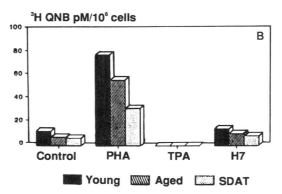

Figure 3. The response of peripheral blood lymphocytes from young, aged individuals and patients with Alzheimer's disease (SDAT) to various stimuli. PHA = phytohemagglutinin, TPA = phorbol ester, H7 = protein kinase C inhibitor.

Some immune functions and lymphocyte proliferation are declining during aging.[14,15,18] The data presented in Figure 3 show that the response (measured by ^{3}H-thymidine incorporation) of PBLs from aged and SDAT groups to various stimuli was significantly decreased as compared to the healthy young group. Similar changes occur in mAChR expression as it was shown earlier.[19] mAChR expression was increased after 72 hours due to phytohemagglutinin (PHA) or antibody (OKT3) stimulation, while treatment of PBLs with phorbol ester (TPA) resulted in the disappearance of mAChR from the cell surface (Figure 3B). This was possibly due to an enhanced phosphorylation by protein kinase C (PKC) which is activated specifically by TPA. Therefore, we were interested to study whether the decreased mAChR expression observed in the aged and SDAT groups was also connected to an enhanced phosphorylation. Inhibition of PKC by H7 (a known specific inhibitor of PKC) resulted in a slight increase in ^{3}H-thymidine incorporation in the aged and SDAT groups and restored the decreased mAChR binding almost completely. The higher PKC activity of PBLs in elderly could also be shown directly and was accompanied by an increased membrane/cytosol ratio (M/C=0.36 for young and 0.57 for the aged group) (Figure 4).

Muscarinic receptors are known to be coupled to a G-protein which interacts in the signal-transduction due to ligand binding. Studying the GTP binding and G-protein function of PBLs from the young, aged and SDAT groups, an increased GTP binding was observed in both the aged and the SDAT group which was highly significant in the latter (Figure 5). Alteration of the G-protein function in these

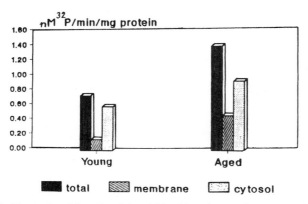

Figure 4. Protein kinase C activity of peripheral blood lymphocytes in young and aged humans.

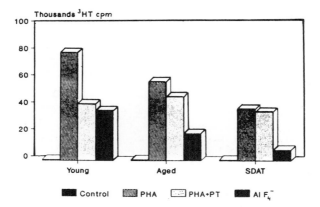

Figure 5. Altered G protein function in lymphocytes from aged individuals and from patients with Alzheimer's disease (SDAT). PHA = phytohemagglutinin, PHA-PT = phytohemagglutinin and pertussis toxin, ALF_4^- = non-hydrolysing analogue of GTP.

groups could be revealed by the facts that the G-protein modulating agent, pertussis toxin (PT), failed to inhibit PHA stimulation in the aged and SDAT groups, while AlF4, the non-hydrolyzing analogue of GTP, did not stimulate PBLs from these groups in comparison to PBLs from young individuals (Figure 6).

These data support our hypothesis that signal transduction is altered by the mAChR in aged and SDAT PBLs, which is possibly due to a permanently activated protein kinase C. It is likely that the PKC isoenzyme pattern is altered in these cases towards a less calcium sensitive enzyme. Perhaps, minor differences in membrane-lipid composition[20] could also contribute to keeping a higher part of PKC attached to the membrane.

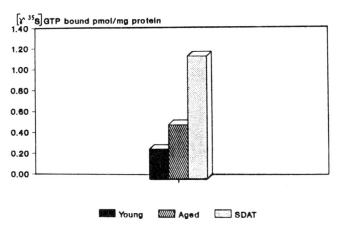

Figure 6. GTP binding by G proteins in lymphocytes from young and aged individuals and from patients with Alzheimer's disease (SDAT).

CONCLUSIONS

Alteration of cholinergic structures on PBLs during normal aging or in dementia of the Alzheimer type (SDAT) can be demonstrated, which is associated with altered responses to various stimuli. mAChR is downregulated by activation of PKC with TPA in PBLs. Inhibition of PKC restores slightly the response to stimulation and completely the expression of mAChR in the aged and SDAT groups. These findings suggest that PKC is constitutively activated in lymphocytes of aged individuals and of patients suffering from SDAT. Indeed, it was found that in these cases mAChR is coupled to a functionally altered G-protein. It was also demonstrated that the alteration of G-protein is caused by the permanent activation of PKC. This constitutively higher PKC activity (possibly due to some change in its isoenzyme pattern) resulted in higher GTP binding and possibly increased GTP hydrolysis accompanied by an altered α-subunit dissociation. This PKC abnormality might be the specific defect underlying the abnormal signal-transduction of PBLs in healthy aged individuals, and in a more pronounced form in lymphocytes of SDAT patients.

REFERENCES

1. T.B. Strom, A. Deisseroth, J. Morganroth, C.B. Carpenter, and J.P. Merrill, Alteration of cytotoxic action of sensitized lymphocytes by cholinergic agents and activators of adenylate cyclase, *Proc. Natl. Acad. Sci.* 69:2995 (1972).
2. D.P. Richman and B.G.W. Arnason, Nicotinic acetylcholine receptor: evidence for functionally distinct receptor on human lymphocytes, *Proc. Natl. Acad. Sci.* 76:4632 (1979).
3. S.J. Zalcman, L.M. Neckers, O. Kayaalp, and R.J. Wyatt, Muscarinic cholinergic binding sites on intact human lymphocytes, *Life Sci.* 29:69 (1983).
4. J.G. Szelenyi, E. Bartha, and S.R. Hollan, Acetylcholinesterase activity of lymphocytes: an enzyme characteristic of the T-cells, *Brit. J. Haematol.* 50:241 (1982).
5. G.F. Schreiner and E.R. Unanue, The modulation of spontaneous and anti IgG stimulated motility of lymphocytes by cyclic nucleotides and adrenergic and cholinergic agents, *J. Immunol.* 114:802 (1975).
6. S. Sipka, P. Dioszeghy, F. Melcher, K. Lukacs, J. Szelenyi, and G. Szegedi, Decreases in the carbamoylcholine-induced chemotaxis of monocytes in myasthenia gravis, *Eur. Arch. Psychiatr. Clin. Neurosci.* 240:279 (1991).

7. M.H. Grieco, I. Siegel, and Z. Goel, Modulation of human T lymphocyte rosette formation by autonomic agonists and cyclic nucleotides, *J. Allergy Clin. Immunol.* 58:149 (1976).

8. A. Alito, H.E. Romeo, R. Baler, H.E. Chuluyan, M. Braun, and D.P. Cardinali, Autonomic nervous system regulation of murine immune responses as assessed by local surgical sympathetic and parasympathetic dennervation, *Acta Physiol. Pharmacol. Latinoam.* 37:305 (1987).

9. R.K. Chandra and K. Madhavankutty, Effect of cholinesterase inhibition by eserine and phospholipase D on human T-cell rosetting, *Experientia* 31:858 (1975).

10. I. Rinner and K. Schauenstein, The parasympathetic nervous system takes part in the immuno-neuroendocrine dialogue, *J. Neuroimmunol.* 34:165 (1991).

11. T.B. Strom, M.A. Lane, and K. George, The parallel, time-dependent bimodal change in lymphocyte cholinergic binding activity and cholinergic influence upon lymphocyte-mediated cytotoxicity after lymphocyte activation, *J. Immunol.* 127:705 (1981).

12. W. Maslinski, E. Grabczewska, H. Laskowska-Bozek, and J. Ryzewski, Expression of muscarinic cholinergic receptors during T-cell maturaiton in the thymus, *Eur. J. Immunol.* 17:1059 (1987).

13. P. Paldi-Haris and J. Szelenyi, Changes in the expression of the cholinergic structures of human T lymphocytes due to maturation and stimulation, *Thymus* 16:119 (1990).

14. W. Augener, G. Cohnen, A. Reuter, and G. Brittinger, Decrease of T lymphocytes during ageing, *Lancet* 1:1164 (1974).

15. A. Ben-Zvi, U. Galili, A. Russel, and M. Schlesinger, Age associated changes in subpopulations of human lymphocytes, *Clin. Immunol. Immunopathol.* 7:139 (1977).

16. J.M. Rabey, L. Schenkman, and G.M. Gilad, Cholinergic muscarinic binding by human lymphocytes: changes with ageing, antagonist treatment, and senile dementia of the Alzheimer type, *Ann. Neurol.* 20:628 (1986).

17. A. Adem, A. Nordberg, G. Bucht, and B. Winblad, Extraneuronal cholinergic markers in Alzheimer's and Parkinson's disease, *Prog. Neuro-Psychopharmacol. Biol. Psychiat.* 10:247 (1986).

18. L. Bonomo, S. Antonici, and E. Jiriollo, Cell mediated immune response in the elderly, *Bull. Inst. Pasteur* 81:347 (1983).

19. J.G. Szelenyi, P. Paldi-Haris, and S. Hollan, Changes in the cholinergic system of lymphocytes due to mitogenic stimulation, *Immunol. Lett.* 16:49 (1989).

20. G. Freund, T.R. Brophy, and J.O. Scott, Membrane fluidity increases low-affinity muscarinic receptor binding in brain: changes with ageing, *Exp. Geront.* 21:37 (1986).

REGULATION OF NK CELLS DURING ACUTE PSYCHOLOGICAL STRESS BY NORADRENALINE

Reinhold E. Schmidt [1], Roland Jacobs [1], Thomas O.F. Wagner [2], Uwe Tewes [3] and Manfred Schedlowski [3]

Divisions of [1] Immunology, [2] Endocrinology, [3] Medical Psychology
Hannover Medical School
30623 Hannover, FRG

INTRODUCTION

Stress has been suggested for a long time to be associated with a decrease in natural killer (NK) activity. There is also accumulating evidence from psychoneuroimmunological research[1] that stress situations such as examination[2,3] or bereavement[4] may alter a wide range of immunological functions, especially NK cell function. Moreover, increased incidence of viral infections or cancer have been reported in stressed populations[5,6].

NK cells are believed to act early in the immune response before specificity can be generated, and represent approximately 12-15% of peripheral blood lymphocytes[7-9]. They mediate first line defense by direct cytotoxicity against various types of target cells without apparent prior immunization[10]. More recently they have been characterized phenotypically as coexpressing the cell surface antigens CD16 and CD56 and lacking the CD3/TCR complex[11]. NK cells have also been demonstrated to play an important role in the immune surveillance against tumors and virally infected cells[9,10]. Other reports have shown that NK cells represent a subset of peripheral blood lymphocytes which quickly respond to activation signals such as IL-2 via the respective cytokine receptors in vivo and in vitro[7,12]. In vitro stimulation of NK cells with high concentrations of adrenaline causes a decrease of NK function. In contrast, lower concentrations appear to increase NK activity[13]. In vivo administration in humans also induces an increase of NK activity with a rapid decrease during follow-up[14,15]. Other in vivo data in humans during physical stress reported an increase of NK activity after exercise[16-18]. These observations together with data demonstrating that lymphocytes express receptors for a variety of hormones[19] suggest that various neuroendocrine parameters may influence NK activity in peripheral blood. Whether psychological stress is influencing peripheral blood lymphocytes and in particular NK cells and which hormones are involved has not been established yet, since the manipulation of the sympathetic-adrenal system in vivo has remained difficult.

Advances in Psychoneuroimmunology, Edited by
I. Berczi and J. Szélenyi, Plenum Press, New York, 1994

279

In order to analyze physiological parameters such as hormones and alterations in the cellular immune system in an acute psychological stress situation a model with a high degree of experimental control was chosen[20-22]. First-time tandem parachutists were continuously monitored for their heart rate and plasma concentrations of the sympathetic-adrenal hormones cortisol and catecholamines before, during and after jumping. At three defined time points lymphocytes and their respective subsets were characterized for cell surface antigens and their functional capacity.

MATERIALS AND METHODS:

Subjects and Procedure

45 male subjects aged 19-39 years (Mean=25,4) participated in this study after having given informed consent. All subjects had passed an intensive health examination before. Subjects with drug or alcohol abuse, medication or infections within the last two weeks were excluded. All subjects performed tandem-jumps, i.e. the novice was secured in front of an experienced tandem-master. Jumps were performed between 10:00 and 12:00 am. Boarding time was 30 minutes before exit. During ascent of the plane to exit altitude (3500m) subjects sat in the plane without being physically active. Jumpers had 45 seconds free-fall time until the parachute was released and approximately 5 minutes under the open parachute until landing. Subjects completed the State-Trait-Anxiety-Inventory (STAI)[23] to determine the state-anxiety at baseline and immediately before exit. Heart rate was continuously recorded with Ag-AgCl electrodes in the lead II configuration by using a portable recording system[24]. For continuous blood sampling a catheter (certofix, Braun Melsungen, Melsungen, Germany) was inserted into a brachialis vein. Two to three hours before jumping, the catheter was connected to a heparinized silicon tubing (1mm ø). Blood was continuously drawn by a small portable pump (Fresenius, Homburg, Germany) and collected in lithium heparin tubes containing glutathione. Blood flow was adjusted to 0.8 ml/min. Every 10 minutes the tubes were changed and samples for the endocrine variables were fractionated. The samples were centrifuged at 4°C, and stored frozen at -70°C until assayed. In addition, blood samples were drawn to determine immunological parameters: two hours before the jump (baseline), immediately after and one hour after the jump. All analyses were performed immediately after the last sample was collected.

Endocrine Analyses

All samples of each subject were analyzed in the same assay. Cortisol was determined by radioimmunoassay (Biermann GmbH, Bad Nauheim, Germany). Catecholamine plasma levels were analyzed by high pressure liquid chromatography (HPLC, electrochemical detector).

Monoclonal Antibodies

The monoclonal antibodies used have been described in detail previously[25]: CD2 (Leu5b), CD3 (Leu4), CD8 (Leu2a), CD16 (Leu11b), CD56 (Leu19), and CD25 were purchased from Becton Dickinson (Heidelberg, Germany). CD3 (OKT3), CD4 (OKT4), CD8 (OKT 8) were from Ortho (Neckargemünd, Germany). CD2R (T11.3), CD26, and CD56 (NKH-1A) were generously provided by Drs. S.F. Schlossmann, E.L. Reinherz and J. Ritz (Dana Farber Cancer Institute, Boston, USA). Tu27 (anti-p75) was generously given by Dr. K. Sugamura (Department of Microbiology, Tohuko University School of Medicine, Sendai, Japan).

Phenotypic Analyses

Phenotypic analyses were performed by indirect immunofluorescence using fluorescein conjugated goat anti mouse F(ab') Ig (GM-FITC, Dianova, Hamburg, Germany). For two colour analyses directly labeled antibodies were utilized. The staining procedure has been described in detail previously[26]. Briefly, 1 to 3×10^5 cells/well were incubated with murine monoclonal antibodies at an optimal dilution for 30 min. Nonspecific binding was eliminated by mixing the samples with a 1:5 solution of a commercial human IgG (Intraglobin, Biotest, Frankfurt, Germany). Samples were incubated for another 30 min with GM-FITC, washed three times and 10 000 cells were then analyzed using a FACScan (Becton Dickinson, Heidelberg, Germany).

Cell Lines

Two continuously growing cell lines were used: the K562 erythromyeloid human line, and the L1210 murine lymphoma cell line.

Cytotoxicity Assays

All cytotoxicity assays were performed in triplicates at various effector to target (E/T) ratios using V-bottom microtiter plates with 5×10^3 ^{51}Cr-labeled target cells per well as previously described in detail[26]. Medium for cytotoxicity assays was RPMI 1640 supplemented with 5% FCS and 1% penicillin-streptomycin. Specific cytotoxicity was measured following 4 hrs of incubation at 37^0C by determining the ^{51}Cr release. Spontaneous release was determined by incubating target cells without effector cells. Maximal release was obtained by lysing target cells with the detergent NP40. The specific lysis was calculated as described[27].

Statistical Analyses

Statistical analyses of the data were performed by using ANOVAs with repeated measures. All repeated measures analyses of variance were corrected for nonsphericity by the Geiser-Greenhouse method[28]. For each repeated measure we report the uncorrected degrees of freedom and the epsilon-corrected p value.

RESULTS

State-anxiety was within normal range at baseline (M = 34±7.4) and significantly increased immediately before exit (M = 45±9.4) (df=1, F=63.43, P<0.0001). The heart rate as representative parameter for cardiovascular reactivity was recorded throughout 3 hours, i.e. during preparation, at exit and after landing (Fig. 1A). Heart rate increased before boarding (30 minutes before exit) and reached 152 beats/min during exit. The frequency normalized 20 minutes after the jump (df=18, F=79.92, P<0.0001). In parallel to the heart rate, plasma concentrations of cortisol and catecholamines in 10 minutes intervals from 120 minutes before to 60 minutes after the jump were determined (Figures 1 B-D). Adrenaline peaked at 400 ng/l synchronously with the heart rate during the jump itself (df=18, F=60.03, P<0.0001)

(Fig. 1 C). Cortisol and noradrenaline peak levels appeared slightly delayed when compared to adrenaline levels (Fig. 1 B and D). Peak values for cortisol were 23 mg/dl and did not reach baseline levels within 1 hour after exit (df = 18, F = 17.72, P < 0.0001). In contrast, noradrenaline (df = 18, F = 19.00, P < 0.0001) and adrenaline quickly decreased within 20 minutes after exit and reached normal plasma concentrations. No significant alteration could be observed for plasma levels of dopamine 120 minutes before to 60 minutes after the jump (df = 18, F = 1.27; n.s.) (data not shown).

When peripheral blood lymphocytes and their respective subsets were measured before, immediately after and 1 hour after jumping, a significant increase of $CD2^+$ cells (T and NK cells) (df = 2, F = 20.48, P < 0.0001) and also of $CD3^+$ (df = 2, F = 10.92, P < 0.0001), $CD4^+$ (df = 18, F = 5.86, P < 0.01), and $CD8^+$ lymphocytes (df = 18, F = 28.64, P < 0.0001) was observed (Fig. 2 A).

The T lymphocytes 1 hour after jumping were significantly decreased compared to baseline values before exit (P < 0.01). The most impressive changes were found for NK cells. Immediately after the jump the absolute number of $CD16^+$ cells increased by more than 100 %. One hour after jumping a decrease below baseline levels (280 NK cells/μl) was observed (df = 2, F = 41.58, P < 0.0001). Similar numbers were obtained when $CD56^+$ cells (df = 2, F = 36.56, P < 0.0001) were determined (Fig. 2 B).

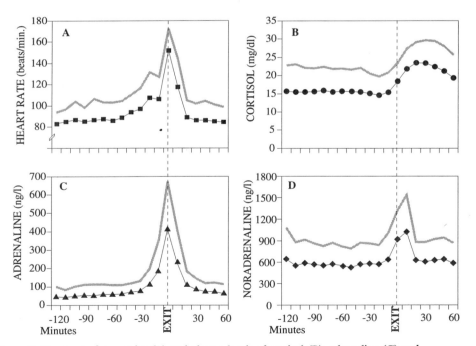

Figure 1. Heart rate frequencies (**A**) and plasma levels of cortisol (**B**), adrenaline (**C**) and noradrenaline (**D**) before, during and after the jump in 10 minutes intervals from 120 minutes before to 60 minutes after the jump (data are expressed as means ± standard deviation ▦).

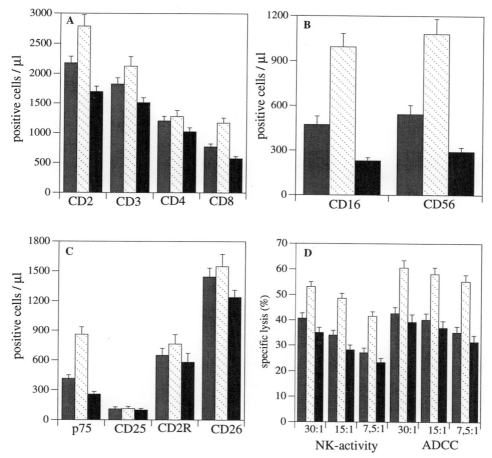

Figure 2. Lymphocyte subsets, NK-activity and ADCC before, immediately after and one hour after the jump. In panels **A** + **B** T and NK cells are depicted. Panel C presents various activation antigens and Panel D shows NK-activity and ADCC. Heparinized blood was drawn two hours before (▨), immediately after (▢), and one hour after the jump (■). Data are presented as means ± standard errors (STE).

In parallel to these enormous changes of the NK cells alterations of activation antigens occured (Fig. 2 C). The number of p75[+] lymphocytes increased significantly immediately after the jump and dropped thereafter (df=2, F=39.34, P<0.0001). The pattern was completely parallel to the changes of NK cells. p75 is known to be mainly expressed on NK cells (11). Using two colour fluorescence it was demonstrated that nearly all CD56[+] NK cells express p75[+] IL-2Rβ chain (data not shown). In addition, there was a significant change in CD26[+] cells (df=2, F=5.49, P<0.01), while statistical analysis revealed no significant differences for CD25[+] and CD2R[+] cells.

NK activity and ADCC were determined in parallel to the phenotypic studies to test the NK lymphocytes for their functional capacity. Significant changes for both types of cytotoxic activities were observed demonstrating that the increase of absolute NK cell numbers and the relatively enhanced frequency in peripheral blood was paralleled by an increase of specific cytotoxicity (NK and ADCC at an E:T ratio of 15:1 / df=2, F=64.77, P<0.0001 / df=2, F=51.01, P<0.0001) (Fig. 2 D).

In order to analyze which of the psychological, cardiovascular and endocrine variables studied correspond to the observed alterations in immune functions, Pearson correlations of anxiety measures, heart rate, endocrine variables and immunological parameters were calculated. There were no significant correlations between the anxiety measures, heart rate, cortisol and immunological variables. However, catecholamines especially noradrenaline plasma levels in 10 minute intervals correlated significantly with $CD8^+$ cells, NK cell numbers ($CD16^+$ and $CD56^+$) and functions (NK cytotoxicity) immediately after the jump and one hour after the jump (Tables 1 and 2).

Table 1: Matrix of significant Pearson correlations between plasma noradrenaline in 10 minute intervals prior to and including exit on the one hand and on the other $CD8^+$, $CD16^+$, $CD56^+$ cells and the NK activity immediately after the jump.

Intervals prior to and including Exit (min.)				
120'	---	---	---	---
110'	.42*	.42**	.39*	.62**
100'	.43**	.46**	.42*	.57**
90'	.44**	.49**	.46**	.59**
80'	.38*	.45**	---	.39*
70'	.37*	.46**	.34*	.43**
60'	.40*	.42**	.38*	.49**
50'	.35*	.39*	---	---
40'	.44*	.42**	.41**	.48**
30'	.46**	.42**	.40**	.42**
20'	.47**	.45**	.39*	.36*
10'	.35*	.35*	.33*	.41**
EXIT	---	---	---	.33*
immediately after jumping	$CD8^+$	$CD16^+$	$CD56^+$	NK-activity

(---) not-significant / * $p<.05$ / ** $p<.01$

In contrast, the plasma concentrations of adrenaline significantly correlated only in the sample 30 minutes before the jump with the $CD56^+$ cells immediately after jumping (r=0.37, P<0.05) and adrenaline plasma levels collected 50 minutes after the jump correlated significantly with CD16$^+$ cells (r=0.39, P<0.05), CD56$^+$ cells (r=0.37, P<0.05) and the NK activity (r=0.34, P=0.05) one hour after jumping.

Table 2: Matrix of significant Pearson correlations between plasma noradrenaline in 10 minute intervals after the jump and CD8$^+$, CD16$^+$, CD56$^+$ cells and the NK-activity one hour after the jump.

Intervals after jumping (min.)				
10'	.39*	---	.33*	---
20'	.39*	---	.38*	---
30'	.50**	.45**	.47**	.32*
40'	.45**	.46**	.41**	---
50'	.52**	.43**	.49**	.35*
60'	.47**	.47**	.47**	.35*
1 hr after jumping	CD8 $^+$	CD16 $^+$	CD56 $^+$	NK-activity

(---) not-significant   / * p<.05/ ** p<.01

DISCUSSION

An increase of NK cells and their functions has been well documented during physical exercise. Possible alterations of the immune system during psychological stress and the hormones involved are not clear [29].

A parachute jump has been employed previously as a model of well controlled intensive psychological stress showing that the increase in cardiovascular activity and hormonal response is principally caused by the psychophysiological arousal prior to and during a jump [20-22]. The present study, using this model demonstrated that emotional strain induces significant changes not only of various hormones but also of leukocytes and their subsets [30]. An increase of granulocytes and a significant change in absolute numbers of lymphocytes were observed. Amongst the various circulating lymphocyte subsets the most significant increases were determined for CD8 $^+$ cells and in particular for NK cells (Fig. 2 B). The latter quickly more than double in the stress situation, drop thereafter and subsequently return to levels significantly below baseline. In addition examination of activation antigens revealed a significantly enhanced expression of the intermediate affinity IL-2 receptor p75 which reflects the NK cell increase. Functional studies demonstrated that increased numbers of NK cells are accompanied by significantly enhanced NK and ADCC activity. The decrease of NK cell numbers after the jump is paralleled by a functional loss of activity (Fig. 2 D). Studies in individuals during physical exercise have also reported increased NK cells and activity [15-18] followed by a decrease after exercise [17,31]. Here we demonstrate that psychological stress also alters the distribution of circulating lymphocytes to the same extent. It remains to be determined whether this is a general post stress phenomenon. In contrast to the phenotypic changes observed in increased NK cells after bone marrow transplantation or in vivo IL-2 application [32,33], the surface antigen characteristics of the enhanced NK cells immediately after the jump are not changed to high density CD56 expression suggesting quick mobilisation of NK lymphocytes from marginal pools in the spleen and the lung.

The enhanced secretion of cortisol, adrenaline and noradrenaline represents the sympathetic-adrenal activation during psychophysiological arousal. While suppressive in vitro effects of glucocorticoids on NK cell functions have been described [34], in our study there is no correlation between plasma levels of cortisol and lymphocyte subpopulations or NK-cell functions at any time. However, the changes in NK cells and activity are significantly correlated with preceding plasma levels of noradrenaline. This analysis is of course restricted to the first hour after jumping and may therefore be biased towards short term effects caused by noradrenaline.

For physical exercise it has been postulated that catecholamines may cause increased NK activity by direct interaction with these lymphocytes. This hypothesis is supported by the observation that in vivo administration of adrenaline in man increased both NK cell numbers and activity [14,15,35]. On NK cells and T cells β-adrenergic receptors have been identified with the highest receptor density on NK cells [36-38]. However, it is still controversial whether the observed increases in lymphocyte subsets are mediated via β_1 or β_2 adrenergic receptors on these lymphocytes. While one study found no changes in the distribution of T and NK cells after infusion of β_1-selective agonist noradrenaline [36] another study reported a significant increase in NK activity after noradrenaline administration and a subsequent decrease 30 minutes later [39]. Correlations between exercise-induced increases in circulating lymphocytes and adrenaline [38] and the noradrenaline associated increase of T and NK cells after moderate exercise [31] are further evidence for an adrenergic stimulation of lymphocytes via β_1 and β_2 receptors.

According to other studies NK cell trafficking could be due to vascular contraction in the spleen and other marginal pools for NK cells which may be predominantly induced by noradrenaline [40]. We suggest that the observed changes of cell numbers in our study most likely must be attributed to cell mobilization from respective reservoirs such as spleen and lung [37]. Alternatively, some of the other sympathetic-adrenal hormones and neuropeptides [41,42] or their interaction with cytokines such as IFN-γ or TNF-α could be responsible. Other studies reported no association between catecholamine levels and increased NK activity during brief psychological stress [43]. Moreover it needs to be considered that exercise induced increase of NK activity was attenuated by prior administration of naloxone, an opioid antagonist [44]. Cytokines and hormones might also alter expression of adhesion structures on NK or endothelial cells allowing the fast NK cell release and most likely a fast pooling thereafter [45]. In contrast to T and B cells, even from animal studies there is not much known on migration and recirculation of NK cells [46]. Further research is needed to clarify these mechanisms. The biological meaning of the increase and subsequent decrease of circulating lymphocytes, particularly of NK cells after acute stress may be interpreted as an adaptation to environmental stimuli. Since evidence is provided here for diminished NK cells after psychological stress, it is of further interest whether this is a transient or long lasting effect which might enhance susceptibility for infections.

In summary, these data suggest that short term psychological stress leads to an enhanced secretion of sympathetic-adrenal hormones with a rapid increase and subsequent decrease of NK cell numbers and their respective functions. These changes are significantly correlated with plasma levels of noradrenaline indicating a central-nervous-immune system interaction during phases of intense emotional stress.

SUMMARY

Emotional stress is often followed by increased susceptibility to infections. One major role in the immediate immune response to infection play natural killer (NK) cells. This study was designed to establish whether acute psychological stress influences cellular immune functions and to elucidate the role of endocrine parameters as potent mediators of stress induced alterations of the immune system. 45 first-time tandem parachutists were monitored continuously for their plasma concentrations of cortisol and catecholamines from 120 minutes before to 60 minutes after jumping. Lymphocyte subsets, NK activity, and ADCC were determined two hours before, immediately after, and one hour after jumping. There was a significant increase of sympathetic-adrenal hormones during and shortly after jumping. Lymphocyte subsets and the functional capacity of NK cells revealed an increase immediately after jumping followed by a decrease significantly below baseline one hour later. These changes were significantly correlated to plasma concentrations of noradrenaline.

ACKNOWLEDGEMENTS

This work was supported by the Volkswagen-Stiftung Germany, grant number I/66 077 and by funds of the Deutsche Forschungsgemeinschaft DFG Schm 596/3-2.

REFERENCES

1. R. Ader, D.L. Felten and N. Cohen. "Psychoneuroimmunology", Academic Press, San Diego (1991).
2. R. Glaser, J. Rice, C.E. Speicher, J.C. Stout and J.K. Kiecolt-Glaser, Stress depresses interferon production by leucocytes concomitant with a decrease in natural killer cell activity, *Behav. Neurosci.* 100:675 (1986).
3. J.K. Kiecolt-Glaser, W. Garner, C. Speicher, G.M. Penn, J. Holliday and R. Glaser, Psychological modifiers of immunocompetence in medical students, *Psychosom. Med.* 46: 7 (1984).
4. M. Irwin, M. Daniels, T.L. Smith, E. Bloom, and H. Weiner, Impaired natural killer activity during bereavement, *Brain Behav. Immun.* 1:98 (1987).
5. S. Cohen, D.A.J. Tyrrell, and A.P. Smith, Psychological stress and susceptibility to the common cold, *N. Engl. J. Med.* 325:606 (1991).
6. S. Levy, R. Herberman, M. Lippman, and T.d´Angelo, Correlation of stress factors with sustained depression of natural killer cell activity and predicted prognosis in patients with breast cancer, *J. Clin. Oncol.* 5:348 (1987)
7. T. Hercend and R.E. Schmidt, Characteristics and uses of natural killer cells, *Immunol. Today* 9:291 (1988).
8. R.E. Schmidt, J. Michon, H. MacMahon, J. Woronicz, S.F. Schlossman, E.L. Reinherz and J. Ritz, Enhancement of natural killer (NK) function through activation of the T11/E rosette receptor, *J. Clin. Invest.* 79:305 (1987).
9. G. Trinchieri, Biology of natural killer cells, *Adv. Immunol.* 46:181 (1989).
10. R. Herberman and J. Ortaldo, Natural killer cells: Their role in defenses against disease, *Science* 214:24 (1981).
11. J. Ritz, T.J. Campen, R.E. Schmidt, H.D. Royer, T. Hercend, R.E. Hussey and E.L. Reinherz, Analysis of T cell receptor gene rearrangement and expression in human natural killer (NK) clones. *Science* 228:1540 (1985).

12. L.L. Lanier, R. Testi, J. Bindl and J.H. Phillips, Identity of Leu-19 (CD56) leukocyte differentation antigen and neural cell adhesion molecule, *J. Exp. Med.* 169:2233 (1989).

13. K. Hellstrand, S. Hermodsson and Ö. Strannegard, Evidence for a beta-adrenoceptor-mediated regulation of human natural killer cells, *J. Immunol.* 134:4095 (1985).

14. E. Tonnesen, N.J. Christensen and M.M. Brinklov, Natural killer cell activity during cortisol and adrenaline infusion in healthy volunteers, *Eur. J. Clin. Invest.* 17:497 (1987).

15. M. Kappel, N. Tvede, H. Galbo, P.M. Haahr, M. Kjaer, M. Linstow, K. Klarlund and B.K. Pedersen, Evidence that the effect of physical exercise on NK cell activity is mediated by epinephrine, *J. Appl. Physiol.* 70:2530 (1991).

16. Z. Brahmi, J.E. Thomas, M. Park, M. Park and I.R.G. Dowdeswell, The effect of acute exercise on natural killer-cell activity of trained and sedentary subjects, *J. Clin. Immunol.* 5:321 (1985).

17. B.K. Pedersen, N. Tvede, F.R. Hansen, V. Andersen, T. Bendix, G. Bendixen, K. Bendtzen, H. Galbo, P.M. Haahr, K. Klarlund, J. Sylvest, B.S. Thomsen and J. Halkjaer-Kristensen, Modulation of natural killer cell activity in peripheral blood by physical exercise, *Scand. J. Immunol.* 27:673 (1988).

18. R.M.A. Landmann, F.B. Müller, C.H. Perini, M. Wesp, P. Erne and F.R. Bühler, Changes of immunoregulatory cells induced by psychological and physical stress: relationship to plasma catecholamines, *Clin. Exp. Immunol.* 58:127 (1984).

19. M. Plaut, Lymphocyte hormone receptor, *Annu. Rev. Immunol.* 5:621-669, 1987

20. H. Ursin, E. Baade and S. Levine. "Psychobiology of Stress. A study of coping men", Academic Press, New York (1987).

21. M. Schedlowski, D. Wiechert, T.O.F. Wagner and U. Tewes, Acute psychological stress increases plasma levels of cortisol, prolactin and TSH, *Life Sci.* 50:1201 (1992).

22. M. Schedlowski and U. Tewes, Physiological arousal and perception of bodily state during parachute jumping, *Psychophysiology* 29:95 (1992).

23. L. Laux, P. Glanzmann, P. Schaffner and C.D. Spielberger. "State-Trait-Angst Inventar (STAI)", Beltz-Testgesellschaft, Weinheim (1984).

24. R.v. Kiparski and H.P. Steffens, Possibilities of wireless recording of biosignals by means of digital memories, *Biomed. Tech.* 30:76 (1985).

25. W. Knapp, B. Dörken, W.R. Gilks, E.P. Rieber, R.E. Schmidt, H. Stein and A.E.G. Kr. von dem Borne. "Leukocyte Typing IV", England, Oxford, University Press (1989).

26. R.E. Schmidt, G. Bartley, H. Levine, S.F. Schlossmann and J. Ritz, .Functional characterization of LFA-1 antigens in the interaction of human NK clones and target cells, *J. Immunol.* 135:1020 (1985).

27. R.E. Schmidt, R.P. Mac Dermott, G.T. Bartley, M. Bertovich, D.A. Amato, K.F. Austen, S.F. Schlossman, R.L. Stevens and J. Ritz, Specific release of proteoglycans from human natural killer cells during target lysis. *Nature* 318:289 (1985).

28. M.W. Vasey and J.F. Thayer, The continuing problem of false positives in repeated measures ANOVA in psychophysiology: A multivariate solution, *Psychophysiology* 24:479 (1987).

29. D.N. Khansari, A.J. Murgo and R.E. Faith, Effects of stress on the immune system, *Immunol. Today* 11:170 (1990).

30. M. Schedlowski, R. Jacobs, G. Stratmann, S. Richter, A. Hädicke, U. Tewes, T.O.F. Wagner and R.E. Schmidt, changes of natural killer cells during acute psychological stress, *J. Clin. Immunol.* 13:119 (1993).

31. D.C. Nieman, S.L. Nehlsen-Cannarella, K.M. Donohue, D.B.W. Chritton, B.L. Haddock, R.W. Stout and J.W. Lee, The effect of acute moderate exercise on leukocyte and lymphocyte subpopulations, *Med. Sci. Sport Exer.* 23:578 (1991).

32. R. Jacobs, M. Stoll, G. Stratmann, R. Leo, H. Link and R.E. Schmidt, CD16⁻ CD56⁺ NK cells after bone marrow transplantation, *Blood* 79:3239 (1992).

33. T. Witte, K. Wordelmann and R.E. Schmidt, Heterogeneity of human NK cells in the spleen, *Immunology* 69:166 (1989).

34. G. Gatti, R. Cavallo, M.L. Sartori, D. Del Ponte, R. Masera, A. Salvadori, R. Carignola and A. Angeli, Inhibition by cortisol of human natural killer (NK) cell activity, *J. Steroid Biochem.* 26:49 (1987).

35. B. Crary, S.L. Hauser, M. Borysenko, I. Kutz, C. Hoban, K.A. Ault, H.L. Weiner and H. Benson, Epinephrine-induced changes in the distribution of lymphocyte subsets in peripheral blood of humans, *J. Immunol.* 131:1178 (1983).

36. M.M. Khan, P. Sansoni, E.D. Silverman, E.D. Engleman and K.L. Melmon, Beta-adrenergic receptors on human suppressor, helper, and cytolytic lymphocytes, *Biochem. Pharmacol.* 35:1137 (1986).

37. L.J.H. Van Tits, M.C. Michel, H. Grosse-Wilde, M. Happel, F.W. Eigler, A. Soliman A, O.E. Brodde, Catecholamines increase lymphocyte beta$_2$-adrenergic receptors via a beta$_2$-adrenergic, spleen-dependent process, *Am. J. Physiol.* 258:E191 (1990).

38. A.S. Maisel, T. Harris, C.A. Rearden and M.C. Michel, b-adrenergic receptors in lymphocyte subsets after exercise, *Circulation* 82:2003 (1990).
39. S. Locke, L. Kraus, I. Kutz, S. Edbril, K. Phillips and H. Benson, Altered natural killer activity during norepinephrine infusion in humans, *in:* "Neuroimmunomodulation," N.H. Spector, ed. Proceedings of the first international workshop on neuroimmunomodulation, Bethesda, Maryland, p. 297 (1984).
40. D.L. Felten, S.Y. Felten, D.L. Bellinger, S.L. Carlson, K.D. Ackerman, K.S. Madden, J.A. Olschowki and S. Livnat, Noradrenergic sympathetic neural interactions with the immune system: Structure and function, *Immunol. Rev.* 100:225 (1987).
41. S.A. Williamson, R.A. Knight, S.L. Lightman and J.R. Hobbs, Differential effects of b-endorphin fragments on human natural killing, *Brain, Behav. Immun.* 1:329 (1987).
42. C.J. Heijnen, A. Kavelaars and R.E. Ballieux, b-endorphin: Cytokine and neuropeptide, *Immunol. Rev.* 119:41 (1991).
43. B.D. Naliboff, D. Benton, G.F. Solomon, J.E. Morley, J.L. Fahey, E.T. Bloom, T. Makinodan and S.L. Gilmore, Immunological changes in young and old adults during brief laboratory stress, *Psychosom. Med.* 53: 121 (1991).
44. M.A. Fiatarone, J.E. Morley, E.T. Bloom, D. Benton, T. Makinodan and G.F. Solomon, Endogenous opioids and the exercise-induced augmentation of natural killer cell activity, *J. Lab. Clin. Med.* 112: 544 (1988).
45. Y. Shimizu, W. Newman, Y. Tanaka and S. Shaw, Lymphocyte interaction with endothelial cells, *Immunol. Today* 13: 106, (1992).
46. J. Westermann and R. Pabst, Distribution of lymphocyte subsets and natural killer cells in the human body, *Clin. Invest.* 70:539 (1992).

PSYCHOLOGICAL AND NEURAL REGULATION
OF INTESTINAL HYPERSENSITIVITY

Veljko J. Djurić
John Bienenstock
Mary H. Perdue

Intestinal Disease Research Programme
McMaster University
Hamilton, Ontario L8N 3Z5

INTRODUCTION

This chapter describes research related to the integration of three diverse areas of science: mucosal immunology, transport physiology and experimental psychology. It is obvious that these three scientific disciplines emerged unaware of one another. However, the situation changed once it was realized that the immune system does not function independently from the influences of the regulatory systems of the body. The interdisciplinary approach advocated by psychoneuroimmunology[1] was welcomed especially by those interested in immediate hypersensitivity. Allergies are by far the most common immunological disorders in humans, and there is also extensive clinical and experimental evidence linking immediate hypersensitivity to environmental and psychological factors.

Interestingly, immediate hypersensitivity was the first experimental model used in the study of central nervous system - immune system interactions. At the beginning of this century, when there was little evidence for a functional link between the two systems, early students of immunology were amazed by the numerous neurological symptoms exhibited by individuals experiencing anaphylactic reactions. Indeed, shortly after the initial report on anaphylaxis by Portier and Richet[2], it was recognized that general anaesthesia protects animals from fatal anaphylactic shock[3] and that the functioning of the brain may be affected during an anaphylactic reaction.[4] These early studies were supported by reports of pronounced disturbances of EEG activity during anaphylactic shock[5,6] even in the absence of any visible signs of shock.[7]

The interest in this area of research persisted regardless of the fact that the central nervous system is not the sole target organ in anaphylaxis since most signs of anaphylactic shock could be observed even in decerebrated animals.[8,9] Later studies reported that intraventricular administration of antigen produced anaphylactic hypersensitivity in the cat[10] and that local cerebral anaphylaxis occurred following

Advances in Psychoneuroimmunology, Edited by
I. Berczi and J. Szélenyi, Plenum Press, New York, 1994

injection of antigen into the brain of a sensitized monkey,[11] indicating that the central nervous system could be used both as a route for immunization and as a route for the induction of active anaphylaxis.

An important contribution was made by Szentivanyi and his coworkers who pointed out that bilateral focal lesions of the tuberal region of the hypothalamus have anti-anaphylactic effects in guinea pigs.[12,13] The clear implication was that the intensity of immediate hypersensitivity reactions can be modified by brain functioning.

This conclusion went hand in hand with numerous clinical observations that asthmatic and hay fever attacks may be provoked by nonallergic stimuli[14,15] emphasizing the possibility that at least some components of the human allergic response can be modified through learning. Indeed, putative conditioning of an immune response was often used as an interpretation for the occurrence of 'allergic reactions' in the absence of any antigen stimulation.[16]

From its very beginnings, psychosomatic medicine[17] considered allergies - and asthma in particular - to fall in its proper domain. Thus, psychological factors such as introversion,[18] suggestion,[19] neuroticism[20] and hypnosis[21,22] were repeatedly related to the susceptibility, and the outcome of asthma. The gastrointestinal tract may be another organ system in which psychological factors can influence allergic reactions.

INTESTINAL HYPERSENSITIVITY

IgE-mediated immediate hypersensitivity to food and other environmental antigens is a contributing factor in the pathogenesis of certain gastrointestinal diseases/syndromes as well as of asthma, rhinitis, urticaria and eczema.[23-25] The intestine is continuously exposed to antigens contained in ingested material. This antigenic stimulation initiates an intricate sequence of events in a complex local microenvirnoment consisting of lymphoid and nonlymphoid cells.[26] Immediate hypersensitivity involves a primary exposure to an antigen that crosses the epithelial barrier and activates B lymphocytes to produce reaginic IgE antibodies (also IgG_4 in humans[27]) that bind to mast cells. Upon subsequent exposure, antigen cross links the antibodies causing mast cells to secrete pre-formed mediators (biogenic amines, proteases, and proteoglycans, etc.) and newly produced mediators, both rapidly (eicosanoids) and slowly (cytokines) synthesized. In the gastrointestinal tract, these mediators cause the physiological changes such as fluid secretion, and increased motility.[28] Mast cell factors also recruit other inflammatory cells, such as eosinophils and neutrophils, which contribute to these reactions. Other hypersensitivities result in pathophysiology via immune complex formation,[29-31] complement deposition,[32] or cell mediated immune reactions.[25,33-35] Immediate hypersensitivity (allergy) is the main reaction focused on in this review.

Neuronal Involvement in Intestinal Hypersensitivity

Activation of mast cells is considered to be the main contributing factor for the structural and functional changes leading to clinical manifestations of gastrointestinal food hypersensitivity.[28] Mediators that are released following the mast cell degranulation act either directly on the epithelium, endothelium or muscle or indirectly via nerves and mesenchymal cells. Neuroimmune interactions in the intestinal mucosa involve close contacts between neurons and mast cells.[36-38] We believe that an appreciation of communication between nerves and mast cells is necessary for an explanation of the regulation of the gut mucosal defence.[39]

Mast cells are found in proximity to substance P-containing nerves in a variety of tissues[40] and ultrastructural evidence for bouton formation between mast cells and enteric nerves has been reported.[41] In patients diagnosed with ulcerative colitis, Yonei[42] demonstrated an anatomical association between nerves and mast cells in the colonic mucosa. Morphometric analysis in normal and nematode infected rats[36] and human appendix[43] demonstrated that the majority of mast cells are juxtaposed to enteric nerves. Close proximity of mast cells to neurons[44] is ideal for the exchange of chemical messages leading to functional effects. This mutual affinity of nerves and mast cells suggests some regular and purposeful communication.

FUNCTIONAL STUDIES

Manifestations of anaphylaxis in rats are the result of the liberation of small molecular weight mediators, such as histamine and serotonin, which primarily act in the small intestine[45] resulting in alterations of gut function that include changes in intestinal ion transport, permeability and motility. Ussing chambers are used to study the effects of mast cell activation on ion transport across the local intestinal epithelium due primarily to secretion of chloride ions.[46] Earlier, we have demonstrated that intestinal anaphylaxis results in increased epithelial permeability.[47] *In vitro* secondary exposure to antigen creates a state of on-going Cl ion secretion resulting in an increase in intestinal short-circuit current (Isc), an indication of the transport tone of the tissue. However, this change is not observed in nonsensitized rats or in sensitized rats challenged with nonspecific antigen. Larger Isc values indicate net transport of large numbers of ions across the membrane and suggest that the tissue is in a more active secretory state.[48]

Using this model of intestinal hypersensitivity, we have repeatedly shown that the effects of antigen-induced mast cell degranulation on the secretory response of jejunal tissue are dependent on intact nerves.[49-50] Immediate responses to luminal antigen challenge (measured by an increase of short-circuit-current) were inhibited by: N-phenylanthranilic acid (chloride channel blocker), doxantrazole (mast cell stabilizer), and ketanserin (5-HT antagonist), and by tetrodotoxin (neurotoxin) and capsaicin (depletes substance P containing nerves). It was concluded that mast cells and nerves interact in the gastrointestinal mucosa and influence epithelial transport during immediate hypersensitivity reactions. Similar observations to these described in the rat intestine were made in the rat trachea.[51]

Inhibition of antigen-induced secretion was observed in substance P-depleted tissues (from neonatally capsaicin treated rats) but not in atropine pretreated tissues.[52] This finding strongly suggests the role of peptidergic fibres in the control of intestinal fluid secretion in reaction to secondary antigen exposure. Nerve involvement in antigen-induced intestinal changes was documented in permeability studies, as well.[53] Sensitized rats exhibited increased intestinal permeability both at baseline and in response to antigen challenge. However, tetrodotoxin applied directly to ligated jejunal segments inhibited antigen-induced uptake of the inert probe ^{51}Cr-EDTA. Since neural blockade inhibited the increases in uptake of the probe and the antigen that are normally seen after secondary antigen exposure, it seems that neural amplification is an important part of antigen-induced alterations of epithelial permeability.

Studies comparing genetically mast cell-deficient WBB6F$_1$-W/W^v with congenic normal WBB6F$_1$-+/+ mice and/or their reconstituted controls emphasize the importance of mast cells in various reactions.[54,55] Our data indicate that mast cell-dependant mechanisms are primarily responsible for the ion secretion associated with intestinal anaphylaxis, but also implicate the involvement of other cell types since response to antigen challenge was not completely abolished in mast cell-deficient mice. In the absence of any identifiable mast cells, isolated intestinal preparations were still capable of generating a small but detectable response to antigen indicating that at least some of physiological changes during anaphylaxis are independent of mast cell activation. Anaphylactic response to antigen was significantly inhibited by neural blockade with tetrodotoxin in congenic normal but not in mast cell-deficient mice. In addition, stimulation of intestinal nerves in the absence of mast cells resulted in a reduced epithelial response. These findings demonstrate that mast cells and nerves interact in the gastrointestinal mucosa and influence epithelial transport.[56]

Psychological Factors in Intestinal Hypersensitivity

The onset and severity of anaphylactic reactions have so far been related to, previous exposure to stress[57] and pretreatment with opioid agonists[58] and antagonists[59] and Pavlovian conditioning. In Pavlovian conditioning, some initially neutral stimulus is paired with an unconditioned stimulus (UCS) that invariably elicits an unconditioned response (UCR). Usually, it takes several pairings of a neutral stimulus with an UCS to form an association between them. Consequently, the initially neutral stimulus becomes a signal (conditioned stimulus, CS) for the occurrence of the UCS; exposure of the CS alone elicits a response (conditioned response, CR) that often mimics the UCR.

Existing knowledge about cellular interactions between mast cells and nerves favours the hypothesis that information processed and stored by the central nervous system could be used by the organism in order to modulate mast cell function. Indeed, there are many reports about putative Pavlovian conditioning of an allergic reaction. However, most of the studies of conditioned responses to stimuli associated with allergens reported conditioning of immuno-nonspecific responses. Exposure to a conditioned stimulus alone, was shown to elicit symptoms such as: decrease in vital capacity,[60,61] increased airway resistance,[62] increased respiratory distress,[63,64] itching response,[65,66] increased erythema,[67] and taste aversion.[68] Conditioned release of histamine was reported,[69] as well and it was claimed that previous exposure to stress or pretreatment with dexamethasone enhances this type of learning.[70] However, since histamine is produced by many tissues and cells other than mast cells, these studies can not be regarded as a conclusive evidence for conditioning of a hypersensitivity response. Only in one experiment[71] was the associative effect of mast cell stimulation reported, as evidenced by the conditioned increase of rat mast cell protease II (RMCP II), an enzyme that is restricted to mucosal mast cells.

Recently we have started to directly relate behavioral variables to mechanisms involved in amplification and inhibition of mucosal reactions in rodent models of intestinal anaphylaxis and inflammation. Comparing two different strains of rats derived from the same genetic pool, we have observed a strong linear relationship between measures of overt behavior (horizontal and vertical activity recorded in an activity box) of rats before the challenge and subsequent antigen-induced changes in gut function.[72] In the tradition of Selye, Saunders and his coworkers, who have demonstrated that cold restraint stress stimulated ion secretion and increased intestinal epithelial permeability in the rats,[73] we hope that these studies will provide further evidence for the involvement of psychological variables in gastrointestinal dysfunction triggered by or resulting from immediate hypersensitivity.

PAVLOVIAN CONDITIONING OF INTESTINAL IMMUNOPHYSIOLOGICAL ABNORMALITIES

Here, we report data from our preliminary study indicating that intestinal immunophysiology can be directly modulated by a Pavlovian conditioning process. The study was conceived as a logical extension of our previous experiment[71] which demonstrated that exposure of an audiovisual cue (CS) alone elicited a release of a specific mucosal mast cell mediator i.e. the response that mimics the presentation of antigen (UCS) itself. In this study, we went one step further and looked for effects of Pavlovian conditioning on functional changes in the small intestine resulting from mast cell activation.

For that purpose, adult male Sprague-Dawley rats were sensitized by a subcutaneous injection of alum-precipitated egg albumin and an intraperitoneal injection of 1 ml Bordatella pertussis vaccine.[52] Fourteen days later, the rats were injected with 3000 larvae of the nematode parasite *Nippostrongylus brasiliensis* to further stimulate reagenic antibody production and induce mast cell hyperplasia in the intestinal mucosa. The conditioning training procedure was initiated 20 days after infection at which time the *Nippostrongylus brasiliensis* had been spontaneously eliminated. Two groups of conditioned rats (Paired and Positive control) learned to associate an audiovisual cue (CS) with a subcutaneous injection of egg albumine (UCS) in the following manner: the rats were removed from the colony room and placed in a plastic cage inside a sound attenuated chamber where a light flashed at an alteration rate of 300 ms, and a background noise was provided by a ventilation fan.[71] The rats were exposed to the cues in the isolation chamber for 15 min, removed and injected with 300 μ g of egg albumin, then returned to the chamber for further 15 min. This protocol was followed once per week for 3 weeks. At the same 3 times, rats in another group (Unpaired) were placed in the isolation chamber for 30 minutes but received antigen challenge in their home cages on the following day. Thus, rats from the Unpaired group were exposed to the same amount of CS and UCS stimulation but in a noncontigent manner to avoid any learned association. In the fifth week, a test trial was performed. Rats from Paired and Unpaired groups were exposed to CS alone, whereas animals from the Positive control group were exposed to both CS and UCS as before. Thirty minutes later, all rats were returned to their home cages and then anaesthetized with intramuscular urethane in preparation for the studies of intestinal immunophysiology. Data were analyzed using one-way analysis of variance. *Post hoc* comparisons among the groups were performed using Student-Newman-Keuls criterion set at α =.05.

Intestinal anaphylaxis (UCR) was indicated by 1) evidence of intestinal mucosal mast activation as determined by reduced intensity of mast cell granule staining, 2) increased baseline secretory tone (Isc) of jejunal tissues, 3) reduced ability of the tissues to respond to *in vitro* antigen with a rise in Isc. Our data show evidence for all three aspects of the expected UCR in the Paired group when it was exposed only to the CS.

Mast cells. There was no significant difference in the absolute numbers of mast cells in the mucosa of jejunal sections in the different groups of rats [$F(2,24)$=.89; p=.425] (Table 1). However, there were changes in mast cell granule stain intensity [$F(2,24)$ = 4.69; p=.019], with tissues from Positive and Paired groups showing reduced granule density compared to those from Unpaired group. This indicates that mast cell activation was a component of the CR.

Table 1. Mast cells in jejunal tissues

		Mast Cells	
Group	n	number/vcu	granule stain intensity
Unpaired	9	37.0±9.6	4.9±1.8
Positive	9	32.8±9.9	2.6±0.9[*]
Paired	9	32.0±5.4	3.7±2.4[*]

Mast cells were counted and granule stain intensity (fully granulated mast cell = 5.0) determined in coded sections of jejunum stained with Alcian blue; values were obtained from at least 10 villus-crypt units (vcu) per section. Jejunal tissues were obtained approximately 45 min after *in vivo* challenge of rats. Values represent the means ± SEM. [*] Statistically significant from the Unpaired group at .05 level

Baseline secretory tone. The Isc data from intestinal tissues from Positive and Unpaired group were consistent within each group. Analysis of variance indicated between group differences [$F(2,24)=4.47$; $p=.022$] (Table 2). The UCR expressed as an increased baseline Isc, was significantly different in Positive versus Unpaired group. Data from the Paired group were less consistent and the overall group mean was not significantly different from the other groups. However, the distribution appeared to be bimodal, and if the data were analyzed as such and compared to controls [$F(2,15=7.13$; $p=.007$] a subgroup of 5 rats appeared to be responders having a significantly elevated Isc, with 4 rats being non-responders.

Table 2. Baseline electrophysiological parameters of jejunal tissues

		Isc
Group	n	($\mu A/cm^2$)
Unpaired	9	64.8±12.9
Positive	9	85.0±10.5[*]
Paired	9	71.1±20.1
Paired		
responders	5	88.5±12.3[*]
non-responders	4	64.0± 9.0

Short-circuit current (Isc) was measured *in vitro* in jejunal tissues obtained from rats approximately 45 min after *in vivo* challenge as described in the text. Values represent the means ± SEM; n indicates the number of rats; * Statistically significant from the Unpaired group at .05 level.

Response to *in vitro* antigen. Previous studies have shown that tissues from sensitized unchallenged rats respond to *in vitro* antigen challenge with a rapid rise in Isc which peaks and then remains elevated above the original baseline;[47,52] this response is reduced in tissues from rats challenged *in vivo* with antigen. This aspect of the UCR was evident in tissues from the Positive control group compared with the

Unpaired group [F(2,24)=11.21; p<.001] (Table 3). However, tissues from Paired group rats demonstrated Isc responses to *in vitro* egg albumin antigen that were significantly increased in magnitude above those in both Unpaired and Positive control groups. Thus, in this case the CR was in the opposite direction to the UCR. This finding may be an example of a priming event, in which mast cells that are exposed first to a neuroendocrine factor or cytokine become more responsive to a second stimulus. In this case, the CS may have resulted in the release of a neuropeptide which primed the mast cells as well as inducing mediator release. Another possibility is that the amplification may have occurred at the level of epithelium. Such priming may have important implications for human conditions in which perception of environmental cues for antigen precedes an actual encounter with antigen itself leading to a situation of potential hyper-responsiveness.

At the end of each experiment, the response (increase in Isc) to electrical transmural stimulation (10 mA, 10 Hz, 5 ms for a total time of 5s) of mucosal nerves in the tissue was determined. Transmural stimulation releases neurotransmitters from enteric nerves in tissue preparations and transiently stimulates luminally-directed anionic secretion resulting in a transient increase in Isc.[49] Transmural responses were completely abolished by treatment of tissues with the neurotoxin, tetrodotoxin. Tissues from Positive control and Paired groups had significantly greater Isc responses to transmural stimulation compared to those from the Unpaired group [F(2,24)=5.26; p=.013) (Table 3).

Taken together these data support the notion that associative learning can result in changes in intestinal immunophysiology. There was evidence for all three aspects of the expected UCR in the rats from the conditioned Paired group: mast cell activation indicated by reduced granule stain intensity, abnormal transport tone expressed as an elevated baseline Isc, and altered responses to antigen *in vitro*.

Table 3. Short-circuit current (Isc) responses of jejunal tissues *in vitro* to antigen and transmural stimulation (TS).

Rat Group	n	Δ Isc (μA/cm^2)	
		EA (100 μg/ml)	TS (10Hz,10μA)
Unpaired	9	40.0±18.0	36.8± 9.6
Positive	9	23.8±12.6[a]	54.7±10.1[a]
Paired	9	58.5±15.6[ab]	49.7±15.6[a]

Values represent the means ± SEM; n indicates the number of rats; [a] Statistically significant from the Unpaired group, [b] from the Positive group at .05 level. Short-circuit current (Isc) responses were determined as maximum increase above baseline in response to addition of egg albumin (EA) antigen and in response to transmural stimulation (TS). *In vitro* tissues were obtained approximately 45 min after *in vivo* challenge.

Nevertheless, this preliminary study suffers from some obvious shortcomings. We feel that two additional control groups need to be included before any definite conclusions could be made: (a) a group that is exposed to UCS only on each conditioning occasion and the test trial (so called 'UCS only' control); and (b) a group that is exposed to CS alone during both the conditioning trials and a test trial (so called 'CS only' or Negative control group). Restrictions in the numbers of the rats studied were due to the length of time for the entire experiment (including handling, sensitization, infection, training and antigen challenge) the requirement for isolation housing, and that only 3 rats could be studied in the Ussing chamber in any one day. Furthermore, the conditioning effect on the Isc baseline was found in only 5 of the 9 conditioned rats (Paired group). Although this finding of variability in the numbers of animals responding to a CS or in the magnitude of the CR had been previously reported in rats and mice, and occurs even within inbred lines,[74,75] we believe that in our future studies we will be able to make an *a priori* distinction between responders and non-responders to our conditioning treatment.

Moreover, we believe that our preliminary findings reported here suggest that signals from the central nervous system may result in immunophysiological responses in the intestine. Such conditioned reactions may play a role in producing symptoms in human diseases such as food allergies, asthma, inflammatory bowel disease and possibly irritable bowel syndrome.

CONCLUSION

In this chapter, we have focused on the study of psychological and neural regulation of intestinal hypersensitivity. Recent evidence indicates that the interaction between enteric nerves and mast cells has an important role in immediate hypersensitivity. Although we do not think that all fixed tissue mast cells necessarily communicate with nerves, we do believe that there is suggestive evidence for regular and adaptively beneficial bi-directional interaction between the neurons and mast cells. Mast cells may serve a sensory function in detecting antigens. Strategically positioned at mucosal surfaces in the intestine and respiratory tracts, mast cells could act as sensitive and specific sensory receptors for environmental antigens and other noxious stimuli.[76]

Neural regulation of intestinal hypersensitivity is only a singular instance of the functional relationship between the immune and the central nervous system. It shares the intellectual appeal of other lines of research of neuroimmune interaction, one that stems from its relevance to the mind-body relationship. After examining the available anatomical, histological, electrophysiological and behavioral evidence relating intestinal immediate hypersensitivity to the functioning of the nervous system, it seems safe to conclude that some crucial information pathways are still unclear and some key components of the system have yet to be identified. The existence of bi-directional mast cell-nerve interactions is unquestionable; however its relevance in health and disease is not clear. Our present research is directed towards better understanding the biochemical and anatomical pathways and stress-related determinants of IgE-dependent and IgE-independent mechanisms of mast cell activation. Obviously, we have been inspired by the pivotal ideas of Hans Selye[77] and Andor Szentivanyi.[78]

ACKNOWLEDGEMENTS

The authors' research is supported by grants from Medical Research Council of Canada, the National Institutes of Health, and the Canadian Foundation for Ileitis and Colitis.

REFERENCES

1. R. Ader, ed., "Psychoneuroimmunology", Academic Press, New York (1981).
2. P. Portier, and C. Richet, De l'action anaphylactique de certain venins, *CR Soc Biol* 6:170 (1902).
3. E. J. Banzhaf, and L.W. Famulaner, The influence of chloral hydrate on serum anaphylaxis, *J. Infect. Dis.* 7:577 (1910).
4. A. Besredka, "Anaphylaxis and Anti-Anaphylaxis and Their Experimental Foundations", C.V. Mosby, St. Louis (1919).
5. A. Brandon, Electrocorticographic and electromyelographic findings in anaphylactic shock, *EEG Clin. Neurophysiol.* 7:371 (1955).
6. E.M. Heimlich, C.W. Dunlop, and R.E. Smith, Central nervous activity in guinea pig anaphylaxis, *J.Allergy* 31:497 (1960).
7. G. Eriksson, and U. Soderberg, Abnormalities in EEG recordings from actively sensitized animals without close correlation to anaphylactic response, *Acta Allerg.* 18:110 (1963).
8. J. Auer, and P. Lewis, A demonstration of the cause of acute anaphylactic death in guinea pig, *Proc. Soc. Exp. Biol. Med.* 7:29 (1910).
9. E. Zunz, and J. La Barre, Modification of coagulability and surface tension in plasma following serum anaphylaxis in tension decerebrated guinea-pigs, *Compt. Rend. Soc. Biol.* 95:858 (1926).
10. B.D. Janković, Lj., Rakić, R. Veskov, and J. Horvat, Anaphylactic reaction in the cat following intraventricular and intravenous injection of antigen, *Experientia* 25:864 (1969).
11. N. Kopelloff, L.M. Davidoff, and L.M. Kopellof, General and cerebral anaphylaxis in the monkey, *Proc. Soc. Exp. Biol. Med.* 34:155 (1936).
12. A. Szentivanyi, and G. Fillip, Anaphylaxis and the nervous system. Part II, *Ann. Allergy* 16:143 (1958).
13. A. Szentivanyi, and J. Szekely, J. Anaphylaxis and the nervous system. Part IV, *Ann. Allergy* 16:389 (1958).
14. J.N. MacKenzie, The production of the so-called "rose cold" by means of an artificial rose, *Am J. Med. Sci.* 91:45 (1886).
15. G.H Smith, and R. Salinger, Hypersensitivities and the conditioned reflex, *Yale J. Biol. Med.* 5:387 (1933).
16. D. Hanson, A psychosomatic theory of allergic sensitization: Allergy as a conditioned response, *Zeit. Psychosomat. Med. Psychoanal.* 27:143 (1981).
17. F. Alexander, "Psychosomatic Medicine", Norton, New York (1950).
18. E. Teiramaa, Psychosocial and psychic factors in the course of asthma, *J. Psychosomat. Res.* 22:121 (1978).
19. T.J. Luparello, M. Stein, and C.D. Park, Effect of hypothalamic lesions on rat anaphylaxis, *Am. J. Physiol.* 207:911 (1964).
20. N.C. Oswald, N.E. Waller, and J. Drinkwater, Relationship between breathlessness and anxiety in asthma and bronchitis: a comparative study, *Br. Med. J.* 2:14 (1970).
21. S. Black, Inhibition of immediate-type hypersensitivity response by direct suggestion under hypnosis, *Br. Med. J.* 1: 925 (1963).
22. T.C. Ewer, and D.E. Stewart, Improvement in bronchial hyper-responsiveness in patients with moderate asthma after treatment with a hypnotic technique: a randomised controlled trial, *Br. Med. J.* 293:1129 (1986).
23. M.H. Lessof, D.G. Wraith, T.G. Merrett, J. Merrett, and Buisseret, P.D. Food allergy and intolerance in 100 patients - local and systemic effects, *Q. J. Med.* 49:259 (1980).
24. H.A. Sampson, R.H. Buckley, and D.D. Metcalfe, Food allergy, *J. Am. Med. Assoc.* 258:2886 (1987).
25. C. Collins-Williams, The role of pharmacological agents in the prevention or treatment of allergic food disorders. *Ann. Allergy* 57:53 (1986).
26. D.M. McKay, and M.H. Perdue, Intestinal epithelial function: The case for immunophysiological regulation. Cells and mediators, *Digestive Dis. Sci.* 38:1377 (1993).
27. G.M. Halpern, and J.R. Scott, Non-IgE antibody mediated mechanisms in food allergy, *Ann. Allergy* 58:14 (1987).
28. S.E. Crowe, and M.H. Perdue, Gastrointestinal food hypersensitivity: Basic mechanisms of pathophysiology, *Gastroenterology* 103:1075 (1992).
29. C. Carcini, J. Brostoff, and D.G. Wraith, IgE complexes in food allergy. *Ann. Allergy* 59:110 (1987).
30. R. Paganelli, G. Cavagni, and F. Pallone, The role of antigenic absorption and circulating immune complexes in food allergy, *Ann. Allergy* 57:330 (1986).

31. F. Shakib, Is IgE-mediated hypersensitivity an autoimmune disease? *Allergy* 45:1 (1990).
32. M. Shiner, J. Ballard, and M.E. Smith, The small-intestinal mucosa in cow's milk allergy, *Lancet* 1:136 (1975).
33. J.A. Walker-Smith, Food sensitive enteropathies, *Clin. Gastroenterol.* 15:55 (1986).
34. W. Strober, Gluten-sensitive enteropathy: a nonallergic immune hypersensitivity of the gastrointestinal tract, *J. Allergy Clin. Immunol.* 78:202 (1986).
35. S.R. Targan, M.F. Kagnoff, M.D. Brogan, and F. Shanahan, Immunologic mechanisms in intestinal diseases, *Ann. Internal Med.* 106:853 (1987).
36. R.H. Stead, M. Tomioka, G. Quinonez, G.T. Simon, S.Y. Felten, and J. Bienenstock, Intestinal mucosal mast cells in normal and nematode-infected rat intestines are in intimate contact with peptidergic nerves, *Proc. Nat. Acad. Sci.* 84:2975 (1987).
37. J. Bienenstock, M.H. Perdue, and R.H. Stead, Neuronal interaction with mast cells, *in:* "New Trends in Allergy III," J. Ring, and B. Przybilla, eds., Springer Verlag, Berlin (1991).
38. N. Arizono, S. Matsuda, T. Hattori, Y. Kojima, T. Maeda, and S.J. Galli, Anatomical variation in mast cell nerve associations in the rat small intestine, heart, lung, and skin: Similarities of distances between neural processes and mast cells, eosinophils, or plasma cells in the jejunal lamina propria, *Lab. Invest.* 62:626 (1990).
39. D.M. McKay, V.J. Djurîc, M.H. Perdue, and J. Bienenstock, Regulating factors affecting gut mucosal defence, in: "Gastrointestinal and Hepatic Immunology," R.V. Heatly ed., Cambridge University Press, Cambridge UK (in press)
40. G. Skofitsch, J.M. Savitt, and D.M. Jacobowitz, Suggestive evidence for a functional unit between mast cells and substance P fibres in the rat diaphragm and mesentery, *Histochemistry* 82:5(1985).
41. B. Newson, A. Dahlstrom, L. Enerback and N. Ahlman, Suggestive evidence for a direct innervation of mucosal mast cells. An electron microscopic study, *Neuroscience* 10:565 (1983).
42. Y. Yonei Y. Autonomic nervous alterations and mast cell degranulation in the exacerbation of ulcerative colitis. *Jap. J. Gastro.* 84:1045-1987.
43. R.H. Stead, A.J. Franks, C.H. Goldsmith, J. Bienenstock, and M.F. Dixon, Mast cells, nerves and fibrosis in the appendix: a morphological assessment, *J. Path.* 161:209 (1990).
44. M.G. Blennerhassett, M. Tomioka and J. Bienenstock Formation of contacts between mast cells and sympathetic neurons in vitro, *Cell Tiss. Res.* 265:121-1991.
45. R.K. Sanyal, and G.B. West, Anaphylactic shock in the albino rat, *J.Physiol.* 141:571 (1958).
46. M.H. Perdue, J. Marshall, and S. Masson, Ion transport abnormalities in inflamed rat jejunum, *Gastroenterology* 98:561 (1990).
47. M.H. Perdue, and D.G. Gall, Rat jejunal mucosal response to histamine and anti-histamines in vitro. Comparison with antigen-induced changes during intestinal anaphylaxis, *Agents Actions* 19:5 (1986).
48. D.M. McKay, and M.H. Perdue, Intestinal epithelial function: The case for immunophysiological regulation. Implications for disease, *Digestive Dis. Sci.* 38:11735 (1993).
49. M.H. Perdue, and J.S. Davison, Response of jejunal mucosa to electrical transmural stimulation and two neurotoxins, *Am. J. Physiol.* 251: G642-G648 (1986).
50. M.H. Perdue, R. Galbraith, and J.S. Davison, Evidence for substance P as a functional neurotransmitter in guinea pig small intestinal mucosa, *Reg. Peptides* 18:63 (1987).
51. P. Sestini, M. Dolovich, C. Vancheri, R.H. Stead, J.S. Marshall, M.H. Perdue, J. Gauldie, and J. Bienenstock, Antigen-induced lung solute clearance in rats is dependent on capsaicin-sensitive nerves, *Am. Rev. Resp. Dis.* 139:401 (1989).
52. S.E. Crowe, P. Sestini, and M.H. Perdue, Allergic reactions of rat jejunal mucosa. Ion transport responses to luminal antigen and inflammatary mediators, *Gastroenterology* 99:74 (1990).
53. S.E. Crowe, K. Soda, A.M. Stanisz, and M.H. Perdue, Intestinal permeability in allergic rats: nerve involvement in antigen-induced changes, *Am. J. Physiol.* 264: G617 (1993).
54. Y. Kitamura, S. Go, and K. Hatanaka, Decrease of mast cells in W/W^V mice and their increase by bone marrow transplantation, *Blood* 52:447 (1978).
55. S.J. Galli, and Y. Kitamura, Genetically mast-cell-deficient W/W^V and Sl/Sl^d mice, *Am J of Pathol* 127:191 (1987).
56. M.H. Perdue, S. Masson, B.K. Wershil, and S. J. Galli, Role of mast cells in ion transport abnormalities associated with intestinal anaphylaxis. Correction of the diminished secretory response in genetically mast cell-deficient W/W^V mice by bone marrow transplantation, *J. Clin. Invest.* 87:687 (1991).
57. I. Marîc, V.J. Djurîc, M. Dimitrijevîc, M. Bukilica, S. Djordjevîc, B.M. Markovîc, and B.D. Jankovîc, B.D. Stress-induced resistance to anaphylactic shock, *Int. J. Neurosci.* 59:159 (1991).
58. B.D. Jankovîc, and D. Marîc, Enkephalins and anaphylactic shock: Modulation and prevention of

shock in the rat, *Immunol. Lett.* 15:153 (1987).

59. V.J. Djurľc, I. Marľc, M. Dimitrijevľc, B.M. Markovľc, and B.D. Jankovľc, Naloxone and anaphylactic shock in the rat, *Ann. New York Acad. Sci.* 650:128 (1992).

60. E. Dekker, H.E. Pelser, J.J. Groen, Conditioning as a cause of asthmatic attacks, *J Psychosomat Res* 1957; 2: 84-98.

61. D.J. Horton, W.L. Suda, R.A. Kinsman, J. Souhrada, S. Spector, Bronchoconstrictive suggestion in asthma: A role for airways hyperactivity and emotions, *Am. Rev. Resp. Dis.* 117:1029 (1987).

62. E.R. Jr. McFadden, T. Luparello, H.A. Lyons, E. Bleecker, The mechanism of action of suggestion in the induction of asthma attacks, *Psychosomat Med* 31:134 (1969).

63. P. Ottenberg, M. Stein, J. Lewis, C. Hamilton, Learned asthma in guinea pig, *Psychosomat Med* 1958; 20: 395 (1958).

64. D.R. Justesen, E.W. Braun, R.G. Garrison, R.B. Pendleton, Pharmacological differentiation of allergic and classically conditioned asthma in the guinea pig, *Science* 170:864 (1970).

65. J.M. Jordan, and F.A. Whitlock, Emotions and the skin: The conditioning of scratch responses in cases of atopic dermatitis, *Dermatology* 86:574 (1972).

66. J.M. Jordan, and F.A. Whitlock, Atopic dermatitis: Anxiety and conditioned scratch responses, *J. Psychosomat. Res.* 18:297 (1974).

67. G.R. Jr. Smith, and S.M. McDaniel, Psychologically mediated effect on the delayed hypersensitivity reaction to tuberculin in humans, *Psychosomat. Med.* 45:65 (1983).

68. B.M. Markovľc, V.J. Djurľc, M. Lazarevľc, B.D.Jankovľc, Anaphylactic shock-induced conditioned taste aversion. I. Demonstration of the phenomenon by means of three modes of CS-US presentation, *Brain. Behav. Immun.* 2: 11 (1988).

69. M. Russell, K.A. Dark, R.W. Cummins, G. Ellman, E. Callaway, and H.V.S. Peeke, Learned histamine release, *Science* 225:733 (1984).

70. H.V.S. Peeke, G. Ellman, K. Dark, M. Salfi, and V.I. Reus, Cortisol and behaviorally conditioned histamine release, *Ann. New York Acad. Sci.* 496:583 (1987).

71. G. MacQueen, S. Siegel, J. Marshall, M. H. Perdue, and J. Bienenstock, Pavlovian conditioning of rat mucosal mast cells to secrete rat mast cell protease II, *Science* 243:83 (1989).

72. V.J. Djurľc, U. Kosecka, J. Bienenstock, and M.H. Perdue, Strain-related difference in susceptibility to anaphylactic shock correlates with measures of spontaneous activity, (submitted)

73. P.R. Saunders, U. Kosecka, D.M. McKay, and M.H. Perdue, Acute stressors stimulate ion secretion and increase epithelial permeability in the rat (submitted)

74. R.M. Gorczynski, S. McRae, and M. Kennedy, Conditioned immune response associated with allogeneic skin grafts in mice, *J. Immunol.* 129:704 (1982).

75. R.L. Elkins, Separation of taste-aversion-prone and taste-aversion-resistant rats through selective breeding: implications for individual differences in conditionality and aversion-therapy alcoholism treatment, *Behav. Neurosci.* 100:121 (1986).

76. J. Bienenstock, Cellular communication networks: Implications for our understanding of gastrointestinal physiology, *Ann. New York Acad. Sci.* 664:1 (1992).

77. H. Selye, The general adaptation syndrome and the diseases of adaptation, *J. Clin. Endocrinol.* 6:117 (1946).

78. A. Szentivanyi, The beta adrenergic theory of atopic abnormality in bronchial asthma. *J. Allergy* 42:203 (1968).

THE CERVICAL SYMPATHETIC TRUNK-SUBMANDIBULAR GLAND AXIS IN THE REGULATION OF INFLAMMATORY RESPONSES

Ronald Mathison[1], Joseph S. Davison[1] and A. Dean Befus[2]

[1]Department of Medical Physiology
The University of Calgary
3330 Hospital Drive N.W.
Calgary, Alberta, Canada, T2N 4N1

[2] Pulmonary Research Group
The University of Alberta,
Edmonton, Alberta, Canada, T6G 2S2

INTRODUCTION

Elaborate homeostatic mechanisms have been developed for regulating the main physiological variables of body temperature, heart rate, blood pressure, water balance and availability of nutrients. Major threats to an organism whether through a stress response or other insults such as infection (viral, fungal, bacterial) and noninfectious pathological causes (e.g. pancreatitis, ischemia, multiple trauma and tissue injury, haemorrhagic shock, immune-mediated organ dysfunction) elicit homeostatic responses in the major body systems; nervous, endocrine, immune, cardiovascular, respiratory, liver, kidneys and gastrointestinal tract.

In this review we present some of our recent findings on the role of the "cervical sympathetic trunk-submandibular gland axis" (CST-SMG axis) in regulating immunological and cardiovascular responses to insults to the homeostatic mechanisms. Initially, however, we will develop the concept that the CST-SMG axis functions closely and in an interrelated manner with two other major system involved in stress and shock responses: the hypothalamo-pituitary axis and the autonomic nervous system. These three systems, involving nervous, endocrine and neuroendocrine pathways of communication, form a functional network that by interacting with the immune system attempt to assure maintenance of body homeostasis.

Advances in Psychoneuroimmunology, Edited by
I. Berczi and J. Szélenyi, Plenum Press, New York, 1994

NEUROENDOCRINE MECHANISMS AND HOMEOSTASIS

Autonomic and Hypothalamic Mechanisms

The two well studied mechanisms by which the central nervous system regulates the homeostatic responses to stress, shock and inflammation involve modification of sympathetic outflow, and controlling the release of endocrine hormones from the pituitary. The hypothalamic neuroendocrine systems, principally involved in homeostatic stress responses, originate in the parvocellular and magnocellular regions of the hypothalamus, which respectively synthesize and release two important endocrine hormones corticotrophin releasing factor (CRF) and vasopressin (AVP). Although the hypothalamo-pituitary axis is under the control of cortical structures (e.g. hippocampus and basal forebrain), which integrate neurally transmitted and blood born sensory messages from peripheral tissues, it has been recognized recently that projections of sympathetic nerves in the cervical sympathetic trunk to the median eminence constitute another pathway by which the brain can communicate with the endocrine system.[1]

The sympathetic nervous system has been considered to play a role in the stress and shock response by primarily modulating heart rate, the distribution of blood to essential organs, and the release of certain stress hormones such as glucagon, renin, etc. However, the studies of Cardinali & Romeo[1] emphasized a fundamental role for the cervical sympathetic trunk in regulating the hypothalamo-pituitary axis to certain stress conditions. For example, they found that the secretion of adrenocorticotropic hormone (ACTH) in response to immobilization and ether stress in rats was markedly reduced by removal of the superior cervical ganglia (SCG) during the phase of Wallerian degeneration, as well as in chronically denervated rats (7 days after SCG ganglionectomy; SCGx).[2]

Interactions of Submandibular Glands with Neuroendocrine Systems

Recently, our studies have shed light on another role for the sympathetic nervous system in regulating cardiovascular and immunological responses to shock and inflammatory stimuli. This novel neuroendocrine system comprises the "cervical sympathetic trunk-submandibular gland axis", and discussion of this system constitutes the major focus of this review.

Nevertheless, it should initially be emphasized that the influence of the submandibular glands on homeostatic mechanisms extends beyond regulation of digestive and immunological functions to include effects on hypothalamic function. Removal of the submandibular glands (sialadenectomy; SMGx) decreases hypothalamic content of thyrotropin releasing hormone (TRH) without affecting plasma levels of thyroid stimulating hormone (TSH),[3] whereas plasma levels of luteinizing hormone (LH), but not those in the hypothalamus, were decreased by sialadenectomy. In this context it is interesting to note that epidermal growth factor (EGF) elicits a rapid and pronounced release of LH from isolated, perfused hypothalamo-pituitary segments,[4] but not from the pituitaries alone. Thus, EGF appears to activate hypothalamic neurons secreting LH-RH.

The mechanisms involved in submandibular gland regulation of hypothalamic function are somewhat problematic as growth factors, such as EGF, probably cannot cross the blood-brain barrier. Several possibilities need to be considered, such as: 1) activation of nervous structures outside the blood-brain barrier (e.g. in the median eminence), 2) alteration of the functional properties of cervical sympathetic nerves projecting to the hypothalamus-pituitary axis and/or the pineal.[5] The pineal releases several substances, including melatonin, which inhibit the release of thyrotropin releasing hormone (TRH) from the hypothalamus.[6,7]

Figure 1 summarizes the major lines of communication among the cervical sympathetic trunk, submandibular glands, hypothalamo-pituitary axis and the cardiovascular and immune systems. Nerve growth factor (NGF) is used as an example of a salivary gland[8] and hypothalamic hormone[9] that exerts effects on all the systems under consideration.

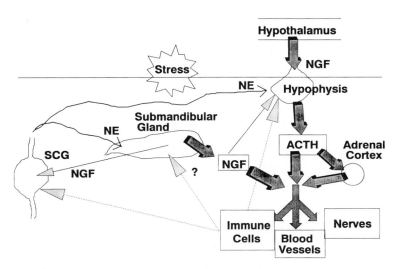

Figure 1. Some of the neuroendocrine relationships involved in homeostatic responses to stress, shock and inflammation. SCG = superior cervical ganglion, NE = norepinephrine, NGF = nerve growth factor, ACTH = adrenocorticotropic hormone.

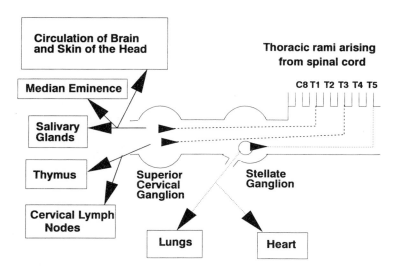

Figure 2. Schematic drawing of the origins and field of innervation of the cervical sympathetic trunk.

THE CERVICAL SYMPATHETIC TRUNK

The sympathetic ganglia receive innervation from preganglionic neurons that reside in the spinal cord (Fig. 2). The axons from spinal cord neurons reach the cervical ganglia by way of the ventral routes to form the thoracic and cervical sympathetic trunks. The axons that project in the cervical trunk from the ganglion cells in the inferior cervical (stellate) and superior cervical ganglia (SCG) innervate smooth muscle, glands, and blood vessels. The post-ganglionic nerves emanating from the stellate ganglia innervate predominately the lungs and heart, whereas those projecting from the SCG innervate the upper thoracic and head regions. Structures of particular relevance to the modulation of inflammatory responses are the hypophysis, the cervical lymph nodes and the submandibular glands.

THE SUBMANDIBULAR SALIVARY GLANDS

Several studies have shown that the submandibular glands, classically considered to have exclusively exocrine functions, also possess endocrine functions. Both of these hormone release processes are, nevertheless, involved in the reestablishing homeostasis to stress, tissue injury and shock.

The presence of 100 to 1000 times greater amounts of some growth factors in the submandibular glands of male, than in female, mice and rats has led to the suggestion that these hormones function in the wound healing.[8, 10-14] Aggressive behaviour provokes an increase circulating levels of NGF [15] and in salivary and plasma renin.[16, 17] Sham surgeries on the submandibular glands increase dramatically the levels of circulating EGF.[18] From this perspective the proposition can be made that the salivary gland factors also participate in the stress response (even those unrelated to fighting) by regulating the functions of the hypothalamopituitary axis and the cardiovascular and immune systems.

Immunomodulatory Factors in the Submandibular Glands

Although definitive molecular characterization has not yet been completed for many activities, the submandibular gland contains both colony-stimulating (CSF), colony-inhibitory (CIF) factors,[19] and an inhibitor of interleukin 1.[20] Two isoforms of transforming growth factor beta (TGFBβ1 and TGFBβ2),[21-23] both potent immunoregulators, are also found in the submandibular glands. NGF is also a potent degranulator of mast cells.[24, 25]

Factors found in the submandibular gland probably contribute to the modulation of neutrophil function. Several of them, such as insulin-like growth factor (IGF-1),[26] βNGF,[27] as well as 7S-NGF,[28] and kallikreins[29] prime neutrophils for superoxide production in response to lipopolysaccharide (LPS). NGF is also a chemotactic factor for neutrophils.[30] TGFβ has suppressive effects on neutrophils and macrophages. Saito and coworkers[31] described a factor in human parotid saliva which inhibited phagocytosis, bacterial killing and generation of hydrochlorous acid by neutrophils. The factor has a molecular weight (MW) of less than 12,500. Our studies (see below) also suggest that the submandibular salivary glands contain factors that modulate neutrophil function.

THE CERVICAL SYMPATHETIC TRUNK-SUBMANDIBULAR GLAND (CST-SMG) AXIS AND THE REGULATION OF INFLAMMATORY RESPONSES

We have investigated the role of the CST-SMG axis in the regulation of inflammatory cell functions that relate specifically to anaphylaxis in *Nippostrongylus brasiliensis* (Nb)-<u>sensitized</u> rats and endotoxemia or ischemia-reperfusion injury in normal <u>unsensitized</u> rats.

The experiments, performed on male Sprague-Dawley rats, involved several manipulations on the CST-SMG axis 1 week prior to experimentation (Fig. 3). These include:

1) bilateral decentralization of the superior cervical ganglia (SCG) by cutting the cervical sympathetic trunk between the stellate and SCG ganglia (Decent),

2) bilateral removal of the SCG (SCGx),

3) extirpation of the submandibular glands (SMGx),

4) a combination of Decent or SCGx and SMGx, or

5) sham surgery by freeing the submandibular glands of their connective tissue but the innervation is left intact.

Using these models we have found that the CST-SMG axis exerts complex effects on inflammatory cell function in *N. brasiliensis*-sensitized and unsensitized rats.

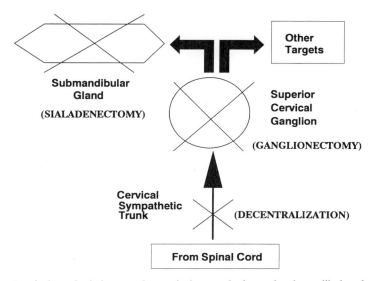

Figure 3. Surgical manipulations on the cervical sympathetic trunk-submandibular gland axis.

N. brasiliensis-Sensitized Rats

Decentralization or ganglionectomy reduces late-phase (4 to 8 h) pulmonary inflammation (neutrophil and macrophage influx into alveolar spaces) following challenge with antigen (Ag)[32,33] in *N. brasiliensis*-sensitized rats. Furthermore, the chemotactic and phagocytic/reductive capacity of neutrophils are down-regulated by decentralization.[34] The early components of anaphylaxis, such as respiratory and cardiovascular perturbations,[35] the release of rat mast cell protease II (RMCP II)[36] and Ag-stimulated histamine release from mast cells[37] are not affected by these operations.

The anti-inflammatory effect of SCG denervations is probably the result of modified neutrophil and macrophage function. Since the attenuated late-phase (8 h) pulmonary inflammation seen in decentralized or ganglionectomized rats was abolished by the removal of the submandibular glands,[33] we proposed that cervical sympathetic nerves exert an inhibitory constraint on the release from the glands of a factor which down-regulates the ability of neutrophils and macrophages to migrate into the lungs (see Fig. 6). Denervation of the glands removes this inhibitory constraint, the factor is released with consequent down-regulation of inflammatory cells such that fewer of them accumulate in the lungs. Removal of the submandibular glands removes the source of the inhibitory factor, and thus abrogates the anti-inflammatory effects of decentralization and ganglionectomy.

Superoxide Anion Production by Neutrophils. In contrast to the role played by the submandibular glands in the attenuation of the late-phase pulmonary inflammation induced by Ag-challenge in cervically sympathectomized rat, the production of superoxide anion (O_2^-) by circulating neutrophils is regulated differently. In Table 1 it can be seen that neutrophils obtained from sham-operated (Sham) and decentralized (Decent), *N. brasiliensis* sensitized rats exhibited increased O_2^- production in response to fMLP, whereas those obtained from sialadenectomized (SMGx) rats had a markedly attenuated oxidative response to this chemotactic factor. This reduced reactivity of the neutrophils from sialadenectomized rats was not affected by decentralization of the SCG (SMGx/Decent).

These observations suggest that a factor released from salivary glands increases the ability of circulating neutrophils to respond to a fMLP with an oxidative burst. The release of the factor is apparently not under sympathetic nervous system control since cervical sympathetic denervations did not modify the oxidative burst of neutrophils (see Fig. 6). However, upon removal of the glands the source of the factor is eliminated and neutrophils exhibit reduced responsiveness to fMLP. This model is substantially different from that proposed for the regulation of Ag-induced pulmonary

Table 1. Superoxide production by circulating neutrophils obtained from *N. brasiliensis*-sensitized rats.

| Operation | Concentration of fMLP (nmoles) | |
	62.5	250
Sham	20.2±0.8	40.5±4.5[*,#]
Decent	17.0±2.0	40.3±5.6[*,#]
SMGx	15.8±1.3	23.7±3.3
SMGx/Decent	17.5±3.5	24.9±3.4

Sham = sham-operated, Decent = decentralized, SMGx = sialadenectomized,
SMGx/Decent = sialadenectomized and decentralized.
[#] > 62.5 nmoles fMLP,
[*] > than SMGx and SMGx/Decent at 250 nmoles fMLP.

inflammation by the CST-SMG axis (discussed above), where the cervical sympathetic nerves inhibit the release of a factor which down-regulates the ability of neutrophils to migrate into the alveolar spaces. Given the variety of growth factors found in the submandibular glands that can regulate neutrophil function (see above), differential regulation of distinct biochemical activities in the neutrophils should be expected. The challenge that remains is to relate the actions of these factors on neutrophils with the regulation of their release from submandibular glands by autonomic nerves.

TNFα Release from Alveolar Macrophages. Table 2 shows the effects of LPS stimulated release of TNFα from alveolar macrophages isolated prior to and at various times after challenge of *N. brasiliensis* sensitized rats with antigen. The macrophages were stimulated with 1μ g/ml of LPS for 24 h and the quantity of TNFα released into the incubation medium was measured by evaluating its cytotoxicity for WEHI cells.

Table 2. Release of TNFα from alveolar macrophages.

Time (h) after antigen challenge in vivo	Decentralized	Sham-operated
0	78.8±11.9[*]	27.9± 5.4[**] [#]
0.25	82.7±13.3[*]	70.9±10.3
1	54.4± 9.5	82.4± 9.1
4	31.4± 4.5	71.0±11.6[#]
8	40.0± 8.2	91.7±12.5[#]
24	65.2±3.0	92.1±17.6

In vitro release of TNFa from LPS-stimulated alveolar macrophages obtained from decentralized and sham-operated N. brasiliensis sensitized rats. TNFa is expressed in units/10E6 adherent cells. * indicates > 1, 4 and 8 h for decentralized rats. ** indicates < 0.25, 1, 4, 8 and 24 h for sham-operated rats. # indicates < decentralized at these times.

The release of this cytokine was markedly different in decentralized and sham-operated rats. 1) Macrophages obtained from unchallenged (time = 0 h) decentralized rats released approximately 2½ times as much TNFα as did those obtained from sham-operated rats. 2) Antigen challenge apparently sensitized the macrophages from sham-operated rats to release TNFα , since as early as 15 min after challenge TNFα release increased approximately 2½ fold and remained elevated even with cells harvested 24 after the challenge. In contrast, the macrophages from challenged, decentralized rats released less TNFα than both unchallenged, decentralized and sham-operated rats.

These observations suggest that cervical sympathetic trunk can dramatically alter cytokine release from inflammatory cells. We have yet to determine if the submandibular glands play a role in this response.

Normal, Unsensitized Rats

A role for the CST-SMG axis in regulating the cardiovascular and immunological responses of unsensitized rats have been studied using models of endotoxemia and ischemia-reperfusion injury.

Hypotensive responses to intravenous lipopolysaccharide. In normal rats ganglionectomy or sialadenectomy markedly aggravates the hypotensive responses to intravenous injection of LPS in sham-operated, unsensitized rats (Fig. 4).[38] We have proposed[38] that the submandibular glands release a factor(s) that reduces the hypotensive effect of LPS, possibly through actions on the cells of the immune and vascular systems. Although the mechanism by which submandibular gland factor(s)

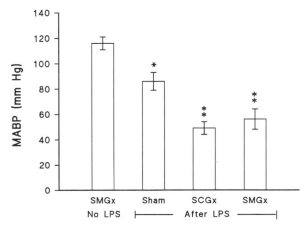

Figure 4. Average decrease in mean arterial blood pressure (MABP) over 60 minutes after injecting 10 mg/kg lipopolysaccharide (LPS) into normal, unsensitized rats that have received either a sham-operation (Sham), ganglionectomy (SCGx) or sialadenectomy (SMGx). The controls show changes in MABP over 60 minutes of sialadenectomized rats not injected with LPS. * < than control, ** < less than control and Sham.

alters the response to LPS remains to be identified, we speculate that it down-regulates monocytes and neutrophils so that these cells release reduced amounts of vasoactive mediators in response to LPS. With removal of the submandibular glands and the associated factor(s), the release of vasoactive mediators by inflammatory cells is increased. The release of the factor(s) is positively controlled by the cervical sympathetic nerves since bilateral removal of the SCG (ganglionectomy) also potentiates the acute hypotensive response to LPS (see Fig. 6).

Presently, we are characterizing submandibular gland factors that modulate the acute hypotensive response to endotoxin. Our initial studies have shown that the increased hypotensive response to LPS in sialadenectomized rats can be significantly reduced by pretreatment with submandibular gland extract. Using extracts of the SMG, prepared in the presence of serine protease inhibitors, we have partially characterized several of the factors. They are: 1) water soluble, 2) heat labile, and 3) have apparent molecular weights of less than 3000 daltons. They must be given at least 1 h prior to administration of endotoxin to exert their anti- hypotensive actions, and on purification with C18-HPLC a peptidergic and a non-peptidergic factor were isolated. Further characterization of these biological activities is in progress.

Accumulation of pulmonary neutrophils following reperfusion-ischemia injury. Neutrophil accumulation in the lungs, determined by measuring myeloperoxidase (MPO) activity, was studied after subjecting the intestine to an ischemia/reperfusion (I/R) insult. Ischemia was induced by clamping the superior mesenteric artery (SMA) for 45 min and reperfusion was allowed to proceed for 3 h. In response to the I/R insult rats that had received no prior surgery had MPO levels of ~40 units of MPO/g wet tissue, whereas those experiencing 'sham-sialadenectomy' 1 week prior to the experiment showed doubling of the neutrophil accumulation in their lungs (Fig. 5). Removal of the submandibular glands (SMGx), decentralization of the SCG (Decent) and removal of the SCG (ganglionectomy, SCGx) reduced the neutrophil accumulation in the lungs induced by 'sham-sialadenectomy (Sham).

These observations suggest that neutrophil function is modified as late as 1 week after a surgical manipulation of the submandibular glands. We propose that a factor(s) released from the submandibular glands of "sham-operated" rats up-regulates neutrophil function such that they accumulate more extensively in the lungs after an I/R insult. Since the factor(s) is no longer present in sialadenectomized rats and less readily released after sympathetic denervation of the glands, neutrophil function is less dramatically modified and thus, fewer accumulate in the lungs following an I/R episode. The release of this submandibular gland factor would be positively regulated by the sympathetic nerves (see Fig. 6); i.e. they promote the release of the factor and when they are removed less of the factor would be released. It cannot be excluded at the present time that the enhanced release of a neutrophil modifying factor due to the trauma of sham-sialadenectomy results in part from trauma (activation) of the sympathetic nerves innervating the submandibular gland.

In concordance with this hypothesis, Hwang et al.[18] reported a massive and prolonged (at least 72 h) increase in circulating EGF following sham-sialadenectomy. They also reported that removal of the submandibular glands increases circulating EGF for up to 24 h. Thus, a possible explanation for the enhanced neutrophil accumulation in the lungs after "true-I/R" in sham-sialadenectomized rats is the sustained release of a neutrophil modifying factor (not necessarily EGF) from the submandibular glands. Further studies are required to identity the factor(s) and determine its circulating levels after sham-sialadenectomy.

SUMMARY OF IMMUNOMODULATORY EFFECTS OF THE CERVICAL SYMPATHETIC TRUNK-SUBMANDIBULAR GLAND (CST-SMG) AXIS

The regulation of inflammatory cell function by the cervical sympathetic trunk-submandibular gland axis appears to be exerted at two different levels (Fig. 6):

1) the effects of submandibular gland factors on inflammatory cells, and

2) the control of the release of the factors from submandibular glands by the sympathetic nerves.

Using *N. brasiliensis* sensitized-rats we have found that neutrophil migration, chemotaxis and reductive/phagocytic activities are down-regulated by a submandibular gland factor(s), whereas the oxidative responses of these cells are up-regulated. With unsensitized rats we speculate that a submandibular gland factor down-regulates

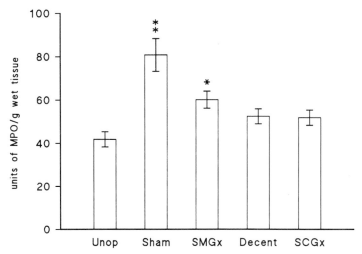

Figure 5. Effects of no operation (Unop), sham-operation (Sham), sialadenectomy (SMGx), decentralization of the superior cervical ganglia (Decent) and superior cervical ganglia removal (SCGx) on neutrophil accumulation in the lungs induced by ischemia-reperfusion. * > than unoperated rats, ** > than all other groups.

inflammatory cells so that they release reduced amounts of vasoactive mediators upon exposure to LPS. On the other hand, the propensity of neutrophils to accumulate in the pulmonary vasculature following an ischemia/reperfusion injury is up-regulated by a submandibular gland factor.

Concerning the regulation of the release of these factors from the submandibular glands by the cervical sympathetic nerves three modes of control have been identified:

1) Inhibitory control, as exerted on the factor controlling neutrophil migration, chemotaxis and reductive/phagocytotic activities in *N. brasiliensis*-sensitized rats.

2) No sympathetic control as observed with the oxidative response of neutrophils obtained from *N. brasiliensis*-sensitized rats.

3) Stimulatory control that is seen with ischemia-reperfusion injury induced pulmonary neutrophil accumulation, and the possible release of vasoactive mediators from inflammatory cells in response to LPS.

THE CERVICAL SYMPATHETIC TRUNK-SUBMANDIBULAR GLAND AXIS: A NOVEL NEUROENDOCRINE SYSTEM WITH IMMUNOMODULATORY ACTIONS

With this survey of some of our results on the role of the CST-SMG axis in the regulation of inflammatory responses several important observations emerge:

1) The cervical sympathetic trunk exerts powerful immunomodulatory effects on inflammatory cells.

2) Some, if not all, of the immunomodulatory effects of the cervical sympathetic trunk involve the intermediacy of the submandibular salivary glands.

3) In view of the variety of actions of the submandibular gland activities (i.e. unidentified factors) on inflammatory cell functions and the complexity of the neural mechanisms regulating the release of these activities, the neuroendocrinology of the submandibular glands can be considered to be in its infancy.

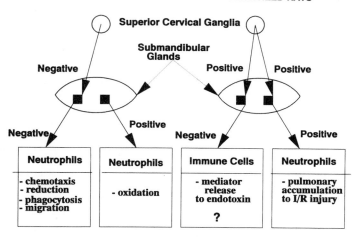

Figure 6. Summary of the effects of the cervical sympathetic trunk-submandibular gland axis on inflammatory cell responses in *N. brasiliensis*-sensitized rats and unsensitized rats.

Another important aspect related to the neuroendocrinology of the cervical sympathetic trunk-submandibular gland axis that needs further study pertains to potential interactions with the hypothalamo-pituitary system.

ACKNOWLEDGMENTS

This work was supported by the Council for Tobacco Research, New York, USA and Medical Research Council of Canada. JSD is an Alberta Heritage Foundation for Medical Research (AHFMR) Honourary Professor, and ADB is an AHFMR Scholar. We thank David Kirk for performing the surgeries on the cervical sympathetic trunk.

REFERENCES

1. D.P. Cardinali, H.E. Romeo, Peripheral neuroendocrine interrelationships in the cervical region, *NIPS* 5:100 (1990).
2. H.E. Romeo, E. Spinedi, F. Vacas, F. Estivariz, D.P. Cardinali, Increase in adrenocortictropin release during wallerian degeneration of peripheral sympathetic neurons after superior cervical ganglionectomy of rats, *Neuroendocrinol.* 51:213 (1990).
3. R. Boyer, L. Tapia-Arancibia, G. Alonso, S. Arancibia, Decrease of hypothalamic TRH levels but not plasmic TSH levels after ablation of submandibular salivary glands in the rat, *J. Biol. Buccale* 16:69 (1988).
4. A. Miyake, K. Tasaka, S. Otsuka, H. Kohmura, H. Wakimoto, T. Aonot, Epidermal growth factor stimulates secretion of rat pituitary lutenizing hormone in vitro, *Acta Endocrinol.* 108:175 (1985).
5. R. Boyer, F. Jame, S. Arancicia, Une fonction non exocrine de la glande sous-maxillaire, *Annales d'Endocrinologie* (Paris) 52:307 (1991).
6. R. Relkin, Effect of pinealectomy and constant light and darkness on thyrotropin level in pituitary and plasma of the rat, *Neuroendocrinol.* 10:46 (1972).

7. J. Vriend, P.M. Hinkle, K.M. Knigge, Evidence for TRH inhibitor in the pineal gland. *Endocrinology* 107:1791 (1980).
8. R. Levi-Montalcini, The nerve growth factor 35 years later, *Science* 286:1154 (1987).
9. M.G. Spillantini, L. Aloe, E. Alleva, R. De Simone, M. Guedert, R. Levi-Montalcini, Nerve growth factor mRNA and protein increase in hypothalamus in a mouse model of aggression, *Proc. Natl. Acad. Sci. USA* 86:8555 (1989).
10. U. Wingren, T.H. Brown, B.M. Watkins, G.M. Larson, Delayed gastric ulcer healing after extirpation of submandibular glands is sex-dependent, *Scand. J. Gastroenterol.* 24:1102 (1989).
11. B.L. Tepperman, B.D. Soper, Effect of sialadenectomy on ethanol-induced gastric mucosal damage in the rat: role of neutrophils, *Can. J. Physiol. Pharmacol.* 68:207 (1990).
12. M. Laato, J. Heino, V.-L. Kahari, J. Nhinikoski, B. Gerdin, Epidermal growth factor (EGF) prevents methylprednisolone-induced inhibition of wound healing, *J. Surg. Res.* 47:354 (1989).
13. Y. Kamei, O. Tsutsumi, Y. Kuwabara, Y. Taketani, Intrauterine growth retardation and fetal losses are caused by epidermal growth factor deficiency in mice. *Am. J. Physiol.* 264:R597 (1993).
14. Noguchi, S., Ohba, Y., and Oka, T., Effect of salivary epidermal growth factor on wound healing of tongue in mice. *Am. J. Physiol.* 260:E620 (1991).
15. Aloe, L., Alleva, E., Bohm, A., and Levi-Montalcini, R., Aggressive behaviour induces release of nerve growth factor from mouse salivary gland into the blood stream. *Proc. Natl. Acad. Sci. USA* 83:6184 (1986).
16. Poulsen, K., and Pedersen, E.B., Increase in plasma renin in aggressive mice originates from kidneys, submaxillary and other salivary glands, and bites. *Hypertension* 5:180 (1983).
17. Pedersen, E.B., and Poulsen, K., Aggression-provoked huge release of submaxillary mouse renin to saliva. *Acta Endocrinol.* 104:510 (1983).
18. Hwang, D.L., Wang, S., Chen, R. C-R., and Lev-Ran, A., Trauma, especially of the submandibular glands, causes release of epidermal growth factor into bloodstream of mice. *Reg. Peptides* 34:133 (1991).
19. Shimizu, M., Sato, J., Ishi, T., Kanada, T. and Shinoda, M., Androgen-induced production of colony-stimulating factor (CSF) and colony-inhibitory factor (CIF) in the submandibular gland in female mice. *J. Pharmacobio-Dyn.* 12: 352 (1989).
20. Kemp, A., Mellow, L. and Sabbadini, E., Inhibition of interleukin 1 activity by a factor in submandibular glands of rats. *J. Immunol.* 237:2245 (1986).
21. Amand, O., Tsuji, T., Nakamura, T., and Iseki, S., Expression of transforming growth factor ß1 in the submandibular gland of the rat. *J. Histochem. Cytochem.* 39:1707 (1991).
22. Miller, D.A., Lee, A., Pelton, R.W., Chen, E.Y., Moses, H.L., and Derynck, E.Y., Murine transforming growth factor ß2 cDNA sequence and expression in adult tissues and embryos. *Mol. Endocrinol.* 3:1108 (1989).
23. Salido, E.C., Yen, P.H., Shapiro, L.J., Fisher, D.A. and Barajas, L., In situ hybridization of nerve growth factor mRNA in the mouse submandibular gland. *Lab. Invest.* 59:625 (1989).
24. Banks, B.E.C., Vernon, C.A., & Warner, J.A., Nerve growth factor has anti-inflammatory activity in the rat hind-paw oedema test. *Neurosci. Lett.* 47:41 (1984).
25. Sugiyama, K., Suzuki, Y., and Furuta, H., Histamine-release induced by 7S nerve-growth factor of mouse submandibular salivary glands. *Arch. Oral Biol.* 30:93 (1985).
26. Fu, Y-K., Arkins, S., Wang, B.S., and Kelley, K.W., A novel role of growth factor and insulin-like growth factor-1. Priming neutrophils for superoxide anion secretion. *J. Immunol.* 146:1602 (1991).
27. Kannan, Y., Ushio, H., Koyama, H., Okada, M., Oikawa, M., Yoshihara, T., Kaneko, M. and Matsuda, H., 2.5S nerve growth factor enhances survival, phagocytosis, and superoxide production by murine neutrophils. *Blood* 77:1320 (1991).
28. Gruber, D.F., O'Halloran, K.P., D'Alesandro, S., and Farese, A.M., Hypermetabolic priming of canine neutrophils by 7-S nerve growth factor. *Am. J. Vet. Res.* 51:921 (1990).
29. Zimmerli, W., Huber, I., Bouma, B.N., and Lammle, B., Purified human plasma kallikrein does not stimulate but primes neutrophils for superoxide production. *Thromb. Haemost.* 29:1221 (1989).
30. Boyle, M.D., Lawman, M.J.P., Gee, A.P., and Young, M., Nerve growth factor: a chemotactic factor for polymorphonuclear leukocytes in vivo. *J. Immunol.* 134:564 (1985).
31. Saito, K., Kato, C., and Teshigawara H., Saliva inhibits chemiluminescence response, phagocytosis and killing of *Staphylococcus epidermidis* by polymorphonuclear leukocytes. *Infect. Immunity* 56:2125 (1988).
32. Ramaswamy, K., Mathison, R., Carter, L., Kirk, D., Green, F., Davison, J.S., and Befus, D., Marked antiinflammatory effects of decentralization of the superior cervical ganglia. *J. Exp. Med.* 172:1819 (1990).

33. Mathison, R.D., Hogan, A., Helmer, D., Bauce, L., Woolner, J., Davison, J.S., Schultz, G., and Befus, D., Role for the submandibular gland in modulating pulmonary inflammation following induction of systemic anaphylaxis. *Brain Behav. Immun.* 6:117 (1992).
34. Carter, L., Ferrari, J.K., Davison, J.S., and Befus, D., Inhibition of neutrophil chemotaxis and activation following decentralization of the superior cervical ganglia. *J. Leukoc. Biol.* 51:597 (1992).
35. Mathison, R.D., Davison, J.S., De Sanctis, G., Green, F., and Befus, A.D., Decentralization of the superior cervical ganglia and the immediate hypersensitivity response. *Proc. Soc. Exp. Biol. Med.* 200:542 (1992).
36. Mathison, R.D., Carter, L., Mowat, C., Bissonnette, E., Davison, J.S., and Befus, D., Temporal analysis of the anti-inflammatory effect of decentralization of the rat superior cervical ganglia. *Am. J. Physiol.* (in press) (1993).
37. Bissonnette, E.Y., Mathison, R.D., Carter, L., Davison, J.S., and Befus, A.D., Decentralization of the superior cervical ganglia inhibits mast cell mediated TNFa-dependent cytotoxicity. 1. Potential role of salivary glands. *Brain Behav. Immun.* (in press) (1993).
38. Mathison, R.D., Befus, D., and Davison, J.S., Removal of the submandibular glands increases the acute hypotensive response to endotoxin. *Circ. Shock* 39:52 (1993).

NEUROENDOCRINE ASPECTS OF LYMPHOCYTE MIGRATION

C.A. Ottaway

Department of Medicine
University of Toronto
Toronto, Ontario, Canada

INTRODUCTION

Lymphocyte migration plays a central role in adaptive responses because it permits the continuous redeployment of specialist lymphocytes between different compartments of the body. Through this process, the immune system undergoes dynamic remodelling in response to physiological and antigenic challenges. There have been rapid advances in our understanding of the cellular and molecular features that contribute to migration of lymphocytes in vivo, which are well reviewed elsewhere.[1,2] It is now traditional to consider the first step in lymphoid migratory behaviour to be the exit of the cells from the blood stream. In lymph nodes, Peyer's patches and virtually all other tissues of the body with the exception of the spleen, lymphocytes leave the cardiovascular system via the postcapillary venules (PCV),[3] and the attachment of migrating lymphocytes to the endothelium depends upon the reciprocal interaction of adhesion molecules at the surface of these cells.[4,5]

Four major groups of molecular determinants have been implicated in these adhesive interactions.[2,4,5] Those on lymphocytes include integrins, a leukocyte specific selectin, and specific carbohydrate determinants (Table 1). Those on endothelial cells include members of the immunoglobulin-like superfamily of cell adhesion molecules, site-specific mucin-like glycoprotein adhesins or addressins, and endothelial selectins. The adhesive events in which the cells use these molecules are usually considered to be influenced only by regulatory signals arising from within the immune apparatus itself, as occurs, for example, when the organism responds to antigenic challenge. Substantial evidence has accumulated, however, demonstrating that neuroendocrine signals can influence lymphocyte traffic and migration *in vivo* and evidence is now accumulating that these signals can influence lymphocyte-endothelial interactions *in vitro*. A number of pathways by which the neuroendocrine system can regulate lymphocyte migration have been described,[6] but the two central neural pivots of general adaptational responses are the activation of the sympathetic autonomic

Advances in Psychoneuroimmunology, Edited by
I. Berczi and J. Szélenyi, Plenum Press, New York, 1994

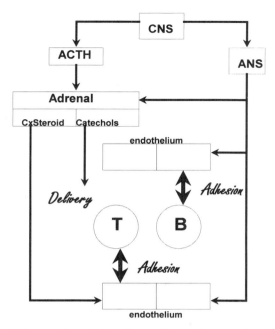

Figure 1. A schematic integration of the effects of central sympathoadrenal activation and hypothalamaic-pituitary axis activation on the interaction of lymphocytes with endothelium in PCV.

Table 1. Some lymphocyte surface molecules involved in attachment to PCV endothelium.

Molecule	Other Names or designations	Endothelial Ligand(s)	Present on
$\alpha_L\beta_2$ Integrin	LFA-1 CD11a /CD18	ICAM-1 ICAM-2	T cells B cells
$\alpha_4\beta_1$ Integrin	VLA4	VCAM-1 MadCAM-1	T cellls B cells
$\alpha_4\beta_7$ Integrin	LPAM	VCAM-1 MadCAM-1	T cells B cells
L-Selectin	Mel-14, Leu-8	MECA-79 GlyCAM-1	T cells B cells
Sialyl Lewis X	CLA	E-Selectin	T cells

nervous system (ANS), and the activation of nerves in the paraventricular nucleus which contin corticotropin-releasing hormone (CRH). I wish here to briefly consider the potential role of these pathways as regulators of the migration of lymphocytes *in vivo* (Figure 1).

SYMPATHOADRENAL ACTIVATION

Lymphoid organs are extensively innervated by noradrenergic autonomic nerves.[7,8] This neural connection permits the potential translation of ANS signals into effects on migration in a number of ways. First, a substantial portion of the innervation of lymphoid tissues is associated with the vasculature, and the perfusion of lymph nodes is very responsive to both α and β adrenergic signals.[9] Sympathetic activation would change the delivery of cells and the turbulence in the PCV, thus altering the opportunities for interaction of lymphocytes with the endothelium. Second, the lymphoid corridors immediately surrounding the vasculature in lymphoid tissues are also innervated, and these so-called T cell areas are important venues through which both T and B cells travel during their migration through the tissues. A third level of interaction may occur through more immediate short range interactions mediated by the synapse-like connections between nerve endings and lymphocytes which have been identified.[10]

Another means by which sympathetic activation affects migration is through the release of catecholamines from the adrenal medulla into the central circulation. It has been known for some time that direct infusion of catechols via the blood stream markedly affects the output of lymphocytes from spleen and lymph nodes,[11,12] and studies with sympathetic disruption in mice treated with 6-hydroxydopamine have demonstrated that both the migratory ability of lymphocytes and the lymphocyte-collecting abilities of lymphoid tissues are affected by chemical ablation of the sympathetic innervation of lymph nodes.[13] Thus, catecholamines can influence aspects of lymphocyte distribution and migration through direct neural effects, local paracrine effects, as well as endocrine pathways.

Two behavioural paradigms that affect sympathetic activation and disrupt the distribution of lymphocytes in various compartments are acute physical exertion and acute psychological or emotional stress. Acute exercise activates the sympathoadrenal system in humans as well as experimental animals and regularly leads to alterations in the availability of lymphocytes in the blood stream, but the relative effect differs for different subpopulations of lymphocytes. In humans, during acute physical exertion there is an incresae in the concentration of $CD4^+$ and $CD8^+$ T cells, B cells and large granular lymphocytes (LGL) in the peripheral blood, which depends upon both the work-dependent catecholamine elevation and the training state, or adaptation, of the subject.[14,15] The relative increase in concentration tends to be largest for LGL, 2intermediate for $CD8^+$ T cells and lowest for $CD4^+$ T cells, but $CD4^+$ T cell concentrations in the blood can increase from approximately 750 cells/mm^3 at rest to 1000 cells/mm^3 after 10 min of moderate exercise, and rise as high as 1300 cells/mm^3 in response to 1 min of exertion at 100% of maximal ventilatory capacity.[16] Such effects are at least in part inhibitable by administering β -adrenergic antagonists and are believed to represent rapid redistribution of sequestered lymphocytes into the central circulation. For T cells and B cells much of this may arise from spleen and other lymphoid tissues, but the mechanism for LGL is presumed to be different. Many LGL appear to be marginated within the vascular space and their rapid recentralization may reflect adrenergic alterations of their interaction with endothelial cells in this marginated pool. In vitro, the presentation of catecholamines leads to the detachment of LGL from cultured endothelial cells.[17] Whether more prolonged adrenergic activation can alter the presentation of adhesion molecules by endothelial cells is not known, but *in vitro* adrenergic agents alter the responsiveness of endothelial cultures to interferon[18] and cyclic AMP activation of endothelial cells inhibits their production of E-selectin and VCAM in response to TNF stimulation.[19]

Another challenge that can activate the sympathoadrenal system and alter the availability of lymphocytes in the blood is acute short-term stressors such as performance of a frustrating laboratory task[20] or rehearsal of disturbing emotional experiences.[21] The changes observed in these circumstances tend to be more complex and variable than those seen with acute physical exertion. With frustration, circulating concentrations of LGL and CD8$^+$ T cells are elevated immediately after the task, but CD4$^+$ T cell concentrations are decreased.[20] With emotional challenge, no alterations could be found in LGL or T cell concentrations during the challenge, but there was an unexpected but regular increase in CD4$^+$ T cell concentrations during recovery from the rehearsal.[21] Meta-analytical review of a variety of naturalistic (real-life) stress paradigms has demonstrated a consistent negative relationship between these stressors and absolute numbers of circulating LGL, CD4$^+$ and CD8$^+$ T cells and B cells.[22]

HYPOTHALAMIC-PITUITARY-ADRENAL ACTIVATION

CRH activation of the hypothalamic-pituitary-adrenal axis provides another major pathway through which lymphocyte distributions in various body compartments can be alered (Figure 1). Corticosteroids have been recognized for many years to profoundly alter the total numbers and subpopulation pattern of lymphocytes in the blood[23,24] and play a crucial role in this cellular response. These effects are dose-dependent, occur within minutes of acute infusion of corticosteroids in humans, affect T cells more than B cells, and have a more profound effect on CD4$^+$ T cells than on CD8$^+$ T cells.[25,26] Similar effects are seen in experimental animals. Bolus infusions of either ACTH or prednisolone in rats results in a transient redistribution of blood-borne lymphocytes into the bone marrow.[27] More prolonged infusion of prednisolone, however, disrupts the accumulation of lymphocytes in many compartments such that the entry of labelled lymphocytes into lymphoid tissues is decreased and the retention of those that do arrive is prolonged.[27] Both of these disruptions are reversible, however, and when the corticosteroid infusions are stopped the subsequent migration behaviour of the previously transferred cells is normal.[27]

An important part of this effect of corticosteroids is mediated through alterations in the adhesiveness of endothelial cells. In mice, dexamethasone treatment of recipient animals also changes the migration patterns of labelled lymphocytes. Assays of the ability of lymphocytes to attach to the endothelium in lymphoid tissue sections have shown that lymphocytes from untreated animals adhere poorly to sections from dexamethasone treated animals, while lymphocytes from treated animals adhere well to lymph node sections from untreated animals.[28] Thus, endothelial cells appear to be central to this effect of corticosteroids. Recently, it has been shown that addition of corticosteroids to endothelial cell cultures inhibits the synthesis and expression of endothelial E-Selectin and ICAM-1 adhesion molecules that normally occurs in response to cytokine stimulation.[29]

Not all of the effects of corticosteroids on lymphocyte distribution are the result of changes in endothlial adhesiveness, however. When acute bolus infusions of methylprednisolone are given to humans, circulating concentrations of CD4$^+$ T cells can decrease by as much as 80% within a few hours.[26] The kinetics of this reaction are consistent with a modest reduction in the rate of clearance of the CD4[+] T cells from the blood stream, but what is most markedly decreased is the rate of re-entry of cells into the vascular compartment.[26] This probably reflects a steroid induced immobilization of the cells throughout the lymphoid compartments of the body similar to that identified in rodents.[27]

CONCLUSION

The ability of stress paradigms and both psychological and physiological challenges to modulate immune responses in intact animals is now well established. The information reviewed here supports the notion that the redistribution of lymphocyte subpopulations within the body is an integral part of adaptive responses. Sympathoadrenal activation and hypothalamic-pituitary activation have direct effects both on the intravascular behaviour of lymphocytes and the adhesion of lymphocytes with the endothelium. The molecular and cellular aspects of these processes require further understanding, however, and exploration of these issues can give important insights into the interactions of the neuroendocrine and immune systems *in vivo*.

REFERENCES

1. M. Dustin and T. Springer, Role of lymphocyte adhesion receptors in transient interactions and cell locomotion, *Annu. Rev. Immunol.* 9:27 (1991).
2. L. Picker and E. Butcher, Physiological and molecular mechanisms of lymphocyte homing, *Annu. Rev. Immunol.* 10:561 (1992).
3. C.A. Ottaway, Dynamic aspects of lymphoid cell migration, in: "Migration and Homing of Lymphoid Cells," vol. II, A.J. Husband, ed., CRC Press, Boca Raton, FL (1988).
4. R.O. Haynes, Integrins: versatility, modulation and signaling in cell adhesion, *Cell* 69:11 (1992).
5. C.R. Mackay and B. Imhof, Cell adhesion in the immune system, *Immunol. Today* 14:99 (1993).
6. C.A. Ottaway and A.J. Husband, Central nervous system influences on lymphocyte migration, *Brain Behav. Immun.* 6:97 (1992).
7. D. Felten, K. Ackerman, S. Weigand and S. Felten, Noradrenergic sympathetic innervation of the spleen. I. Nerve fibers associate with lymphocytes and macrophages in specific compartments of the white pulp, *J. Neurosci. Res.* 18:28 (1987).
8. S. Felten, D. Felten, D. Bellinger, S. Carlson, K. Ackerman and K. Madden, Noradrenergic sympathetic innervation of lymphoid organs, *Prog. Allergy* 43:14 (1988).
9. O. Lundgren and I. Wallentin, Local chemical and nervous control of consecutive sections in the mesenteric lymph node of the cat, *Angiologica* 1:284 (1964).
10. S. Felten and J. Olschowka, Noradrenergic sympathetic innervation of the spleen. II. Tyrosine hydroxylase positive nerve terminals form synaptic-like contacts on lymphocytes in the splenic white pulp, *J. Neurosci. Res.* 18:37 (1987).
11. T. Moore, Modification of lymphocyte traffic by neurotransmitter substances, *Immunology* 52:511 (1984).
12. M. Loeper and O. Crouzon, L'action de l'adrenaline sur le sang, *Arch. Med. Exp.* 16:83 (1904).
13. D. Felten, S. Felten, D. Bellinger, S. Carlson, K. Ackerman and K. Madden, Noradrenergic sympathetic neural interactions with the immune system: structure and function, *Immunol. Rev.* 100:225 (1987).
14. D. McCarthy and M. Dale, The leukocytosis of exercise, *Sports Med.* 6:333 (1988).
15. A. Kendall, L. Hoffman-Goetz, M. Houston, B. MacNeil and Y. Arumugam, Exercise and blood lymphocyte subset responses: intensity, duration, and subject fitness effects, *J. Appl. Physiol.* 69:251 (1990).
16. A. Gray, Y. Smart, R. Telford, M. Weidemann and T. Roberts, Anaerobic exercise causes transient changes in leukocyte subsets and IL-2R expression, *Med. Sci. Sports Exerc.* 24:1332 (1990).
17. R. Benschop, F. Oostveen, C. Heijnen and R. Ballieux, Beta-adrenergic stimulation causes

detachment of natural killer cells from cultured endothelium, *Eur. J. Immunol.* 23:3242 (1993).

18. S. Bourdoulous, O. Durieu-Trautmann, A. Strosberg and P. Courand, Catecholamines stimulate MHC Class I, Class II and invariant chain gene expression in brain endothelium through different mechanisms, *J. Immunol.* 150:1486 (1993).

19. J. Pober, M. Slowik, L. De Luca and A. Ritchie, Elevated cyclic AMP inhibits endothelial cell synthesis and expression of TNF-induced endothelial leukocyte adhesion molecule-1, and vascular cell adhesion molecule-1, but not intracellular adhesion molecule-1, *J. Immunol.* 150:5114 (1993).

20. E. Bachen, S. Manuck, A. Marsland, S. Cohen, S. Malkoff, M. Muldoon and B. Rabin, Lymphocyte subset and cellular immune responses to a brief experimental stressor, *Psychosom. Med.* 54:673 (1992).

21. P. Knapp, E. Levy, R. Giorgi, P. Black, B. Fox and T. Heeren, Short-term immunological effects of induced emotion, *Psychosom. Med.* 54:133 (1992).

22. T. Bennet-Herbert and S. Cohen, Stress and immunity in humans: a meta-analytic review, *Psychosom. Med.* 55:364 (1993).

23. A. Fauci and D. Dale, The effect of hydrocortisone on the kinetics of normal human lymphocytes, *Blood* 46:235 (1975).

24. A. Fauci, Mechanisms of corticosteroid action on lymphocyte subpopulations, *Immunology* 28:669 (1975).

25. J. Slade and B. Hepburn, Prednisone induced alterations of circulating human lymphocyte subsets, *J. Lab. Clin. Med.* 101:479 (1983).

26. L. Fisher, E. Ludwig, J. Wald, R. Sloan, E. Middleton and W. Jusko, Pharmacokinetics and pharmacodynamics of methylprednisolone when administered at 8 a.m. versus 4 p.m., *Clin. Pharmacol. Ther.* 51:677 (1992).

27. J. Cox and W. Ford, The migration of lymphocytes across specialized vascular endothelium IV. Prednisolone acts at several points on the recirculation pathway of lymphocytes, *Cell. Immunol.* 66:407 (1982).

28. H. Chung, W. Samlowski and R. Daynes, Modification of the murine immune system by glucocorticosteroids: alterations of the tissue localization properties of circulating lymphocytes, *Cell. Immunol.* 101:571 (1986).

29. B. Cronstein, S. Kimmel, R. Levin, F. Martiniuk and G. Weissmann, A mechanism for the anti-inflammatory effects of corticosteroids: the glucoccorticoid receptor regulates leukocyte adhesion to endothelial cells and expression of endothelial-leukocyte adhesion molecule 1 and intercellular adhesion molecule, *Proc. Natl. Acad. Sci. USA* 89:9991 (1992).

MECHANISMS OF STRESS-INDUCED HOST DEFENSE ALTERATIONS

Alex Kusnecov and Bruce S. Rabin

The Brain, Behavior, and Immunity Center and the
Department of Pathology
The University of Pittsburgh School of Medicine
Pittsburgh, PA 15213

INTRODUCTION

The initial processing and subsequent physiological response to an aversive situation or environmental cues previously associated with an aversive situation (conditioned stimuli) occurs within the brain. Using experimental models, many of the areas of the brain which respond to a stressor have been identified. The role of neurohormones within the brain and the participation of the sympathetic nervous system in altering immune function is also beginning to be clarified. The primary interest of our research laboratory is to understand the mechanisms which are activated after a subject experiences a stressor and the resultant alteration of immune system function.

We have found that stressor induced immune alteration of spleen lymphocyte function, in rodents, is related to catecholamines while alteration of natural killer cell function is related to opioids. The alteration of function of peripheral blood lymphocytes is an adrenal dependent function[1]. Conditioning rats to an aversive stimulus produces the same degree of immunologic alteration as does the actual aversive event[2,3]. Thus, a suitable model is available which allows studies for determining the mechanism of stressor induced immune alteration.

I. Identification of Areas of the Brain Which are Activated by Stress

The perception of the existence of an aversive situation or a situation which has been conditioned to be aversive, must occur within the brain. To identify the areas of the brain which are involved in stressor-induced immune alteration, a variety of approaches can be taken. For example, to mimic neurochemical changes due to stress, activation of selected brain regions by the infusion of agonists can be used. Conversely, the influence of lesioning of discrete brain regions prior to exposure of

Advances in Psychoneuroimmunology, Edited by
I. Berczi and J. Szélenyi, Plenum Press, New York, 1994

animals to a stressful situation, may also aid in the identification of central brain regions which trigger subsequent immune alteration. However, these procedures must, of necessity, focus on one area at a time, thereby failing to provide a comprehensive overview of all areas activated by a stressor. An early marker of cell activation, the c-fos proto-oncogene can be used to identify individual neurons within the brain which are activated by stress. Anti-sera which have specific reactivity for c-fos have been used along with immunohistochemical procedures to assist in such identification.

In interpreting the c-fos localization studies, it is important to remember that the presence of c-fos within an area of the brain does not indicate that the activated cells are involved in the activation of pathways which suppress immune function. Some areas which are activated may be involved with the transmission of signals which activate the H-P-A axis or the sympathetic nervous system while other areas may inhibit the transmission of signals which activate the H-P-A and sympathetic pathways. Further, there may be areas of the brain which do not become c-fos positive following a stressor and which are involved in the pathways of immune system alteration. Given these qualifications, we have sacrificed male Lewis rats immediately after either a single 60 minute session of electric shock or reexposure to the shock box (but no shock) that 14 days previously had been associated with electric shock[4,5]. We then used immunohistochemical procedures to identify the c-fos areas of the brain. These areas were:

Brain stem nuclei activated by a conditioned aversive stimulus:
Locus Coeruleus
Nucleus of the solitary tract
Ventral lateral medulla
A5
A7
Dorsal and ventral subdivisions of the periaqueductal gray area
Serotonergic neurons of the dorsal raphe nuclei

Forebrain nuclei activated by a conditioned aversive stimulus:
CRH containing neurons of the paraventricular nucleus of the hypothalamus
Ventral lateral septal nuclei
Medial amygdaloid nuclei
Sensorimotor cortex
Basal ganglia
Thalamic nuclei

Although a number of areas are activated, it is not yet possible to determine the relevance of each of these to the eventual alteration of immune function following a stressor. However, some preliminary data does indicate a central role of the locus coeruleus. Rats were conditioned to an electric shock stimulus and then re-exposed to the conditioned stimulus (the box) after being injected with diazepam. The mitogenic function of spleen lymphocytes was not suppressed in the diazepam treated animals (Fowler, et al, Manuscript in preparation). When the use of the c-fos proto-oncogene was employed to identify the activated brain areas in the diazepam-treated animals, all of the areas which were activated in saline injected animals, other than the locus coeruleus and some neurons in the A5 area were activated (unpublished data). This suggests that prevention of activation of the locus coeruleus prevents the functional alteration of spleen lymphocyte mitogenic activity. Thus, a central role for the locus coeruleus is suggested.

The pre-ganglionic neurons of the sympathetic nervous system which are involved with innervation of the spleen originate within the intermediolateral cell column of the spinal cord. It is likely that the areas of the central nervous system which are capable of activating the intermediolateral cell column must be activated by a stressor. Using retrograde tracing from the spleen with a herpes virus that is transmitted transsynaptically, preliminary data suggests that the locus coeruleus has synaptic connections with the intermediolateral cell column (Ackerman, et al, unpublished). This is further supported by preliminary studies which suggest that CRH activation of the locus coeruleus leads to suppressed mitogenic function of spleen lymphocytes (Rassnick, et al, unpublished data).

It will be important to identify the brain areas which promote as well as inhibit the transmission of activation signals to the neurons which innervate the spleen. As lesioning of the paraventricular nucleus is associated with an increase in the amount of stressor-induced suppression of spleen lymphocyte mitogenic function[6], an interaction between the locus coeruleus and the paraventricular nucleus may be involved in regulating the amount of activation which converges upon a common pathway which leads to alteration of the amount of sympathetic outflow in the spleen.

Another study, which evaluated the effect of stress in hypophysectomized rats[7], found that the suppression of spleen lymphocytes to non-specific mitogens was greater in hypophysectomized animals than in intact or sham treated animals. Both the PVN lesioning and hypophysectomy studies suggest that there are areas of the brain which, when activated by a stressor, interfere with the transmission of the neuronal signals to the spleen. Removal of these areas results in enhanced catecholamine release in the spleen. Thus, these regulatory areas of the brain may serve to prevent excessive immune suppression when a subject experiences a stressor, and therefore, may serve an adaptive function.

In addition, there have been interesting studies, performed in humans, which indicate that the ability of an individual to cope with stress may influence the extent of stressor-induced immune alteration[8]. These studies would further suggest that there are interactions between various areas of the brain which modify the final outflow pathway to the spleen.

II.   Sympathetic Nervous System Activation

Studies in both rodents and humans indicate that activation of the sympathetic nervous system is associated with suppression of the response of lymphocytes to non-specific mitogenic stimulation. However, there are differences which are found between species as well as between different compartments of the immune system. In addition, finding an association between activation of the sympathetic nervous system and the functional alteration of lymphocyte mitogenic function does not causally identify a sympathetic mechanism for lymphocyte alteration.

Receptors for catecholamines are present on both mononuclear phagocytes and lymphoid cells. The concentration of catecholamine receptors differs on different lymphocyte subpopulations with the highest numbers being present on mononuclear phagocytes. As there is extensive interaction between mononuclear phagocytes and lymphocyte subpopulations in the generation of the immune response, it is possible that alteration of the function of one of the components of mononuclear cell interaction will lead to a functional alteration of the immune system. For example, if the non-specific mitogen, phytohemagglutinin (PHA) primarily stimulates the CD4 population of lymphocytes to divide, an increase in the functional activity of the CD8 lymphocyte population may suppress the ability of the CD4 cells to respond to mitogen. It is also possible that mononuclear phagocytic cells will be activated by catecholamines and then release factors which alter the ability of lymphocytes to respond to non-specific mitogens.

Our laboratory has used β-adrenergic receptor antagonists, in rats, to study the role of catecholamines in the alteration of lymphocyte mitogenic function[9]. We find that injection of rats with nadolol, which does not cross the blood brain barrier prior to stressor exposure, prevents the decrease of lymphocyte mitogenic function of spleen lymphocytes. Spleen lymphocyte mitogenic function is decreased in animals which have been pretreated with saline. This data suggests that the decreased function of spleen lymphocytes is mediated by catecholamines. However, as indicated above, it does not identify the mechanism of the altered function and does not indicate that catecholamines reacting with receptors on lymphocytes are responsible for the altered function.

In animals which are pretreated with nadolol, the mitogenic response of peripheral blood lymphocytes remains suppressed suggesting that catecholamines are not associated with stressor-induced alteration of the responsiveness of peripheral blood lymphocytes to non-specific mitogen. A possible explanation for the differences between the role of catecholamines modifying the function of spleen and peripheral blood lymphocytes is related to the architecture of the spleen. There is a dense sympathetic innervation in the spleen which may produce very high localized concentrations of catecholamines which are not achieved in the blood; further, the catecholamines may effect macrophages which then release substances which modify lymphocyte function. The lack of close interaction between macrophages and lymphocytes in the blood may be related to the lack of effective catecholamines on the mitogenic function of peripheral blood lymphocytes.

Natural killer cell function, in the spleen of rats exposed to a stressor, is also suppressed[10]. When experimental animals are pretreated with nadolol, natural killer cell function remains suppressed. However, pretreating animals with naltrexone, prevents the stressor-induced modulation of natural killer cell function[9]. Thus, the sympathetic nervous system does not appear to be involved in modulating the activity of natural killer cells which are functionally altered, in rats, by the opioid system.

Studies in humans do not have the luxury of studying the effect of stress on spleen lymphocyte function but are capable of evaluating changes in peripheral blood function. Experimental models of stress can be performed in humans who are exposed to a frustrating mental task. Such tasks produce an increase in sympathetic nervous system activity and both quantitative and qualitative changes in peripheral blood lymphocytes. There is a fairly rapid increase in the number of CD8 T lymphocytes and natural killer cells which are present in the blood. The responsiveness of lymphocytes to non-specific mitogenic function decreases[11]. Those individuals with the greatest increase in sympathetic activity show the largest changes in immune function. As cortisol does not increase during the time that the test is being administered, the changes which occur in the immune system are independent of glucocorticoid binding to receptors. Both male and female subjects show stressor-induced immune alterations. Thus, there are no apparent sex differences which differentiate immune alterations in response to stress which are influenced by sex hormones. However, the mechanism by which catecholamines alter immune function in humans has not been identified.

A study in our laboratory added different concentrations of catecholamines to mononuclear cells that had been purified on a density gradient or to cultures of whole blood. Although purified mononuclear cells did show suppression of mitogenic function in the presence of added catecholamines, whole blood cultures had greater decrease of mitogenic function and showed suppressed mitogenic function with low concentrations of catecholamines that failed to have an effect on isolated mononuclear cells. This data suggests that an accessory cell is involved in the alteration of mitogenic activity.

Another way to determine an interrelationship between sympathetic nervous system activity and immune function would be to evaluate the effect of the stresses associated with life events. We have found that there is a highly significant decrease of the numbers of CD8 lymphocytes in healthy adults with high levels of stress in comparison to healthy adults with low levels of stress. This is exactly the opposite finding of the effect of acute stress on CD8 lymphocytes. Individuals with high levels of stress have been reported to be more susceptible to viral infection[12], and possibly the development and/or exacerbation of autoimmune disease. These observations indicate that it will be important to differentiate the effects of acute or chronic stress on immune system function.

III.   Adrenergic Receptors on Lymphocytes

In order to implicate catecholamines as having an effect upon immune function, receptors for catecholamines must be present on lymphocytes. Quantitation of β-adrenergic receptors indicates that natural killer cells have the largest number of adrenergic receptors followed respectively by B lymphocytes, CD8 lymphocytes and CD4 lymphocytes. This is of interest as the lymphocyte population which often is most greatly elevated following a stressor is natural killer cells. If, as has been suggested that stress causes a release of lymphocytes from the spleen, it is possible that adrenergic receptors are involved with the adherence of lymphocyte populations to the vascular endothelium. In this regard, it has recently been reported that treating lymphocytes with either epinephrine or norepinephrine decreases the ability of lymphocytes with the phenotypic characteristics of NK cells to adhere to vascular endothelial cells, in vitro. The altered adherence was mediated by intracellular changes of the concentration of cAMP in the lymphocytes.[23] However, as B lymphocytes have the second largest number of β-adrenergic receptors, and there is no change in B lymphocyte numbers occurring in association with stress-induced elevation of catecholamines, it is possible that there may be a difference in coupling of the adrenergic receptor in B lymphocytes to second messenger systems in comparison to second messenger coupling in other lymphocyte populations.[13]

Studies in humans[14] have shown that two signals which consist of a ligand binding to the β-adrenergic receptor occurring simultaneously with activation of the T cell receptor, produces a marked increase in cyclic AMP in lymphocytes and a decrease of mitogenic function. Whether ligands such as NPY, ACTH, or CRH for example, are necessary to participate in stressor-induced immune alteration of lymphocytes and therefore the number of receptors for these other ligands is a factor in immune changes which occur, has not yet been determined.

Injection of humans with catecholamines has helped to implicate the role of these substances in quantitative and qualitative immune changes. When epinephrine is injected into healthy humans, there is a transient increase in the number of lymphocytes in the peripheral blood and a reduction in the response of lymphocytes to non-specific mitogenic stimulation[15]. As occurs with acute stress, catecholamine injection produces an increase in the numbers of CD8 and NK lymphocytes[16]. However, when patients who have had a splenectomy are injected with catecholamines, there is no change in the composition of lymphocyte subsets in the peripheral blood[17]. In addition, there is no change in the response of peripheral blood lymphocytes to mitogen in splenectomized patients while there is a decrease of mitogenic responsiveness which occurs in normal individuals. These studies suggest that the spleen participates in the catecholamine modulated alteration of lymphocyte numbers and function.

IV.   Cortisol

Short term studies of stress on immune function are terminated before cortisol becomes elevated[11] in humans as cortisol becomes elevated approximately 30 minutes after stress onset.   Cortisol changes in chronic stress have not adequately been related to the lymphocytic changes which are present in chronically stressed individuals and, as has been indicated, the changes in acute and chronic stress differ.

There are studies, which suggest that glucocorticoids can have an effect on lymphocytes.   Cortisol receptors are present in the cytoplasm of cells and when receptors bind to glucocorticoids they move to the nucleus where they bind to DNA. The receptors are of two types and are termed Type 1 and Type 2.   In rats, ligands for the Type 1 and Type 2 receptor significantly decrease the white blood cell count and the number of lymphocytes and monocytes which are in the peripheral blood. Binding of a ligand to the Type 1 glucocorticoid receptor decreases the number of NK cells in the circulation while a ligand binding to the Type 2 receptor decreases the numbers of all lymphocyte subsets present. As there are different concentrations of Type 1 and Type 2 receptor in different lymphocyte subpopulations, the effect of glucocorticoid elevation which occurs in chronic stress may have a varied effect upon the immune system, depending on the lymphocyte subpopulation which is being studied[18].   It is important because of the possible effect of glucocorticoids on lymphocyte circulation, that these factors be taken into consideration when studying either acute or chronic stress.

V.   The Effect of Development on Stress Related Alterations of Immune Function

If the central nervous system and adrenergic receptors on leukocytes are involved in the functional regulation of immune system activity, events which alter the ability of the central nervous system to regulate the sympathetic nervous system or which alter quantitative aspects of the peripheral sympathetic nervous system, including adrenergic receptor numbers on mononuclear cells, can have an influence on the impact of stress on immune function.   It is possible that during the developmental stage of the fetus, the number of receptors for various ligands which are available on the external surface of a cell may be permanently altered depending upon the concentration of ligands which are present during the early stages of the organization of the fetus. If such alterations occur in the developing fetus, there may be subsequent alterations in various biochemical processes which have an influence on behavior or the effect of stress on immune function. In this regard, there are studies which have demonstrated that exposing a pregnant animal to a stressor will alter the manner in which an offspring responds to a stressor [19].   In particular, alterations in the function of the H-P-A axis, particularly in female offspring of stressed animals have been reported[20]. After birth, there are studies which indicate that postnatal handling of young animals can alter the development of the glucocorticoid receptor system in the hippocampus and the frontal cortex[21]. These studies suggest that differences in the responsiveness of the immune system to a stressor in an adult animal, in comparison to other animals, may partially be due to in utero or early developmental events.

In non-human primates, it has been shown that the response of lymphocytes to mitogenic stimulation is significantly different when comparing nursery reared infants to maternally reared infants[22].   The differences in the lymphocyte responsiveness persists for at least two years. Whether such differences are related to the lack of immune factors transferred from mother to child through breast milk, or to an altered psychosocial environment of the separated infant, has not been determined. Thus, it is not only in rodents where such developmental modifications of immune system function occur, but also in non-human primates.

These studies are relevant to determining whether an individual response to a stressor with immune alterations, and the magnitude of the response. Early events which alter the concentrations of receptors on cells with a concentration of neuropeptides, may have an effect on the magnitude and duration of the immune alterations which are produced by stress.

VI. Conclusions

Once an event is perceived by the brain as being stressful, a series of coordinated reactions occurs. Specific areas within the brain become activated and interact with each other to either activate the release of catecholamines or activate the H-P-A axis. The various neurohormones which are produced bind to their receptors which then may alter the adherence of lymphocytes to endothelial walls, result in the production of immune regulatory substances from mononuclear phagocytic cells, change the functional relationship of lymphocyte subpopulations, or directly alter the ability of a lymphocyte to participate in an immune reaction. As a result, the ability of an individual to resist infectious disease becomes altered. In addition, susceptibility to the development of, of exacerbation of, autoimmune disease results. Once the detailed mechanisms of interaction between the brain, the multiple components of the immune system, and immune function are understood, procedures to ameliorate the effect of stress on immune system function can be initiated.

REFERENCES

1. B.S. Rabin, J.E. Cunnick, and D.T. Lysle, Stress-induced alteration of immune function, *Progress in NeuroEndocrinImmunology.* 3:116 (1990)
2. D.T. Lysle, J.E. Cunnick, H. Fowler, and B.S. Rabin, Pavlovian conditioning of shock induced suppression of lymphocyte reactivity: Acquisition, extinction, and preexposure effects, *Life Sciences.* 42:2185 (1988).
3. D.T. Lysle, J.E. Cunnick, B.J. Kucinski, H. Fowler, and B.S. Rabin, Characterization of immune alterations induced by a conditioned aversive stimulus, *Psychobiol.* 18:220 (1990)
4. M.A. Pezzone, W-S Lee, G.E. Hoffman, and B.S. Rabin, Induction of c-Fos immunoreactivity in the rat forebrain by conditioned and unconditioned aversive stimuli, *Brain Research.* 597:41 (1992).
5. M.A. Pezzone, W-S Lee, G.E. Hoffman, K.M. Pezzone, and B.S. Rabin, Activation of brainstem catecholaminergic neurons by conditioned and unconditioned aversive stimuli as revealed by C-Fos immunoreactivity, *Brain Research.* 608:310 (1993).
6. M.A. Pezzone, H. Dohanics, J.G. Verbalis, and B.S. Rabin, Effects of footshock stress upon spleen and peripheral blood lymphocyte mitogenic responses in paraventricular nucleus (PVN) lesioned rats, *Society for Neuroscience Abstracts.* 18:679 (1992).
7. S.E. Keller, S.J. Schleifer, A.S. Liotta, R.N. Bond, N. Farhoody, and M. Stein, Stress-induced alterations of immunity in hypophysectomized rats, *Proc. Natl. Acad. Sci. USA.* 85:9297 (1988).
8. R. Glaser, J.K. Kiecolt-Glaser, J.C. Stout, K.L. Tarr, C.E. Speicher, and J.E. Holliday, Stress-related impairments in cellular immunity, *Psychiatry Research.* 16:233 (1985).
9. J.E. Cunnick, D.T. Lysle, A. Armfield, and B.S. Rabin, Shock-induced   modulation of lymphocyte responsiveness and natural killer cell activity: Differential mechanisms of induction, *Brain, Behavior, and Immunity.* 2:102 (1988).
10. Y. Shavit, F.C. Martin, R. Yirmiya, S. Ben-Eliyahu, G.W. Terman, H. Weiner, R.P. Gale, and

J.C. Liebeskind, Effects of a single administration of morphine or footshock stress on natural killer cell cytotoxicity, *Brain, Behavior, and Immunity.* 1:318 (1987).

11. S.B. Manuck, S. Cohen, B.S. Rabin, M.F. Muldoon, and E.A. Bachen, Individual differences in cellular immune responses to stress *Psychological Science.* 2:1 (1991).

12. S. Cohen, D.A.J. Tyrrell, and A.P. Smith, Psychological stress and susceptibility to the common cold, *NEJM.* 325:606 (1991).

13. A.S. Maisel, P. Fowler, A. Rearden, H.J. Motulsky, and M.C. Michel, A new method for isolation of human lymphocyte subsets reveals differential regulation of b-adrenergic receptors by terbutaline treatment, *Clin. Pharmac. Ther.* 46:429 (1989).

14. S.L. Carlson, W.H. Brooks, and T.L. Roszman, Neurotransmitter-lymphocyte interactions: Dual receptor modulation of lymphocyte proliferation and cAMP production. *Journal of Neuroimmunology.* 24:155 (1989).

15. B. Crary, M. Borysenko, D.C. Sutherland, I. Kutz, J.S. Borysenko, and H. Benson, Decrease in mitogen responsiveness of mononuclear cells from peripheral blood after epinephrine administration in humans, *Journal of Immunology.* 130:694 (1983).

16. A.S. Maisel, K.U. Knowlton, P. Fowler, A. Reardon, M.G. Ziegler, H.J. Motulsky, P.A. Insel, and M.C. Michel, Adrenergic control of circulating lymphocyte subpopulations. Effects of congestive heart failure, dynamic exercise, and terbutaline treatment, *Journal of Clinical Investigation.* 85:462 (1990).

17. L.J.H. van Tits, M.C. Michel, H. Grosse-Wilde, M. Happel, F.W. Eigler, A. Soliman, and O.E. Brodde, Catecholamines increase lymphocyte b2-adrenergic receptors via a b2-adrenergic, spleen dependent process, *American Journal of Physiology.* 258:E191 (1990).

18. A. Miller, Presented at the Fourth Research Perspectives in Immunology Meeting, Boulder, Colorado, (1993).

19. E. Fride, Y. Han, J. Feldon, G. Halevy, and M. Weinstock, Effects of prenatal stress on vulnerability to stress in prepubertal and adult rats, *Psychological Behavior.* 37:681 (1986).

20. M. Weinstock, E. Matlina, E.I. Maor, H. Rosen, and B.S. McEwen, Prenatal stress selectively alters the reactivity of the hypothalamic-pituitary adrenal system in the female rat, *Brain Research.* 595:195 (1992).

21. A. Wakshlak, and M. Weinstock, Neonatal handling reverses behavioral abnormalities induced in rats by prenatal stress, *Physiological Behavior.* 48:289 (1990).

22. C.L. Coe, G.R. Lubach, W.B. Ershler, and R.G. Klopp, Influence of early rearing on lymphocyte proliferation responses in juvenile Rhesus monkeys, *Brain, Behavior, and Immunity.* 3:47 (1989).

23. R.L. Benschop, F.G. Oostveen, C.J. Heijnen and R.E. Ballieux, b2-Adrenergic stimulation causes detachment of natural killer cells from cultured endothelium, *Eur. J. Immunol.* 23:3242 (1993).

PLASTICITY OF NEUROENDOCRINE-IMMUNE INTERACTIONS DURING AGING

N. Fabris,[1,2] E. Mocchegiani[1] and M. Provinciali[1]

[1]Immunology Center, Gerontological Research Department; Italian Natl. Res. Ctrs. on Aging (INRCA), Ancona, Italy and [2]Inst. of Haematology, Medical Faculty, University of Pavia, Italy

INTRODUCTION

A good body of experimental evidences now supports the existence of numerous interactions among the nervous, endocrine and immune systems. Communication between these networks is mediated by humoral mediators, such as hormones, neurotransmitters and immune-derived cytokines, which are to a large extent shared by the different homeostatic systems. Common to nervous, neuroendocrine and immune cells are also the receptor sites sensitive to such signals. Hormones and neurotransmitters, in addition to regulating various target tissues in the body, reach also lymphoid organs and cells through blood circulation or through direct autonomic-nervous-system (ANS) connections between the nervous tissue and the organs of the lymphoid system itself (for review see[1,2]). The neuro-endocrine-immune interactions supported by circulating humoral mediators are mainly due to and mediated by the hypothalamus-pituitary axis; this may influence the immune system either releasing by various hormones and neuropeptides into the blood with direct modulatory action on the immune effectors, or regulating the hormonal secretion of peripheral endocrine glands, which also exert an immuno-modulating action.

In spite of the great understanding reached in recent years on these kinds of interconnections, little effort has been made to confront and whenever possible to specifically adopt the notions, that we have on the more classical neuro-endocrinological pathways. Hormones, neuro-transmitters and immune cytokines may exert both developmental actions, related to the structural and functional organisation of target organ or cell and play roles in the actual performance of mature cells, such as those required to counteract stressful conditions, antigenic insults included (for review see[3,4,5]). The findings related to neuroendocrine-immune interactions responsible for developmental steps should, therefore, be clearly distinguished from those related to emergency events in fully matured systems. The stimuli required to activate a given pathway as well as the end effects, may be quite different, either quantitatively or qualitatively, according to the functional demand, "morphogenic" or "of actual performance" of the organism.

These considerations have suggested to distinguish at least two levels of neuroendocrine-immune interrelationships[4]. The first level (Fig. 1A) is based on the

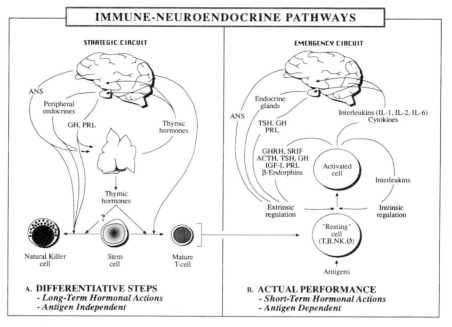

Fig. 1. Schematic representation of immune-neuroendocrine pathways. In: A) the strategic circuit, B) the emergency circuit. For explanation see text.

interactions between the neuroendocrine system and the thymus, where the proliferation and differentiation of stem cells into mature T-lymphocytes is induced. Such interactions should take into account the fact that the thymus is synthesising and secreting various hormone-like peptides with differentiation properties on the T-cell lineage[6]. The second level of interaction (Fig. 1B) is at the periphery, between neuroendocrine signals and the humoral products which are secreted by immune cells during specific reactions to various antigens[1,7].

The rationale for discriminating these two levels is based on various orders of considerations. The first level of interactions is primarily involved in maturation steps of both immune and neuroendocrine systems, which occur, in absence of pathological events, independently of antigenic stimulation (strategic circuit): in fact, neuroendocrine-thymus interactions are observable also in animals maintained under germ-free conditions. The second level of interaction requires the presence of fully differentiated immune cells and the occurrence of a specific antigenic or stress-mediated hormonal stimulus. The main role played by these regulatory interactions appears to be the normalization of the neuroendocrine and immune balance after a sudden alteration by a stressful cognitive or non-cognitive event[1,4-5] (emergency circuit).

The present paper aims to summarise the data available on the aging of both neuroendocrine and immune systems and on possible interaction between them in favouring the maintenance of discrete immune functions in old age.

NEUROENDOCRINE AND IMMUNE AGING

A sizeable body of experimental and clinical evidence has demonstrated that both the neuroendocrine and the immune system undergo, with advancing age, a progressive deterioration in their efficiency.

Immunological Aging

With regard to the immune system: the stem cell compartment, the efficiency of B-cells, and, with regard to nonspecific immunity, the function of macrophages and of antigen-presenting cells are not significantly affected by age.

Other components, such as the whole T-cell compartment and the efficiency of NK cells are more or less altered during aging.

The decline of T-cell function depends largely on the involution of the thymus, which is responsible, also through the production of various hormone-like peptides[6], for the proliferation and differentiation of the different subpopulations of T-cells.

The thymus attains its maximum size at puberty, after which it starts to progressively involute and to be replaced by fat. This process is nearly complete in man by the 5th to 6th decade: the cortical areas are depleted of lymphoid cells and the epithelial cells show cystic changes and reduction of intracellular granules[8].

Measurement of the circulating level of thymic factors has demonstrated, both in animals and humans, that the plasma level of one of the best known thymic peptides, i.e. thymulin, declines progressively from birth to old age and is virtually undetectable in humans past 60 years of age. According to a recent more precise determination of thymulin, which takes into account the interference due to a marginal zinc deficiency present with advancing age, the decline of thymulin levels is less pronounced than previously reported. Even in very old age, a residual production of thymulin is detectable[9].

The age-associated decline of thymic endocrine activity seems to be one of the major causes for the deterioration, observed in old age, of the peripheral T-cell compartment, in the efficiency of T-cells with cytotoxic properties, as well as of T-cells with helper and suppressor activity. In fact, treatments of old mice with thymic hormone preparations have demonstrated the reversibility of age-associated defects in peripheral T-cell functions[10].

The effector mechanisms of nonspecific immunity, and especially the cytotoxicity of natural killer (NK) cells are also altered with advancing age. Several abnormalities occur in advanced age both in animals and humans at the level of either endogenous NK activity or of lymphokine-induced cytotoxicity. In old mice and rats, basal spleen cell NK activity is nearly undetectable; the sensitivity to IL-2 is generally maintained, though at a reduced level when compared to middle age; and the responsiveness to IFN is strongly diminished[11]. The changes occurring in human basal NK cell activity, with advancing age, are still not well defined. In fact, no change, or a decrement, or an increment of basal NK activity all have been reported in elderly humans. With regard to the responsiveness of NK cells to the boosting action of IFN or IL-2, a progressive decline in both IFN and IL-2 induceable NK activities appears with increasing age, the defect being more evident at the level of IFN rather than of IL-2 responsiveness of NK cells[12,13].

Neuroendocrine Aging

The neuroendocrine system is a complex network, which functions to connect various body apparatuses among themselves and with the central nervous system, in order to adapt end-organ responses to external and internal demands.

Aging may affect all steps of the endocrine cascade and the alterations observed, regardless of the endocrine gland specificity, in general follow these characteristics[14,15]: a) the basal hormonal blood levels are generally found unmodified by aging, whereas the secretory response to an appropriate stimulus is frequently decreased; b) many protein hormones (particularly those secreted by the pituitary gland) show, even at a young age, a biochemical and antigenic polymorphism, which is generally accentuated with advancing age: the functional activity of the different forms and the significance of the increased polymorphism with age remain to be established; c) secretory and clearance rates appear diminished; d) since a given hormone may act on various cell types, it is not infrequent to find that aging may affect the response of one or more cell types, leaving others unaffected, this phenomenon being due to the different rate at which aging may affect receptor availability and hormone-receptor binding[15].

With regard to the "master" glands, i.e. the pineal and the pituitary, no significant intrinsic alterations seem to be present in old age, in many species examined. In addition, in these species, the pituitary hormonal content and basal blood levels of pituitary hormones do not show significant alterations in old age with the exception of prolactin, which is usually increased in the elderly, and of gonadotropins, whose pituitary release is greatly augmented particularly in females as a result of alterations in ovarian function. With regard to protein hormones, other than the pituitary content or blood basal levels, the circadian rhythmicity and the secretory response to specific stimuli are altered in old age[15-17].

The pineal gland produces various hormones, in particular melatonin, derived from the neurotransmitter serotonin. A major function of melatonin is the control of cyclicity of daily as well as seasonal rhythms. With advancing age, the gland undergoes calcification with loss of pinealocytes, which is not associated with major alterations of the secretory capacities. The night and time amplitude of peak secretion is, however, generally altered with aging and such an alteration may be responsible for the modified night-time profile of other hormones such as adrenocorticoltropic hormone (ACTH), growth hormone (GH), prolactin (PRL) and thyrotropic hormone (TSH). A major consequence of this phenomenon is that the 24 hr total secretion rate of these hormones, with the exception of PRL, is generally decreased[17].

NEUROENDOCRINE-IMMUNE INTERACTIONS DURING AGING

Since it has been demonstrated that modifications induced by experimental manipulation of the neuroendocrine system may alter the immune system, it may also be expected that the physiological decline with advancing age of the immune or the neuroendocrine system and vice-versa may be in part responsible for the alterations observed in the other partner.

As a matter of fact, it has been demonstrated that regrowth of the thymus, with recovery of plasma level of zinc-thymulin (for review see[4,5]), may be achieved in old animals by treatment with thyroid hormones, with growth hormone, with analogues of LH-RH, and with either melatonin or intrathymic young pineal graft[18].

Interestingly, similar thymic "rejuvenation" was obtained in mice by treatment with arginine which has a secretagogue activity on pituitary GH, and with zinc salts which certainly may have a direct action on the thymus, but are also capable of acting on pituitary-thyroid axis[4,5].

In humans, very few trials have been done, but, at least with regard to thymic endocrine activity, both arginine and zinc are effective in old age[4]. On the other hand, the capacity of thyroid hormones and of growth hormone to restore thymic function plasma level of some hormones, such as an increased insulin and a decreased triiodo-thyronine (T_3) level and reduction of the adaptive reaction to beta-adrenergic stimuli. This latter change is probably due to a drecreased density of beta-adrenoceptors on cell membranes of various tissues, including the submandibular gland and some parts of the brain[20,21]. All of these deficits are fully restored in old animals by neonatal thymus grafts.

The majority of the neuroendocrine changes demonstrated in these studies are strictly age-dependent, i.e. they display a linear progression starting early in life. Their interconnection with the thymus, which shows a quite similar age-dependent progressive deterioration, does not seem , therefore, to be merely coincidental.

Focusing on the "strategic" circuit reported in Fig. 1A, there does not seem to be any single intrinsic and critical event with advancing age, which can account for the progressive deterioration of the neuroendocrine-thymus functions in old age.

Quite similar considerations apply to the emergency circuit. In the case of neurohormonal modulation of peripheral immune functions, some, although not exhaustive, data suggest that the age-related deterioration of T-cell and of NK-cell functions do not represent irreversible processes, but rather defects which can be corrected "in vivo" and in some cases even "in vitro" by appropriate hormonal treatments.

Thus it has been demonstrated that lymphocyte proliferation may be enhanced by treatments with growth hormone, with thyrotropic hormone, with thyroid hormones, with enkephalins, with melatonin, with arginine and with zinc salts. With regard to NK activity, growth hormone, endorphins and melatonin are able to increase basal NK activity, whereas treatments with thyrotropic hormone increase only IL-2 induced NK cytotoxicity and with T3 IFN-boosted NK activity (for review see[4,22]).

Do immunocyte-derived immunotransmitters play a role in the age-related deterioration of some neuroendocrines? This might be deduced from the known effects of some interleukins upon hypothalamic and pituitary functions, but to our knowledge, no investigations have been carried out to answer this question.

The finding discussed above clearly demonstrate that at least some age-related alterations of the neuroendocrine and the immune networks are not "per se" intrinsic and irreversible and that the site of age-related deterioration cannot easily be assigned to one or another homoestatic apparatus.

These considerations suggest that we need to revise, at a theoretical level, both the "nervous-neuroendocrine"[23] and the "immune" hypotheses of aging[24,25], especially the assumption that there is a genetically determined hierarchy among the three homeostatic systems of the body and that the deterioration of one of them, according to each single theory, is a primary, intrinsic and irreversible event. If this were true, the age-associated alterations of the other systems would necessarily follow.

As an alternative to both the "neuroendocrine" and the "immune" theories of aging, we hypothesise that, due to the continuos interactions existing among the nervous, the neuroendocrine and the immune systems during the entire life of the organism, it is the disruption of such interactions in old age which is responsible for most of the age-associated dysfunctions[4] (Fig. 2). These interactions are, in fact, continuously modified during life by external stressors which may be either cognitive (psycho-emotional, social, etc.) or non-cognitive (chemical, antigenic, etc.). The individual diversity of life experience may account for variations of a single homeostatic mechanism, both in early ontogeny and in aging in different individuals. The interaction, however, with the other networks, may explain the global alterations usually observed in these situations. In other words, priority in the appearance of age-related phenomena among nervous, neuroendocrine and immune systems is not a strictly genetically determined phenomenon, but develops in an individual according to personal life experience. This assumption can explain the increased incidence of diseases in ontogeny as due to defective development of the interactions among those homeostatic systems and/or to abnormal stressor events. In aging, such increases in diseases may be due to the accumulation of individually different "collages" of various non cognitive and/or cognitive stress effects.

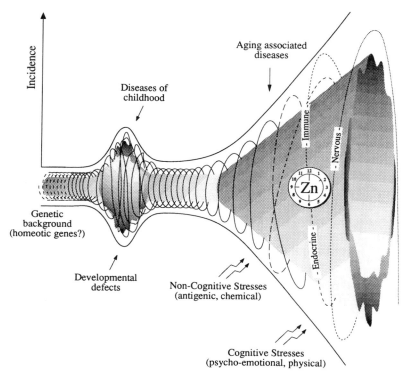

Fig. 2. Schematic representation of the neuroendocrine-immune theory of aging. For explanation see text.

The demonstrated possibility to correct, by exogenous interventions, various age-related neuroendocrine and immune defects, supports, however, the idea that a relevant plasticity is still present in the old homeostatic systems, or, alternatively, there remains a consistent residual potentiality, which can allow reversibility of both neuroendocrine and immune ageing. Such an interpretation would also be compatible with the observation that some markers, either immunological or neuroendocrinological, are "normal" in super healthy old individuals[16,26]. In fact, it is conceivable that, while in the whole population an immunological or neuroendocrinological manipulation may be able to repair age-associated deterioration, those selected individuals (for genes or life-experience?) are either profiting from better interactions among the homeostatic systems, or are endogenously capable of correcting the reversible alterations due to external or internal noxious stimuli.

The deterioration of neuroendocrine-immune interactions with aging, in addition to depending on age-dependent alterations at the level of one of the homeostatic systems involved, might also depend on modifications of some basic mechanisms capable of influencing all of the homeostatic systems. A putative factor in this context is zinc metabolism, since it has been demonstrated that zinc is required for the functional efficiency of the nervous, the neuroendocrine and the immune systems[18,27].

These findings, together with the observations that: 1) zinc turnover is altered with advancing age both in humans and in animals; 2) that zinc supplementation in old age is able to restore thymic function, restore various immune deficiencies, and also can correct some age-related hormonal defects, suggest a crucial role for zinc in neuroendocrine-immune interactions during aging[27].

SUMMARY

A common and generally accepted assumption is that with advancing age, the thymus undergoes progressive and irreversible involution. This is considered the main cause for the age-related deterioration of various immune functions and, ultimately, for the increased incidence of infectious, neoplastic, and autoimmune diseases in old age. This assumption is no longer tenable because of several clear-cut demonstrations that age-related thymic, and, more generally, immune involution is not an intrinsic and irreversible phenomenon.

Various neuroendocrine or nutritional manipulations can induce a regrowth of the thymus, and recovery of peripheral immune reactivity even when applied in old age.

These data strongly support the idea that thymic involution is a phenomenon secondary to age-related alterations in neuroendocrine-thymus interactions and that it is the disruption of such interactions in old age that is responsible for most of the age-associated dysfunctions.

On the basis of this experimental and clinical evidence and as an alternative to purely immune or neuroendocrine theories of aging, a neuroendocrine-immune hypothesis is proposed.

Further work is required to determine if the age-related disruption of neuroendocrine-immune interactions occurs because of progressive accumulation of stressor-dependent consequences at the level of one or the other system or if it may depend on a single common cause.

REFERENCES

1. J.E. Blalock. "Neuroimmunoendocrinology," Chem. Immunol., Karger, Basel, Switzerland vol. 52 (1992).
2. N. Fabris, B.D. Jankovic, B. Markovic, B. and N.H. Spector, eds., "Ontogenetic and Phylogenetic Mechanisms of Neuroimmunomodulation: from Molecular Biology to Psychosocial Sciences," Ann. N.Y. Acad. Sci. vol. 650 (1992).
3. G. Jasmine and M. Cantin, "Stress Revised," Methods Exp. Path., Karger, Basel, vol. 14 (1991).
4. N. Fabris. "The Neuroendocrine-Immune Domain. Time for a New Theory of Aging? Ann. N.Y. Acad. Sci. 663:335 (1992).
5. N. Fabris, Neuroendocrine/thymus interactions during development and aging, in: "Hormones and Immunity. Bilateral Communication Between the Endocrine and Immune Systems," C. Grossman, ed., Springer Verlag, USA, p. 265-299 (1993).
6. J.W. Hadden, Immunostimulants, Immunology Today 14(6):275 (1993).
7. H.O. Besedovsky and A. Del Rey, Immune-neuroendocrine circuits: integrative role of cytokines, in: "Frontiers in Neuro-endocrinology," Raven Press, New York, 13(1) (1992).
8. J.F. Simpson, E. S. Gray and J. S. Beck., Age involution in the normal human adult thymus, Clin. Exper. Immunol. 19:261 (1975)
9. N. Fabris, E. Mocchegiani, L. Amadio, M. Zannotti, F. Licastro and C. Franceschi, Thymic hormone deficiency in normal aging and Down's syndrome: is there a primary failure of the thymus? Lancet 1:983 (1984).

10. M. M. Zatz and A. L. Goldstein, Thymosins, lymphokines and the immunology of ageing, *Gerontol.* 31:263 (1985).
11. M. Provinciali, M. Muzzioli and N. Fabris, Timing of appearance/ disappearance of IFN and IL-2 induced natural cytotoxicity during ontogenetic development and aging, *Exp. Gerontol.* 24:227 (1989).
12. B.S. Bender, F.J. Chrest and W.H. Adler, Phenotypic expression of natural killer cell associated membrane antigens and cytolytic function of peripheral blood cell from different aged humans, *J. Clin. Lab. Immunol.* 21:31 (1986).
13. R. Krishnaray. Immunosenescence: the activity, subset profile and modulation of human natural killer cells, *Aging, Immunol. Infect. Diseases,* 2(3):127 (1990).
14. G.T. Baker and R.L. Sprott, Biomarkers of aging., *Exp. Gerontol.,* 23:223 (1988).
15. N. Fabris, D. Harman, D.L. Knook, E. Steinhagen-Thiessen and I. Zs. Nagy. "Physyiopathological Processes of Aging: Towards a Multicausal Interpretation", *Ann. N.Y. Acad. Sci. USA* vol. 673 (1992).
16. M.R. Blackman, Pituitary hormones and aging, *in:* "Endocrinology and Aging", B. Sacktor, ed.W. Saunders, London 16(4):981 (1989).
17. K.A., Stokkan, R.J. Reiter, K.O. Nonaka, A. Lerchl, B. Pal Yu and M. K. Vaughan, Food restriction retards aging of the pineal gland, *Brain Res.* 545:66 (1991).
18. N. Fabris, Neuroendocrine-immune aging: the role of the zinc and an integrative view, *Ann. N.Y. Acad. Sci.* 719 (1994) (in press).
19. P. Travaglini, E. Mocchegiani, C. De Min, T. Re and N. Fabris, Modifications of thymulin titers in patients affected with prolonged low or high zinc circulating levels are independent of patients' age, *Arch. Gerontol. Ger.* (3):349 (1992).
20. L. Piantanelli, S. Gentile, P. Fattoretti and C. Viticchi, Thymic regulation of brain cortex beta-adrenoceptors during development and aging, *Arch. Gerontol. Ger.* 4:179 (1985).
21. G. Rossolini, A. Basso, L. Piantanelli, R. Tacconi, D. Amici and G.L. Gianfranceschi, Neuroendocrine thymus and beta-adrenergic responsiveness in aging mice, *Arch. Gerontol. Geriatr.* 3:311 (1992).
22. N. Fabris. Endocrine-immune interactions, *in:* "Immuno-Pharmacology Reviews," J. W. Hadden and A. Szentivanyi, eds., Plenum Pub. Corp., New York (1994) in press.
23. J. Meites, R. Goya and S. Takahashi, Why the neuroendocrine systemis important in aging processes: a review, *Exp. Gerontol.* 22:1 (1986).
24. R.L. Walford. "The immunological theory of aging," Munksgaard, Copenhagen (1969).
25. F.M. Burnet. "Immunological Surveillance," Pergamon Press, Sydney, Australia (1970).
26. C. Franceschi, D. Monti, A. Cossarizza, F. Fagnoni, G. Passeri and P. Sansoni, Aging, longevity, and cancer: studies in Down's syndrome and centenarians, *Ann. N. Y. Acad. Sci, USA* 621:428 (1991).
27. N. Fabris, E. Mocchegiani, M. Muzzioli and M. Provinciali, Role of zinc in neuroendocrine-immune interactions during aging, *Ann. N.Y. Acad. Sci., USA,* 621:314 (1991).

BETA-ENDORPHIN IN PERIPHERAL BLOOD MONONUCLEAR CELLS: EFFECT OF AGING AND PHARMACOLOGICAL TREATMENTS

Paola Sacerdote,[1] Emilio Clementi,[2] Barbara Manfredi,[1] and Alberto E. Panerai[1]

[1]Department of Pharmacology, University of Milano, Italy
[2]C.N.R. Centre of Cytopharmacology, Milano, Italy and
Department of Pharmacology, School of Pharmaci, University of Reggio Calabria, Catanzaro, Italy

SUMMARY

Beta-endorphin is known to affect the immune system when added *in vitro* or when its release is increased *in vivo*, e.g. during stress. Moreover, the synthesis of the beta-endorphin precursor proopiomelanocortin (POMC) has been consistently shown, by now, in several cells of the immune system, while the effect of extracellular beta-endorphin has been widely studied *in vitro* and *in vivo*. We investigated the synthesis of the opioid peptide in immune cells and its release from peripheral blood mononuclear cells obtained from normal subjects of different ages, both under resting conditions and after stimulation with polyclonal mitogens such as PHA or Con-A. The *in vivo* and *in vitro* modulation of intracellular beta-endorphin by pharmacological treatments is also described. Our findings indicate that the concentrations of beta-endorphin in peripheral blood mononuclear cells increase after the age of thirty years, and that the peptide is released by PHA or Con-A stimulation from cells obtained from subjects aged over thirty years, but not from cells of younger donors. Finally, we show that beta-endorphin is under a tonic inhibition exerted by the dopaminergic system, while serotonin exerts a tonic stimulatory effect. Preliminary data indicate that calcium homeostasis might be important for the differences observed in cells obtained from young (<30 years) or old (> 30 years) patients.

INTRODUCTION

By now a compelling body of experimental evidence supports the hypothesis that neuropeptides play a role in the modulation of the immune system.[1] Several immune functions, including lymphocyte proliferation induced by mitogens, antibody production and NK activity were shown to be influenced by neuropeptides.[2,3] There is much information on the effects of neuropeptides on immune reactions, whereas little is known about the synthesis and secretion of neuropeptides and hormones by lymphocytes. In contrast with cytokines, which are always newly synthesized, beta-endorphin (β-END) is constitutively present in immune cells that include splenocytes, macrophages, thymocytes and peripheral blood lymphocytes.[4] However, β-END synthesis can be further stimulated by the infection of cells with viruses such as Newcastle disease virus or human immunodeficiency virus (HIV).[3] The regulation/modulation of β-END production and storage by immune cells has not been studied in great detail, and there is no information at all with regards to the possible influence of aging on this process.

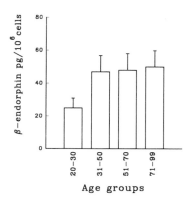

Figure 1. Beta-endorphin concentrations in peripheral blood mononuclear cells otained from normal subjects of different age. Vertical bars are Mean   SD.

Aging is associated with a decline of cognitive[5] and immune functions.[6] For this reason, we studied peripheral blood mononuclear cells (PBMC) from subjects of various age groups for the presence and release of β-END in the resting state and after stimulation in culture. The effect of dopaminergic, serotoninergic and GABAergic drugs has also been examined.[4] Our working hypothesis was that the pharmacological modulation of β-END in PBMC might offer new approaches to the correction of immune abnormalities.

Proopiomelanocortin Derivatives in Immune Cells

The genes of several neuropeptides and hormones are expressed in lymphocytes and in other cells of the immune system after stimulation by PHA, Con-A, LPS or infection with Newcastle disease virus.[7-9] However, the significance of the presence of neuropeptides in PBMC is unknown at the present time. It was suggested that POMC products could not be released from PBMC because of the lack of signalling sequences within exon 1 and 2.[10] Yet the release of β -END has been demonstrated under various experimental conditions.[11] Moreover, recent data indicate that the signalling sequence is not necessary for the release of proteins, and more importantly, of cytokines such as IL-1.[12,13]

β -END During Aging

For the study of β -END concentrations during aging, samples were obtained from healthy volunteers, aged from 20-99 years. The samples were divided into the following age groups: 20-30 (mean ± SD = 24.5±2.7), 31-50 (38.9±5.4), 51-70 (58.6±4.9), 71-99 (83.3±7.8). Figure 1 shows that the β -END content of resting PBMC increases significantly after 30 years of age (p<0.01) and remains constant thereafter. In additional experiments we showed that this increased concentration was due to an increase of the number of cells synthesizing β -END and not to an increase of production by single cells as revealed by the reverse plaque forming cell assay (data are not shown).

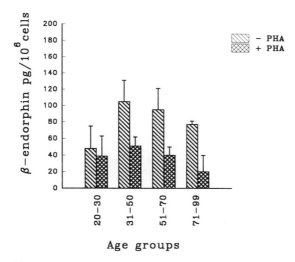

Figure 2. Beta-endorphin concentration in cultured peripheral blood mononuclear cells obtained from normal subjects of different age before and after stimulation with PHA. Vertical bars are Mean ± SD.

Effect of Polyclonal Mitogens on β -END Concentrations at Different Ages

Peripheral blood mononuclear cells from donors either younger or older than 30 years of age were stimulated by PHA or Con-A using standard methods and the β -END concentrations of the cells were measured. In unstimulated cultures the β -END content of cells from older donors was higher than in those from subjects under 30 years of age (Figure 2). Furthermore, stimulation with PHA did not affect the β -END content of PBMC from healthy donors aged 20-30 years, whereas such stimulation decreased significantly the β-END concentration of PBMC from donors of all other age groups examined. There was no correlation between the degree of cell proliferation and β-END content of cells in any of the groups studied. Similar results were obtained after stimulation with Con-A (data are not presented).

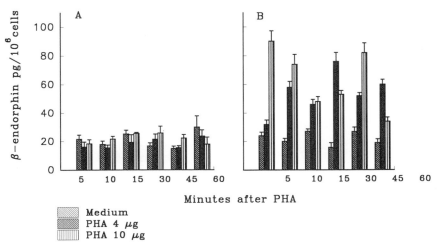

Figure 3. Beta-endorphin concentrations in the medium of cultured peripheral blood mononuclear cells obtained from 20-30 (A) and 30-40 (B) years old subjects with or without PHA stimulation.

Effect of Polyclonal Mitogens on β-END Release from PBMC at Different Ages

The release of β-END from PBMC of donors younger (Figure 3A) or older (Figure 3B) than 30 years of age after stimulation with PHA was markedly different in that no release was observed from cells of young individuals, whereas cells of older individuals released significant amounts of β-END in a dose dependent manner. This experiment was performed in Krebs medium in order to avoid the presence of neuropeptides in the culture which could complicate the results. Similar results were obtained after stimulation with Con-A (data are not shown).

Study of Calcium Homeostasis in PBMC of Young and Old Subjects

PBMC from young or old donors were loaded with the fluorescent Ca^{++} indicator dye fura-2 and stimulated with PHA in the presence of EGTA in order to prevent Ca^{++} influx. PHA induced a greater mobilization of Ca^{++} from the intracellular stores of cells obtained from older subjects and also elicited a greater Ca^{++} influx when the ion was added to the incubation medium. In cells from donors older than 30 years of age (Figure 4) no such differences were observed after stimulation with Con-A under similar experimental conditions (data not shown).

Pharmacological Modulation of β-END in Human and Rat PBMC

Cells from three groups of patients were examined as follows: Patients suffering from epilepsy and treated with the GABAergic drug, sodium valproate (1-2 g/day for 15 days); psychotic patients treated with the antidopaminergic agent, haloperidol (5-9 mg/day for 2-15 days per os); patients suffering from depression and treated with the

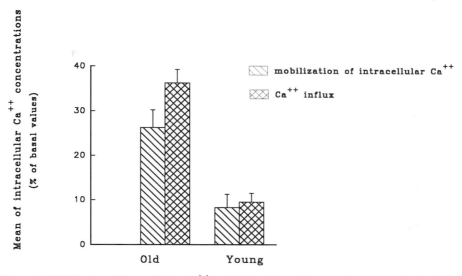

Figure 4. Mobilization of intracellular Ca^{++} and influx induced by PHA in PBMC obtained from young or old donors (see text).

serotoninergic drug, chlorimipramine (100 mg/day for 15 days per os). For *in vitro* studies PBMC samples were obtained from normal volunteers of 20-30 years of age.

Male Sprague-Dawley rats (Charles River Calco, Italy) of 200 g body weight were used, 8 animals in each experiment. The dopaminergic agent, bromocriptine, was administered at 5 mg/kg i.p. and haloperidol at 2 mg/kg i.p. 2 hrs before sampling in acute experiments and twice daily for 2 weeks in chronic studies. The serotonin precursor, 5-hydroxytryptophan, was given at 30 mg/kg i.p.; the serotonin receptor antagonist, metergoline, at 7.5 mg/kg i.p.; and the serotonin reuptake blocker, chlorimipramine, was administered at the dose of 40 mg/kg i.p. in the acute and 20 mg/kg i.p. in the chronic study. The drugs acting on the serotoninergic system were given 60 min before sampling in the acute study and twice daily for 2 weeks in the chronic study. Sodium valproate was given i.p. for 2 weeks at the dose of 200 mg/kg twice daily.

The effect of chronic treatment with agents modifying the serotoninergic tone on the β-END content of PBMC from rat or human donors is shown in Figure 5. 5-Hydroxytryptophan and chlorimipramine increased the β-END concentrations in PBMC in the rat which could be blocked by concomitant treatment with the serotoninergic antagonist, metergoline. Metergoline also decreased the concentration of β-END in PBMC when administered alone. In humam PBMC chlorimipramine treatment increased the content of β-END.

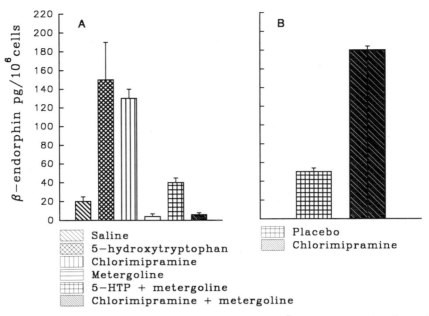

Figure 5. Effect of the modulation of the serotoninergic tone on ß-END concentrations in rat (A) or human (B) PBMC. Vertical bars are Mean ± SD.

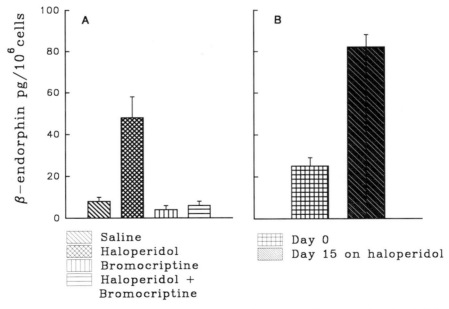

Figure 6. Effect of the modulation of the dopaminergic tone on BE concentrations in rat (A) or human (B) PBMC. Vertical bars are Mean ± SD.

In rats the β-END concentrations of PBMC increased after chronic treatment with haloperiodol and decreased after chronic bromocriptine administration. No change of β-END content was observed when the two agents were administered together (Figure 6A). The treatment of patients with haloperidol for 15 days increased significantly the concentrations of β-END in their PBMC (Figure 6B).

Treatment with sodium valproate decreased the β-END content of PBMC in rats and also in patients (Figure 7).

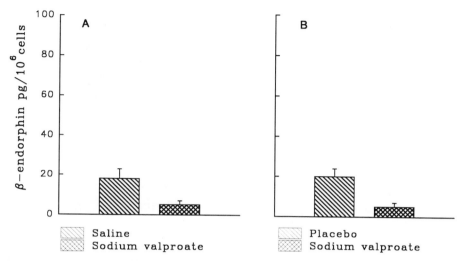

Figure 7. Effect of the modulation of the GABAergic tone on BE concentrations in rat (A) or human (B) PBMC. Vertical bars are Mean ± SD.

In Vitro Studies

Macrocultures of PBMC from normal healthy volunteers were incubated for 24 hrs, which was followed by pharmacological treatments for 4 or 24 hrs. Serotonin, the serotonin receptor agonist quipazine, and the GABA receptor agonist, muscimol, were used at the doses of 10^{-5} M and 10^{-7} M.

Both serotonin and quipazine increased the concentrations of β-END in cultured human PBMC, while muscimol induced a decrease (data not shown).

Dopamine or dopaminergic receptor agonists were ineffective when adminsitered *in vitro*. However, Figure 8 shows that the intracerebroventricular administration of haloperidol, at doses that are ineffective when administered peripherally, increased β-END concentrations in PBMC of rats.

CONCLUSIONS

The data reviewed here indicate that β-END is unequivocally present and synthesized in peripheral blood mononuclear cells. Its function, however, remains largely unknown, although it has been shown that the concentrations in PBMC change during pathological processes in the central nervous system or in the immune system,

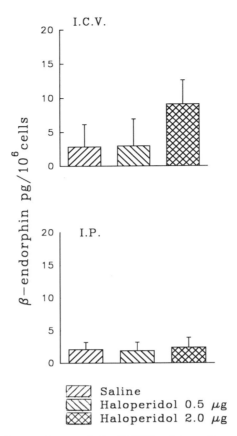

Figure 8. Effect of the intracerebroventricular (I.C.V.) or intraperitoneal (I.P.) administration of haloperidol on BE concentrations in PBMC. Vertical bars area Mean ± SD.

such as migraine, schizophrenia or rheumatoid arthritis.[14-16]

It is interesting to observe that *in vivo* the same pharmacological agents that modify β-END in the nervous system exert a similar effect on β-END concentrations in PBMC. This observation suggests that PBMC could be used as a tool for the study of the modulation of β-END in the brain both in health and disease. On the other hand, modifying the concentrations of β-END may be used to influence the immune response.

Aging is associated with an increase in β-END concentrations in PBMC and with an increased release from mitogen stimulated cells. The time course of changes in β-END concentrations, and release after PHA or Con-A stimulation from PBMC of subjects of different ages does not directly correlate to the time course of immune defects found during normal aging. These changes are already present in age groups that do not show immunological impairments, such as impaired mitogen induced proliferation or IL-2 production.[17]

The results presented here indicate that β-END concentrations are increased in resting cells, and its release is increased after PHA or Con-A stimulation with increasing age. One explanation for this could be the disappearance of a previously existing inhibitory tone after the age of 30/35. A correlation can be envisaged between the time course of β-END changes in PBMC and the age associated atrophy of the cellular matrix of the thymus and the initial decline of plasma thymic hormone levels. In fact, both these phenomena seem to take place with an age pattern that parallels the one we find in β-END concentrations in resting cells or after PHA or Con-A stimulation.[18] Thus, one may speculate on a possible inhibitory effect of thymic factors on β-END in PBMC, or, alternatively, on an inhibitory effect of β-END on the thymus. This hypothesis is now under investigation in nude athymic rats.

A second hypothesis can be put forward for the role of the inhibitory neurotransmitter GABA in β-END concentrations and responsiveness. It was shown, in fact, that in cerebrospinal fluid (CSF) obtained from 30-40 year old subjects, the concentration of GABA is 50% of that present in 20 year old subjects.[19] Therefore, it can be suggested that the increase in β-END concentrations, and increased susceptibility to stimuli we observe in PBMC during aging parallels a decrease of GABA concentrations in CSF.[19]

Finally, the data presented here suggest that changes in Ca^{++} mobilization and/or influx could be an important factor in the different responses observed after PHA or Con-A stimulation in cells obtained from young or old subjects.

REFERENCES

1. J.E. Morley, N.E. Kay, G.E. Solomon, and N.P. Plotnikoff, Neuropeptides: conductors of the immune orchestra, *Life Sci.* 41:527 (1987).
2. P.M. Matthews, C.J. Froelich, W.L. Sibbitt, and A.D. Bankhurst, Enhancement of natural cytotoxicity by B-endorphin, *J. Immunol.* 130:3045 (1983).
3. J.E. Blalock, A molecular basis for bidirectional communication between the immune and neuroendocrine system, *Physiol. Rev.* 69:1 (1989).
4. P. Sacerdote, F. Rubboli, L. Locatelli, I. Ciciliato, P. Mantegazza, and A.E. Panerai, Pharmacological modulation of neuropeptides in peripheral mononuclear cells, *J. Neuroimmunol.* 32:35 (1991).
5. J.W. Rowe and R.L. Kahn, Human aging: usual and successful, *Science* 237:143 (1987).
6. M.W. Weskler and G.W. Siskind, The cellular basis of immune senescence, *Monogr. Dev. Biol.* 17:110 (1984).
7. E.M. Smith and J.E. Blalock, Human lymphocyte production of ACTH and endorphin-like substances: Association with leucocyte itnerferon, *Natl. Acad. Sci. USA* 78:7530 (1981).
8. D. Harbour-McMenamin, E.M. Smith and J.E. Blalock, Production of lymphocyte derived chorionic gonadotropin in a mixed lymphocyte reaction. *Proc. Natl. Acad. Sci. USA* 83:2599 (1986).
9. R. Buzzetti, L. McLoughlin, P.M. Lavender, A.J. Clark and L.H. Rees, Expression of proopiomelanocortin gene and quantitization of adreno-corticotropic hormone like immunoreactivity in human normal peripheral mononuclear cells and lymphoid and myeloid malignancies, *J. Clin. Invest.* 83:733 (1989).
10. A.J.L. Clark, P.M. Lavender, P. Costes, M.E. Johnson and L.H. Rees, In vitro and in vivo analysis of the processing and fate of the peptide products of the short proopiomelanocortin mRNA, *Mol. Endocrinol.* 4:1737 (1990).
11. C.J. Heijnen, A. Kavelaars and R.E. Ballieux, b-endorphin: cytokine and neuropeptide, *Immunol. Rev.* 119:41 (1991).
12. A. Rubartelli, F. Cozzolino, M. Talio and R.A. Sitia, A novel secretory pathway for interleukin-b, a protein lacking a signal sequence, *EMBO J.* 9:1503 (1990).
13. P.C.W. Lord, L.M.G. Wilmoth, S.B. Mizel and C.E. McCail, Expression of interleukin-1a and b genes by human blood polymorphonuclear leukocytes, *J. Clin. Invest.* 87:1312 (1991).
14. G. Panza, E. Monzani, P. Sacerdote, G. Penati and A.E. Panerai, Beta-endorphin, vasoactive intestinal peptide and cholecystokinin in peripheral blood mononuclear cells from healthy

subjects and from drug-free and haloperidol-treaed schizophrenic patients, *Acta Psychiat. Scand.* 85:207 (1982).

15. M. Leone, P. Sacerdote, D. D'Amico, A.E. Panerai and G. Bussone, Beta-endorphin concentrations in the peripheral blood mononuclear cells of migraine and tension-type headache patients, *Cephalagia* 12:155 (1982).

16. C.J. Wiedermann, P. Sacerdote, E. Mur, U. Kinigadner, T. Wicker, A.E. Panerai and H. Braunsteiner, Decreased immunoreactive beta-endorphin in mononuclear leukocytes from patients with rheumatic diseases, *Clin. Exp. Immunol.* 87:178 (1992).

17. W. Barcellini, M.O. Borghi, C. Sguotti, R. Palmieri, D. Frasca, P.L. Meroni and C. Zanussi, Heterogeneity of immune responsiveness in healthy elderly subjects, *Clin. Immunol. Immunopathol.* 47:142 (1988).

18. G. Goldstein and T.K. Audhya, Thymopoietin to thymopeptin: experimental studies, *Survey Immun. Res.* 4, Suppl. 1:1 (1985).

19. S.R. Bareggi, M. Franceschi, L. Bonini, L. Zecca and S. Smirne, Decreased CSF concentrations of homovanillic acid and s-aminobutyric acid in Alzheimer's disease, *Arch. Neurol.* 39:709 (1982).

THE ROLE OF THE ADRENERGIC/CHOLINERGIC BALANCE IN THE IMMUNE-NEUROENDOCRINE CIRCUIT

Konrad Schauenstein,[1] Ingo Rinner,[1] Peter Felsner,[1] Dietmar Hofer,[1] Harald Mangge,[2] Elisabeth Skreiner,[1] Peter Liebmann,[1] Amiela Globerson[3]

[1]Institute of General and Experimental Pathology, [2]Department of Pediatrics, University of Graz, Austria, [3]Department of Cell Biology, The Weizmann Institute of Sciences, Rehovot, Israel

STRESS AND IMMUNE RESPONSE

The concept of an extrinsic regulation of the immune system through neuroendocrine signals is well established, as is the fact that the immune system in turn informs the brain about contacts with antigens via "immunotransmitters", i.e. cytokines and/or hormones with central effects.[1] All these data that have accumulated during the last few years have contributed to the vision of the immune system as "the sixth sense".[2] While there is certainly still more work needed to define the physiology of this concept in all details, strong evidence has been obtained that the immune-neuroendocrine dialogue is of relevance for the homeostasis of the immune response, as defects in the activation of the hypothalamo-pituitary-adrenal (HPA) axis by immune signals were found to be associated with and/or to predispose to spontaneously occurring[3] and experimentally induced autoimmune diseases in animal models.[4,5]

Neuroendocrine immunoregulation in vivo becomes obvious under conditions that elicit somatic stress responses. This has been documented in numerous clinical and experimental studies, in which the effects of physical or psychological stressors on immune functions have been examined.[6] However, the results reported are contradictory and depend on animal species, quantity and/or quality of the stressor applied, and immune function investigated. In one of our earlier studies we have attempted to define some further parameters that may be of relevance to qualitative and quantitative effects exerted by a given stressor to immune reactivity.[7] These were longitudinal studies in cannulated rats, which were subjected to restrain stress, and, using a whole blood stimulation assay requiring 300μ l blood samples, were repeatedly analyzed for the mitogen reactivity of peripheral blood lymphocytes (PBL) at different time points of the stress regimen. It was found that (i) a single 90 minute immobilization leads to a total loss of B or T cell reactivity in peripheral blood that is reversible within 24 hrs after stress, (ii) repeated daily stress treatments does not lead to adaptation, (iii) the effect is not due to shifts in white blood cell counts, and it is confined to PBL leaving spleen cell reactivity unaffected, and (iv) while one minute handling leads to a highly significant and persistent increase of mitogen reactivity, increasing the dose of immobilization entails a gradual decrease. The new aspects from these data are that in one and the same individual different doses of a stressor can have opposite effects on immune functions, and that the reactivity of lymphocytes is differently regulated by neuroendocrine signals in different compartments of the immune system, an observation that will also be dealt with later on in this chapter.

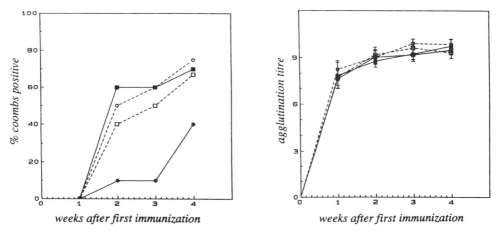

Figure 1. Occurrence of anti-erythrocyte autoantibodies (% of Coombs positive animals, left) and development of specific heteroantibodies (right) in Balb/c mice weekly immunized with rat erythrocytes. For the weekly bleeding animals were anesthesized by barbiturate i.p. (■) or ether inhalation (●), open symbols represent animals that were adrenalectomized 10 days before the experiment.

Another parameter that obviously determines the sensitivity of immune functions towards neuroendocrine influences lies in the complexity of the immune response. This is reflected by the well known sensitivity of immune response kinetics to regulatory factors, where the phase of sensitization and initiation is generally by far more prone to regulatory influence, as compared to the effector phase. Furthermore, it is to be expected that the quantity and/or quality of an immune response determines the resistance against neuroendocrine signals. This notion was confirmed by recent circumstantial observations by us suggesting that an autoimmune response is by far more sensitive to stress effects as compared to a concomittant response to heteroantigen.[8] The data are shown in Figure 1. Balb/c mice weekly immunized (i.p.) with rat erythrocytes (RRBC), besides producing RRBC-specific antibodies, develop with time autoantibodies against autologous erythrocytes that are detectable by a direct Coombs assay.[9] For the weekly blood sampling from the retroorbital plexus the animals were shortly anesthesized, and we observed that ether inhalation significantly delayed the onset of the autoantibody response as compared with animals anesthesized by i.p. barbiturate injection, whereas the "regular" anti-RRBC response was not affected by the stress of anesthesia. Adrenalectomy prior to immunization eliminated the suppressive effect of ether anesthesia, and had no effect on the anti-RRBC response either. As in vivo treatment with adrenergic agonists and/or antagonists did not significantly alter the autoimmune response (not shown), we conclude that this stress effect is mediated by glucocorticoids, which confirms the role of the HPA axis in the prevention of autoimmunity.

One may conclude from this brief section that somatic stress responses do significantly influence immune functions in vivo and ex vivo, whereby the actual outcome is not predictable and depends on various parameters of both the specific stressor and the specific immune response under investigation. Similarly, the "talking back" of the immune system to the brain is expected to be quantitatively determined by several parameters that still remain to be defined.

In the following we review data on the in vivo role of the autonomous nervous system in this context, in order to answer the question if and how disturbances in the adrenergic/cholinergic balance may affect the immune-neuroendocrine interplay, and to better understand functional relationships between the immune and the adrenergic and cholinergic systems.

ADRENERGIC IMMUNOREGULATION

Beginning with the early seventies a vast literature on morphological and functional interrelationships between the sympathetic nervous system and the immune system has accumulated. Lymphoid organs are strongly innervated by adrenergic fibres, the terminals of which come in close "synapsis like" apposition with lymphoid cells.[10] Furthermore, lymphocytes express β-adrenergic receptors which is dependent on lineage and activation stage.[11] The presence of α-adrenergic receptors has been documented for human lymphocytes by radioligand binding.[12] Concerning adrenergic effects on immune functions the data are numerous and conflicting. Both enhancement and suppression of lymphocyte functions have been ascribed to either α- or β-adrenergic mechanisms. These conflicting findings may be due to reasons of complexity as outlined in the previous section, but also to differences in the experimental approach, particularly between in vivo and in vitro studies.

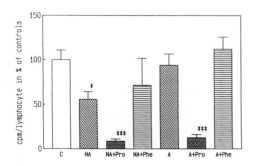 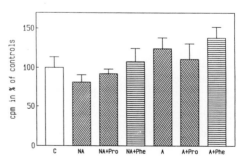

Figure 2. Maximal mitogen (Con A) response of PBL (left) and spleen lymphocytes (right) of Sprague Dawley rats 20 hours after s.c. implantation of retard tablets containing noradrenaline (NA) or adrenaline (A) alone, or together with phentolamine (Phe) or propranolol (Pro). Controls (C) received placebos. *, *** significantly (p<0.05, p<0.001) different from controls.

The aim of our own studies is to define the in vivo effects of chronically (20 hours) increased levels of peripheral catecholamines, such as adrenalin (A) and noradrenalin (NA), on lymphocyte functions in the rat model. The experimental approach consists in the s.c. implantation of retard tablets that continuously release defined amounts of adrenergic agonists or antagonists into the circulation.[13] This technology avoids handling stress effects (see previous section) due to repeated injections that may interfere with effects of the adrenergic treatment proper. The data obtained with this system[14] are summarized as follows (Figure 2): (i) a 20 hours treatment with A (15 mg) or NA (5 mg) has no (A) or only a marginal (NA) suppressive effect on the ex vivo proliferative response of PBL to Concanavalin A, (ii) combination of either A or NA with propranolol (Pro), or other β-adrenergic blockers (not shown), resulted in profound suppression of this T cell responsiveness, whereas concomittant α-adrenergic blockade with phentolamine (Phe) had no effect, neither had β- or α-adrenergic blockade per se induced a measurable effect (not shown), (iii) in analogy to the experiences with immobilization stress, spleen cells were resistant to chronic α- or β-adrenergic treatment.

During our first attempts to elucidate the possible mechanism(s) of this α-adrenergic receptor mediated immunosuppression, we have shown that it does not represent a general stress symptom due to metabolic inhibition, nor is it due to secondarily elicited glucocorticoids, or to quantitative shifts in total T cells (CD3+) or CD4+/CD8+ subsets in PBL. More recently we could confirm this effect with the synthetic α2-adrenergic receptor agonist clonidine, which, again only in combination with β-blockade, led to significant suppression of T cell proliferation,[15] whereas methoxamine (α1-receptor agonist) or isoproterenol (β-receptor agonist) had no effect. Furthermore, evidence was obtained that the in vivo antibody response against a foreign antigen (sheep red blood cells) is influenced in a similar way. Finally, we observed an analogous suppressive effect on the T-cell proliferative response one hour after a single i.p. injection of the indirect sympathomimetic drug tyramine in combination with β-adrenergic receptor blockade. Thus, this phenomenon is not an artefact inherent to treatment with exogenous catecholamines, but is likewise observed under enhanced release of endogenous NA from peripheral terminals. Interestingly, the effect of tyramine was again restricted to PBL, leaving the mitogen reactivity of splenic T-lymphocytes unaffected, which in view of the aforementioned intense sympathetic innervation of the spleen seems puzzling and is presently further investigated.[15]

Taken together, these data are in agreement with earlier studies of Besedovsky et al.[16] and contradict, at least in the rat model, the significance of a β-adrenergic receptor mediated immunosuppression in vivo. They also suggest that in vivo a significant adrenergic effect on immune functions requires not only enhanced levels of agonists, but also an alteration in the α/β receptor status. This may be explained by the fact that α-adrenergic stimulation has a positive feedback on endogenous release of catecholamines,[17] which may activate free β-adrenoceptors and thereby alleviate α-mediated effects. More has to be learned if and how a similar effect could occur under physiological conditions. Attempts to reproduce a_2-adrenergic suppression of mitogen response in vitro have so far failed in our hands, which leaves the possibility open that the phenomena observed in vivo are due to indirect mechanisms. In this context, however, recent data should be mentioned showing the suppression of antigen specific proliferative response of murine lymph node cells by addition of α- but not β-adrenergic agonists in vitro,[18] whereby the responsible α-adrenergic interaction appears to occur at the level of antigen presenting cells.

THE IMMUNE SYSTEM AS PART OF THE CHOLINERGIC SYSTEM?

Much less was known until recently about the interrelationship between the cholinergic and immune systems. The thymus has been reported to be innervated by cholinergic fibres,[19] which, however, was questioned by other authors.[20] Concerning the spleen, there is no evidence of cholinergic innervation, although low activities of both acetylcholine (ACh) and its synthetic enzyme, choline acetyltransferase (ChAT), have been detected in the spleen of several species.[21] Human and rodent lymphocytes from thymus, spleen and peripheral blood were reported to express muscarinic and nicotinic cholinergic receptors,[22] as well as a cellular acetylcholine esterase (AChE).[23] Finally, cholinergic agents were reported to influence immune functions in vivo and in vitro,[24-26] although the physiological source of the specific ligand, i.e. acetylcholine, in immune organs remained obscure.

In our own studies to investigate chronic cholinergic stimulation in the rat model using a similar technology as described for the adrenergic treatment we obtained evidence that the cholinergic system is intrinsically involved in the immuno-neuroendocrine dialogue.[27] This was concluded from the following data: (i) Altering the cholinergic tonus by chronic treatment with the AChE inhibitor physostigmine or the muscarinic antagonist atropine influenced the T cell mitogen response of lymphocytes of different compartments in different ways, suggesting an enhancing cholinergic effect on lymphocytes from thymus and spleen, but not in PBL, (ii) an enhancing effect of cholinergic in vivo treatment was also observed on the antigen specific humoral immune response as measured by the frequency of plaque forming cells in the spleen, although the in vitro addition of carbachol led to an inhibition, (iii) chronic physostigmine treatment prior to immunization with antigen (sheep red blood cells, SRBC) was found to abrogate the activation of the HPA axis due to immunization, and (iv) three days after immunization with SRBC a transient increase in affinity and decreased numbers of muscarinic receptors in the gyrus hippocampus was observed as determined by the binding of ^{3}H N-methyl scopolamine (^{3}H-NMS) (Figure 3).

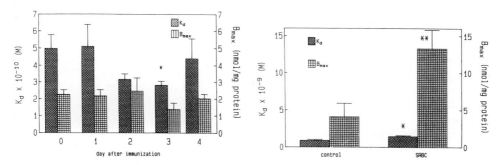

Figure 3. Changes in affinity (K_d) and number (B_{max}) of muscarinic receptors (^{3}H NMS binding) in brain (gyrus hippocampus, left) and spleen (right) membrane preparations after in vivo (i.p.) immunization with sheep red blood cells. Control = animals injected with saline. The values with spleen cells were measured 4 days after immunization. *, ** significantly (p<0.05, p<0.01) different from day 0 or control.

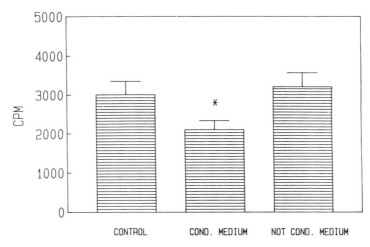

Figure 4. Effect of supernatants from Con A activated spleen cells (COND. MEDIUM), and non activated spleen cells (NOT COND. MEDIUM) on ^{3}H NMS binding to membrane preparations of gyrus hippocampus. * significantly (p<0.05) different from control.

As shown in Figure 3, immunization of rats with SRBC also affected peripheral muscarinic receptors in the spleen, but in an opposite way to what has been found in the brain.[28] In view of the lack of cholinergic innervation of the spleen, the question arose which ligand(s) might be responsible for these changes. These results together with the observation that supernatants from mitogen (Con A) activated spleen cells were able to significantly reduce the in vitro binding of ^{3}H-NMS to membrane preparations of the gyrus hippocampus (Figure 4) triggered our interest to search for ChAT activity in isolated lymphocytes. Using the radioenzymatic method by Fonnum[29] we were indeed able to detect the synthesis of ACh in homogenates of isolated rat lymphocytes from thymus (34.5±2.8 pmol/mg/min), spleen (22.9±3.3 pmol/mg/min) and peripheral blood 13.8±1.6 pmol/mg/min), as well as in several murine and human lymphoid cell lines.[30] The activity found in rat thymus lymphocytes was quantitatively very similar to what has been reported elsewhere with murine thymus tissue homogenates,[31] and we conclude that this activity may be entirely due to lymphocyte derived ChAT, rather than to cholinergic innervation.

CONCLUSIONS

The physiological roles of adrenergic and cholinergic mechanisms in the interaction of the immune-neuroendocrine network has been reviewed. As it emerges from these data, the integration of the immune system with the parasympathetic nervous system may be much closer and of more physiological relevance than the one with the adrenergic system. In our hands adrenergic suppression of immune functions was difficult to reproduce in vitro, and in vivo it required pharmacological intervention to block β-adrenergic receptors. In contrast, direct in vitro cholinergic effects on lymphocyte functions, such as thymocyte proliferation[26] (Rinner et al, unpublished observations) and the formation of antigen-specific plaques by spleen cells[27] could readily be observed. Cholinergic in vivo stimulation was found to directly interfere with the activation of the HPA axis due to immunization, which may facilitate unwanted immune responses. No such effect was found under adrenergic treatment (Felsner et al, unpublished observations). Finally, immune cells not only express functional cholinergic receptors and acetylcholine esterase, but they are also equipped to synthesize the specific ligand to these surface molecules, i.e. acetylcholine, which may be instrumental to communication with the cholinergic part of the autonomous nervous system, as well as contribute to autocrine immunoregulatory mechanisms. Attempts to detect the key enzyme of catecholamine synthesis, i.e. tyrosine hydroxylase, in lymphocytes were unsuccesful (Felsner et al, unpublished observations). Further studies to investigate ChAT activity in lymphocytes in dependence of activation and differentiation are presently under way. In this context recent data from our lab may be of interest as they suggest that cholinergic, but not adrenergic, stimuli can modulate survival and differentiation of thymic lymphocytes via a direct effect on the thymic epithelium.[32]

ACKNOWLEDGMENTS

Supported by the Austrian Science Foundation (projects 7509 and 7038) and by the Austrian National Bank (projects 3556 and 4349).

REFERENCES

1. C.W. Cotman, R.E. Brinton, A. Galaburda, B. McEwen, and D. Schneider. "The Neuro-Immune-Endocrine Connection," Raven Press, New York (1987).
2. J.E. Blalock, The immune system - our sixth sense, *The Immunologist* 2:8 (1994).
3. K. Schauenstein, R. Faessler, H. Dietrich, S. Schwarz, G. Kroemer, and G. Wick, Disturbed immune-endocrine communication in autoimmune disease. Lack of corticosterone response to immune signals in obese strain chickens with spontaneous autoimmune thyroiditis, *J. Immunol.* 139:1830 (1987).
4. E.M.Sternberg, J.M. Hill, G.P. Chrousos, T. Kamilaris, S.J. Listwak, P.W. Gold, and R.L. Wilder, Inflammatory mediator-induced hypothalamic-pituitary-adrenal axis activation is defective in streptococcal cell wall arthritis-susceptible Lewis rats, *Proc. Natl. Acad. Sci. USA* 86:4771 (1989).
5. D. Mason, I. MacPhee, and F. Antoni, The role of the neuroendocrine system in determining genetic susceptibility to experimental allergic encephalomyelitis in the rat, *Immunology* 70:1 (1990).
6. S.E. Keller, S.J. Schleifer, and M.K. Demetrikopoulos, Stress-induced changes in immune function in animals: Hypothalamo-pituitary-adrenal influences, *in*: "Psychoneuroimmunology," 2nd edition, R. Ader, D.L. Felten, and N. Cohen, eds., Academic Press (1991).
7. I. Rinner, K. Schauenstein, H. Mangge, S. Porta, and R. Kvetnansky, Opposite effects of mild and severe stress on in vitro activation of rat peripheral blood lymphocytes, *Brain Behav. Immun.* 6:130 (1992).
8. D. Hofer and K. Schauenstein, Enhanced stress sensitivity of an autoimmune response as compared to a simultaneous response against foreign antigen in mice, Abstract, Annual Meeting of the Austrian Society of Allergology and Immunology, Graz, Austria, November 1993, p. 17.
9. J.D. Naysmith, P.M.G. Ortega, and C.J. Elson, Rat erythrocyte-induced anti- erythrocyte autoantibody production and control in normal mice, *Immunol. Rev.* 55:55 (1981).
10. D.L. Felten, S.Y. Felten, D.L. Bellinger, S.L. Carlson, K.D. Ackerman, K.S. Madden, J.A. Olschowski, and S. Livnat, Noradrenergic sympathetic neural interactions with the immune system: structure and function, *Immunol. Rev.* 100:225 (1987).
11. M.M. Khan, P. Sansoni, E.D. Silverman, E.G. Engleman, and K.L. Melmon, Beta-adrenergic receptors on human suppressor, helper and cytolytic lymphocytes, *Biochem. Pharm.* 7:1137 (1986).

12. S. Titinchi and B. Clark, Alpha2-adrenoceptors in human lymphocytes: direct characterization by (^{3}H) yohimbine binding, *Biochm. Biophys. Res. Commun.* 121:1 (1984).
13. W. Korsatko, S. Porta, A. Sadjak, and S. Supanz, Implantation von Adrenalin-retard Tabletten zur Langzeitunzersuchung in Ratten, *Pharmazie* 37:565 (1982).
14. P. Felsner, D. Hofer, I. Rinner, H. Mangge, M. Gruber, W. Korsatko, and K. Schauenstein, Continuous in vivo treatment with catecholamines suppresses in vitro reactivity of rat peripheral blood T-lymphocytes via a-mediated mechanisms, *J. Neuroimmunol.* 37:47 (1992).
15. P. Felsner, D. Hofer, I. Rinner, W. Korsatko, and K. Schauenstein, In vivo immunosuppression by enhanced catecholamines in the rat model is due to activation of peripheral a2-receptors. *J. Neuroimmunol.* (Submitted for publication).
16. H.O. Besedovsky, A. Del Rey, E. Sorkin, M. Da Prada, and H.H. Keller, Immunoregulation mediated by the sympathetic nervous system, *Cell. Immunol.* 48:346 (1979).
17. S. Porta, B. Rangentiner, I. Rinner, U. Ertl, A. Sadjak, J. Nauman, Long-term application of some catecholamines elevates levels of other catecholamines in rats, *Exp. Path.* 28:181 (1985).
18. M. Heilig, M. Irwin, G. Iqbal, and E. Sercarz, Sympathetic regulation of T-helper cell function, *Brain Behav. Immun.* 7:154 (1993).
19. K. Bulloch, A comparative study of the autonomous nervous system innervation of the thymus in the mouse and chicken, *Int. J. Neurosci.* 40:129 (1988).
20. D.M. Nance, D.A. Hopkins, and D. Bieger, Re-investigation of the innervation of the thymus gland in mice and rats, *Brain Behav. Immun.* 1:134 (1987).
21. S.Y. Felten and D.L. Felten, Innervation of lymphoid tissue, *in:* "Psychoneuroimmunology," 2nd edition, R. Ader, D.L. Felten, and N. Cohen, eds., Academic Press (1991).
22. W. Maslinski, Cholinergic receptors on lymphocytes, *Brain Behav. Immun.* 3:1 (1989).
23. J. Szelenyi, P. Palldi-Haris, and S. Hollan, Changes in the cholinergic system due to mitogenic stimulation, *Immunol. Lett.* 16:49 (1987).
24. G. Iliano, G.P.E. Tell, M.I. Segal, and P. Cuatrecasas, Guanosine 3´, 5´-cyclic monophosphate and the action of insulin and acetylcholine, *Proc. Natl. Acad. Sci. USA* 70:2443 (1973).
25. T.B. Strom, A.T. Sytkowski, C.B. Carpenter, and J.B. Merill, Cholinergic augmentation of lymphocyte mediated cytotoxicity. A study of the cholinergic receptor of cytotoxic T lymphocytes, *Proc. Natl. Acad. Sci. USA* 71:1330 (1974).
26. A. Rossi, M.A. Tria, S. Baschieri, G. Doria, and D. Frasca, Cholinergic agonists selectively induce proliferative responses in the mature subpopulation of murine thymocytes, *J. Neurosci. Res.* 24:369 (1989).
27. I. Rinner and K. Schauenstein, The parasympathetic nervous system takes part in the immuno-neuroendocrine dialogue, *J. Neuroimmunol.* 34:165 (1991).
28. I. Rinner, K. Schauenstein, E. Skreiner, and M. Posch, Relationship between the parasympathetic nervous system and the immune system, Abstract No. 313, 1st International Congress ISNIM, Florence, Italy, May 23-26, 1990, p. 420.
29. F.A. Fonnum, Rapid radiochemical method for the determination of choline acetyltransferase, *J. Neurochem.* 24:407 (1975).
30. I. Rinner and K. Schauenstein, Detection of choline-acetyltransferase activity in lymphocytes, *J. Neurosci. Res.* 35:188 (1993).
31. M. Badamchian, H. Damavandy, T. Radojcic, and K. Bulloch, Choline O-acetyltransferase (ChAT) and muscarinic receptors in the Balb/C mouse thymus, Abstract, Proc. Satellite meeting of the 8th International Congress of Immunology "Advances in Psychoneuroimmunology", Budapest, August 1992, p. 3.
32. I. Rinner, T. Kukulansky, E. Skreiner, A. Globerson, M. Kasai, K. Hirokawa, and K. Schauenstein, Adrenergic and cholinergic regulation of apoptosis and differentiation of thymic lymphocytes, *in:* "In Vivo Immunology: Regulatory Processes During Lymphopoiesis and Immunopoiesis," J. Bonniver, ed., Plenum Pub. Corp., NY, in press.

INDEX

AML, *see* Acute myelogenous leukemia
AMP, *see* Cyclic adenosine monophosphate
Anaphylactic shock, 27, 28, 291
Anaphylaxis, 44, 291, 307
Anchorage, 49
Androgens, 115, 150, 192, 203, 213-216, 223, 226, 227, 234
Androgen-like steroids, 225
Androgen receptors, 192, 215, 227
Androstane, 225
Androsterone, 225
Anemia
   normochromic-normocytic, 111
Angiopexis, 20
Angiotensin, 140, 142
Antibody, 48, 213, 274
   anti-corticotroph, 248
   anti-DNA, 127, 128, 230
   anti-L3T4/CD4, 128
   antinuclear, 131
   anti-pituitary, 248
   combining site, 44
   formation, 227
   forming cells, 150
   pituitary stimulating, 248
   production, 28, 113
   response, 114, 139, 234, 352
Antibody dependent cellular cytotoxicity (ADCC), 283
   activity, 285
Antidromic electrical stimulation of sensory nerves, 22
Antiestrogens, 234
Antigen-antibody bonds, 56
Antigen binding site, 56
Antigen challenge, 295
Antigen-driven selection, 54
Antigen-induced histamine release, 64
Antigen presentation, 153
Antigen presenting cells, 352
Antigen processing, 265
Antigen receptor site, 44
Antigen signal, 117, 118
Antigenic competition, 153
Antiphlogistic corticoids, 9
Apoptosis, 182
Arachidonic acid, 181, 188
Area postrema, 142
Arginase, 46
Arginine, 334, 335
β-Arrestin, 64
Arsenobenzene compounds, 19
Arthritis, 116, 154
Arthus reaction, 28
Asthma, 25, 27, 29, 31, 44, 61-63, 265
   atopic, 62
   bronchial, 60
   chronic, 42
   extrinsic, 63
   intrinsic, 63

Astrocytes, 141, 166
   class II MHC antigen expression, 151
Astroglial proliferation, 60
Asymmetry, 53
Ataxia-telangiectasia, 88
Atopic allergy, 52
Atopy, 44
Atrophy
   of thymus and other lymphatic organs, 7
Atropine, 353
AtT-20 cells, 139
   corticotroph cell line, 246
Autism, 53
Autoantibody formation, 230
Autocrine, 131
Autocrine regulation, 114, 139
Autoimmune diseases, 116, 138, 150, 213, 216, 223, 227, 327
   autoimmune phenomena, 244
   autoimmune progesterone dermatitis, 238
   autoimmune reactions, 32, 154
   autoimmune response, 350
   autoimmune thyroiditis, 154, 230, 245
   autoimmune thyroid disease, 130
   autoimmune uveitis, 130
   human, 130
   model, 228
Autoimmunity, 125, 192, 213
   pituitary, 248
Autonomic imbalance, 43, 44
Autonomic nervous system (ANS), 317, 319, 331, 351
Avidity, 56
AVP, *see* vasopressin

Bacille Calmette-Guerin, 60
Bacterial adherence, 210
Bacterial endotoxin, 151, 234
Bacterial lipopolysaccharide (LPS), 114; *see also* Lipopolysaccharide
Bacterial translocation, 210
Barbituate injection, 350
Basement membrane, 48
Basophilic leukocytes, 6, 47, 61, 63
B cells, *see* B lymphocytes
Behcet's syndrome, 130
Bereavement, 279
Bernard, Claude, 5
   "Claude Bernard Profesors", 5
Bile, 205
   output, 206
Blastogenic transformation, 193
Blood-brain barrier, 57, 142
Blood sugar, 6
Bloom syndrome, 88
B lymphocytes, 54, 57, 85, 114, 118, 126, 128, 234, 255, 272, 319, 320, 333
   differentiation, 244
   growth, 244
   total (surface IgM$^+$), 216

Hepatocytes *(cont'd)*
  secretory component (SC) production, 206
Hershberger test, 227
5-HETE, 46
8,15-diHETE, 47
Hexamethonium, 43
High performance liquid chromatography (HPLC), 64, 208
Hippocampus, 55, 57, 60, 166
  electrical stimulation of, 56
Hippocampal neurons, 57
Histamine, 17, 19, 28, 29, 46, 61, 117, 153
  desensitization, 20
  release, 308
  releasing factors (HRFs), 61, 63
Histamine-release inhibitory factor, 61
Histocompatibility antigen, 118
  class II, 115
HIV, *see* Human immunodeficiency virus
HLA-DR, 62
Homeostasis, 200
Homeostatic control, 45
Hormones, *see* specific types
Hormonal control, 204
Hormone receptors
  class I, 50
  class II, 50
  class III, 50
HPA, *see* Hypothalamic-pituitary-adrenal
^{3}H-Quinuclidinyl benzoate, 273
Human immunodeficiency virus (HIV), 89
Human placental lactogen, 112
Humoral immunity, 44, 114
Hydrocortisone, 28
Hydrogen peroxide, 238
Hydrogenation, 183
6-Hydroxydopamine (6-OHDA), 150, 319
2-Hydroxyestrone, 225
11α-Hydroxynandrolone, 230
17α-Hydroxypregnanes, 229
Hypergammaglobulinemia, 130
Hyperglycemia, 28, 29
Hyperinsulinemia, 60
Hyperprolactinemia, 112, 113, 125, 127, 129, 130, 138, 248
  longstanding, 131
Hypersensitivity, 42
Hypertension, 7
Hyperthyroidism, 335
Hypogammaglobulinemia, 88
  non-lymphopenic, 88
  sex linked lymphopenic, 88
Hypoglycemia, 154, 155
Hypophysectomy, 83, 114, 214, 325
Hypophysis, 7
Hypophysitis, 116
Hypotensive responses, 310
Hypothalamic-pituitary axis, 214, 217
Hypothalamic-pituitary-adrenal (HPA) axis, 46, 137, 153, 154, 244, 245, 320, 328, 349,

Hypothalamic-pituitary-adrenal axis *(cont'd)*
  351, 353
Hypothalamic-pituitary-growth hormone axis, 83
Hypothalamically imbalanced anaphylaxis, 44
Hypothalamo-hypophyseal hormones, 46
Hypothalamopituitary peptides, 59
Hypothalamus, 42, 44, 57, 60, 117, 139, 141, 152, 192
  CRF release, 167
  electrical stimulation of mamillary region, 28
  electrical stimulation of posterior, 56
  median basal, 117
  neurons, 152
    firing, 152
  organum vasculosum of the laminae terminalis, 142
  tuberal lesions, 28
  ventral medial, 60, 167
  ventromedial nucleus, 152
Hypox, *see* Hypophysectomy

^{125}Iα-bungarotoxin, 273
IBD, *see* inflammatory bowel disease
ICAM-1, *see* intercellular adhesion molecule-1
IFN, *see* Interferon
IgA, 85, 203, 213, 215, 255, 257
  monomeric, 209
  plasma cells, 214
  polymeric, 209, 210
  redistribution of, 210
  secretion, 257
    antibodies, 214
  serum levels, 208
  synthesis, 266
  uterine, 207
  vaginal, 207
IgE, 61, 63
  antibodies, 31, 292
  antibody synthesis, 63
  binding factor, 61, 64
  synthesis, 64
IGF, *see* Insulin-like growth factor
IgG, 56, 85, 114, 128, 214, 234, 257
Ig-like domains, 48
IgM, 56, 85, 114, 234, 257
IgSFs, 51
Ig superfamily, 48
IL, *see* Interleukin
IM-9 cell line, 89
Immature T cells, 216
Immediate hypersensitivity, 291
Immune complex, 128, 238
Immune complex glomerulonephritis, 127
Immune homeostasis, 45
Immune-neuroendocrine circuit, 45, 59, 349
Immune-neuroendocrine interactions, 149
Immune response, 45, 61
Immune suppression, 45
Immune surveillance, 279
Immune system, 45, 83

Ischemia/reperfusion (I/R) insult, 311
3-Isobutyl-1-methylxanthine, 215
Isoproterenol, 215, 352

Jancso, Miklos (Nicholas), 17

K562 erythromyeloid human line, 281
K562 target cells, 235
Keratoconjunctivitis sicca, 216
Kidney grafts, 115
Killer cells, 239
Kinases, 55
Kinins, 29, 47
Kupffer cells of the liver, 19

L929 tumor cells, 182
Lacrimal glands, 213-215
    autoimmune disease, 213
    acinar epithelial cells, 214
    epithelial cells, 217
    secretions, 203
Lacrimal tissue
    IgA-containing cells, 214
Lactation, 115
Lactogenic hormone, 7
Learning disorders, 53
Left cerebral neocortex, 53
Left cortical ablation, 54
Leukemia, 87
    acute lymphocytic leukemia (ALL), 87
    acute myelogenous leukemia (AML), 87, 131
    and growth hormone deficiency, 87
Leukemia inhibitory factor, 50
    receptor, 50
Leukemic lymphoblast colony, 87
Leukocyte specific selectin, 317
Leukopenia, 111
Leukotrienes, 46, 47
    B_4, 64
    C_4, 63
Levator ani muscle, 226
Lewis rats, 245
LFA1, *see* Lymphocyte functional antigen
LGL, *see* large granular Lymphocytes
LH, *see* Luteinizing hormone
LHRH, *see* Luteinizing hormone releasing hormone
Life events, 327
Lipid A, 165, 234
Lipid extraction, 183
Lipid saturation, 181
Lipopolysaccharide (LPS), 151, 310; *see also* Bacterial lipopolysaccharide
Listeria monocytogenes, 116
Liver, 7
Locus coeruleus, 166, 324
Low density lipoprotein (LDL), 88
LPS, *see* Bacterial lipopolysaccharide; Lipopolysaccharide
Luteinizing hormone (LH), 112, 115, 126, 130,

Luteinizing hormone *(cont'd)*
    131, 138, 140, 245, 246, 304
Luteinizing hormone releasing hormone (LHRH)
    analogue of, 193
    hypothalamic, 193
Lymphocyte steroid receptors, 192
Lymphocytes, 6, 46, 150, 271, 353; *see also* B lymphocytes; T lymphocytes
    "activated", 63
    class II antigen-positive (Ia⁺), 216
    large granular (LGL), 319, 320
    maturation, 56
    mature, in blood, 192
    migration, 317
    p75⁺, 283
    production of PRL or GH, 114
    rat, 354
    subsets, 85, 283
Lymphocyte blastogenesis, 86
Lymphocyte functional antigen (LFA), 118
Lymphocyte proliferation, 228, 274
Lymphocytic hypophysitis, 248
Lymphocytic infiltration, 216
Lymphoid cell migration, 258, 265
Lymphoid infiltration
    lacrimal gland, 216
    salivary gland, 216
Lymphoid organs, 6, 45, 319
    adrenergic innervation, 351
Lymphoid feedback signals, 117
Lymphokines, 46, 63, 152, 215
    T-cell-derived, 61
Lymphokine activated killer
    activity, 236
    cell mediated cytotoxicity, 237
Lymphoma
    L1210 murine, 281
Lynestrenol, 228
Lysozyme activities, 87

mAChR, *see* Acetylcholine receptor, muscarinic
Macrophages, 6, 46, 62, 88, 116, 126, 152, 258, 309
    alveolar, 64
    induced cytostasis, 225
    murine, 88
Macroprolactinemia, 130
Major histocompatibility complex (MHC), 44
    compatibility at MHC loci, 152
    glycoproteins, 48
    major histocompatibility antigens, 49
        class I and II antigen expression, 151
        class II antigen-positive (Ia⁺) expression, 216
Male, 306
Malignant mesenchymal tumors, 235
Malnutrition, 91
Mammary gland, 115
Man, 117